AF904034

CNS Regeneration

CNS REGENERATION

BASIC SCIENCE AND CLINICAL ADVANCES

SECOND EDITION

EDITED BY

JEFFREY H. KORDOWER
Research Center for Brain Repair
Rush University Medical Center
Chicago, Illinois

MARK H. TUSZYNSKI
Center for Neural Repair
University of California, San Diego
La Jolla, California

ELSEVIER

AMSTERDAM • BOSTON • HEIDELBERG • LONDON • NEW YORK • OXFORD
PARIS • SAN DIEGO • SAN FRANCISCO • SINGAPORE • SYDNEY • TOKYO

Academic Press is an imprint of Elsevier

Academic Press is an imprint of Elsevier
84 Theobald's Road, London WC1X 8RR, UK
Radarweg 29, PO Box 211, 1000 AE Amsterdam, The Netherlands
30 Corporate Drive, Suite 400, Burlington, MA 01803, USA
525 B Street, Suite 1900, San Diego, CA 92101-4495, USA

First edition 1999
Second edition 2008

Copyright © 2008 Elsevier Inc. All rights reserved

No part of this publication may be reproduced, stored in a retrieval system or transmitted in any form or by any means electronic, mechanical, photocopying, recording or otherwise without the prior written permission of the publisher

Permissions may be sought directly from Elsevier's Science & Technology Rights Department in Oxford, UK: phone (+44) (0) 1865 843830; fax (+44) (0) 1865 853333; email: permissions@elsevier.com. Alternatively you can submit your request online by visiting the Elsevier web site at http://elsevier.com/locate/permissions, and selecting *Obtaining permission to use Elsevier material*

Notice
No responsibility is assumed by the publisher for any injury and/or damage to persons or property as a matter of products liability, negligence or otherwise, or from any use or operation of any methods, products, instructions or ideas contained in the material herein. Because of rapid advances in the medical sciences, in particular, independent verification of diagnoses and drug dosages should be made

British Library Cataloguing in Publication Data
A catalogue record for this book is available from the British Library

Library of Congress Cataloging-in-Publication Data
A catalog record for this book is available from the Library of Congress

ISBN: 978-0-12-373994-0

For information on all Academic Press publications
visit our web site at books.elsevier.com

Printed and bound in the United Kingdom
Transferred to Digital Printing, 2011

Working together to grow
libraries in developing countries

www.elsevier.com | www.bookaid.org | www.sabre.org

ELSEVIER BOOK AID International Sabre Foundation

Contents

Part I

Responses to Injury

1

Intrinsic Determinants of Axon Regeneration

Rhona Seijffers and Larry Benowitz

2

AXONAL RESPONSES TO INJURY

MIKE FAINZILBER

3

GLIAL CELLS, INFLAMMATION, AND CNS TRAUMA MODULATION OF THE INFLAMMATORY ENVIRONMENT AFTER INJURY CAN LEAD TO LONG-DISTANCE REGENERATION BEYOND THE GLIAL SCAR

MICHAEL T. FITCH AND JERRY SILVER

PART II

STEM CELLS AND TROPHIC FACTORS IN CNS REPAIR

4

NEUROTROPHIC FACTORS

EUGENE M. JOHNSON, JR AND MARK H. TUSZYNSKI

5

STEM CELL THERAPY FOR BRAIN TUMORS

RAHUL JANDIAL, ALLEN L. HO, MICHAEL L. LEVY, AND EVAN Y. SNYDER

6

STEM CELL THERAPIES FOR PARKINSON'S DISEASE

ANDREAS ANDROUTSELLIS-THEOTOKIS, MARIA A. RUEGER, AND RON D.G. MCKAY

7

ADULT NEURAL PROGENITOR CELLS IN CNS FUNCTION AND DISEASE

SEBASTIAN JESSBERGER, STEFAN AIGNER, JAMES B. AIMONE, AND FRED. H. GAGE

8

NEUROTROPHIC FACTORS IN ALZHEIMER'S DISEASE

ARMIN BLESCH AND MARK H. TUSZYNSKI

PART III

NOVEL THERAPIES FOR PARKINSON'S AND HUNTINGTON'S DISEASE

9

DIRECT DELIVERY OF GDNF INTO THE NON-HUMAN PRIMATE AND HUMAN PARKINSONIAN BRAIN: SUCCESS AND ROAD BLOCKS

R GRONDIN, Z ZHANG, Y AI, J SLEVIN, AB YOUNG, DM GASH, AND GA GERHARDT

10

STIMULATING AND ABLATIVE TREATMENT FOR PARKINSON'S DISEASE

JULIE G. PILITSIS AND ROY A.E. BAKAY

11

GENE AND CELLULAR TRANSPLANTATION THERAPIES FOR HUNTINGTON'S DISEASE

SHILPA RAMASWAMY AND JEFFREY H. KORDOWER

PART IV

NOVEL THERAPIES FOR ALZHEIMER'S DISEASE

12

ANTI-AMYLOID-β IMMUNOTHERAPY AS A TREATMENT FOR ALZHEIMER'S DISEASE

JOHN R. CIRRITO AND DAVID M. HOLTZMAN

PART V

NOVEL THERAPIES FOR SPINAL CORD INJURY AND WHITE MATTER DEMYELINATION

13

AXONAL PLASTICITY AND REGENERATION IN THE INJURED SPINAL CORD

M.H. TUSZYNSKI AND P. LU

14

MODIFYING THE EXTRACELLULAR MATRIX AS A TREATMENT TO IMPROVE FUNCTIONAL RECOVERY AFTER SPINAL CORD INJURY

JAMES W. FAWCETT

15

MYELIN NEUTRALIZATION FOR SPINAL CORD INJURY AND STROKE

A.R. WALMSLEY, D. BOURIKAS, AND A.K. MIR

16

STRATEGIES TO INHIBIT SIGNALING THROUGH NOGO RECEPTOR 1 FOR SPINAL CORD INJURY AND STROKE

JANE K. RELTON AND PAUL H. WEINREB

17

CYCLIC AMP MODULATION OF AXONAL REGENERATION FOLLOWING SPINAL CORD INJURY

SARI S. HANNILA AND MARIE T. FILBIN

18

MULTIPLE SCLEROSIS: REMYELINATION

JEFFERY D. KOCSIS, MASANORI SASAKI, KAREN L. LANKFORD, AND CHRISTINE RADTKE

PART VI

DESIGN OF CLINICAL TRIALS

19

CLINICAL TRIALS IN REHABILITATION WITH NEURAL REPAIR

BRUCE H. DOBKIN

INTRODUCTION

Much has happened in the neurology/neuroscience community since the publication of the First Edition of **CNS Regeneration: Basic Science and Clinical Advances**. Some area of research such as gene therapy have taken large steps forward, with preclinical and translational neuroscience in this field leading to the initiation of multiple clinical trials. Other areas of regeneration research have necessarily taken steps backwards, with lessons learned in clinical trials clearly indicating that more basic and translational science is required. This is the essence of productive translational science: the lab informs the clinic and visa versa. The most obvious area wherein patience has been required is cellular replacement strategies, with the publication of two negative double-blind trials of fetal grafting in Parkinson's disease accompanied by unacceptable side effects such as "off-medication" dyskinesias. The practical difficulties of fetal grafting for large numbers of patients also encourages time for pause, as the stem cell era for cell replacement strategies is emerging and we await data establishing its safety and efficacy.

As stated in our first edition, the dogma that regeneration in the central nervous system cannot occur is a myth and there are currently more examples of this fact than ever. The adult brain and spinal cord are responsive to signals provided by several classes of molecules that can promote neuronal survival, stimulate axon outgrowth and even lead to self-replenishing sets of neural progenitor cells that form new neuronal cells. This capacity for regeneration is most often dependent upon specific interventions, but can be readily elicited by experimental manipulation. The culmination of these interventions is functional recovery, a phenomenon that has now been demonstrated in numerous animal models including nervous system trauma, stroke, degeneration, and inherited degenerative diseases. Based upon preclinical studies, several of these approaches are now undergoing clinical testing.

This second edition will update many of the areas detailed in the first edition, with thought leaders in their respective fields focusing on modern structurally-based and molecularly-based models of neurological disease as well as methods for promoting nervous system regeneration. The organization of this edition reflects the growth in this area of research.

The book opens with the Responses to Injury section which details responses of the nervous system to cellular and axonal insults and identifies salient issues that are critical in mediating nervous system repair. Responses of neurons, axons, and glia to injury will be described, together with the deleterious effects on the nervous system of secondary damaging factors such as free radicals and excitotoxins. The Stem Cells and Trophic Factors in CNS Regeneration section of the book goes on to identify recent advances in neuroscience that offer the tools for promoting regeneration of the central nervous system. These tools include neurotrophic factors, fetal cell grafts, primary and immortalized pro-genitor cells, and genetic engineering. These opening chapters are followed by several sections that detail the application of regeneration-promoting strategies to specific animal models of human disease. Means of promoting neural repair in models of Alzheimer's disease, Parkinson's disease, Huntington's disease and spinal cord injury are described as prototypes for the treatment of neurological disease in general. Specific methods for treating these disorders, including the use of growth factors, gene therapy, and fetal grafting are discussed in a chapter on design of clinical trials.

We are in the midst of a new era in the treatment of neurological disease. The focus of much neurological therapy remains symptomatic, with dramatic benefit found in some procedures such as deep brain stimulation. However, the "holy grail" remains neuroprotection and regeneration of remaining systems and augmentation of existing pathways. This field still grows at an astounding pace and no book can at present be comprehensive. Thus, the present offering is but a sampling of mechanistically enlightening and therapeutically promising progress in a field that has undergone remarkable evolution in a very short period of time.

LIST OF CONTRIBUTORS

Y. Ai (223) Departments of Anatomy & Neurobiology, and Morris K. Udall Parkinson's Disease Research Center of Excellence, University of Kentucky Medical Center, Lexington, KY 40536

Stefan Aigner (181) The Salk Institute for Biological Studies, La Jolla, CA 92037

James B. Aimone (181) The Salk Institute for Biological Studies, La Jolla, CA 92037

Andreas Androutsellis-Theotokis (161) Laboratory of Molecular Biology, National Institute of Neurological Disorders and Stroke, National Institutes of Health, Bethesda, MD 20892

Roy A.E. Bakay (245) Department of Neurosurgery, Rush University Medical Center, Chicago, IL 60612

Larry Benowitz (1) FM Kirby Neurobiology Center and Department of Neurosurgery, Children's Hospital, Department of Surgery and Program in Neuroscience, Harvard Medical School, Boston, MA 02115

Armin Blesch (201) Departments of Neurosciences-0626, University of California-San Diego, La Jolla, CA 92093-0626

D. Bourikas (355) Neuroscience Research, Novartis Institutes for Biomedical Research, 4056 Basel, Switzerland

John R. Cirrito (295) Departments of Psychiatry, Neurology, Hope Center for Neurological Disorders, Alzheimer's Disease Research Center, Washington University School of Medicine, St Louis, MO 63110

Bruce H. Dobkin (437) University of California Los Angeles, Los Angeles, CA 90095

Mike Fainzilber (41) Department of Biological Chemistry, Weizmann Institute of Science, 76100, Rehovot, Israel

James W. Fawcett (337) Cambridge University Centre for Brain Repair, Robinson Way, Cambridge CB2 2PY, UK

Marie T. Filbin (389) Department of Biological Sciences, Hunter College, City University of New York, NY 10021

Michael T. Fitch (59) Department of Emergency Medicine, Wake Forest University School of Medicine, Medical Center Boulevard, Winston-Salem, NC 27157

Fred. H. Gage (181) The Salk Institute for Biological Studies, La Jolla, CA 92037

D.M. Gash (223) Departments of Anatomy & Neurobiology, and Morris K. Udall Parkinson's Disease Research Center of Excellence, University of Kentucky Medical Center, Lexington, KY 40536

G.A. Gerhardt (223) Departments of Anatomy & Neurobiology, Neurology, and Morris K. Udall Parkinson's Disease Research Center of Excellence, University of Kentucky Medical Center, Lexington, KY 40536

R. Grondin (223) Departments of Anatomy & Neurobiology, and Morris K. Udall Parkinson's Disease Research Center of Excellence, University of Kentucky Medical Center, Lexington, KY 40536

Sari S. Hannila (389) Department of Biological Sciences, Hunter College, City University of New York, NY 10021

Allen L. Ho (145) University of California, San Diego, CA

David M. Holtzman (295) Departments of Neurology, Molecular Biology and Pharmacology, Hope Center for Neurological Disorders, and Alzheimer's Disease Research Center, Washington University School of Medicine, St Louis, MO 63110

Rahul Jandial (145) Division of Neurological Surgery – University of California, San Diego, CA

Sebastian Jessberger (181) The Salk Institute for Biological Studies, La Jolla, CA 92037

Eugene M. Johnson, Jr (95) Department of Neurology and of Molecular Biology and Pharmacology, Washington University School of Medicine, St Louis, MO 63110

Jeffery D. Kocsis (413) Department of Neurology, Yale University School of Medicine, and Neuroscience Rehabilitation Research and Development Center, VA Connecticut Healthcare System, West Haven, CT 06516

Jeffrey H. Kordower (267) Department of Neuroscience, Rush University Medical Center, Chicago, IL 60612

Karen L. Lankford (413) Department of Neurology, Yale University School of Medicine, and Neuroscience Rehabilitation Research and Development Center, VA Connecticut Healthcare System, West Haven, CT 06516

Michael L. Levy (145) Department of Pediatric Neurosurgery – Children's Hospital San Diego, CA

P. Lu (319) Department of Neurosciences, University of California-San Diego, La Jolla, CA 92093-0626, and Veterans Administration Medical Center, San Diego, CA 92165

Ron D.G. McKay (161) Laboratory of Molecular Biology, National Institute of Neurological Disorders and Stroke, National Institutes of Health, Bethesda, MD 20892

A.K. Mir (355) Neuroscience Research, Novartis Institutes for Biomedical Research, 4056 Basel, Switzerland

Julie G. Pilitsis (245) Department of Neurosurgery, Rush University Medical Center, Chicago, IL 60612

Christine Radtke (413) Department of Neurology, Yale University School of Medicine, and Neuroscience Rehabilitation Research and Development Center, VA Connecticut Healthcare System, West Haven, CT 06516

Shilpa Ramaswamy (267) Department of Neuroscience, Rush University Medical Center, Chicago, IL 60612

Jane K. Relton (373) Department of Neuropharmacology, Biogen Idec, 14 Cambridge Center, Cambridge, MA 02142

Maria A. Rueger (161) Laboratory of Molecular Biology, National Institute of Neurological Disorders and Stroke, National Institutes of Health, Bethesda, MD 20892

Masanori Sasaki (413) Department of Neurology, Yale University School of Medicine, and Neuroscience Rehabilitation Research and Development Center, VA Connecticut Healthcare System, West Haven, CT 06516

Rhona Seijffers (1) Neural Plasticity Research Group, Department of Anesthesia and Critical Care, Massachusetts General Hospital and Harvard Medical School, Charlestown, MA 02129

Jerry Silver (59) Department of Neurosciences, Case Western Reserve University School of Medicine, 10900 Euclid Avenue, Cleveland, OH 44106

J. Slevin (223) Departments of Neurology, and Morris K. Udall Parkinson's Disease Research Center of Excellence, University of Kentucky Medical Center, Lexington, KY 40536

Evan Y. Snyder (145) Center for Neurosciences and Aging – The Burnham Institute for Medical Research, La Jolla, CA

Mark H. Tuszynski (95, 201, 319) Department of Neurosciences-0626, University of California-San Diego, La Jolla, CA 92093, and Veterans Affairs Medical Center, San Diego, CA 92161

A.R. Walmsley (355) Neuroscience Research, Novartis Institutes for Biomedical Research, 4056 Basel, Switzerland

Paul H. Weinreb (373) Department of Drug Discovery, Biogen Idec, 14 Cambridge Center, Cambridge, MA 02142

A.B. Young (223) Departments of Neurosurgery, and Morris K. Udall Parkinson's Disease Research Center of Excellence, University of Kentucky Medical Center, Lexington, KY 40536

Z. Zhang (223) Departments of Anatomy & Neurobiology, and Morris K. Udall Parkinson's Disease Research Center of Excellence, University of Kentucky Medical Center, Lexington, KY 40536

PART

I

RESPONSES TO INJURY

1

Intrinsic Determinants of Axon Regeneration

Rhona Seijffers* and Larry Benowitz†

*Neural Plasticity Research Group, Department of Anesthesia and Critical Care, Massachusetts General Hospital and Harvard Medical School, Charlestown, MA 02129
†FM Kirby Neurobiology Center and Department of Neurosurgery, Children's Hospital, Department of Surgery and Program in Neuroscience, Harvard Medical School, Boston, MA 02115

ABSTRACT

In the peripheral nervous system (PNS), neurons spontaneously regenerate injured axons to reinnervate skin, muscle, and other targets. In contrast, neurons cannot generally regenerate injured axons within the central nervous system (CNS: brain, spinal cord, eye). While most research in the field has tended to ascribe these differences to cell-extrinsic inhibitors of growth, recent evidence suggests that intrinsic factors are at least as important in governing a neuron's ability to regenerate its axon. Following peripheral nerve injury, dorsal root ganglion (DRG) neurons undergo striking changes in the expression of genes required for cell survival and axon outgrowth, enabling them to regenerate both their peripheral axon branch and their centrally directed axon branch into the CNS. In contrast, following injury to the optic nerve, retinal ganglion cells (RGCs), the projection neurons of the eye, show only modest changes in gene expression and modest levels of terminal sprouting, but no long-distance regeneration. In response to growth factors released by macrophages, however, RGCs undergo changes in gene expression similar to those seen in regenerating DRG neurons and extend lengthy axons through the optic nerve. Here, we discuss what is known about the inductive signals for axon regeneration in DRG neurons and RGCs, the signal transduction pathways and transcription factors involved, and the use of combinatorial treatments

Copyright © 2008 Elsevier Inc.
All rights of reproduction in any form reserved.

to overcome cell-extrinsic growth-inhibitory signals while also activating neurons' intrinsic growth state. A greater understanding of the molecular events that accompany axon regeneration in the CNS and PNS is likely to help us achieve better outcome after neural injury in the clinical setting.

INTRODUCTION

The marked difference between the regenerative capacities of the peripheral nervous system (PNS) and the central nervous system (CNS) is well known. Spinal motorneurons, along with sensory and autonomic ganglionic neurons, are able to re-grow damaged axons through peripheral nerves to reinnervate muscle, skin, and other end organs. In contrast, most neurons within the brain, spinal cord, and optic nerve cannot regenerate injured axons over any appreciable distance. Compounding this problem, only a small percentage of neurons that are irretrievably lost after CNS injury are replaced by endogenous precursor cells, and CNS neurons that escape damage have only a limited ability to extend axon collaterals into areas that have lost their normal inputs. Because of these limitations, victims of spinal cord injury, stroke, closed head trauma, and various neurodegenerative diseases often suffer devastating and permanent losses in sensory, motor, cognitive, and autonomic functioning, depending on the site of damage. Even within the PNS, the situation is far from ideal. Although PNS regeneration is often complete in animal studies, the accuracy and extent of repair are often imperfect in human patients.

To date, most research in the area of axon regeneration has focused on cell-extrinsic factors, and for good reason. When CNS neurons are presented with an opportunity to grow through a peripheral nerve graft, a small percentage of these cells extend lengthy axons but then stop growing when they re-enter the CNS environment (Vidal-Sanz et al., 1987; Aguayo et al., 1991). This growth is likely to be associated with both positive growth signals that are present in the PNS but not in the CNS, e.g., laminin and Schwann cell-derived growth factors, and with an absence of negative cues that are more prevalent in the CNS than in the PNS. The search for such negative signals has led to the discovery of multiple growth-inhibitory proteins that are expressed in the CNS by oligodendrocytes, along with their cognate receptors on neurons. Additional sources of growth inhibition include chondroitin sulfate proteoglycans (CSPGs) and other proteins that are present in the perineuronal net and that accumulate in the scar that forms at the site of injury, as well as growth cone repellants of the Ephrin and semaphorin families. These findings have been the subject of several excellent reviews (Schwab, 2002; Filbin, 2003; Lee et al., 2003; Silver and Miller, 2004; Carmeliet and Tessier-Lavigne, 2005) and will not be covered here in detail.

In spite of these advances, it is becoming increasingly clear that regenerative failure in the CNS is also related to the neurons' intrinsic growth state. Even when grown on permissive substrates, mature CNS neurons show far less ability than peripheral ganglionic neurons to extend axons (Chierzi et al., 2005). One CNS population that has been studied extensively, retinal ganglion cells (RGCs), show a rapid decline in their ability to grow axons in the early postnatal period, when they shift over to begin elaborating dendrites (Goldberg et al., 2002a). The importance of neurons' intrinsic growth state in limiting CNS regeneration is also evident in vivo. Many approaches have been used to interfere with growth-inhibitory proteins, their receptors, or their downstream signaling pathways, and while some of these studies have given promising results, others have shown little or no benefit (Bartsch et al., 1995; Lehmann et al., 1999; Dergham et al., 2002; Fournier et al., 2002; GrandPre et al., 2002; Fournier et al., 2003; Kim et al., 2003; Simonen et al., 2003; Zheng et al., 2003; Fischer et al., 2004a,b; Kim et al., 2004; Song et al., 2004; Zheng et al., 2005). These findings suggest that overcoming inhibition is not sufficient to produce extensive axon regeneration in the CNS (Woolf, 2003). One possible reason for this outcome is that counteracting one or even several inhibitory signals still leaves other ones in place. Another possibility, however, is that even if all inhibitory signals could be overcome, the low intrinsic growth potential of most CNS neurons would still limit the amount of regeneration that can occur. In this case, repair strategies aimed only at overcoming inhibition would be like trying to drive a car by taking one's foot off the brake without stepping on the accelerator (Steeves and Tetzlaff, 1998).

What are the cell-autonomous factors that govern a neuron's ability to extend an axon? To address this question, we will focus primarily on two types of neurons: sensory neurons of the dorsal root ganglia, which regenerate peripheral axons spontaneously, and RGCs, which do not regenerate injured axons under normal circumstances but which can be induced to do so. To date, these are probably the best-studied populations of PNS and CNS neurons, respectively, in terms of the cellular response to injury, involvement of trophic factors, and changes in gene expression associated with axon regeneration.

AXON REGENERATION IN THE PNS

DRG NEURONS AND THE CONDITIONING EFFECT

Peripheral nerve regeneration is defined as the ability of neurons with cell bodies located either in the CNS, i.e., motor neurons, or outside the CNS, i.e., sensory neurons, to regenerate severed axons that span the PNS and innervate peripheral target organs. Our understanding of peripheral nerve regeneration has increased significantly in the last two decades, based in large part upon the many studies that have been carried out in dorsal root ganglion (DRG) neurons.

The DRG neurons are unique in that they extend one axonal branch to peripheral targets through the PNS environment and a second axonal branch that either terminates in the spinal cord or ascends up through the spinal cord in the dorsal columns to reach the brainstem. Whereas the peripheral axon branch of DRG neurons regenerates when injured, the central branch fails to regenerate following injury to either the dorsal roots or the dorsal columns of the spinal cord. This failure is not only dependent in part on the environment encountered in the CNS associated with myelin inhibitory molecules along with astroglia-derived CSPGs that form the glial scar (Filbin, 2003; Schwab, 2004; Silver and Miller, 2004; Schwab et al., 2006), but it also depends upon the intrinsic growth state of the neurons. If the peripheral axonal branches of DRG neurons are injured prior to injuring its central axonal branch, the neurons are primed into a growth state that enables them to regenerate their central branch through a lesion site in the spinal cord (Neumann and Woolf, 1999). Peripheral "conditioning" lesions likewise enhance the ability of the central axonal branch to regenerate into a peripheral nerve graft in the spinal cord (Richardson and Issa, 1984). Thus, a peripheral nerve injury primes DRG neurons into an active growth state, and enhances the ability of the axon to grow through either a permissive or a restrictive extracellular environment in vivo.

A peripheral conditioning lesion also enhances regeneration of DRG neurons' peripheral axon branch: a crush injury of the sciatic nerve performed 3–7 days prior to a second crush injury enhances the rate of peripheral axonal growth (McQuarrie et al., 1977; Bisby and Pollock, 1983; Sjoberg and Kanje, 1990), and this is also seen when cells are placed in culture. Adult DRG neurons readily grow and extend many highly branched neurites when cultured on a permissive substrate such as laminin, but do not grow well on a non-permissive substrate such as CNS myelin. However, if the peripheral axonal branch is injured several days prior to culturing the neurons, DRG neurons extend very long and sparsely branched neurites on laminin and also grow more readily on non-permissive substrates such as CNS myelin. A central axonal injury (dorsal root rhizotomy) performed prior to culture fails to produce these priming effects (Hu-Tsai et al., 1994; Smith and Skene, 1997; Neumann et al., 2002; Qiu et al., 2002).

Activating the intrinsic growth state of neurons is essential for both peripheral and central axonal regeneration. Under normal circumstances, CNS neurons show a small and transient upregulation of growth-associated genes, but lack the ability to maintain their expression. Sustaining the intrinsic growth state by performing two preconditioning injuries of the sciatic nerve consecutively enables injured DRG axons in the dorsal columns to extend further through the hostile CNS environment than a single conditioning lesion (Neumann et al., 2005). This study, though not clinically applicable, indicates that successful regeneration requires inducing and maintaining neurons' intrinsic growth state. Combinatorial treatments that will activate the neurons' intrinsic growth state while also counteracting inhibitory signals may result in much greater recovery than either approach alone (Steinmetz et al., 2005).

Many questions arise as to how a peripheral conditioning lesion, but not injury to the central axon branch of a DRG neuron, enhances axon regeneration. Timing is essential, as only peripheral conditioning lesions performed at a certain interval before a subsequent axonal injury enhance regeneration. This implies that signals emanating from a peripheral injury site produce changes in the cell body that prime the neuron for growth. Transcription plays a significant role, as hindering transcription using RNA polymerase II inhibitors prevents neurite outgrowth (Smith and Skene, 1997; Cai et al., 2002a). More than a thousand genes undergo transcriptional changes after a peripheral injury, whereas the changes seen after a central axonal injury are much more modest, with little change in the expression of growth-associated genes (Figure 1.1) (Costigan et al., 2002; Xiao et al., 2002; Bareyre and Schwab, 2003).

GROWTH-PROMOTING SIGNALS

The signals emanating from the injury site that are responsible for the vast changes in gene expression associated with axon growth have not yet been fully elucidated. Among the many changes that occur at the injury site are the disruption of retrograde axonal transport, loss of target-derived trophic support, local axonal protein synthesis, changes in ion flux with increased Na^+ and Ca^{2+} intake leading to excessive neuronal firing, leukocytes and macrophage recruitment to the injury site, and Wallerian degeneration of the distal nerve stump that helps generate a permissive environment for growth (Makwana and Raivich, 2005; Hanz and Fainzilber, 2006; Twiss and van Minnen, 2006; Chen et al., 2007; Raivich and Makwana, 2007).

Peripheral nerve injury initiates cellular and transcriptional changes of many growth-associated genes that are sustained until trophic support is achieved by reinnervation of the peripheral targets. This observation implies that induction of the growth response depends upon the loss of target-derived neurotrophic support. However, several growth factors, including glial cell line-derived neurotrophic factor (GDNF), nerve growth factor (NGF) and cytokines, are upregulated in Schwann cells at the distal nerve stump after peripheral injury and may play an important role in stimulating neuronal growth through binding to their cognate receptors on axons, which can also be induced by the lesion. Thus, it is thought that the loss of neurotrophic support may contribute to initiating growth, whereas the gradual increase in growth factors and their neuronal receptors are important in supporting and enhancing regrowth following peripheral nerve injury.

The neurotrophin family of growth factors, which includes NGF, brain-derived neurotrophic factor (BDNF), NT-3, and NT4/5, signal through the Trk family of receptor tyrosine kinases and/or the common neurotrophin receptor p75 to influence neuronal survival and axonal growth. Embryonic and neonate DRG neurons depend on neurotrophic support for survival in culture, though adult DRG neurons do not. Therefore, the effects of neurotrophins on growth of adult DRG neurons can be monitored independent of any survival effects. In culture, NGF or BDNF stimulates adult DRG neurons to extend neurites, and NGF combined with BDNF or NT-3 enables more cells to extend neurites than either

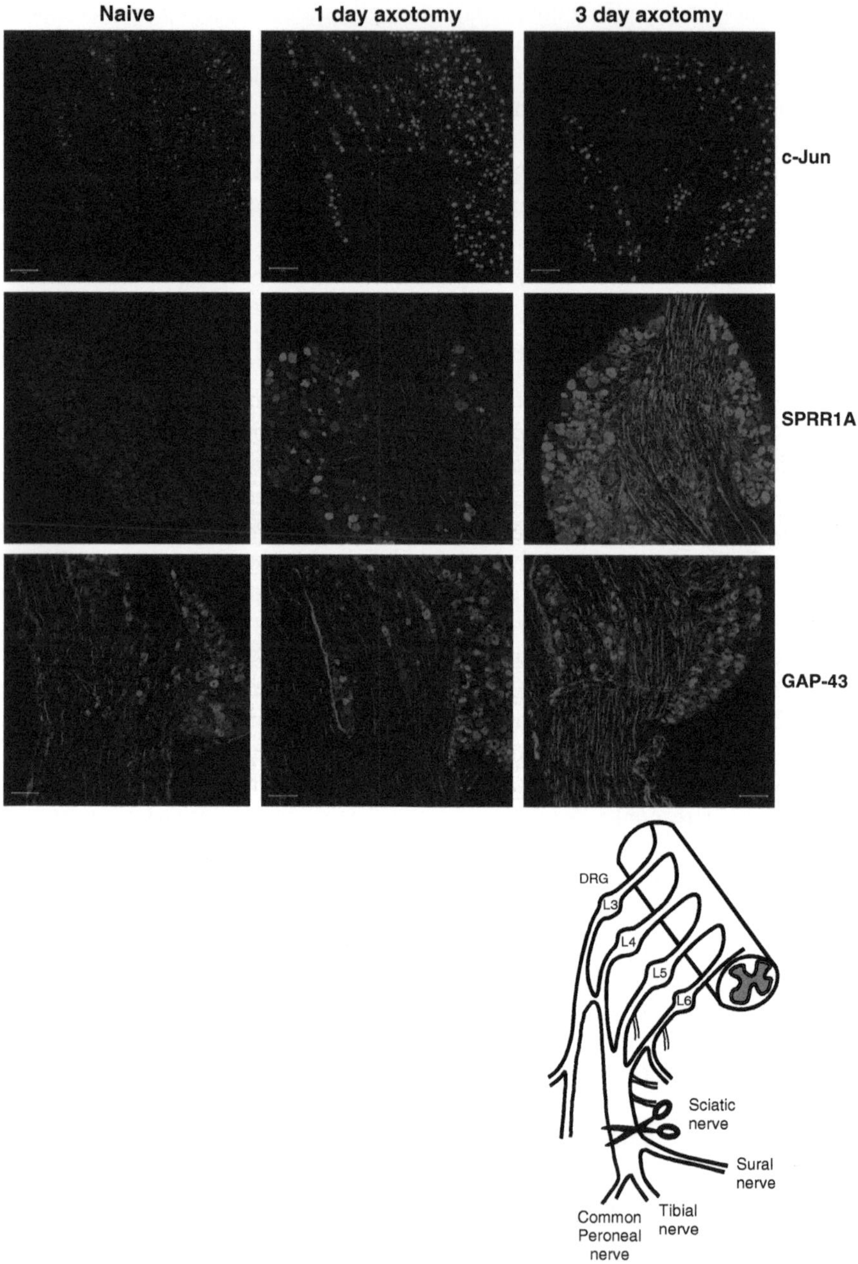

FIGURE 1.1 Peripheral nerve injury induces the expression of growth-associated genes in DRG neurons. Detection by immunostaining of c-Jun, SPRR1A, and GAP-43 in uninjured mouse adult DRG and in DRG neurons 1 and 3 days following sciatic nerve transection (axotomy). The transcription factor c-Jun is localized to the nucleus of the neurons, whereas GAP-43 and SPRR1A are localized to the cytoplasm and massively distributed along the nerve fibers after nerve injury. Scale bar = 100 μm.

one alone (Lindsay, 1988; Hu-Tsai et al., 1994; Gavazzi et al., 1999). NGF promotes axonal elongation, whereas BDNF or NT-3 do not (Kimpinski et al., 1997). However, the effect of NGF in culture does not measure up to the effect of a prior conditioning lesion (Hu-Tsai et al., 1994; Smith and Skene, 1997). A conditioning lesion causes DRG neurons to extend long and sparsely branched neurites which resemble a true regenerative mode, whereas NGF promotes long but highly branched neurites in culture, suggestive of a sprouting mode of growth (Hu-Tsai et al., 1994; Smith and Skene, 1997; Gavazzi et al., 1999). Delivery of neurotrophins promotes some degree of axonal regeneration in vivo. Following injury of centrally directed axons in the dorsal roots (dorsal rhizotomy), delivery of neurotrophins to the injury site enables DRG neurons to project through the PNS–CNS boundary region, the dorsal root entry zone (DREZ), and into the spinal cord. The extent of regeneration is neurotrophin-selective. NGF and NT-3 promote growth in those sub-populations of DRG neurons that express the TrkA and TrkC receptors, respectively, whereas BDNF fails to promote growth (Ramer et al., 2000). Delivery of NGF or NT-3 to the spinal cord using viral vectors, or NT-3 administered intrathecally with or without nerve grafts, also enhances growth through the injury site and into the spinal cord (Oudega and Hagg, 1996; Zhang et al., 1998; Bradbury et al., 1999; Iwaya et al., 1999; Oudega and Hagg, 1999; Romero et al., 2001; Ramer et al., 2002).

The GDNF family of growth factors signal through a receptor complex that includes a member of the GFR family of glycosylphosphatidylinositol (GPI)-linked receptors and the RET tyrosine kinase receptor. GDNF receptors are expressed in subpopulations of small and large DRG neurons (Bennett et al., 2000; Josephson et al., 2001). GDNF, like NGF, promotes neurite outgrowth in adult DRG neurons in culture (Gavazzi et al., 1999). GDNF expressed in vivo in fibroblast grafts promotes the growth of sensory DRG axons into but not through the graft in the spinal cord after dorsal column injury (Blesch and Tuszynski, 2003). In vivo delivery of GDNF intrathecally after a dorsal rhizotomy enables robust growth through the DREZ into the spinal cord and, after dorsal column injury, growth around but not through the lesion site (Bradbury et al., 1999; Ramer et al., 2000). GDNF also seems to augment the growth effect of a preconditioning lesion. Low doses of GDNF coupled with a preconditioning lesion enhance growth after a dorsal column injury; however, higher doses fail and even perturb the effect of a conditioning lesion (Mills et al., 2007). This again points to the importance in balancing neurotrophic support levels, as high doses may mimic a state in which the axon reconnects with the peripheral target and halts regeneration. Among the other GDNF family members, artemin affect neurite outgrowth of cultured DRG neurons, apparently by inducing transcriptional changes in genes involved in regulating actin polymerization (Park and Hong, 2006). Both GDNF and neurturin prevent semaphorin 3A-mediated growth cone collapse in cultured adult DRG neurons (Wanigasekara and Keast, 2006) and promote neurite outgrowth of adult motor neurons in spinal cord explant cultures (Bilak et al., 1999).

Insulin and insulin-like growth factors (IGFs: IGF-1, IGF-2) also affect the growth characteristics of sensory neurons. In culture, IGF-1, IGF-2, and insulin stimulate neurite initiation and elongation in DRG neurons and augment the effects of NGF when administered together (Fernyhough et al., 1993; Akahori and Horie, 1997; Kimpinski and Mearow, 2001; Jones et al., 2003). IGFs and insulin signal through insulin and insulin-like growth factor receptors, and in vivo, enhance peripheral nerve regeneration (Kanje et al., 1989; Glazner et al., 1993; Xu et al., 2004; Toth et al., 2006). Fibroblast growth factor (FGFs) also enhance peripheral nerve regeneration. FGF-1 and FGF-2 enhance axonal outgrowth in adult DRG neurons in culture (Mohiuddin et al., 1996) and in vivo, FGF transgenic mice show enhanced early peripheral regeneration and enhanced myelination (Jungnickel et al., 2006). Bridging of the injured sciatic nerve with grafts expressing FGF-2 promotes axonal growth into the graft (Timmer et al., 2003; Haastert et al., 2006). Moreover, delivery of FGF-2 to the spinal cord after dorsal root crush promotes growth through the DREZ with functional recovery (Romero et al., 2001).

The IL-6 family of cytokines, including Interleukin-6 (IL-6), Leukemia inhibitory factor (LIF) and Ciliary neurotrophic factor (CNTF), are secreted by macrophages and Schwann cells and are elevated in DRG neurons after peripheral nerve injury. These cytokines also contribute to nerve regeneration. CNTF can mimic the effect of a conditioning lesion by promoting growth of injured dorsal root axons into the spinal cord when delivered to the injury site (Wu et al., 2007). IL-6 and LIF are also important for axonal regeneration, as the growth-promoting effects of a conditioning lesion on injured central axons fails in IL-6 and LIF knockout mice (Cafferty et al., 2001, 2004). Moreover, intrathecal delivery of IL-6 to DRG neurons enhances central axonal growth after dorsal column injury (Cao et al., 2006).

ROLE OF MONOCYTES

Circulating blood monocytes are rapidly recruited to the injury site and differentiate into activated macrophages after peripheral nerve injury. Macrophages, including resident macrophages, together with Schwann cells help in clearing myelin and axonal debris, thereby creating a permissive environment for growth (Stoll et al., 1989; Lu and Richardson, 1993; Bruck, 1997; Hu and McLachlan, 2003). The macrophages decrease in numbers at about 2 weeks after peripheral injury, which appears to be mediated, at least in part, by the process of remyelination. Myelin expresses myelin-associated glycoprotein (MAG) which binds to the NgR on macrophages and promotes clearance (Fry et al., 2007). The mechanisms responsible for macrophage recruitment are not fully understood. Serum complement is essential, as both in vitro and in vivo studies show that complement depletion inhibits the phagocytic ability of macrophages and delays axonal regeneration after a sciatic nerve crush (Bruck and Friede, 1991; Dailey et al., 1998). The tissue plasminogen activator (tPA) also mediates macrophage recruitment. The serine protease tPA, known for its role in lysis of blood clots, is rapidly upregulated in Schwann cells after peripheral injury. Mice lacking tPA show increased axonal degeneration and demyelination, delayed axonal regeneration, and reduced recruitment of macrophages (Akassoglou et al., 2000; Siconolfi and Seeds, 2001; Ling et al., 2006). Exogenous delivery of tPA to crushed sciatic nerve results in an increase in macrophage infiltration and enhanced peripheral nerve regeneration. tPA induces an increase in the expression of matrix metalloproteinase-9 (MMP-9) in macrophages, suggesting that the macrophages induce MMP-9 to help clear myelin debris and to prevent collagen scar formation (Ling et al., 2006; Zou et al., 2006). Schwann cells also attract macrophages by secreting the chemokine MCP-1, which acts through the CCR2 receptor (Siebert et al., 2000; Tofaris et al., 2002). Oxidized galectin-1 and PAP-III are two other chemoattractant proteins produced by Schwann cells that are important in recruiting macrophages and enhancing peripheral nerve regeneration (Horie et al., 2004; Namikawa et al., 2006). Macrophage activation within the DRG enhances the ability of sensory neurons to regenerate axons through a crush site in the dorsal roots (Lu and Richardson, 1991) and, when combined with methods to overcome CSPGs, enables axons in the dorsal root to regenerate into the dorsal horn of the spinal cord and reestablish appropriate connections (Steinmetz et al., 2005). Macrophages seem to enhance axonal growth by secreting growth-promoting factors, as activated macrophage-conditioned medium administered in vitro to DRG explants promotes axonal growth and Schwann cell migration (Luk et al., 2003; Horie et al., 2004). Macrophages secrete several cytokines and neurotrophic factors including IL-1, IL-6, NGF, oncomodulin (Yin et al., 2006) and more, though it is still unclear which of these are the most relevant in stimulating axon growth after peripheral nerve injury (Review: Kiefer et al., 2001).

SIGNALING CASCADES

What are the downstream signaling cascades that prime DRG neurons for growth? IL-6 and related cytokines bind to cell-surface receptor complexes that include the signal transducing receptor gp130 subunit (glycoprotein 130), and this leads to the activation of the JAK (Janus Kinase) tyrosine kinase. JAK phosphorylates STAT3 (signal transducer and activator of transcription-3), which in turn dimerizes and translocates into the nucleus to alter gene expression. Inhibition of the JAK–STAT transduction pathway, using JAK2 inhibitors prevents neurite outgrowth of cultured sensory neurons (Liu and Snider, 2001), attenuates the effect of a conditioning lesion on the growth of sensory axons after a dorsal column injury and blocks STAT3 phosphorylation after sciatic nerve transection (Qiu et al., 2005). Activated STAT3 promotes neurite outgrowth, whereas delivery of the suppressor of STAT3 (SOCS3) inhibits neurite outgrowth by blocking nuclear translocation of STAT3 (Miao et al., 2006). Thus, the JAK–STAT pathway is activated after peripheral nerve injury and is necessary for axonal growth.

As mentioned above, peripheral nerve injury leads to increased expression of IL-6 and related cytokines in non-neuronal cells and in DRG neurons. The induction of IL-6, CNTF, and LIF mRNA in DRG neurons is cAMP-dependent (Cao et al., 2006; Wu et al., 2007). In general, cAMP levels play an integral part in the regenerative potential of neurons after injury. cAMP levels are elevated in DRG neurons by peripheral nerve injury and by neurotrophins, but are suppressed by MAG and myelin, which inhibit the activity of adenylate cyclase (Spencer and Filbin, 2004). DRG neurons treated with dibutyril cAMP, a cAMP analog, can overcome myelin inhibition in culture (Cai et al., 2001) and regenerate severed axons through a lesion site in the spinal cord (Neumann et al., 2002; Qiu et al., 2002). This effect is similar to, though not as robust, as what is seen after a conditioning lesion (Neumann and Woolf, 1999). Also, although cAMP enhances regeneration through the non-permissive environment of the CNS, it fails to enhance growth through a peripheral nerve graft (Han et al., 2004). cAMP effects are transcription-dependent, and involve signaling through protein kinase A (PKA) to activate the transcription factor CREB (Cai et al., 1999; Gao et al., 2004). Activated CREB, like cAMP delivery, is sufficient to overcome myelin inhibition in vitro and to promote regeneration of lesioned dorsal column axons (Gao et al., 2004). CREB upregulates Arginase I, resulting in subsequent synthesis of polyamines (Cai et al., 2002b; Gao et al., 2004). Either overexpression of Arginase I or addition of the polyamine putrescine is sufficient to allow neurons to overcome the inhibitory effects of myelin in culture (Cai et al., 2002a). Polyamines are thought to affect axonal growth through interaction with cytoskeletal tubulin. Activated PKA, in addition to its effects on gene transcription, acts locally to prevent myelin-induced growth cone collapse by inhibiting the activity of RhoA (Dong et al., 1998; Snider et al., 2002). IL-6 and cAMP pathways seem to converge as cAMP induces IL-6 cytokines expression in neurons, though inhibition of the JAK–STAT pathway in culture does not perturb growth on a non-permissive substrate and only partially perturbs the effect of cAMP on axonal growth when cells are cultured on a permissive substrate. These results suggest that both signaling cascades are required to stimulate neurons' intrinsic growth potential and overcome myelin inhibition (Cao et al., 2006; Wu et al., 2007).

Besides elevating cAMP levels, neurotrophins, acting through their respective Trk receptors, exert multiple other effects on DRG neurons. The Ras–Raf–MAPK/ERK pathway is important for neurotrophin-dependent survival and axonal growth during development (Markus et al., 2002; Zhong et al., 2007), and there is compelling evidence that this pathway is also essential for nerve regeneration. Phosphorylated ERK is detected in DRG neurons, satellite cells, and axons after peripheral nerve injury (Sheu et al., 2000; Obata et al., 2003; Doya et al., 2005; Agthong et al., 2006), and inhibition of ERK prevents adult DRG neurons from spontaneously initiating neurites in culture (Chierzi et al., 2005). Inhibition of MEK, the upstream activator of ERK, suppresses neurotrophin-induced neurite outgrowth and the robust outgrowth that occurs when DRG neurons are placed in culture after a conditioning lesion (Sjogreen et al., 2000; Wiklund et al., 2002). Activation of the ERK induces its retrograde transport and nuclear translocation (Reynolds et al., 2001; Perlson et al., 2006). ERK activation in turn activates many transcription factors including CREB and STAT3, both of which are important components of the intrinsic growth response (Gao et al., 2004; Qiu et al., 2005). Besides its effects on transcription, activation of the ERK pathway exerts local effects on microtubule assembly.

The PI3K/Akt signaling cascade also contributes to peripheral nerve regeneration. Akt is highly phosphorylated in motor neurons, and delivery of activated Akt to hypoglossal motor neurons enhances axonal regeneration in vivo (Namikawa et al., 2000). Conversely, in culture, inhibition of PI3K diminishes neurite outgrowth and elongation in response to growth factors (Kimpinski and Mearow, 2001; Edstrom and Ekstrom, 2003), although neurite outgrowth in adult DRG neurons evoked by a preconditioning injury is unaffected (Liu and Snider, 2001). This suggests that the PI3K/Akt pathway may act locally to regulate the cytoskeleton rather than initiating intrinsic cell body changes in response to nerve injury. In contrast, the SAPK/JNK signaling pathway, which is also activated following peripheral nerve injury, induces transcriptional changes that contribute to axonal growth (review (Waetzig et al., 2006). JNK inhibitors reduce neurite outgrowth from adult explants of dorsal root and nodose ganglia and decrease both c-Jun phosphorylation and ATF-3 expression (Lindwall et al., 2004). Loss of target-derived NGF and GDNF following nerve injury induces the expression of the transcription factors c-Jun and ATF3 (Gold et al., 1993; Averill et al., 2004) and treatment with NGF reduces c-Jun activation (Lindwall and Kanje, 2005b). Furthermore, inhibition of either axonal transport or JNK activation in vitro results in reduction in c-Jun and ATF3 expression (Lindwall and Kanje, 2005a), implying that both activation and retrograde transport of JNK to the nucleus are required for the transcriptional activation that leads to enhanced axonal growth. Figure 1.2 summarizes some of the signaling pathways that have been implicated in peripheral nerve repair.

TRANSCRIPTIONAL CHANGES

The transcription factors that lie downstream of these signaling cascades and that are essential for the regenerative cell body response are only partly known. Several transcription factors are elevated and/or activated in DRG and motor neurons following peripheral nerve injury. These include c-Jun, ATF3, STAT3, CREB, C/EBP-β, sox11 and JunD; other transcription factors such as ATF2 and Islet-1 are downregulated (Jenkins and Hunt, 1991; Leah et al., 1991; Herdegen et al., 1992; Martin-Villalba et al., 1998; Hol et al., 1999; Schwaiger et al., 2000; Tsujino et al., 2000; Tanabe et al., 2003; Lee et al., 2004; Nadeau et al., 2005; Jankowski et al., 2006). Neuron-specific knockout of c-Jun decreases the rate of regeneration of injured facial motor neuron axons and reduces the expression of growth-associated genes (Raivich et al., 2004). Forced expression of ATF3

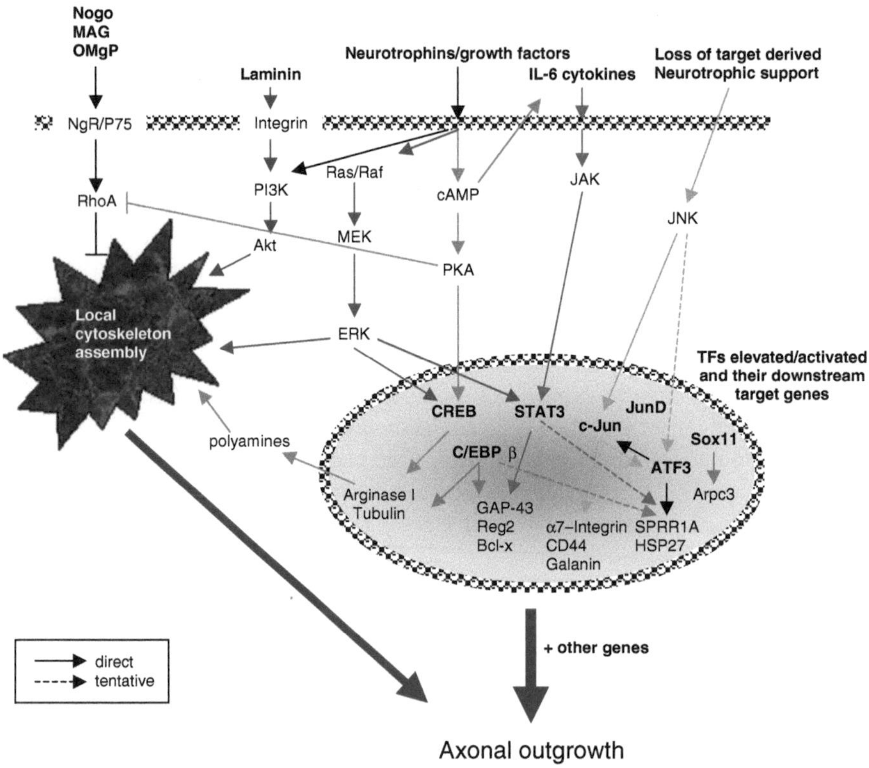

FIGURE 1.2 Signaling pathways implicated in peripheral nerve regeneration. Loss of target-derived neurotrophins and GDNF after injury activates the SAPK/JNK pathway. Activated JNK translocates to the nucleus and induces the expression of ATF3 and phosphorylates c-Jun. ATF3 also activates c-Jun expression and the growth-associated genes HSP27 and SPRR1A. c-Jun may reciprocally elevate ATF3 transcription and also controls the expression of α7-Integrin, CD44, and galanin. The IL-6 family of cytokines acts through the JAK/STAT pathway, inducing dimerization and translocation of STAT3, which leads to the upregulation of GAP-43, Reg2, Bcl-x, and possibly SPRR1A. Neurotrophins elevate cAMP, which activates PKA and also elevates levels of the IL-6 family of cytokines. PKA activates CREB, stimulating transcription of tubulins and Arginase I. Arginase I directs synthesis of polyamines that regulate the cytoskeleton. PKA also inhibits RhoA activation thus preventing growth cone collapse induced by inhibitory myelin proteins. Neurotrophins and growth factors activate the MAPK/ERK pathway. ERK mediates activation of CREB and STAT3. Neurotrophins, laminins, and growth factors activate the PI3K pathway, which acts locally to mediate cytoskeleton assembly. The transcription factors JunD, Sox11, and C/EBP-β are also elevated in DRG neurons after peripheral nerve injury and some of their target genes have been identified including GAP-43, tubulin, and Arpc3 and possibly SPRR1A. This diagram should be considered as a working model. Locally synthesized axonal proteins implicated in axonal growth and the machinery responsible for retrograde transport of many of the signals that are generated in the injured axon are not shown here (Hanz and Fainzilber, 2006; Perlson et al., 2006; Twiss and van Minnen, 2006). Many more transcriptional changes occur after peripheral nerve injury that are likely to be essential in promoting axonal growth.

in DRG neurons in vivo enhances peripheral nerve regeneration by inducing the expression of growth-associated genes, but is not sufficient to promote regeneration of DRG central axons after a dorsal column injury (Seijffers et al., 2007). c-Jun and ATF3 both enhance neurite outgrowth in vitro, and when co-expressed, have synergistic effects on axonal growth (Nakagomi et al., 2003; Pearson et al., 2003; Seijffers et al., 2006). ATF3, c-Jun, CREB, ATF2, and JunD belong to the leucine zipper family of transcription factors that can form homo- or heterodimers, and bind to AP1 and CRE/ATF DNA promoter elements, thereby activating or repressing the expression of target genes (Hai et al., 1999; Hai and Hartman, 2001). Only some of the target genes regulated by ATF3 and c-Jun after injury have been identified. Forced ATF3 expression in non-injured adult DRG neurons in vivo is sufficient to induce expression of the growth-associated genes Hsp27, SPRR1A, and c-Jun, but not GAP-43, CAP-23, STAT3, or α7-integrin, genes that are normally induced after peripheral injury (Seijffers et al., 2007). Absence in neuronal c-Jun inhibits the axotomy-induced expression of the neuropeptide galanin and the receptors CD44 and α7-integrin (Raivich et al., 2004). In addition to ATF3 regulating c-Jun expression, it is likely that c-Jun regulates ATF3 expression, as both their promoters contain functional AP1 sites and JNK inhibition results in loss of c-Jun activation and reduced ATF3 expression levels (Morooka et al., 1995; Cai et al., 2000; Lindwall and Kanje, 2005b; Seijffers et al., 2007). The expression of the injury-induced growth-associated genes GAP-43 and tubulin-α1 is inhibited in C/EBP-β knockout mice (Nadeau et al., 2005). C/EBP-β may also be responsible for the expression of the growth-associated gene SPRR1A, as a functional C/EBP-β binding site is located in the SPRR1A promoter (Pradervand et al., 2004). The gene encoding the growth-associated protein GAP-43 is a downstream target of STAT3 (Cafferty et al., 2004; Qiu et al., 2005). In addition, STAT3 regulates the expression of Reg-2 and the anti-apoptotic gene Bcl-x in motor neurons following nerve injury, and may also regulate the expression of SPRR1A (Schweizer et al., 2002; Wu et al., 2007). CREB, in addition to inducing the expression of Arginase I (Cai et al., 2002a; Gao et al., 2004), may also regulate tubulin expression (Han et al., 2004). Undoubtedly, injury-induced transcription factors control the expression of many more target genes than are currently identified. To date, most studies in this field have evaluated the effects of transcription factors one at a time, using either loss- or gain-of-function techniques. However, transcription factors usually do not act alone to control gene transcription, but either dimerize with other transcription factors, interact through co-activators or co-repressors, or bind to adjacent binding sites to synergistically regulate transcription. Therefore, it is highly possible that the signaling cascades initiated by peripheral nerve injury converge through their known and still to be identified downstream transcription factors. These transcription factors such as ATF3, c-Jun, CREB and STAT3 probably interact or act in concert to modulate gene transcription of many target genes, leading to both enhanced axonal growth and a diminished response to inhibitory cues. Transcription factors that have been implicated in peripheral nerve repair are indicated in Figure 1.2.

GROWTH-ASSOCIATED GENES

Microarray analysis reveals that peripheral nerve injury in DRG or SCG neurons alters the expression of a thousand or more genes encoding cell adhesion molecules, cytoskeletal proteins, survival factors, growth-associated genes, ion channels, receptors, neuropeptides, transcription factors, and other proteins (Costigan et al., 2002; Xiao et al., 2002; Tanabe et al., 2003; Boeshore et al., 2004). Some of the striking changes in gene expression that accompany peripheral nerve regeneration are shown in Figure 1.1. Growth-associated genes or RAGs (regeneration associated genes) are genes that are expressed when neurons are actively growing; the significance of only a few of these for nerve regeneration is established. GAP-43 and CAP-23 are considered prototypic RAGs (Skene, 1989; Benowitz and Routtenberg, 1997; Caroni, 2001). These two proteins are abundant, plasmalemma-associated PKC substrates that are enriched in lipid rafts of growing axons, especially in growth cones. These proteins are thought to be involved in transducing extracellular signals to modulate actin accumulation and dynamics through interaction with the phosphoinositide lipid PI(4,5)P2 (Benowitz and Routtenberg, 1997; Caroni, 1997; He et al., 1997; Walsh et al., 1997; Frey et al., 2000; Laux et al., 2000). Transgenic mice that overexpress GAP-43 and CAP23 in neurons show enhanced axonal sprouting in vivo (Aigner et al., 1995; Caroni et al., 1997), whereas knockout mice show defects in axonal pathfinding, sprouting, and brain organization (Aigner and Caroni, 1995; Maier et al., 1999; Frey et al., 2000; Shen et al., 2002). GAP-43 and CAP-23 differ in sequence but share similar motifs and can in part functionally substitute for one another (Frey et al., 2000). Either GAP-43 or CAP-23 alone is not sufficient to induce regeneration of DRG axons after dorsal column injury, but co-expression of the two enables DRG neurons to extend axons into a peripheral nerve graft in the spinal cord to nearly the same extent that as a peripheral conditioning lesion (Bomze et al., 2001). Integrins are another group of proteins that are upregulated in DRG and motor neurons after nerve injury. Integrins are membrane-bound receptors for basement membrane proteins, including laminins, and signal through the PI3K/AKT pathway to facilitate axonal growth. Mice lacking the $\alpha7$ integrin receptor show delayed motor axon regeneration after facial nerve injury (Werner et al., 2000). Conversely, delivery of $\alpha7$ integrin to adult DRG primary cultures promotes neurite outgrowth, and overexpression of $\alpha1$ integrin enables growth on inhibitory substrates containing CSPGs (Condic, 2001). Another growth-associated gene encoding the small proline-rich repeat 1A (SPRR1A) protein is not expressed in naïve uninjured DRG and motor neurons, but is highly induced after peripheral nerve injury, though not after injury of the dorsal roots. Overexpression of SPRR1A in adult DRG cultures enhances growth on both laminin and CNS myelin, and depletion of SPRR1A in preconditioned adult DRG neurons significantly inhibits growth. This protein is enriched along the nerve fibers and at the leading edge of growth cones, and its localization with F-actin structures, suggesting that SPRR1A stimulates axonal elongation by regulating actin dynamics at growth cone ruffles (Bonilla et al., 2002).

SUMMARY

Unfortunately, successful peripheral nerve regeneration in humans, in contrast to rodents, is a milestone that is yet to be met. Deciphering the signals and transcriptional downstream cascades that underlie this process are important in achieving this goal. Therefore, seeking "harmless" strategies that will mimic the conditioning effects to enhance and maintain the intrinsic growth state of DRG neurons is of great value. It remains to be seen whether the pathways that promote growth on a favorable substrate are the same or independent from pathways that enable DRG neurons to overcome inhibitory CNS myelin cues and grow in the CNS. Moreover, defining the signals that activate the intrinsic growth state and that maintain it is clearly important. Is it a matter of loss of target-derived support that initiates the growth? What are the optimal levels of neurotrophins, cytokines, and other signals required to maintain the growth state? Defining what initiates and primes neurons for enhanced peripheral regeneration is an active area of investigation. Whether the same mechanisms can enable other neurons to regenerate axons in the CNS also remains to be determined.

AXON REGENERATION IN THE CNS

THE OPTIC NERVE AS A MODEL SYSTEM

The optic nerve has long been viewed as a paradigm of regenerative failure in the CNS. RGCs, like other cells in the retina, originate in the diencephalon, and their axons are ensheathed by CNS oligodendrocytes. Although axons in the optic nerve do not normally regenerate if injured, this situation can be reversed if RGCs are appropriately stimulated. Because of the anatomical simplicity of the optic nerve, its well-defined projections, and its accessibility, this system is in many ways ideal for understanding mechanisms that inhibit or promote regeneration in the CNS. Over the past 20 years, factors that determine the death, survival, and regenerative potential of RGCs have received a great deal of attention, and are thought to be representative of parallel phenomena occurring in other CNS neurons after injury (Isenmann et al., 2003).

OPTIC NERVE REGENERATION IN LOWER VERTEBRATES

Unlike mammals, fish and amphibia are able to regenerate their optic nerves and certain other CNS connections throughout life. Following injury to the optic nerve, RGCs increase in size and show a rapid increase in the number of polysomes associated with the endoplasmic reticulum, as overall levels of mRNA and protein synthesis increase several fold (Grafstein, 1986). The expression of multiple genes increases significantly over and above this general increase, including genes encoding components of the cytoskeleton (Burrell et al., 1978; Heacock and Agranoff, 1982; Hall and Schechter, 1991; Jian et al., 1996), cell-surface proteins, and proteins that are transported down the axon in membranous vesicles. This latter group undergoes some of the most striking changes, and includes GAP-43 (Benowitz et al., 1981; Skene and Willard, 1981). Other rapidly transported proteins that undergo striking increases during regeneration include glycoproteins of the N-CAM/L1 family (Bastmeyer et al., 1990; Blaugrund et al., 1990; Vielmetter et al., 1991; Bernhardt et al., 1996) and lipid raft proteins (flotillins) (Schulte et al., 1997). Changes in neuropilin expression have also been reported (Fujisawa et al., 1995). Most of these changes return to baseline levels when axons reach their appropriate targets (Benowitz et al., 1983).

INJURY RESPONSE IN MAMMALIAN RGCs

Compounding the problem of regenerative failure in the mammalian optic nerve, RGCs begin to die a few days after injury through both apoptotic and necrotic mechanisms. If the optic nerve is damaged within 1–2 mm of the eye, RGCs begin showing classic signs of apoptosis after about 5 days, i.e., DNA fragmentation, nuclear condensation, apoptotic bodies, and cell shrinkage (Berkelaar et al., 1994; Quigley et al., 1995). By the end of the second week, approximately 90% of RGCs have died. At the electron microscopic level, some RGCs show signs of necrotic cell death (Bien et al., 1999). Levels of the anti-apoptotic Bcl family members Bcl-2 and Bcl-x decrease after injury, while levels of Caspase-3 (Laquis et al., 1998) and Cytochrome C release (He et al., 2004) increase. Axotomy leads to upregulation of several transcription factors that are known to be associated with execution of the apoptotic program, including c-jun, MafK, and Fos-related antigen (Hull and Bahr, 1994; Fischer et al., 2004b), though these same transcription factors, in combination with others, may also contribute to cell survival and axon outgrowth. Cell death can be delayed somewhat, but not prevented, by inhibitors of Caspase-3 (Kermer et al., 1998; Laquis et al., 1998) and its upstream activator, Caspase-9 (Kermer et al., 1998). Overexpression of Bcl-2, on the other hand, enables the majority of RGCs to survive for long periods (Bonfanti et al., 1996; Chierzi et al., 1999). An excellent recent review has covered research in this area in detail (Isenmann et al., 2003).

Both the time of onset and the extent of RGC death depend on the distance between the injury site and the eye. Proximal injury causes more rapid and more extensive damage than distal injury (Villegas-Perez et al., 1993); disconnecting RGCs from their targets does not even cause cell death if axons are severed intracranially (Carpenter et al., 1986). The exact mechanism responsible for RGC death remains uncertain. One possibility is that RGC death may depend upon the entry (or local generation) of a death-inducing signal at the cut end, whose effect is attenuated if the injury is far from the soma (Isenmann et al., 2003). Another possible cause could be the loss of one or more retrogradely transported trophic agents, perhaps from the surrounding glial cells, though not from synaptic targets.

In culture, the survival of neonate RGCs depends upon a combination of growth factors plus physiological activity (Meyer-Franke et al., 1995). Early postnatal RGCs require a combination of a TrkB ligand (i.e., BDNF or NT-4/-5), a member of the CNTF family, and an insulin-like growth factor, along with elevation of intracellular cAMP. cAMP levels may normally be regulated by physiological activity: depolarization causes an influx of Ca^{2+}, which in turn activates the Ca^{2+}-calmodulin-sensitive adenylate cyclase, leading to an elevation in intracellular cAMP concentration (Meyer-Franke et al., 1995). Following optic nerve injury, RGCs begin to withdraw their distal dendrites (Bahr et al., 1988), and it is possible that the decline in synaptic activation that would be expected to follow from this leads to decreased intracellular cAMP.

TROPHIC FACTORS ENHANCE CELL SURVIVAL AFTER AXOTOMY

A number of trophic factors, including the ones that affect survival in early postnatal RGCs, increase the survival of mature RGCs after optic nerve injury. The most widely studied of these factors is BDNF (Mey and Thanos, 1993; Mansour-Robaey et al., 1994; Di Polo et al., 1998; Koeberle and Ball, 2002; Nakazawa et al., 2002). In one study, for example, BDNF enabled approximately 50% of RGCs to survive 2 weeks after optic nerve injury, compared to 15% in controls (Koeberle and Ball, 2002). NT-4/5, an alternate ligand to TrkB, also enhances RGC survival (Nakazawa et al., 2002). BDNF and NT-4/5 activate several downstream pathways, including those involving ras-MAP kinase, PI3 kinase-Akt, and phospholipase C-γ (Segal and Greenberg, 1996; Kaplan and Miller, 2000). BDNF can block apoptosis via both the PI3K and the MAPK signaling pathways, leading to the phosphorylation and inhibition of the pro-apoptotic Bcl-2 family member, Bad, and of Caspase-9, and to the expression of pro-survival genes (Datta et al., 1997; Cardone et al., 1998; Bonni et al., 1999; Brunet et al., 1999). In one study, blockade of MAPK signaling, but not of the PI3 kinase pathway, fully abrogated the effects of BDNF on survival (Cheng et al., 2002), whereas in another study, blockade of either the MAP kinase pathway or the PI3 kinase pathway only partially blocked the effects of BDNF (Nakazawa et al., 2002). Infecting RGCs with viruses expressing a constitutively active form of MEK-1, a key kinase in the MAP kinase signaling cascade, enhanced RGC survival after axotomy to a lesser extent than BDNF overexpression (Cheng et al., 2002).

With time after injury, RGCs lose their ability to respond to BDNF (Mansour-Robaey et al., 1994; Di Polo et al., 1998; Cheng et al., 2002). Much of this loss can be attributed to the loss of the cognate receptor, TrkB. TrkB expression levels decline rapidly in RGCs after optic nerve injury (Cheng et al., 2002), as does the portion of the receptor pool trafficked to the cell surface (Meyer-Franke et al., 1998). Infecting RGCs with adeno-associated virus (AAV) expressing TrkB increases RGC survival appreciably, and addition of exogenous BDNF on top enables approximately 75% of RGCs to survive 2 weeks after injury (Cheng et al., 2002). As mentioned above, elevation of intracellular [cAMP] enables developing RGCs to respond to growth factors in culture, and the same is true in vivo: following axotomy, the ability of RGCs to respond to BDNF is enhanced by elevation of intracellular [cAMP], which leads to the translocation of TrkB from the intracellular compartment to the plasma membrane (Meyer-Franke et al., 1998).

CNTF is at least as effective as BDNF in protecting axotomized RGCs from cell death (Mey and Thanos, 1993; Watanabe et al., 2003). As with BDNF, however, the response of RGCs to CNTF declines with time, and this correlates with diminished expression of CNTFR-α, an essential part of the receptor complex (Miotke et al., 2007). Combining BDNF and CNTF is no more effective than using either one alone (Watanabe et al., 2003). Unlike the situation with BDNF described above, elevation of [cAMP] does not increase the effects of CNTF on RGC survival when a peripheral nerve graft is sutured to the cut end of the optic nerve (Cui et al., 2003).

The RGCs also express receptors for the TGF-β superfamily members GDNF and neurturin (i.e., GFRα-1 and GFRα-2, respectively), along with the co-receptor Ret, and both of these factors markedly enhance RGC survival after axotomy. Unlike CNTF, GDNF and neurturin augment the effects of BDNF on the survival of injured RGCs, bringing survival levels up to 80% (Yan et al., 1999; Koeberle and Ball, 2002). GDNF and XIAP, a caspase inhibitor, also have additive effects on RGC survival (Schmeer et al., 2002).

A number of other trophic factors have lesser effects on RGC survival after ON injury, including vascular endothelial growth factor (VEGF) (Kilic et al., 2006), Erythropoietin (Kretz et al., 2005), FGF-2 (Cheng et al., 2002), and IGF-1 (Kermer et al., 1998). It should be noted that many of these studies have not distinguished between direct effects on RGCs and indirect effects via activation of another cell type and subsequent release of other trophic agents. RGC survival is also enhanced when a peripheral nerve fragment is grafted to the cut end of the optic nerve (Aguayo et al., 1991) or when macrophages are activated intravitreally (Leon et al., 2000; Yin et al., 2003). The factors responsible for increasing RGC survival are not known in either instance. Combining macrophage activation with BDNF augments RGC survival to very high levels (Pernet and Di Polo, 2006).

ROLE OF MICROGLIA

Microglia, the resident immune cells of the nervous system, are distributed in a regular mosaic array in the ganglion cell layer and inner nuclear layer of the retina and become activated a few days after optic nerve injury (Thanos, 1992; Zhang and Tso, 2003). Microglia are antigen-presenting cells and can have both neuroprotective and neurotoxic effects when activated (Sobrado-Calvo et al., 2007). Their role in phagocytosing dying RGCs is readily demonstrated by dye-transfer studies (Thanos, 1992). Microglia produce nitric oxide-free radicals and other neurotoxic agents, as well as a multiplicity of trophic factors (Koprivica et al., 2005). This raises the question of whether microglial activation contributes to RGC death or whether their activation is a consequence of RGCs dying for other reasons. BDNF or intravitreal macrophage activation, which are neuroprotective to RGCs, prevent microglial activation (Leon et al., 2000), but this could be due to the prevention of cell death by other means. Some studies have reported that suppressing microglial activation with minocycline or tetracycline have a small, transitory effect in protecting RGCs (Baptiste et al., 2005), whereas others report that some anti-inflammatory cytokines have an appreciable, though far from complete, protective effect after axotomy (Boyd et al., 2003; Koeberle et al., 2004). Since anti-inflammatory cytokines can act directly on other cell types (Ledeboer et al., 2002; Boyd et al., 2003), it is possible that the protective effects that have been reported were not mediated via microglial suppression. Overall, the literature would suggest that microglial activation is likely to be an effect of RGC death rather than a cause of it.

AXON REGENERATION AFTER OPTIC NERVE INJURY

The intrinsic ability of rat RGCs to extend axons declines sharply shortly after birth. Throughout the embryonic period when RGCs are extending axons in vivo, they exhibit rapid axon extension when placed in culture, but this declines at around postnatal day 2, when RGCs switch to a dendritogenic mode of growth. This switch in RGCs' growth mode does not occur if cells from prenatal retinas are left for extended times in culture, suggesting that a cell-extrinsic signal causes RGCs to change their growth mode. Amacrine cells, which normally come into contact with RGCs at about this time, appear to play a pivotal role in this regard (Goldberg et al., 2002a). The ability of RGCs to regenerate axons through the optic nerve, like the ability of RGCs to extend axons in dissociated cell culture, similarly declines with age. Within the first day or two after birth, RGCs can extend axons through the optic tract in hamsters, but this ability is lost within the first postnatal week.

The RGCs are fully dependent upon target-derived growth factors for survival in the early postnatal period. However, overexpression of the anti-apoptotic gene Bcl-2 enables RGCs to survive axotomy in this period (Bonfanti et al., 1996), and to regenerate their axons (Chen et al., 1997). Although this increased regenerative ability of mice overexpressing Bcl-2 has been attributed to a role of Bcl-2 in axon outgrowth, an alternate interpretation is that Bcl-2 overexpression enables RGCs to survive growth factor deprivation at a time when their intrinsic growth state is developmentally high. Abundant other evidence demonstrates that Bcl-2 overexpression and enhanced survival are not sufficient for regeneration. RGCs that overexpress Bcl-2 survive well in culture in the absence of BDNF, but do not extend axons unless BDNF is added (Goldberg et al., 2002b). In addition, Bcl-2 overexpression does not promote the regeneration of axons after optic nerve injury in adult animals, even when additional measures are taken to partially overcome the inhibitory effects of myelin (Chierzi et al., 1999).

In vivo, mature RGCs revert to a limited growth state when their axons are injured. This can be seen by examining axon outgrowth from retinal explants in culture. Retinal fragments placed in culture several days after injuring the optic nerve show considerably more axon outgrowth than retinas from normal controls (Meyer and Miotke, 1990); this outgrowth is seen only if axotomized RGCs are allowed to remain in vivo for 3–4 days before being explanted (Fischer et al., 2004b). RGCs from growth-enabled explants express higher levels of GAP-43 than controls (Meyer et al., 1994). A recent study used microarrays to examine gene expression in RGCs purified by fluorescence-activated cell sorting (FACS: Table I). These studies showed that 3.5% of RGC genes undergo significant changes in expression after axotomy (Fischer et al., 2004b). Some of these genes encode proteins associated with axon growth (e.g., SPRR1A, GAP-43, galanin, Fn14, galectin 3), neuroprotection (e.g., heat shock protein (HSP)-27, GADD-45), and apoptosis (cell-death activator CIDE-A); several of the transcription factors that are upregulated, including MafK, Fos-related antigen, ATF-3, and c-jun, could conceivably be related to either apoptosis or to survival and outgrowth (Hull and Bahr, 1994; Fischer et al., 2004b). In vivo, methods that selectively stain a small fraction of RGCs show that, after optic nerve injury, axons first withdraw from the region of damage but then extend back toward the lesion site over the first 2 weeks and show turning, looping, and branching (Ramon y Cajal, 1991). EphB3 produced by macrophages acts as an attractant for these growth cones (Liu et al., 2006).

Regenerative growth involves not only changes in gene expression at the cell body level, but also protein synthesis and degradation in the axon (Chierzi et al., 2005). Regenerating retinal axons contain ribosomes and elements of the translational machinery, and show evidence of local protein translation; inhibitors of protein translation block elongation of isolated axons. Levels of protein synthesis in isolated RGC axons are considerably lower, however, than in DRG neurons. RGC axons also contain the machinery for protein degradation, including the proteosome core, ubiquitin itself, and ubiquitinated proteins (Chierzi et al., 2005).

As mentioned above, when presented with the opportunity, 5–10% of axotomized RGCs will extend lengthy axons through a segment of peripheral nerve grafted to the cut end of the optic nerve, and even form synapses if the far end of the graft is sutured to the superior colliculus (Carter et al., 1989; Aguayo et al., 1991). This growth has generally been attributed to the permissive environment of the PNS, but it is also likely to require the partial activation of RGCs' intrinsic growth state following axotomy and by the transport of growth factors provided by cells in the graft, e.g., Schwann cells and macrophages. The ability of RGCs to regenerate axons through peripheral nerve grafts was first observed by Tello, a student of Ramon y Cajal, and represented the first demonstration that CNS neurons retain an intrinsic capacity to extend axons (Ramon y Cajal, 1991).

AXON REGENERATION THROUGH THE OPTIC NERVE

Although RGCs are normally unable to regenerate injured axons into the optic nerve, this situation can be reversed to some extent. Implanting a fragment of peripheral nerve into the vitreous enables many RGCs to extend lengthy axons through the optic nerve (Berry et al., 1996). These implants contain viable Schwann cells and fibroblasts, and are decorated with numerous macrophages, any of which could be the source of the factors that stimulate axon regeneration. Equally strong regeneration can be achieved by simply injuring the lens (Leon et al., 2000; Fischer et al., 2001). Lens injury leads to an influx of blood-borne macrophages into the eye, and Zymosan, a pro-inflammatory agent, produces similar or even stronger regeneration when injected intravitreally. These results suggest that macrophage activation may be the key to stimulating extensive regeneration (Leon et al., 2000; Yin et al., 2003). Proteins secreted by macrophages, when tested on mature rat RGCs in culture, produced more extensive outgrowth than any other factors known to affect RGCs, e.g., BDNF, CNTF, GDNF, or bFGF. These findings suggested that axon-promoting effects of macrophages are linked to something other than these factors (Leon et al., 2000).

When isolated by column chromatography and sequenced, the principal macrophage-derived growth factor was identified as Oncomodulin. Oncomodulin is a small Ca^{2+} protein not previously known to act as a trophic factor and not related to other identified trophic factors. Oncomodulin binds with high affinity to a cell-surface receptor on RGCs and stimulates more extensive outgrowth than any other growth factor tested. The binding of oncomodulin to its receptor requires elevation of [cAMP], similar to what has been shown for BDNF and TrkB (Meyer-Franke et al., 1998). Immune-depletion of oncomodulin from macrophage-conditioned media eliminated the axon-promoting activity of the conditioned media. When delivered into the eye along with a cAMP analog, oncomodulin stimulated nearly as much growth as macrophage activation (Figure 1.3) (Yin et al., 2006). It should be noted, however, that intravitreal macrophage activation stimulates extensive outgrowth even without adding a cAMP analog. This observation suggests that some other factors may be involved that are secreted by macrophages (but which cannot stimulate outgrowth on their own) or that are secreted from another cell type in the eye secondary to macrophage activation. A mixture of NGF, BDNF, and basic FGF, but not any one alone, stimulate RGCs to extend lengthy axons through the optic nerve, though the direct cellular targets of these factors are unknown (Logan et al., 2006).

Besides stimulating axon regeneration through the optic nerve, lens injury (Fischer et al., 2000) and intravitreal macrophage activation (Yin et al., 2003) strongly augment the amount of axon regeneration that occurs when a peripheral

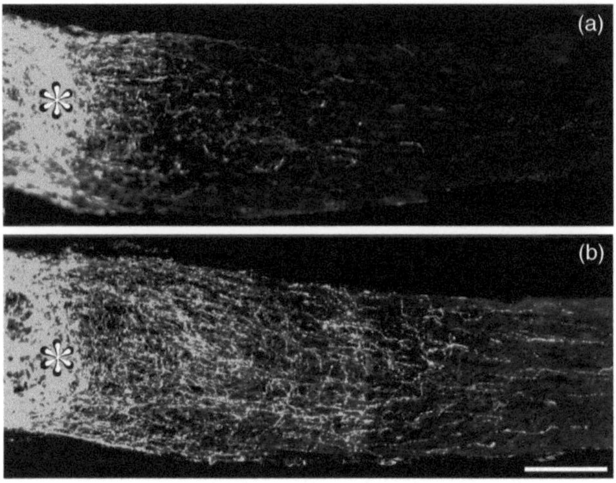

FIGURE 1.3 Oncomodulin promotes optic nerve regeneration in vivo. Rats with optic nerve injury were injected intraocularly with slow-release polymer beads that released buffer alone (a) or oncomodulin plus a cAMP analog (b). Regenerating axons (*green fluorescence*) were visualized using an antibody to GAP-43. The *asterisk* indicates the injury site; D represents the distal optic nerve. Scale bar, 250 μm (Yin et al., 2006).

nerve fragment is sutured to the cut end of the optic nerve. In this case, a large fraction of the surviving RGCs regenerate their axons all the way through a lengthy graft after a few weeks (Yin et al., 2003). Calculating the rate of growth, i.e., the distance traversed by RGC axons divided by the time elapsed, reveals that many axons grew at 700—800 μm/day (Yin et al., 2003), comparable to the rate of axon growth seen in embryonic and early postnatal RGCs (Goldberg et al., 2002a). Similar numbers of axons grow through peripheral nerve grafts when RGCs are exposed to CNTF and a cAMP analog (Cui et al., 2004). However, a recent study has shown that the axon-promoting effects of CNTF in vivo are due to the chemotropic effects of CNTF on macrophages, and an agent that suppresses macrophages in the eye (clodrinate liposomes) blocks the axon-promoting effects of CNTF (Cen et al., 2007).

CHANGES IN GENE EXPRESSION ASSOCIATED WITH OPTIC NERVE REGENERATION

Approximately 4 days after injuring the optic nerve close to the orbit, RGCs make a decision to begin dying or, if appropriately stimulated, to survive and regenerate their axons. Following axotomy in vivo, RGCs which had been exposed to macrophage-derived factors for 3–4 days in vivo show far greater outgrowth in culture than ones which had not been exposed to these factors (Fischer et al., 2004b); this observation suggests that the "priming" of the growth response requires factors that are only available in vivo, but not in culture and presumably include oncomodulin and other growth factors (Yin et al., 2003). As mentioned above, the first clear signs of apoptosis are seen in vivo around 5 days after axotomy when RGCs are not exposed to appropriate growth factors (Fischer et al., 2004b).

What are the molecular differences between RGCs that go on to survive and grow axons vs. ones that fail to regenerate and undergo apoptosis? This question was investigated using FACS and microarrays to compare gene expression profiles between RGCs in the two experimental conditions 4 days after nerve injury, i.e., the time that RGCs first switch into a strong growth state. Surprisingly, only about 0.6% of the genes represented on the microarrays were found to be differentially expressed between the two conditions. Many of the genes that are differentially expressed in the regenerative state are upregulated to an even greater extent after exposure to appropriate growth factors. For example, with lens injury, SPRR1A expression increased several thousand-fold and GAP-43 expression increased 9.3-fold (Table I). It is likely that the genes which show this pattern of differential expression contribute to the greater survival and outgrowth seen in axotomized RGCs after exposure to appropriate growth factors. Other genes are upregulated to an equal extent after axotomy whether or not RGCs are induced to regenerate their axons, and these include several transcription factors (MafK, ATF-3, c-jun, fos-related antigen). Genes in this group may be related to injury *per se*, though some of these may help set the stage for the more extensive growth that occurs if RGCs are exposed to appropriate factors. A third group of genes were expressed to a lesser extent if axotomized RGCs went into an active growth state, while a fourth group showed a lesser degree of downregulation after exposure to appropriate factors after axotomy. These latter two groups may include genes related to cell death (Table I: Fischer et al., 2004b).

SIMILAR CHANGES IN GENE EXPRESSION UNDERLIE REGENERATION IN THE CNS AND PNS

The pattern of molecular changes that accompanies axon regeneration in the optic nerve strongly resembles the pattern seen during PNS repair. Approximately 50% of the genes that are strongly upregulated in axotomized RGCs in response to macrophage-derived factors also undergo significant increases in peripheral neurons regenerating their axons after sciatic nerve injury (Fischer et al., 2004b). RGCs and DRG neurons also show similar groups of genes that are downregulated during regeneration, including several ion channels and transmitter receptors (Fischer et al., 2004b). Some of the changes in DRG neurons during sciatic nerve regeneration are discussed in the section on PNS repair (Shewan et al., 1995; Zigmond and Sun, 1997; Aldskogius and Kozlova, 1998; Magnusson et al., 2000; Nakagomi et al., 2000; Tsujino et al., 2000; Bonilla et al., 2002; Costigan et al., 2002; Xiao et al., 2002; Tanabe et al., 2003; Zhelyaznik et al., 2003; Seijffers et al., 2006). More exact comparisons between RGCs and peripheral neurons are limited, however, by the fact that microarray results are reported as changes in the expression level of individual genes relative to a baseline state of the cell, and the baseline levels of gene expression are likely to be quite different between different neural populations. In addition, whereas the DRG as a whole contains several types of neurons that differ in their response to injury as well as non-neuronal cells, the RGCs that were studied represent a relatively homogeneous population of neurons. We also know very little thus far about the equivalent time points at which the growth state becomes activated in DRG neurons vs. RGCs. Nevertheless, to a first approximation, the resemblance in programs of gene expression activated during axon regeneration in the one CNS population and one PNS population that have been studied in greatest detail is striking. This resemblance suggests the possibility that CNS and PNS neurons, and perhaps other types of neurons as well, share a common molecular program for axon regeneration that may reflect similar signaling pathways and similar transcriptional cascades.

INTRACELLULAR SIGNALING PATHWAYS

The intracellular signaling pathways that underlie axon regeneration in the optic nerve are not known. In vivo, the effects of lens injury have been reported to be unaffected by an inhibitor of Trk receptors or an antibody to the receptor complex used by CNTF, IL-6, and LIF (Lorber et al., 2002). Similarly, in culture, the effects of oncomodulin, mannose, and forskolin, which all presumably contribute to axon regeneration in vivo, are unaffected by inhibitors of these same pathways (Yin et al., 2006). The activity of oncomodulin is blocked, however, by an inhibitor of calcium-calmodulin kinase II (Yin et al., 2006). Another signaling molecule that is important in the activity of oncomodulin is Mst3b. Mst3b is a neuron-specific homolog of Ste20, a kinase that controls budding in yeast. The kinase activity of Mst3b is rapidly activated in neurons exposed to growth factors (Irwin et al., 2006), and can be inhibited with the purine nucleoside analog 6-thioguanine (6-TG). In culture, interfering with the activity of Mst3b or with its expression levels blocks axon outgrowth in embryonic cortical neurons and PC12 cells. In vivo, suppression of Mst3b expression in RGCs using an shRNA approach blocks axon regeneration following optic nerve injury and intravitreal inflammation (Lorber, Howe, Benowitz and Irwin SNS Abstract, 2007). Conversely, activation of Mst3b with inosine promotes brain plasticity and functional recovery in the adult brain (see below).

In addition to promoting axon regeneration, intravitreal macrophage activation strongly promotes RGCs survival after axotomy (Leon et al., 2000; Yin et al., 2003). Combining intravitreal macrophage activation with virally induced BDNF overexpression results in the survival of nearly all RGCs 2 weeks after optic nerve damage, but eliminates axon regeneration (Pernet and Di Polo, 2006). This observation reinforces the idea that the cellular and molecular pathways activated by macrophage-derived factors and by BDNF are distinct: in the case of survival, they have additive effects, whereas for axon regeneration, they interfere with one another. Interfering with BDNF signaling, as well as with signaling pathways downstream from CNTF-related cytokines, actually *enhances* RGC survival in a peripheral nerve graft paradigm (Park et al., 2004). It is clear that there is a great deal more to be learned about the signaling pathways that underlie axon regeneration and cell survival in vivo.

FURTHER ENHANCEMENT OF AXON REGENERATION REQUIRES COMBINATORIAL THERAPIES

The optic nerve, like most other CNS pathways, contains many molecules that suppress axon growth. Among these are the myelin proteins Nogo, MAG, and oligodendrocyte-myelin glycoprotein (OMgp), which act, in part, through the Nogo receptor. Cells in the optic nerve also express semaphorins (Goldberg et al., 2004), ephrins (Liu et al., 2006), and, following injury, CSPGs and other inhibitory molecules that accumulate at the glial scar (Silver and Miller, 2004). The effects of overcoming these inhibitors on axon regeneration have been controversial: in the optic nerve, antibodies to Nogo have been reported to induce a small amount (Weibel et al., 1994) or no regeneration, even when RGC survival is maintained via Bcl-2 overexpression (Chierzi et al., 1999). Likewise, expression of a dominant-negative form of the Nogo receptor is ineffective in promoting axon regeneration (Fischer et al., 2004a), as is deletion of the MAG gene (Bartsch et al., 1995). Many inhibitory molecules exert their effects via the small GTPase RhoA, and therefore blocking RhoA activity represents a comprehensive strategy to counteract many inhibitory signals at once (Lehmann et al., 1999; Niederost et al., 2002; Oertle et al., 2003; Schweigreiter et al., 2004). Blocking RhoA function can be accomplished by using pharmacological blockers to ROCK, a kinase that lies downstream from RhoA, or by introducing C3 ribosyltansferase, an enzyme that irreversibly inactivates RhoA, into cells. RhoA inactivation by either of these means causes only a modest amount of axon regeneration after optic nerve injury (Lehmann et al., 1999; Fischer et al., 2004b).

Although counteracting inhibitory signals is insufficient to promote extensive optic nerve regeneration, it greatly augments regeneration if RGCs' intrinsic growth state is activated. Using AAV2 to express either a dominant-negative form of the Nogo receptor or C3 ribosyltransferase in RGCs increases regeneration after optic nerve damage and lens injury approximately threefold and fivefold, respectively (Fischer et al., 2004a,b). Recent studies demonstrate the efficacy of using similar combinatorial treatments to promote regeneration of RGC axons through a peripheral nerve graft (Hu et al., 2007). Together, these studies show that various signals that have been shown to inhibit axon growth in culture also suppress regeneration in vivo, but overcoming these signals is insufficient to promote extensive regeneration unless neurons' intrinsic growth state is activated.

THE CELL BODY RESPONSE IN OTHER CNS NEURONS

In rubrospinal neurons, axotomy close to the cell body causes an upregulation of GAP-43 and other growth-related proteins, whereas more distal injury does not. Among the proteins that are upregulated following proximal injury are actin and particular tubulin isoforms; most intermediate filament proteins are downregulated (Fernandes and Tetzlaff, 2001). These changes correlate with the ability of these neurons to regenerate their axons through a peripheral nerve graft, but they are transient, and with time the neurons appear to atrophy (Fernandes and Tetzlaff, 2001). Rubrospinal neurons downregulate certain trophic factor receptors, e.g., TrkB and the FGFR1, and thus become less able to respond to trophic signals. Treatment with BDNF increases cell survival, the expression of regeneration-associated genes, and axon regeneration into a peripheral nerve graft (Fernandes and Tetzlaff, 2001). These effects can be seen even if BDNF treatment is delayed for a year after injury (Kwon et al., 2002). CNTF also promotes the survival of neurons in the red nucleus after axotomy (Ye and Houle, 1997).

Layer 5 pyramidal cells in the sensorimotor cortex undergo massive cell death if injured close to their somata, but can be rescued by elevating BDNF. Surprisingly, enhanced survival can also be achieved by blocking endogenous NT-3 signaling (Giehl, 2001). Deletion of the NT-3 gene or use of a neutralizing anti-NT-3 antibody increases the survival of upper motorneurons after axotomy even without introducing exogenous BDNF. Exogenous BDNF increases expression of regeneration-related genes in upper motorneurons and, after a unilateral transection of the corticospinal tract (CST), promotes sprouting of axon collaterals from the intact CST to the denervated side of the spinal cord (Hiebert et al., 2002; Vavrek et al., 2006). Injured CST axons show only local sprouting, however, and cannot regenerate beyond the site of the damage or through a peripheral nerve graft, even when stimulated. Paradoxically, high levels of exogenous NT-3 can also rescue upper motorneurons, which has been attributed to stimulating BDNF expression in other cells (Giehl, 2001). Exogenous NT-3 stimulates CST sprouting after spinal cord hemisections, and this is accompanied by improved locomotor behavior (Grill et al., 1997). Other studies suggest, however, that NT-3 antagonizes collateral growth (Hagg et al., 2005). NT-3 used in combination with the anti-NogoA antibody IN-1 was found to produce greater sprouting of CST axons after spinal cord injury than either approach alone (Schnell et al., 1994).

An alternative way to enhance CST sprouting after unilateral brain injury is with the purine nucleoside inosine. Inosine directly activates the protein kinase Mst3b, and stimulates axon outgrowth in several types of neurons in culture (Benowitz et al., 2002; Irwin et al., 2006). When delivered into the brain after focal brain damage, inosine enhances the ability of undamaged neurons to sprout axon collaterals that innervate brain regions that have lost their normal inputs. This regrowth is accompanied by significant improvements in behavior mediated by the affected limbs (Chen et al., 2002).

CONCLUSIONS

Research into the molecular bases of axon regeneration is in a very dynamic phase. Recent data suggest strong commonalities in the molecular programs that underlie axon regeneration in the PNS and in the CNS. Because regeneration occurs spontaneously in the PNS, this system is expected to continue providing insights into the molecular signaling pathways that enable regeneration to occur. In addition, because PNS repair is still imperfect in humans, there is still a need to discover ways to enhance the rate and accuracy of this process. In the CNS, we are just beginning to see some hope that regeneration through the native CNS environment may be possible some day. However, the amount of regeneration that has been achieved so far is still insufficient to restore function. It is hoped that future research will provide insights into better methods to activate neurons' intrinsic growth state, to overcome inhibitory signals without interfering with axon guidance, and to steer axons to their appropriate destinations. At one time, the mere possibility of getting axons to grow in the CNS seemed remote: now that this is becoming a reality, we need to get ready for the next set of challenges.

ACKNOWLEDGEMENTS

We are grateful for the research support of our laboratories, including funding from the National Institutes of Health (EY05690), Dr. Miriam and Sheldon G. Adelson Medical Research Foundation, Boston Life Sciences, Inc., (Alseres), Paralyzed Veterans of America, the DFG, Christopher Reeve Foundation, The Glaucoma Research Foundation, the DDRC Core of Children's Hospital (NIH P30 HD018655), MDA, IsrALS and the MGH/Neuromuscular Day Laboratory (courtesy of Dr. Robert H Brown Jr.).

REFERENCES

Agthong S, Kaewsema A, Tanomsridejchai N et al. (2006). Activation of MAPK ERK in peripheral nerve after injury. *BMC Neurosci* 7: 45.

Aguayo AJ, Rasminsky M, Bray GM et al. (1991). Degenerative and regenerative responses of injured neurons in the central nervous system of adult mammals. *Philos Trans R Soc Lond B Biol Sci* 331: 337–343.

Aigner L and Caroni P (1995). Absence of persistent spreading, branching, and adhesion in GAP-43-depleted growth cones. *J Cell Biol* 128: 647–660.

Aigner L, Arber S, Kapfhammer JP et al. (1995). Overexpression of the neural growth-associated protein GAP-43 induces nerve sprouting in the adult nervous system of transgenic mice. *Cell* 83: 269–278.

Akahori Y and Horie H (1997). IGF-I enhances neurite regeneration but is not required for its survival in adult DRG explant. *Neuroreport* 8: 2265–2269.

Akassoglou K, Kombrinck KW, Degen JL et al. (2000). Tissue plasminogen activator-mediated fibrinolysis protects against axonal degeneration and demyelination after sciatic nerve injury. *J Cell Biol* 149: 1157–1166.

Aldskogius H and Kozlova EN (1998). Central neuron-glial and glial-glial interactions following axon injury. *Prog Neurobiol* 55: 1–26.

Averill S, Michael GJ, Shortland PJ et al. (2004). NGF and GDNF ameliorate the increase in ATF3 expression which occurs in dorsal root ganglion cells in response to peripheral nerve injury. *Eur J Neurosci* 19: 1437–1445.

Bahr M, Vanselow J, and Thanos S (1988). In vitro regeneration of adult rat ganglion cell axons from retinal explants. *Exp Brain Res* 73: 393–401.

Baptiste DC, Powell KJ, Jollimore CA et al. (2005). Effects of minocycline and tetracycline on retinal ganglion cell survival after axotomy. *Neuroscience* 134: 575–582.

Bareyre FM and Schwab ME (2003). Inflammation, degeneration and regeneration in the injured spinal cord: insights from DNA microarrays. *Trends Neurosci* 26: 555–563.

Bartsch U, Bandtlow CE, Schnell L et al. (1995). Lack of evidence that myelin-associated glycoprotein is a major inhibitor of axonal regeneration in the CNS. *Neuron* 15: 1375–1381.

Bastmeyer M, Schlosshauer B, and Stuermer CA (1990). The spatiotemporal distribution of N-CAM in the retinotectal pathway of adult goldfish detected by the monoclonal antibody D3. *Development* 108: 299–311.

Bennett DL, Boucher TJ, Armanini MP et al. (2000). The glial cell line-derived neurotrophic factor family receptor components are differentially regulated within sensory neurons after nerve injury. *J Neurosci* 20: 427–437.

Benowitz LI and Routtenberg A (1997). GAP-43: an intrinsic determinant of neuronal development and plasticity. *Trends Neurosci* 20: 84–91.

Benowitz LI, Shashoua VE, and Yoon MG (1981). Specific changes in rapidly transported proteins during regeneration of the goldfish optic nerve. *J Neurosci* 1: 300–307.

Benowitz LI, Yoon MG, and Lewis ER (1983). Transported proteins in the regenerating optic nerve: regulation by interactions with the optic tectum. *Science* 222: 185–188.

Benowitz LI, Goldberg DE, and Irwin N (2002). Inosine stimulates axon growth in vitro and in the adult CNS. *Prog Brain Res* 137: 389–399.

Berkelaar M, Clarke DB, Wang YC et al. (1994). Axotomy results in delayed death and apoptosis of retinal ganglion cells in adult rats. *J Neurosci* 14: 4368–4374.

Bernhardt RR, Tongiorgi E, Anzini P et al. (1996). Increased expression of specific recognition molecules by retinal ganglion cells and by optic pathway glia accompanies the successful regeneration of retinal axons in adult zebrafish. *J Comp Neurol* 376: 253–264.

Berry M, Carlile J, and Hunter A (1996). Peripheral nerve explants grafted into the vitreous body of the eye promote the regeneration of retinal ganglion cell axons severed in the optic nerve. *J Neurocytol* 25: 147–170.

Bien A, Seidenbecher CI, Bockers TM et al. (1999). Apoptotic versus necrotic characteristics of retinal ganglion cell death after partial optic nerve injury. *J Neurotrauma* 16: 153–163.

Bilak MM, Shifrin DA, Corse AM et al. (1999). Neuroprotective utility and neurotrophic action of neurturin in postnatal motor neurons: comparison with GDNF and persephin. *Mol Cell Neurosci* 13: 326–336.

Bisby MA and Pollock B (1983). Increased regeneration rate in peripheral nerve axons following double lesions: enhancement of the conditioning lesion phenomenon. *J Neurobiol* 14: 467–472.

Blaugrund E, Bartsch U, Martini R et al. (1990). Immunological evidence that the neural adhesion molecule L1 is expressed in fish brain and optic nerve: possible association with optic nerve regeneration. *Brain Res* 530: 239–244.

Blesch A and Tuszynski MH (2003). Cellular GDNF delivery promotes growth of motor and dorsal column sensory axons after partial and complete spinal cord transections and induces remyelination. *J Comp Neurol* 467: 403–417.

Boeshore KL, Schreiber RC, Vaccariello SA et al. (2004). Novel changes in gene expression following axotomy of a sympathetic ganglion: a microarray analysis. *J Neurobiol* 59: 216–235.

Bomze HM, Bulsara KR, Iskandar BJ et al. (2001). Spinal axon regeneration evoked by replacing two growth cone proteins in adult neurons. *Nat Neurosci* 4: 38–43.

Bonfanti L, Strettoi E, Chierzi S et al. (1996). Protection of retinal ganglion cells from natural and axotomy-induced cell death in neonatal transgenic mice overexpressing bcl-2. *J Neurosci* 16: 4186–4194.

Bonilla IE, Tanabe K, and Strittmatter SM (2002). Small proline-rich repeat protein 1A is expressed by axotomized neurons and promotes axonal outgrowth. *J Neurosci* 22: 1303–1315.

Bonni A, Brunet A, West AE et al. (1999). Cell survival promoted by the Ras-MAPK signaling pathway by transcription-dependent and -independent mechanisms. *Science* 286: 1358–1362.

Boyd ZS, Kriatchko A, Yang J et al. (2003). Interleukin-10 receptor signaling through STAT-3 regulates the apoptosis of retinal ganglion cells in response to stress. *Invest Ophthalmol Vis Sci* 44: 5206–5211.

Bradbury EJ, Khemani S, Von R et al. (1999). NT-3 promotes growth of lesioned adult rat sensory axons ascending in the dorsal columns of the spinal cord. *Eur J Neurosci* 11: 3873–3883.

Bruck W (1997). The role of macrophages in Wallerian degeneration. *Brain Pathol* 7: 741–752.

Bruck W and Friede RL (1991). The role of complement in myelin phagocytosis during PNS wallerian degeneration. *J Neurol Sci* 103: 182–187.

Brunet A, Bonni A, Zigmond MJ et al. (1999). Akt promotes cell survival by phosphorylating and inhibiting a Forkhead transcription factor. *Cell* 96: 857–868.

Burrell HR, Dokas LA, and Agranoff BW (1978). RNA metabolism in the goldfish retina during optic nerve regeneration. *J Neurochem* 31: 289–298.

Cafferty WB, Gardiner NJ, Gavazzi I et al. (2001). Leukemia inhibitory factor determines the growth status of injured adult sensory neurons. *J Neurosci* 21: 7161–7170.

Cafferty WB, Gardiner NJ, Das P et al. (2004). Conditioning injury-induced spinal axon regeneration fails in interleukin-6 knock-out mice. *J Neurosci* 24: 4432–4443.

Cai D, Shen Y, De Bellard M et al. (1999). Prior exposure to neurotrophins blocks inhibition of axonal regeneration by MAG and myelin via a cAMP-dependent mechanism. *Neuron* 22: 89–101.

Cai Y, Zhang C, Nawa T et al. (2000). Homocysteine-responsive ATF3 gene expression in human vascular endothelial cells: activation of c-Jun NH(2)-terminal kinase and promoter response element. *Blood* 96: 2140–2148.

Cai D, Qiu J, Cao Z et al. (2001). Neuronal cyclic AMP controls the developmental loss in ability of axons to regenerate. *J Neurosci* 21: 4731–4739.

Cai D, Deng K, Mellado W et al. (2002a). Arginase I and polyamines act downstream from cyclic AMP in overcoming inhibition of axonal growth MAG and myelin in vitro. *Neuron* 35: 711–719.

Cai D, Deng K, Mellado W et al. (2002b). Arginase I and polyamines act downstream from cyclic AMP in overcoming inhibition of axonal growth MAG and myelin in vitro. *Neuron* 35: 711.

Cao Z, Gao Y, Bryson JB et al. (2006). The cytokine interleukin-6 is sufficient but not necessary to mimic the peripheral conditioning lesion effect on axonal growth. *J Neurosci* 26: 5565–5573.

Cardone MH, Roy N, Stennicke HR et al. (1998). Regulation of cell death protease caspase-9 by phosphorylation. *Science* 282: 1318–1321.

Carmeliet P and Tessier-Lavigne M (2005). Common mechanisms of nerve and blood vessel wiring. *Nature* 436: 193–200.

Caroni P (1997). Intrinsic neuronal determinants that promote axonal sprouting and elongation. *Bioessays* 19: 767–775.

Caroni P (2001). New EMBO members' review: actin cytoskeleton regulation through modulation of PI(4,5)P(2) rafts. *EMBO J* 20: 4332–4336.

Caroni P, Aigner L, and Schneider C (1997). Intrinsic neuronal determinants locally regulate extrasynaptic and synaptic growth at the adult neuromuscular junction. *J Cell Biol* 136: 679–692.

Carpenter P, Sefton AJ, Dreher B et al. (1986). Role of target tissue in regulating the development of retinal ganglion cells in the albino rat: effects of kainate lesions in the superior colliculus. *J Comp Neurol* 251: 240–259.

Carter DA, Bray GM, and Aguayo AJ (1989). Regenerated retinal ganglion cell axons can form well-differentiated synapses in the superior colliculus of adult hamsters. *J Neurosci* 9: 4042–4050.

Cen LP, Luo JM, Zhang CW et al. (2007). Chemotactic effect of ciliary neurotrophic factor on macrophages in retinal ganglion cell survival and axonal regeneration. *Invest Ophthalmol Vis Sci* 48: 4257–4266.

Chen DF, Schneider GE, Martinou JC et al. (1997). Bcl-2 promotes regeneration of severed axons in mammalian CNS. *Nature* 385: 434–439.

Chen P, Goldberg DE, Kolb B et al. (2002). Inosine induces axonal rewiring and improves behavioral outcome after stroke. *Proc Natl Acad Sci U S A* 99: 9031–9036.

Chen ZL, Yu WM, and Strickland S (2007). Peripheral regeneration. *Annu Rev Neurosci* 30: 209–233.

Cheng L, Sapieha P, Kittlerova P et al. (2002). TrkB gene transfer protects retinal ganglion cells from axotomy-induced death in vivo. *J Neurosci* 22: 3977–3986.

Chierzi S, Strettoi E, Cenni MC et al. (1999). Optic nerve crush: axonal responses in wild-type and bcl-2 transgenic mice. *J Neurosci* 19: 8367–8376.

Chierzi S, Ratto GM, Verma P et al. (2005). The ability of axons to regenerate their growth cones depends on axonal type and age, and is regulated by calcium, cAMP and ERK. *Eur J Neurosci* 21: 2051–2062.

Condic ML (2001). Adult neuronal regeneration induced by transgenic integrin expression. *J Neurosci* 21: 4782–4788.

Costigan M, Befort K, Karchewski L et al. (2002). Replicate high-density rat genome oligonucleotide microarrays reveal hundreds of regulated genes in the dorsal root ganglion after peripheral nerve injury. *BMC Neurosci* 3: 16.

Cui Q, Yip HK, Zhao RC et al. (2003). Intraocular elevation of cyclic AMP potentiates ciliary neurotrophic factor-induced regeneration of adult rat retinal ganglion cell axons. *Mol Cell Neurosci* 22: 49–61.

Cui Q, Cho KS, So KF et al. (2004). Synergistic effect of Nogo-neutralizing antibody IN-1 and ciliary neurotrophic factor on axonal regeneration in adult rodent visual systems. *J Neurotrauma* 21: 617–625.

Dailey AT, Avellino AM, Benthem L et al. (1998). Complement depletion reduces macrophage infiltration and activation during Wallerian degeneration and axonal regeneration. *J Neurosci* 18: 6713–6722.

Datta SR, Dudek H, Tao X et al. (1997). Akt phosphorylation of BAD couples survival signals to the cell-intrinsic death machinery. *Cell* 91: 231–241.

Dergham P, Ellezam B, Essagian C et al. (2002). Rho signaling pathway targeted to promote spinal cord repair. *J Neurosci* 22: 6570–6577.

Di Polo A, Aigner LJ, Dunn RJ et al. (1998). Prolonged delivery of brain-derived neurotrophic factor by adenovirus-infected Muller cells temporarily rescues injured retinal ganglion cells. *Proc Natl Acad Sci USA* 95: 3978–3983.

Dong JM, Leung T, Manser E et al. (1998). cAMP-induced morphological changes are counteracted by the activated RhoA small GTPase and the Rho kinase ROKalpha. *J Biol Chem* 273: 22554–22562.

Doya H, Ohtori S, Takahashi K et al. (2005). Extracellular signal-regulated kinase mitogen-activated protein kinase activation in the dorsal root ganglion (DRG) and spinal cord after DRG injury in rats. Spine 30: 2252–2256.

Edstrom A and Ekstrom PA (2003). Role of phosphatidylinositol 3-kinase in neuronal survival and axonal outgrowth of adult mouse dorsal root ganglia explants. *J Neurosci Res* 74: 726–735.

Fernandes KJL and Tetzlaff W (2001). Gene expression in axotomized neurons: identifying the intrinsic determinants of axon growth. In NA Ingoglia and M Murray, eds. *Axonal Regeneration in the Central Nervous System.*) Marcel Dekker, New York, p. 47.

Fernyhough P, Willars GB, Lindsay RM et al. (1993). Insulin and insulin-like growth factor I enhance regeneration in cultured adult rat sensory neurones. *Brain Res* 607: 117–124.

Filbin MT (2003). Myelin-associated inhibitors of axonal regeneration in the adult mammalian CNS. *Nat Rev Neurosci* 4: 703–713.

Fischer D, Pavlidis M, and Thanos S (2000). Cataractogenic lens injury prevents traumatic ganglion cell death and promotes axonal regeneration both in vivo and in culture. *Invest Ophthalmol Vis Sci* 41: 3943–3954.

Fischer D, Heiduschka P, and Thanos S (2001). Lens-injury-stimulated axonal regeneration throughout the optic pathway of adult rats. *Exp Neurol* 172: 257–272.

Fischer D, He Z, and Benowitz LI (2004a) Counteracting the Nogo receptor enhances optic nerve regeneration if retinal ganglion cells are in an active growth state. *J Neurosci* 24: 1646–1651.

Fischer D, Petkova V, Thanos S et al. (2004b). Switching mature retinal ganglion cells to a robust growth state in vivo: gene expression and synergy with RhoA inactivation. *J Neurosci* 24: 8726–8740.

Fournier AE, Gould GC, Liu BP et al. (2002). Truncated soluble Nogo receptor binds Nogo-66 and blocks inhibition of axon growth by myelin. *J Neurosci* 22: 8876–8883.

Fournier AE, Takizawa BT, and Strittmatter SM (2003). Rho kinase inhibition enhances axonal regeneration in the injured CNS. *J Neurosci* 23: 1416–1423.

Frey D, Laux T, Xu L et al. (2000). Shared and unique roles of CAP23 and GAP43 in actin regulation, neurite outgrowth, and anatomical plasticity. *J Cell Biol* 149: 1443–1454.

Fry EJ, Ho C, and David S (2007). A role for Nogo receptor in macrophage clearance from injured peripheral nerve. *Neuron* 53: 649–662.

Fujisawa H, Takagi S, and Hirata T (1995). Growth-associated expression of a membrane protein, neuropilin, in Xenopus optic nerve fibers. *Dev Neurosci* 17: 343–349.

Gao Y, Deng K, Hou J et al. (2004). Activated CREB is sufficient to overcome inhibitors in myelin and promote spinal axon regeneration in vivo. *Neuron* 44: 609–621.

Gavazzi I, Kumar RD, McMahon SB et al. (1999). Growth responses of different subpopulations of adult sensory neurons to neurotrophic factors in vitro. *Eur J Neurosci* 11: 3405–3414.

Giehl KM (2001). Trophic dependencies of rodent corticospinal neurons. *Rev Neurosci* 12: 79–94.

Glazner GW, Lupien S, Miller JA et al. (1993). Insulin-like growth factor II increases the rate of sciatic nerve regeneration in rats. Neuroscience 54: 791–797.

Gold BG, Storm-Dickerson T, and Austin DR (1993). Regulation of the transcription factor c-JUN by nerve growth factor in adult sensory neurons. *Neurosci Lett* 154: 129–133.

Goldberg JL, Klassen MP, Hua Y et al. (2002a). Amacrine-signaled loss of intrinsic axon growth ability by retinal ganglion cells. *Science* 296: 1860–1864.

Goldberg JL, Espinosa JS, Xu Y et al. (2002b). Retinal ganglion cells do not extend axons by default: promotion by neurotrophic signaling and electrical activity. *Neuron* 33: 689–702.

Goldberg JL, Vargas ME, Wang JT et al. (2004). An oligodendrocyte lineage-specific semaphorin, Sema5A, inhibits axon growth by retinal ganglion cells. *J Neurosci* 24: 4989–4999.

Grafstein B (1986). The retina as a regenerating organ. In R Adler and DB Farber, eds. *The Retina: A Model for Cell Biology Studies, Part II*). Academic Press, New York, NY, pp. 275–335.

GrandPre T, Li S, and Strittmatter SM (2002). Nogo-66 receptor antagonist peptide promotes axonal regeneration. *Nature* 417: 547–551.

Grill R, Murai K, Blesch A et al. (1997). Cellular delivery of neurotrophin-3 promotes corticospinal axonal growth and partial functional recovery after spinal cord injury. *J Neurosci* 17: 5560–5572.

Haastert K, Lipokatic E, Fischer M et al. (2006). Differentially promoted peripheral nerve regeneration by grafted Schwann cells over-expressing different FGF-2 isoforms. *Neurobiol Dis* 21: 138–153.

Hagg T, Baker KA, Emsley JG et al. (2005). Prolonged local neurotrophin-3 infusion reduces ipsilateral collateral sprouting of spared corticospinal axons in adult rats. *Neuroscience* 130: 875–887.

Hai T and Hartman MG (2001). The molecular biology and nomenclature of the activating transcription factor/cAMP responsive element binding family of transcription factors: activating transcription factor proteins and homeostasis. *Gene* 273: 1–11.

Hai T, Wolfgang CD, Marsee DK et al. (1999). ATF3 and stress responses. *Gene Expr* 7: 321–335.

Hall CM and Schechter N (1991). Expression of neuronal intermediate filament proteins ON1 and ON2 during goldfish optic nerve regeneration: effect of tectal ablation. *Neuroscience* 41: 695–701.

Han PJ, Shukla S, Subramanian PS et al. (2004). Cyclic AMP elevates tubulin expression without increasing intrinsic axon growth capacity. *Exp Neurol* 189: 293–302.

Hanz S and Fainzilber M (2006). Retrograde signaling in injured nerve – the axon reaction revisited. *J Neurochem* 99: 13–19.

He Q, Dent EW, and Meiri KF (1997). Modulation of actin filament behavior by GAP-43 (neuromodulin) is dependent on the phosphorylation status of serine 41, the protein kinase C site. *J Neurosci* 17: 3515–3524.

He MH, Cheung ZH, Yu EH et al. (2004). Cytochrome c release and caspase-3 activation in retinal ganglion cells following different distance of axotomy of the optic nerve in adult hamsters. *Neurochem Res* 29: 2153–2161.

Heacock AM and Agranoff BW (1982). Protein synthesis and transport in the regenerating goldfish visual system. *Neurochem Res* 7: 771–788.

Herdegen T, Fiallos-Estrada CE, Schmid W et al. (1992). The transcription factors c-JUN, JUN D and CREB, but not FOS and KROX-24, are differentially regulated in axotomized neurons following transection of rat sciatic nerve. *Brain Res Mol Brain Res* 14: 155–165.

Hiebert GW, Khodarahmi K, McGraw J et al. (2002). Brain-derived neurotrophic factor applied to the motor cortex promotes sprouting of corticospinal fibers but not regeneration into a peripheral nerve transplant. *J Neurosci Res* 69: 160–168.

Hol EM, Schwaiger FW, Werner A et al. (1999). Regulation of the LIM-type homeobox gene islet-1 during neuronal regeneration. *Neuroscience* 88: 917–925.

Horie H, Kadoya T, Hikawa N et al. (2004). Oxidized galectin-1 stimulates macrophages to promote axonal regeneration in peripheral nerves after axotomy. *J Neurosci* 24: 1873–1880.

Hu P and McLachlan EM (2003). Distinct functional types of macrophage in dorsal root ganglia and spinal nerves proximal to sciatic and spinal nerve transections in the rat. *Exp Neurol* 184: 590–605.

Hu Y, Cui Q, and Harvey AR (2007). Interactive effects of C3, cyclic AMP and ciliary neurotrophic factor on adult retinal ganglion cell survival and axonal regeneration. *Mol Cell Neurosci* 34: 88–98.

Hull M and Bahr M (1994). Regulation of immediate-early gene expression in rat retinal ganglion cells after axotomy and during regeneration through a peripheral nerve graft. *J Neurobiol* 25: 92–105.

Hu-Tsai M, Winter J, Emson PC et al. (1994). Neurite outgrowth and GAP-43 mRNA expression in cultured adult rat dorsal root ganglion neurons: effects of NGF or prior peripheral axotomy. *J Neurosci Res* 39: 634–645.

Irwin N, Li YM, O'Toole JE et al. (2006). Mst3b, a purine-sensitive Ste20-like protein kinase, regulates axon outgrowth. *Proc Natl Acad Sci U S A* 103: 18320–18325.

Isenmann S, Kretz A, and Cellerino A (2003). Molecular determinants of retinal ganglion cell development, survival, and regeneration. *Prog Retin Eye Res* 22: 483–543.

Iwaya K, Mizoi K, Tessler A et al. (1999). Neurotrophic agents in fibrin glue mediate adult dorsal root regeneration into spinal cord. *Neurosurgery* 44: 589–595; discussion 595–586.

Jankowski MP, Cornuet PK, McIlwrath S et al. (2006). SRY-box containing gene 11 (Sox11) transcription factor is required for neuron survival and neurite growth. *Neuroscience* 143: 501–514.

Jenkins R and Hunt SP (1991). Long-term increase in the levels of c-jun mRNA and jun protein-like immunoreactivity in motor and sensory neurons following axon damage. *Neurosci Lett* 129: 107–110.

Jian X, Szaro BG, and Schmidt JT (1996). Myosin light chain kinase: expression in neurons and upregulation during axon regeneration. *J Neurobiol* 31: 379–391.

Jones DM, Tucker BA, Rahimtula M et al. (2003). The synergistic effects of NGF and IGF-1 on neurite growth in adult sensory neurons: convergence on the PI 3-kinase signaling pathway. *J Neurochem* 86: 1116–1128.

Josephson A, Widenfalk J, Trifunovski A et al. (2001). GDNF and NGF family members and receptors in human fetal and adult spinal cord and dorsal root ganglia. *J Comp Neurol* 440: 204–217.

Jungnickel J, Haase K, Konitzer J et al. (2006). Faster nerve regeneration after sciatic nerve injury in mice over-expressing basic fibroblast growth factor. *J Neurobiol* 66: 940–948.

Kanje M, Skottner A, Sjoberg J et al. (1989). Insulin-like growth factor I (IGF-I) stimulates regeneration of the rat sciatic nerve. *Brain Res* 486: 396–398.

Kaplan DR and Miller FD (2000). Neurotrophin signal transduction in the nervous system. *Curr Opin Neurobiol* 10: 381–391.

Kermer P, Klocker N, Labes M et al. (1998). Inhibition of CPP32-like proteases rescues axotomized retinal ganglion cells from secondary cell death in vivo. *J Neurosci* 18: 4656–4662.

Kiefer R, Kieseier BC, Stoll G et al. (2001). The role of macrophages in immune-mediated damage to the peripheral nervous system. *Prog Neurobiol* 64: 109–127.

Kilic U, Kilic E, Jarve A et al. (2006). Human vascular endothelial growth factor protects axotomized retinal ganglion cells in vivo by activating ERK-1/2 and Akt pathways. *J Neurosci* 26: 12439–12446.

Kim JE, Li S, GrandPre T et al. (2003). Axon regeneration in young adult mice lacking nogo-a/b. *Neuron* 38: 187–199.

Kim JE, Liu BP, Park JH et al. (2004). Nogo-66 receptor prevents raphespinal and rubrospinal axon regeneration and limits functional recovery from spinal cord injury. *Neuron* 44: 439–451.

Kimpinski K and Mearow K (2001). Neurite growth promotion by nerve growth factor and insulin-like growth factor-1 in cultured adult sensory neurons: role of phosphoinositide 3-kinase and mitogen activated protein kinase. *J Neurosci Res* 63: 486–499.

Kimpinski K, Campenot RB, and Mearow K (1997). Effects of the neurotrophins nerve growth factor, neurotrophin-3, and brain-derived neurotrophic factor (BDNF) on neurite growth from adult sensory neurons in compartmented cultures. *J Neurobiol* 33: 395–410.

Koeberle PD and Ball AK (2002). Neurturin enhances the survival of axotomized retinal ganglion cells in vivo: combined effects with glial cell line-derived neurotrophic factor and brain-derived neurotrophic factor. *Neuroscience* 110: 555–567.

Koeberle PD, Gauldie J and Ball AK (2004). Effects of adenoviral-mediated gene transfer of interleukin-10, interleukin-4, and transforming growth factor-beta on the survival of axotomized retinal ganglion cells. *Neuroscience* 125: 903–920.

Koprivica V, Cho KS, Park JB et al. (2005). EGFR activation mediates inhibition of axon regeneration by myelin and chondroitin sulfate proteoglycans. *Science* 310: 106–110.

Kretz A, Happold CJ, Marticke JK et al. (2005). Erythropoietin promotes regeneration of adult CNS neurons via Jak2/Stat3 and PI3K/AKT pathway activation. *Mol Cell Neurosci* 29: 569–579.

Kwon BK, Liu J, Messerer C et al. (2002). Survival and regeneration of rubrospinal neurons 1 year after spinal cord injury. *Proc Natl Acad Sci U S A* 99: 3246–3251.

Laquis S, Chaudhary P, and Sharma SC (1998). The patterns of retinal ganglion cell death in hypertensive eyes. *Brain Res* 784: 100–104.

Laux T, Fukami K, Thelen M et al. (2000). GAP43, MARCKS, and CAP23 modulate PI(4,5)P(2) at plasmalemmal rafts, and regulate cell cortex actin dynamics through a common mechanism. *J Cell Biol* 149: 1455–1472.

Leah JD, Herdegen T and Bravo R (1991). Selective expression of Jun proteins following axotomy and axonal transport block in peripheral nerves in the rat: evidence for a role in the regeneration process. *Brain Res* 566: 198–207.

Ledeboer A, Breve JJ, Wierinckx A et al. (2002). Expression and regulation of interleukin-10 and interleukin-10 receptor in rat astroglial and microglial cells. *Eur J Neurosci* 16: 1175–1185.

Lee DH, Strittmatter SM, and Sah DW (2003). Targeting the Nogo receptor to treat central nervous system injuries. *Nat Rev Drug Discov* 2: 872–878.

Lee N, Neitzel KL, Devlin BK et al. (2004). STAT3 phosphorylation in injured axons before sensory and motor neuron nuclei: potential role for STAT3 as a retrograde signaling transcription factor. *J Comp Neurol* 474: 535–545.

Lehmann M, Fournier A, Selles-Navarro I et al. (1999). Inactivation of Rho signaling pathway promotes CNS axon regeneration. *J Neurosci* 19: 7537–7547.

Leon S, Yin Y, Nguyen J et al. (2000). Lens injury stimulates axon regeneration in the mature rat optic nerve. *J Neurosci* 20: 4615–4626.

Lindsay RM (1988). Nerve growth factors (NGF, BDNF) enhance axonal regeneration but are not required for survival of adult sensory neurons. *J Neurosci* 8: 2394–2405.

Lindwall C and Kanje M (2005a). Retrograde axonal transport of JNK signaling molecules influence injury induced nuclear changes in p-c-Jun and ATF3 in adult rat sensory neurons. *Mol Cell Neurosci* 29: 269–282.

Lindwall C and Kanje M (2005b). The Janus role of c-Jun: cell death versus survival and regeneration of neonatal sympathetic and sensory neurons. *Exp Neurol* 196: 184–194.

Lindwall C, Dahlin L, Lundborg G et al. (2004). Inhibition of c–Jun phosphorylation reduces axonal outgrowth of adult rat nodose ganglia and dorsal root ganglia sensory neurons. *Mol Cell Neurosci* 27: 267–279.

Ling C, Zou T, Hsiao Y et al. (2006). Disruption of tissue plasminogen activator gene reduces macrophage migration. *Biochem Biophys Res Commun* 349: 906–912.

Liu RY and Snider WD (2001). Different signaling pathways mediate regenerative versus developmental sensory axon growth. *J Neurosci* 21: RC164.

Liu X, Hawkes E, Ishimaru T et al. (2006). EphB3: an endogenous mediator of adult axonal plasticity and regrowth after CNS injury. *J Neurosci* 26: 3087–3101.

Logan A, Ahmed Z, Baird A et al. (2006). Neurotrophic factor synergy is required for neuronal survival and disinhibited axon regeneration after CNS injury. *Brain* 129: 490–502.

Lorber B, Berry M, Logan A et al. (2002). Effect of lens lesion on neurite outgrowth of retinal ganglion cells in vitro. *Mol Cell Neurosci* 21: 301–311.

Lu X and Richardson PM (1991). Inflammation near the nerve cell body enhances axonal regeneration. *J Neurosci* 11: 972–978.

Lu X and Richardson PM (1993). Responses of macrophages in rat dorsal root ganglia following peripheral nerve injury. *J Neurocytol* 22: 334–341.

Luk HW, Noble LJ, and Werb Z (2003). Macrophages contribute to the maintenance of stable regenerating neurites following peripheral nerve injury. *J Neurosci Res* 73: 644–658.

Magnusson S, Ekstrom TJ, Elmer E et al. (2000). Heme oxygenase-1, heme oxygenase-2 and biliverdin reductase in peripheral ganglia from rat, expression and plasticity. *Neuroscience* 95: 821–829.

Maier DL, Mani S, Donovan SL et al. (1999). Disrupted cortical map and absence of cortical barrels in growth-associated protein (GAP)-43 knockout mice. *Proc Natl Acad Sci U S A* 96: 9397–9402.

Makwana M and Raivich G (2005). Molecular mechanisms in successful peripheral regeneration. *FEBS J* 272: 2628–2638.

Mansour-Robaey S, Clarke DB, Wang YC et al. (1994). Effects of ocular injury and administration of brain-derived neurotrophic factor on survival and regrowth of axotomized retinal ganglion cells. *Proc Natl Acad Sci USA* 91: 1632–1636.

Markus A, Patel TD, and Snider WD (2002). Neurotrophic factors and axonal growth. *Curr Opin Neurobiol* 12: 523–531.

Martin-Villalba A, Winter C, Brecht S et al. (1998). Rapid and long-lasting suppression of the ATF-2 transcription factor is a common response to neuronal injury. *Brain Res Mol Brain Res* 62: 158–166.

McQuarrie IG, Grafstein B, and Gershon MD (1977). Axonal regeneration in the rat sciatic nerve: effect of a conditioning lesion and of dbcAMP. *Brain Res* 132: 443–453.

Mey J and Thanos S (1993). Intravitreal injections of neurotrophic factors support the survival of axotomized retinal ganglion cells in adult rats in vivo. *Brain Res* 602: 304–317.

Meyer RL and Miotke J (1990). Rapid initiation of neurite outgrowth onto laminin from explants of adult mouse retina induced by optic nerve crush. *Exp Neurol* 107: 214–221.

Meyer RL, Miotke JA, and Benowitz LI (1994). Injury induced expression of growth-associated protein-43 in adult mouse retinal ganglion cells in vitro. *Neuroscience* 63: 591–602.

Meyer-Franke A, Kaplan MR, Pfrieger FW et al. (1995). Characterization of the signaling interactions that promote the survival and growth of developing retinal ganglion cells in culture. *Neuron* 15: 805–819.

Meyer-Franke A, Wilkinson GA, Kruttgen A et al. (1998). Depolarization and cAMP elevation rapidly recruit TrkB to the plasma membrane of CNS neurons. *Neuron* 21: 681–693.

Miao T, Wu D, Zhang Y et al. (2006). Suppressor of cytokine signaling-3 suppresses the ability of activated signal transducer and activator of transcription-3 to stimulate neurite growth in rat primary sensory neurons. *J Neurosci* 26: 9512–9519.

Mills CD, Allchorne AJ, Griffin Robert S et al. (2007). GDNF selectively promotes regeneration of injury-primed sensory neurons in the lesioned spinal cord. *Mol Cell Neurosci* 36: 185–194.

Miotke JA, MacLennan AJ, and Meyer RL (2007). Immunohistochemical localization of CNTFRalpha in adult mouse retina and optic nerve following intraorbital nerve crush: evidence for the axonal loss of a trophic factor receptor after injury. *J Comp Neurol* 500: 384–400.

Mohiuddin L, Fernyhough P, and Tomlinson DR (1996). Acidic fibroblast growth factor enhances neurite outgrowth and stimulates expression of GAP-43 and T alpha 1 alpha-tubulin in cultured neurones from adult rat dorsal root ganglia. *Neurosci Lett* 215: 111–114.

Morooka H, Bonventre JV, Pombo CM et al. (1995). Ischemia and reperfusion enhance ATF-2 and c-Jun binding to cAMP response elements and to an AP-1 binding site from the c-jun promoter. *J Biol Chem* 270: 30084–30092.

Nadeau S, Hein P, Fernandes KJ et al. (2005). A transcriptional role for C/EBP beta in the neuronal response to axonal injury. *Mol Cell Neurosci* 29: 525–535.

Nakagomi S, Suzuki Y, Namikawa K et al. (2003). Expression of the activating transcription factor 3 prevents c-Jun N-terminal kinase-induced neuronal death by promoting heat shock protein 27 expression and Akt activation. *J Neurosci* 23: 5187–5196.

Nakazawa T, Tamai M, and Mori N (2002). Brain-derived neurotrophic factor prevents axotomized retinal ganglion cell death through MAPK and PI3K signaling pathways. *Invest Ophthalmol Vis Sci* 43: 3319–3326.

Namikawa K, Honma M, Abe K et al. (2000). Akt/protein kinase B prevents injury-induced motoneuron death and accelerates axonal regeneration. *J Neurosci* 20: 2875–2886.

Namikawa K, Okamoto T, Suzuki A et al. (2006). Pancreatitis-associated protein-III is a novel macrophage chemoattractant implicated in nerve regeneration. *J Neurosci* 26: 7460–7467.

Neumann S and Woolf CJ (1999). Regeneration of dorsal column fibers into and beyond the lesion site following adult spinal cord injury. *Neuron* 23: 83–91.

Neumann S, Bradke F, Tessier-Lavigne M et al. (2002). Regeneration of sensory axons within the injured spinal cord induced by intraganglionic cAMP elevation. *Neuron* 34: 885–893.

Neumann S, Skinner K, and Basbaum AI (2005). Sustaining intrinsic growth capacity of adult neurons promotes spinal cord regeneration. *Proc Natl Acad Sci U S A* 102: 16848–16852.

Niederost B, Oertle T, Fritsche J et al. (2002). Nogo-A and myelin-associated glycoprotein mediate neurite growth inhibition by antagonistic regulation of RhoA and Rac1. *J Neurosci* 22: 10368–10376.

Obata K, Yamanaka H, Dai Y et al. (2003). Differential activation of extracellular signal-regulated protein kinase in primary afferent neurons regulates brain-derived neurotrophic factor expression after peripheral inflammation and nerve injury. *J Neurosci* 23: 4117–4126.

Oertle T, van der Haar ME, Bandtlow CE et al. (2003). Nogo-A inhibits neurite outgrowth and cell spreading with three discrete regions. *J Neurosci* 23: 5393–5406.

Oudega M and Hagg T (1996). Nerve growth factor promotes regeneration of sensory axons into adult rat spinal cord. *Exp Neurol* 140: 218–229.

Oudega M and Hagg T (1999). Neurotrophins promote regeneration of sensory axons in the adult rat spinal cord. *Brain Res* 818: 431–438.

Park S and Hong YW (2006). Transcriptional regulation of artemin is related to neurite outgrowth and actin polymerization in mature DRG neurons. *Neurosci Lett* 404: 61–66.

Park K, Luo JM, Hisheh S et al. (2004). Cellular mechanisms associated with spontaneous and ciliary neurotrophic factor-cAMP-induced survival and axonal regeneration of adult retinal ganglion cells. *J Neurosci* 24: 10806–10815.

Pearson AG, Gray CW, Pearson JF et al. (2003). ATF3 enhances c-Jun-mediated neurite sprouting. *Brain Res Mol Brain Res* 120: 38–45.

Perlson E, Michaelevski I, Kowalsman N et al. (2006). Vimentin binding to phosphorylated Erk sterically hinders enzymatic dephosphorylation of the kinase. *J Mol Biol* 364: 938–944.

Pernet V and Di Polo A (2006). Synergistic action of brain-derived neurotrophic factor and lens injury promotes retinal ganglion cell survival, but leads to optic nerve dystrophy in vivo. *Brain* 129: 1014–1026.

Pradervand S, Yasukawa H, Muller OG et al. (2004). Small proline-rich protein 1A is a gp130 pathway- and stress-inducible cardioprotective protein. *EMBO J* 23: 4517–4525.

Qiu J, Cai D, Dai H et al. (2002). Spinal axon regeneration induced by elevation of cyclic AMP. *Neuron* 34: 895–903.

Qiu J, Cafferty WB, McMahon SB et al. (2005). Conditioning injury-induced spinal axon regeneration requires signal transducer and activator of transcription 3 activation. *J Neurosci* 25: 1645–1653.

Quigley HA, Nickells RW, Kerrigan LA et al. (1995). Retinal ganglion cell death in experimental glaucoma and after axotomy occurs by apoptosis. *Invest Ophthalmol Vis Sci* 36: 774–786.

Raivich G and Makwana M (2007). The making of successful axonal regeneration: genes, molecules and signal transduction pathways. *Brain Res Rev* 53: 287–311.

Raivich G, Bohatschek M, Da Costa C et al. (2004). The AP-1 transcription factor c-Jun is required for efficient axonal regeneration. *Neuron* 43: 57–67.

Ramer MS, Priestley JV, and McMahon SB (2000). Functional regeneration of sensory axons into the adult spinal cord [see comments]. *Nature* 403: 312–316.

Ramer MS, Bishop T, Dockery P et al. (2002). Neurotrophin-3-mediated regeneration and recovery of proprioception following dorsal rhizotomy. *Mol Cell Neurosci* 19: 239–249.

Ramon y Cajal S (1991). *Degeneration and Regeneration of the Nervous System*. Oxford University Press, New York.

Reynolds AJ, Hendry IA, and Bartlett SE (2001). Anterograde and retrograde transport of active extracellular signal-related kinase 1 (ERK1) in the ligated rat sciatic nerve. *Neuroscience* 105: 761–771.

Richardson PM and Issa VM (1984). Peripheral injury enhances central regeneration of primary sensory neurones. *Nature* 309: 791–793.

Romero MI, Rangappa N, Garry MG et al. (2001). Functional regeneration of chronically injured sensory afferents into adult spinal cord after neurotrophin gene therapy. *J Neurosci* 21: 8408–8416.

Schmeer C, Straten G, Kugler S et al. (2002). Dose-dependent rescue of axotomized rat retinal ganglion cells by adenovirus-mediated expression of glial cell-line derived neurotrophic factor in vivo. *Eur J Neurosci* 15: 637–643.

Schnell L, Schneider R, Kolbeck R et al. (1994). Neurotrophin-3 enhances sprouting of corticospinal tract during development and after adult spinal cord lesion. Nature 367: 170–173.

Schulte T, Paschke KA, Laessing U et al. (1997). Reggie-1 and reggie-2, two cell surface proteins expressed by retinal ganglion cells during axon regeneration. Development 124: 577–587.

Schwab ME (2002). Repairing the injured spinal cord. Science 295: 1029–1031.

Schwab ME (2004). Nogo and axon regeneration. *Curr Opin Neurobiol* 14: 118–124.

Schwab JM, Tuli SK, and Failli V (2006). The Nogo receptor complex: confining molecules to molecular mechanisms. *Trends Mol Med* 12: 293–297.

Schwaiger FW, Hager G, Schmitt AB et al. (2000). Peripheral but not central axotomy induces changes in Janus kinases (JAK) and signal transducers and activators of transcription (STAT). *Eur J Neurosci* 12: 1165–1176.

Schweigreiter R, Walmsley AR, Niederost B et al. (2004). Versican V2 and the central inhibitory domain of Nogo-A inhibit neurite growth via p75NTR/NgR-independent pathways that converge at RhoA. *Mol Cell Neurosci* 27: 163–174.

Schweizer U, Gunnersen J, Karch C et al. (2002). Conditional gene ablation of Stat3 reveals differential signaling requirements for survival of motoneurons during development and after nerve injury in the adult. *J Cell Biol* 156: 287–297.

Segal RA and Greenberg ME (1996). Intracellular signaling pathways activated by neurotrophic factors. *Annu Rev Neurosci* 19: 463–489.

Seijffers R, Allchorne AJ, and Woolf CJ (2006). The transcription factor ATF-3 promotes neurite outgrowth. *Mol Cell Neurosci* 32: 143–154.

Seijffers R, Mills CD, and Woolf CJ (2007). ATF3 increases the intrinsic growth state of DRG neurons to enhance peripheral nerve regeneration. *J Neurosci* 27: 7911–7920.

Shen Y, Mani S, Donovan SL et al. (2002). Growth-associated protein-43 is required for commissural axon guidance in the developing vertebrate nervous system. *J Neurosci* 22: 239–247.

Sheu JY, Kulhanek DJ, and Eckenstein FP (2000). Differential patterns of ERK and STAT3 phosphorylation after sciatic nerve transection in the rat. *Exp Neurol* 166: 392–402.

Shewan D, Berry M and Cohen J (1995). Extensive regeneration in vitro by early embryonic neurons on immature and adult CNS tissue. *J Neurosci* 15: 2057–2062.

Siconolfi LB and Seeds NW (2001). Mice lacking tPA, uPA, or plasminogen genes showed delayed functional recovery after sciatic nerve crush. *J Neurosci* 21: 4348–4355.

Siebert H, Sachse A, Kuziel WA et al. (2000). The chemokine receptor CCR2 is involved in macrophage recruitment to the injured peripheral nervous system. *J Neuroimmunol* 110: 177–185.

Silver J and Miller JH (2004). Regeneration beyond the glial scar. *Nat Rev Neurosci* 5: 146–156.

Simonen M, Pedersen V, Weinmann O et al. (2003). Systemic deletion of the myelin-associated outgrowth inhibitor nogo-a improves regenerative and plastic responses after spinal cord injury. *Neuron* 38: 201–211.

Sjoberg J and Kanje M (1990). The initial period of peripheral nerve regeneration and the importance of the local environment for the conditioning lesion effect. *Brain Res* 529: 79–84.

Sjogreen B, Wiklund P, and Ekstrom PA (2000). Mitogen activated protein kinase inhibition by PD98059 blocks nerve growth factor stimulated axonal outgrowth from adult mouse dorsal root ganglia in vitro. *Neuroscience* 100: 407–416.

Skene JH (1989). Axonal growth-associated proteins. *Annu Rev Neurosci* 12: 127–156.

Skene JH and Willard M (1981). Changes in axonally transported proteins during axon regeneration in toad retinal ganglion cells. *J Cell Biol* 89: 86–95.

Smith DS and Skene JH (1997). A transcription-dependent switch controls competence of adult neurons for distinct modes of axon growth. *J Neurosci* 17: 646–658.

Snider WD, Zhou FQ, Zhong J et al. (2002). Signaling the pathway to regeneration. *Neuron* 35: 13–16.

Sobrado-Calvo P, Vidal-Sanz M, and Villegas-Perez MP (2007). Rat retinal microglial cells under normal conditions, after optic nerve section, and after optic nerve section and intravitreal injection of trophic factors or macrophage inhibitory factor. *J Comp Neurol* 501: 866–878.

Song XY, Zhong JH, Wang X et al. (2004). Suppression of p75NTR does not promote regeneration of injured spinal cord in mice. *J Neurosci* 24: 542–546.

Spencer T and Filbin MT (2004). A role for cAMP in regeneration of the adult mammalian CNS. *J Anat* 204: 49–55.

Steeves JD and Tetzlaff W (1998). Engines, accelerators, and brakes on functional spinal cord repair. *Ann N Y Acad Sci* 860: 412–424.

Steinmetz MP, Horn KP, Tom VJ et al. (2005). Chronic enhancement of the intrinsic growth capacity of sensory neurons combined with the degradation of inhibitory proteoglycans allows functional regeneration of sensory axons through the dorsal root entry zone in the mammalian spinal cord. *J Neurosci* 25: 8066–8076.

Stoll G, Griffin JW, Li CY et al. (1989). Wallerian degeneration in the peripheral nervous system: participation of both Schwann cells and macrophages in myelin degradation. *J Neurocytol* 18: 671–683.

Tanabe K, Bonilla I, Winkles JA et al. (2003). Fibroblast growth factor-inducible-14 is induced in axotomized neurons and promotes neurite outgrowth. *J Neurosci* 23: 9675–9686.

Thanos S (1992). Sick photoreceptors attract activated microglia from the ganglion cell layer: a model to study the inflammatory cascades in rats with inherited retinal dystrophy. *Brain Res* 588: 21–28.

Timmer M, Robben S, Muller-Ostermeyer F et al. (2003). Axonal regeneration across long gaps in silicone chambers filled with Schwann cells overexpressing high molecular weight FGF-2. *Cell Transplant* 12: 265–277.

Tofaris GK, Patterson PH, Jessen KR et al. (2002). Denervated Schwann cells attract macrophages by secretion of leukemia inhibitory factor (LIF) and monocyte chemoattractant protein-1 in a process regulated by interleukin-6 and LIF. *J Neurosci* 22: 6696–6703.

Toth C, Brussee V, Martinez JA et al. (2006). Rescue and regeneration of injured peripheral nerve axons by intrathecal insulin. *Neuroscience* 139: 429–449.

Tsujino H, Kondo E, Fukuoka T et al. (2000). Activating transcription factor 3 (ATF3) induction by axotomy in sensory and motoneurons: A novel neuronal marker of nerve injury. *Mol Cell Neurosci* 15: 170–182.

Twiss JL and van Minnen J (2006). New insights into neuronal regeneration: the role of axonal protein synthesis in pathfinding and axonal extension. *J Neurotrauma* 23: 295–308.

van Adel BA, Kostic C, Deglon N et al. (2003). Delivery of ciliary neurotrophic factor via lentiviral-mediated transfer protects axotomized retinal ganglion cells for an extended period of time. *Hum Gene Ther* 14: 103–115.

Vavrek R, Girgis J, Tetzlaff W et al. (2006). BDNF promotes connections of corticospinal neurons onto spared descending interneurons in spinal cord injured rats. *Brain* 129: 1534–1545.

Vidal-Sanz M, Bray GM, Villegas-Perez MP et al. (1987). Axonal regeneration and synapse formation in the superior colliculus by retinal ganglion cells in the adult rat. *J Neurosci* 7: 2894–2909.

Vielmetter J, Lottspeich F, and Stuermer CA (1991). The monoclonal antibody E587 recognizes growing (new and regenerating) retinal axons in the goldfish retinotectal pathway. *J Neurosci* 11: 3581–3593.

Villegas-Perez MP, Vidal-Sanz M, Rasminsky M et al. (1993). Rapid and protracted phases of retinal ganglion cell loss follow axotomy in the optic nerve of adult rats. *J Neurobiol* 24: 23–36.

Waetzig V, Zhao Y, and Herdegen T (2006). The bright side of JNKs-multitalented mediators in neuronal sprouting, brain development and nerve fiber regeneration. *Prog Neurobiol* 80: 84–97.

Walsh FS, Meiri K, and Doherty P (1997). Cell signalling and CAM-mediated neurite outgrowth. *Soc Gen Physiol Ser* 52: 221–226.

Wanigasekara Y and Keast JR (2006). Nerve growth factor, glial cell line-derived neurotrophic factor and neurturin prevent semaphorin 3A-mediated growth cone collapse in adult sensory neurons. *Neuroscience* 142: 369–379.

Watanabe M, Tokita Y, Kato M et al. (2003). Intravitreal injections of neurotrophic factors and forskolin enhance survival and axonal regeneration of axotomized beta ganglion cells in cat retina. *Neuroscience* 116: 733-742.

Weibel D, Cadelli D, and Schwab ME (1994). Regeneration of lesioned rat optic nerve fibers is improved after neutralization of myelin-associated neurite growth inhibitors. *Brain Res* 642: 259–266.

Werner A, Willem M, Jones LL et al. (2000). Impaired axonal regeneration in alpha7 integrin-deficient mice. *J Neurosci* 20: 1822–1830.

Wiklund P, Ekstrom PA, and Edstrom A (2002). Mitogen-activated protein kinase inhibition reveals differences in signalling pathways activated by neurotrophin-3 and other growth-stimulating conditions of adult mouse dorsal root ganglia neurons. *J Neurosci Res* 67: 62–68.

Woolf CJ (2003). No Nogo: now where to go? *Neuron* 38: 153–156.

Wu D, Zhang Y, Bo X et al. (2007). Actions of neuropoietic cytokines and cyclic AMP in regenerative conditioning of rat primary sensory neurons. *Exp Neurol* 204: 66–76.

Xiao HS, Huang QH, Zhang FX et al. (2002). Identification of gene expression profile of dorsal root ganglion in the rat peripheral axotomy model of neuropathic pain. *Proc Natl Acad Sci U S A* 99: 8360–8365.

Xu QG, Li XQ, Kotecha SA et al. (2004). Insulin as an in vivo growth factor. *Exp Neurol* 188: 43–51.

Yan Q, Wang J, Matheson CR et al. (1999). Glial cell line-derived neurotrophic factor (GDNF) promotes the survival of axotomized retinal ganglion cells in adult rats: comparison to and combination with brain-derived neurotrophic factor (BDNF). *J Neurobiol* 38: 382–390.

Ye JH and Houle JD (1997). Treatment of the chronically injured spinal cord with neurotrophic factors can promote axonal regeneration from supraspinal neurons. *Exp Neurol* 143: 70–81.

Yin Y, Cui Q, Li Y et al. (2003). Macrophage-derived factors stimulate optic nerve regeneration. *J Neurosci* 23: 2284–2293.

Yin Y, Henzl MT, Lorber B et al. (2006). Oncomodulin is a macrophage-derived signal for axon regeneration in retinal ganglion cells. *Nat Neurosci* 9: 843–852.

Zhang C and Tso MO (2003). Characterization of activated retinal microglia following optic axotomy. *J Neurosci Res* 73: 840–845.

Zhang Y, Dijkhuizen PA, Anderson PN et al. NT-3 delivered by an adenoviral vector induces injured dorsal root axons to regenerate into the spinal cord of adult rats. *J Neurosci Res* 54: 554–562.

Zhelyaznik N, Schrage K, McCaffery P et al. (2003). Activation of retinoic acid signalling after sciatic nerve injury: up-regulation of cellular retinoid binding proteins. *Eur J Neurosci* 18: 1033–1040.

Zheng B, Ho C, Li S et al. (2003). Lack of enhanced spinal regeneration in nogo-deficient mice. *Neuron* 38: 213–224.

Zheng B, Atwal J, Ho C et al. (2005). Genetic deletion of the Nogo receptor does not reduce neurite inhibition in vitro or promote corticospinal tract regeneration in vivo. *Proc Natl Acad Sci U S A* 102: 1205–1210.

Zhong J, Li X, McNamee C et al. (2007). Raf kinase signaling functions in sensory neuron differentiation and axon growth in vivo. *Nat Neurosci* 10: 598–607.

Zigmond RE and Sun Y (1997). Regulation of neuropeptide expression in sympathetic neurons. Paracrine and retrograde influences. *Ann NY Acad Sci* 814: 181–197.

Zou T, Ling C, Xiao Y et al. (2006). Exogenous tissue plasminogen activator enhances peripheral nerve regeneration and functional recovery after injury in mice. *J Neuropathol Exp Neurol* 65: 78–86.

2

Axonal Responses to Injury

MIKE FAINZILBER

Department of Biological Chemistry, Weizmann Institute of Science, 76100 Rehovot, Israel

ABSTRACT

A successful axonal response to injury requires retrograde signaling to induce changes in the cell body response, and mobilization of outgrowth programs while integrating growth-promoting and growth-inhibiting signals from the environment. Central nervous system (CNS) and peripheral nervous system (PNS) axons differ in their capacity for useful regeneration, most likely due to differences in intrinsic growth capacity coupled with differential composition of extracellular growth-modulating agents in central versus peripheral environments. Very recent studies have demonstrated marked increases in growth rates of injured CNS axons using combinatorial manipulations of the environment together with re-activation of intrinsic growth programs. This chapter reviews progress on understanding the different mechanistic aspects of the axonal response to injury, with particular focus on mechanisms of activation of intrinsic growth programs, and on the gaps in knowledge that must be bridged for stimulating effective axon re-growth.

Axonal injury to peripheral neurons elicits a sequence of molecular, ultrastructural, and cellular responses that play a vital role in the mounting of a successful regenerative response, and the ensuing recovery of function. Injured nerve fibers must overcome inhibitory influences in the environment (Yamashita et al., 2005) and mobilize intrinsic capacity for neurite outgrowth (Goldberg, 2004; Rossi et al., 2007) to achieve functional regeneration. In the injured neurons, arrival of signals for cellular injury and stress is followed by the induction

Copyright © 2008 Elsevier Inc.
All rights of reproduction in any form reserved.

of transcription factors, adhesion molecules, growth-associated proteins, and structural components needed for axonal elongation. These molecular changes are accompanied by shifts in cellular organization: the appearance of growth cones at the proximal tip of the lesioned axons, swelling of the neuronal cell body associated with a strong increase in cellular metabolism and protein synthesis, and the augmentation and regional dispersion of Nissl body areas of rough endoplasmic reticulum. The neuronal response is also associated with the expression of growth factors, cytokines, neuropeptides, and other secreted molecules involved in cell-to-cell communication, which may be involved in the activation of neighboring non-neuronal cells around the cell body of the injured neuron and in the distal nerve fiber tracts (Hanz and Fainzilber, 2006; Raivich and Makwana, 2007). In the adult mammalian nervous system, there is a striking dichotomy in the ability to regenerate, where peripheral nerves are capable of far more vigorous regeneration than that seen in axons injured inside central nerve tracts. In this chapter, I will focus on the axonal response to injury in "regeneration-competent" systems including mammalian peripheral nerve and non-mammalian model organisms. The chapter will discriminate between two aspects of axonal response – on the one hand, those mechanisms by which the injured axon communicates its distress to the cell body; and on the other hand, the processes that allow functional re-growth of the axon. Some (but not all) of the latter are dependent on cell body responses to retrograde injury signalling. Finally, I will briefly describe processes that impact on survival and maintenance of distal axonal fragments, and whether it might be feasible to address these in a different approach to axonal repair. Wherever relevant throughout the text, I will add comparisons to our (lack of) understanding of these processes in mammalian central neurons.

COMMUNICATION OF INJURY INFORMATION BY RETROGRADE SIGNALING FROM INJURED AXONS TO THE CELL BODY

RAPID SIGNALING – THE ELECTROPHYSIOLOGICAL RESPONSE

Several types of signals in the axon might be involved in generating cell body responses after axonal injury, ranging from very rapid electrophysiological mechanisms such as lesion-induced depolarization to much slower mechanisms propagated by molecular motors. Rapid ion fluxes emanating from the lesion site are likely to be the first indication of breach of axonal integrity. Axotomy of Aplysia neurons in culture elevates intra-axonal calcium concentrations to levels above 1 mM near the tip of the cut axon and to hundreds of micromolars along the axon (Ziv and Spira, 1995). A wave of increased calcium then propagates at a rate of approximately 1 mm/min from the point of transection towards the intact portions of the cell. Calcium recovery in these invertebrate neurons is fairly rapid (within minutes) once the cut ends are resealed, and calcium levels recover as a retreating front traveling back towards the lesion site (Ziv and Spira, 1993). Such increases in calcium are likely to be important for cytoskeletal rearrangement and growth cone formation (see below), in addition to their contribution to retrograde signaling. In mammalian systems, axotomy of postnatal cortical neurons from day-old rats caused an increase of axonal calcium, propagated to the soma via a mechanism dependent on voltage-dependent sodium channels (Mandolesi et al., 2004). Injury was followed by vigorous spiking activity that caused a sodium load and the activation of transient calcium currents that were opened by each action potential. This was followed by inversion of the sodium–calcium exchange pump, thus providing an additional and prolonged means of entry for calcium (Mandolesi et al., 2004). Stretch-induced injuries in neuron-like cell lines or in embryonic cortical neurons in vitro induced similar changes via activation of tetrodotoxin-sensitive sodium channels and inversion of sodium–calcium exchange (Wolf et al., 2001; Iwata et al., 2004). The stretch injury paradigm caused a continued increase in axonal calcium for at least 60 min (Iwata et al., 2004). This increase was facilitated by calcium-dependent proteolysis of an intra-axonal domain in tetrodotoxin-sensitive sodium channels, providing an additional mechanism for propagation of the injury signal. In vivo, resealing of lesioned mammalian axons can take hours, and is dependent on axon diameter and on calcium in the extracellular environment (Howard et al., 1999), thus the rapid changes in axonal calcium levels following injury may be sustained over time to a degree that is proportional to severity of the injury and the resealing capacity of the axon. Finally, there is preliminary evidence for differences in the resealing capacity of central versus peripheral axons (Ahmed et al., 2001), and initial studies on induction of regeneration in vivo by electrical stimulation revealed positive effects on peripheral motor neurons (Al-Majed et al., 2004), but not in the central rubrospinal tract (Harvey et al., 2005). Thus, although most of the findings described above are from invertebrate or embryonic mammalian neurons in culture, taken together they highlight the potential importance of early changes in axonal calcium in determining subsequent responses of the injured neuron.

SIGNALING COORDINATED BY MOLECULAR MOTORS

Some time after arrival of the calcium fluxes described above, signals dependent on motor-driven transport systems start to affect the cell body. This phase includes both an interruption of the normal supply of retrogradely transported molecules such as trophic factor signals, and arrival of new signals elicited at the injury site (Ambron and Walters, 1996; Perlson et al., 2004; Hanz and Fainzilber, 2006). Definitive early studies on retrogradely transported injury signals were carried out by Richard Ambron and colleagues in Aplysia nerves. First they demonstrated that a cross-linked nuclear localization signal (NLS) peptide conveyed microtubule-dependent retrograde transport of heterologous proteins microinjected into the axoplasm of Aplysia neurons in vitro (Ambron et al., 1992). Rhodamine-labeled axoplasm protein fractions were transported retrogradely and accumulated within the nucleus, and microinjection of ligature-concentrated axoplasm from lesioned nerve into cell bodies of uninjured neurons in vitro elicited both growth and survival responses (Ambron et al., 1996; Schmied and Ambron, 1997). Identification of retrograde injury signaling components in Aplysia axoplasm is an ongoing process; studies so far have implicated MAP kinases and protein kinase G as signals activated after injury, while axonal NFκB seems to be inactivated following nerve lesion (Povelones et al., 1997; Sung et al., 2001; Sung et al., 2004). Thus, there is a body of evidence showing the existence of macromoleclar injury signal proteins in molluscan neurons, although the full spectrum of these signals is still not clear.

Macromolecular retrograde injury signals were also inferred to exist in mammalian peripheral nerve, based on the observation that conditioning lesion of the sciatic nerve switches L4/L5 dorsal root ganglia (DRG) neurons from arborizing to elongating axonal growth if the lesion is carried out several days prior to the culture (Smith and Skene, 1997). A number of candidate retrograde injury signals have been suggested, including Erk1/2 (Chierzi et al., 2005; Perlson et al., 2005; Reynolds et al., 2001), p38 MAPK (Zrouri et al., 2004), jun kinase (Jnk); (Kenney and Kocsis, 1998; Lindwall et al., 2004; Cavalli et al., 2005; Lindwall and Kanje, 2005), protein kinase A (PKA) (Chierzi et al., 2005), protein kinase G (Sung et al., 2006), and the transcription factors STAT3 (Lee et al., 2004; Qiu et al., 2005) and ATF2/3 (Lindwall and Kanje, 2005). How might this diversity of cargos link to the retrograde transport machinery?

The evidence in Aplysia that NLS sequences might target injury-signaling proteins to the retrograde transport system suggested that importins, nuclear import proteins from the karyopherin superfamily, might be involved in retrograde transport in axons. A number of importins α were found in axons of both control and injured sciatic nerve in constitutive association with dynein, while importin β protein was present only after injury (Hanz et al., 2003). Interestingly, mRNA for importin β was found in axons, and the upregulation of importin β protein after injury was attributed to local translation in the axon (Hanz et al., 2003). This leads to the formation of importin α/β heterodimers bound to the retrograde motor dynein, which may allow transport of signaling cargos that bind to the importins (Figure 2.1). One such cargo turns out to be the type III intermediate filament vimentin, which is locally translated in axoplasm after injury, and cleaved to soluble fragments by the calcium-activated protease calpain (Perlson et al., 2005). These vimentin fragments bind directly to both phosphorylated Erks and importin β, thereby linking activated Erks to importin-mediated retrograde transport (Perlson et al., 2005). Strikingly, the vimentin–Erk complex protects Erk from dephosphorylation, and since the interaction is calcium dependent, the signal generated may provide information both on the injury and on the degree of damage as reflected by sustained calcium elevation (Perlson et al., 2006) (Figure 2.1).

Most of the candidate signaling cargos listed above lack an obvious NLS and might not directly associate with importins. Although linker molecules such as vimentin provide a solution for importin-mediated transport of activated Erks,

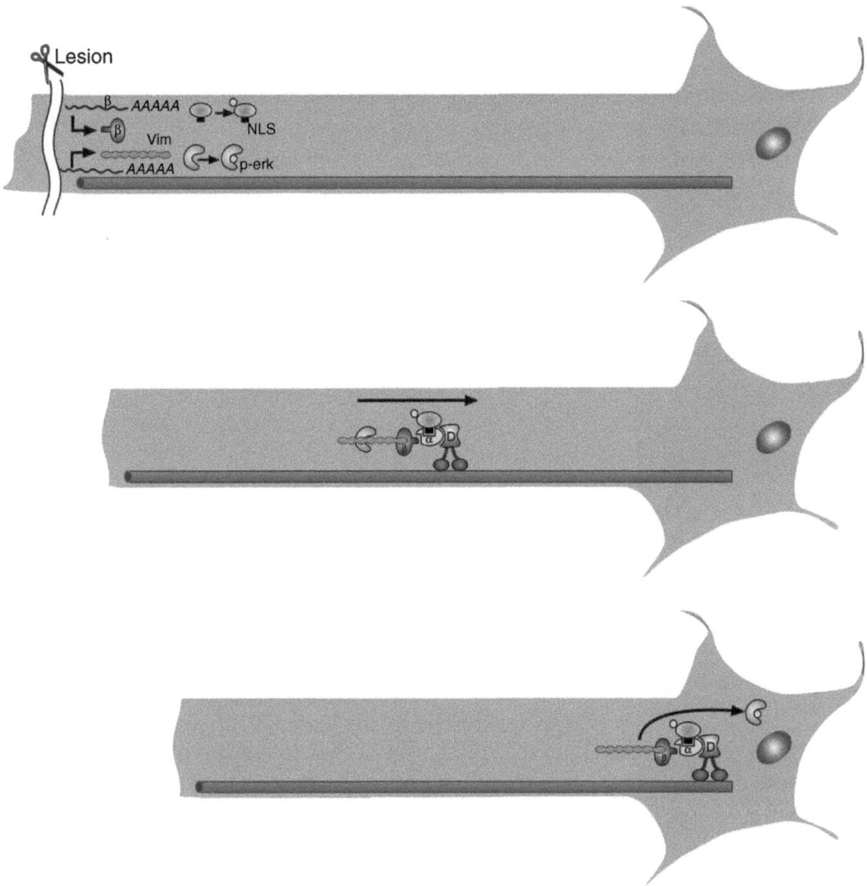

FIGURE 2.1 Retrograde injury signaling in peripheral sensory axons. Upon lesion, importin β and vimentin are translated locally from axonal mRNA, concomitantly with activation of signaling molecules such as Erk. The newly synthesized importin β binds to importin α, which is complexed with dynein, and cleaved fragments of vimentin are then able to interact with the retrograde complex by binding to importin β. Phosphorylated Erk (pErk) links up to the complex via an interaction with vimentin. Upon arrival in the cell body, pErk can dissociate in a calcium-dependent manner, and is then able to activate downstream targets.

additional mechanisms for linking signaling molecules to retrograde transport may be important. For example, the Jnk scaffold protein Sunday driver (Syd) may link activated Jnk to injury signaling (Cavalli et al., 2005). Syd and Jnk3 are present on vesicular structures in axons and are transported in both anterograde and retrograde directions. Nerve injury induces axonal activation of Jnk3, following which the activated Jnk is transported together with Syd, predominantly in the retrograde direction, most likely due to an enhanced interaction between Syd and dynactin after injury. Cavalli et al. (2005) propose that the Jnk–Syd complex acts as a damage surveillance system, and that the direction switch after injury provides a rapid response mechanism for propagation of retrograde injury signals.

The recent studies of Perlson et al. (2005) and Cavalli et al. (2005) suggest that different compositions of retrograde injury signaling complexes are possible, with at least three non-exclusive potential binding sites for signaling molecules identified to date – dynactin-Syd for Jnk and associated molecules, the classical NLS binding site on importin α, and importin β-vimentin for activated Erk (Hanz and Fainzilber, 2006). Variability in subunit composition within the dynein complex might also allow for differential cargo-binding (Pfister et al., 2006), leading to different combinations of signals being transported in different cell types. So far there have been no published studies on retrogradely transported injury signals in adult mammalian neurons from the CNS, and it will be very interesting to find out which aspects of the above-described mechanisms function in the CNS, and which do not.

In addition to the signals described above, a *loss* of normally transported constitutive signals in the axon might also provide retrograde instructions to the neuronal cell body. Neurotrophic factor signaling from axon terminals to the cell body is important in neuronal survival during development, and in maintenance of phenotype and modulation of plasticity in diverse neuronal populations in the adult (Howe and Mobley, 2005; Bronfman et al., 2007). Sciatic nerve axotomy causes a 10-fold decrease in the levels of retrogradely transported NGF (Raivich et al., 1991), and artificial interruption of the supply of NGF induces axotomy-like alterations in gene expression in injured sensory or sympathetic neurons (Shadiack et al., 2001). These observations suggest that reductions in the levels of retrogradely transported trophic factors and their associated signaling complexes might be decoded as an injury signal by the cell body. In support of this notion, intrathecal infusion of NGF delayed GAP 43 induction in DRG neurons and the early phase of peripheral nerve regeneration (Hirata et al., 2002). These findings must, however, be interpreted with caution, since neurotrophins can affect many different aspects of nerve regeneration, from survival responses of injured neurons to guidance effects on re-growing growth cones (Lu et al., 2004, 2007; Lykissas et al., 2007).

THE CELL BODY RESPONSE TO AXONAL INJURY – MOBILIZING GROWTH PROGRAMS

Once information on the injury has been communicated to the cell body, a regeneration-competent neuron must activate transcriptional and translational programs to elicit re-growth of the injured axon. Our mechanistic understanding of these processes is alarmingly limited, and there is no clear understanding to date what a neuron must do in order to grow. The cell body response to injury can be mimicked by inducing changes in intracellular cAMP, and a series of studies have focused on downstream mediators of this signaling, although a clear picture is yet to emerge. Microinjection of cAMP analogs in lumbar DRG markedly increases the regeneration of injured central sensory branches (Neumann et al., 2002; Qiu et al., 2002). Combining stimulation of L4 DRG neuronal cell bodies with cAMP and their injured axons with neurotrophin 3 (NT-3) elicited regeneration of dorsal-column sensory axons into and beyond the lesion (Lu et al., 2004). Regeneration beyond the lesion did not occur after treatment with cAMP or NT-3 alone.

Much effort on the cell body response in recent years has been devoted to microarray approaches to identify regeneration-associated genes (Costigan et al., 2002; Bosse et al., 2006; Rossi et al., 2007). Since neuronal injury is a drastic process, straightforward comparisons of injured versus non-injured neurons in differential gene screens are complicated by the fact that it is difficult to select the genes actually causative for a regeneration response from the many transcriptional events induced by the stress of the injury and the changed connectivity and signaling of the affected cells. Another complicating factor is the fact that most such screens to date have been carried out at the tissue level, hence on heterogenous mixtures of responsive and non-responsive cells. Consequently, differential gene expression screens in the regeneration field have for the most part resulted in ever-lengthening lists of regeneration-*associated* genes, without clear identification of regeneration-*causing* genes. Transgenic and overexpression approaches with a few regeneration-associated genes, primarily GAP-43/CAP-43 or c-jun, have shown modest improvements in regeneration in peripheral neurons, but these results have for the most part not been recapitulated in outgrowth-refractory central neurons (Rossi et al., 2007). Clearly, much additional work will be required to define the critical elements of a neuronal regeneration program.

AXONAL REGENERATION – HOW DO AXONS RE-GROW?

When considering the issue of axonal re-growth after injury, two problems become apparent. The first is that in large mammals, regenerating axons may need to cover very significant distances for full recovery of function – up to 1 meter in the human peripheral nervous system. The second problem is that this re-growth should occur by elongating growth mechanisms, which typically did not evolve to cover such distances in normal development. Most neurons reach their targets during development, when the embryo is still at a relatively small size. Subsequent growth of the organism must therefore be accompanied by interstitial axonal growth in the elongating nerves (Rossi et al., 2007). Both interstitial and elongating regenerative modes of growth may differ mechanistically from embryonic growth cone-driven elongating growth (Zhou and Snider, 2006). Recapitulating a modified form of elongating growth over lengthy distances in a non-embryonic environment is likely to be a daunting challenge for the injured neuron.

Formation of a new growth cone at axonal injury sites has been studied extensively in Aplyisia. Imaging of intracellular calcium, proteolytic activity, and growth cone formation in injured Aplysia neurons showed that localized calpain is required for growth cone formation, and that the process can be blocked by calpeptin, a calpain inhibitor (Gitler and Spira, 1998). This process can be mimicked by microinjection of exogenous proteases (Ziv and Spira, 1998). Endogenous calpain activation causes spectrin cleavage and restructuring of the axonal cytoskeleton near the injury site, forming microtubule enclosed compartments that trap vesicles transported by molecular motors (Spira et al., 2003; Erez et al., 2007). These specialized structures seem to facilitate the fusion of vesicles with the plasma membrane, promoting the extension of the growth cone. Similar cytoskeletal transformations have been described in lesioned spinal axons from Xenopus embryos (Robles et al., 2003), and in the sprouting of collateral branches from mammalian axons (Dent et al., 2003). On the other hand, it should be noted that calcium-dependent activation of proteolytic cleavage of axonal cytoskeleton may also be involved in pathological processes of degeneration after nerve injury (Buki and Povlishock, 2006).

Recent studies have followed up on the mechanisms of new growth cone formation in DRG and retinal explants in vitro. The two neuron subtypes differed in their capacity to form new growth cones after injury. Most DRG axons extended new growth cones within 4 h after injury, while only a small percentage of adult retinal ganglion cells (RGC) responded similarly (Chierzi et al., 2005). Interestingly, RGCs could be induced to extend new growth cones by upregulating their cAMP levels. Inhibition of Erk1/2 or PKA at the time of injury significantly impaired the capacity of DRG axons to extend new growth cones, suggesting that these signaling kinases may connect axotomy to initiation of re-growth at its earliest stages (Chierzi et al., 2005). This process seems to be dependent on localized changes in calcium, and can proceed by solely post-transcriptional events, since adult DRG axons maintained a capacity to form new growth cones even after disconnection from their cell bodies (Verma et al., 2005). This ability was shown to depend on local protein synthesis and degradation within the axon.

Adult DRG axons regenerating in vitro after a conditioning lesion contain ribosomal proteins, translational initiation factors, and rRNA (Zheng et al., 2001). Isolated preparations of such axons separated from their cell bodies maintain the capacity to synthesize proteins in vitro, and blocking such synthesis causes growth cone retraction (Zheng et al., 2001). A comparison of ribosomal protein P0 and phosphorylated translation initiation factor in retinal and sensory axons of different developmental stages revealed high levels in the peripheral sensory axons, lower in embryonic central axons, and absence of these factors in adult retinal axons (Verma et al., 2005). Erk and other signaling proteins activated at the lesion site may regulate local translation via phosphorylation of translation initiation or elongation factors (Zhou and Snider, 2006). Axonally synthesized proteins have been identified by proteomics carried out on injury-conditioned DRG axons, revealing a diverse ensemble of proteins and their encoding mRNAs – including proteins with important roles in cytoskeleton, stress response, endoplasmic reticulum, and general metabolism, as well as proteins associated with neurodegenerative diseases (Willis et al., 2005). Localized application of neurotrophins was found to regulate transport of cytoskeleton protein mRNAs from the cell body into the DRG axons. Taken together, these findings suggest that intra-axonal translation is required for effective growth cone elongation and outgrowth of regenerating sensory axons, and that a deficiency in local translation mechanisms might account for part of the poor regenerative capacity in central neuronal populations (Willis and Twiss, 2006).

After extrusion of a new growth cone and the initiation of an elongating mode of growth, the axon must navigate through an environment that is not necessarily conducive to regeneration. The long-standing notion that adult CNS is not permissive for axon regeneration has been questioned in recent years by studies showing that embryonic CNS neurons and adult PNS neurons are capable of extending axons in adult CNS environments (Zhou and Snider, 2006). The model of the reticulon family member Nogo as a major myelin-derived axon outgrowth inhibitor (Schwab, 2002) has to be modified to accommodate newly described fundamental roles of reticulons in the formation of tubular ER (Voeltz et al., 2006). Moreover, knockouts of Nogo or of the Nogo receptor, that mediates the inhibitory effects described in vitro for Nogo, MAG, and OMgp, do not lead to a marked improvement in axon regeneration in the spinal cords of null animals (Woolf, 2003; Zheng et al., 2005; Steward et al., 2007). These results strongly suggest that axon growth over distance requires both activation of intrinsic elongation programs and a capacity to overcome growth inhibition by the environment. The intracellular signaling within the axon that enables such coordinated action is still largely unknown, although a series of candidate molecules have been implicated at different stages of the process (Filbin, 2006; Zhou and Snider, 2006; Rossi et al., 2007). The GTPase Rho is thought to be a critical junction for both growth-stimulating intrinsic signals and growth-inhibiting environmental signals. Rho activation downstream of inhibitory myelin-derived signals arrests axon growth in vitro, and pharmacological inhibition of Rho promotes axon growth (Yamashita et al., 2005; McKerracher and Higuchi, 2006). A number of recent attempts to apply combinatorial stimuli to elicit axon regeneration after injury have targeted the Rho pathway.

One approach to concomitant activation of intrinsic growth programs together with blockade of inhibitory signals from the environment was demonstrated by attracting growth factor secreting macrophages to the site, concomitantly with transfection of a dominant-negative form of the Nogo receptor to overcome myelin inhibition (Fischer et al., 2004a). This combinatorial strategy resulted in threefold more axon growth by retinal ganglion neurons than that achieved by growth factors alone. A subsequent study coupled activation of the intrinsic growth program with blockade of RhoA-dependent inhibitory responses, leading to enhanced axon regeneration in optic nerve in vivo by almost one order of magnitude more than seen with a single intervention strategy (Fischer et al., 2004b). A similar approach was recently used in a spinal cord injury model. Combining inflammation-induced preconditioning of DRG with chondroitinase ABC (ChABC) digestion of proteoglycans enabled regeneration of sensory axons through the dorsal root entry zone and into the spinal cord after root injury (Steinmetz et al., 2005). Thus, mature neurons can reactivate elongating axon growth programs under appropriate stimuli, and when these are combined with interventions to inhibitory signals from the environment, axon growth can be significantly enhanced (Benowitz and Yin, 2007). If the mechanistic basis for these effects can be elucidated in detail, one might have a firm basis to develop combinatorial approaches to treatment of nerve injuries in humans.

AXONAL MAINTENANCE AND SEGMENT FUSION AS AN ALTERNATIVE TO RE-GROWTH?

So far this chapter has focused on the notion that axon injury induces signaling back to the cell body, which in turn initiates intrinsic growth programs that must propel elongating growth cones to regenerate the axon from the injury site all the way back to its target tissues. From a clinical point of view, it would be highly advantageous if severed axons could be preserved and reconnected, rather than undergoing complete degeneration of the distal segment followed by long distance re-growth and reinnervation. In this last section, I will review the literature pertinent to this very speculative alternative strategy for possible repair.

Under normal conditions, the distal segment of a lesioned axon undergoes rapid Wallerian degeneration after a short latent phase following the injury (Coleman, 2005). Normal peripheral axons appear to degenerate anterogradely after transection and retrogradely after a crush, but in both cases a sharp boundary separates intact and degenerated regions, indicating that Wallerian degeneration progresses as a sequential wave along the axon (Beirowski et al., 2005). A mutant mouse termed "Wlds" (for Wallerian degeneration slow) has a remarkable phenotype of extremely delayed degeneration in distal nerve projections after injury (Perry et al., 1990). In contrast to wild-type axons, once Wlds axons finally degenerate, they differ in the rate, direction, morphology, and abruptness of progression as compared to wild-type (Beirowski et al., 2005; Coleman, 2005). These observations suggest that axon degeneration is an active process intrinsic to the axon itself and potentially reversible in a targeted manner.

Positional cloning identified the Wlds gene product as an in-frame fusion protein combining the N-terminal 70 amino acids of the ubiquitination assembly factor Ube4b with the entire coding sequence of nicotinamide mononucleotide adenyltransferase-1 (Nmnat-1), an essential enzyme in the biosynthesis pathway of nicotinamide adenine dinucleotide (NAD). Transgenic expression of Wlds in diverse genetic backgrounds in both rat and mouse slows Wallerian degeneration after injury, and in various models of neurodegeneration (Coleman, 2005). The fact that Wlds is a fusion of two gene products suggests two possibilities for its mode of action, on the one hand a putative dominant negative effect of the ubiquitination factor fragment, and on the other hand activity of Nmnat1 via NAD biosynthesis (Fainzilber and Twiss, 2006). An interesting twist on the latter possibility was recently raised by Bellen and colleagues, who showed that Drosophila Nmnat maintains neuronal integrity independently of its NAD synthesis activity (Zhai et al., 2006). Perplexingly however, neither the Ube4b or the Nmnat1 domains of Wlds can recapitulate Wlds-like protection in transgenic rodents beyond the first few hours after injury (Conforti et al., 2006), suggesting a novel role for the combined molecule, or targeting or modulatory functions of the 18 amino acid linker region or 3′ or 5′ untranslated sequences (Fainzilber and Twiss, 2006). If the Wlds transcript contains axon-targeting sequences, translation upon injury might upregulate Nmnat1 locally at the lesion site. On the other hand, different studies have localized Wlds in the nucleus, suggesting that it might indirectly prime axons for greater resistance to degeneration by inducing changes in neuronal gene expression or differential acetylation of axoplasmic proteins shipped out from the cell body. Pinpointing the critical features underlying resistance to degeneration in Wlds axons will most likely require a combination of transcriptome and proteome analyses.

Recent studies in Drosophila have provided new insights on Wlds function and on the roles of glia in mediating and modulating axonal degeneration. Draper, a scavenger receptor expressed in fly glia, and ced6, an intracellular transducer/adaptor that may interact with Draper, both function in glial engulfment and clearance of axons during development and after lesion (Awasaki et al., 2006; MacDonald et al., 2006). Strikingly, transgenic expression of mouse Wlds can block degeneration of Drosophila axons for at least 30 days following lesion, and neuronal expression of Wlds can block the attraction of engulfing glia to injured axons (MacDonald et al., 2006). In contrast, Wlds protein did not appear to prevent developmental pruning of axons or dendrites in Drosophila or in mice (Hoopfer et al., 2006). Draper-mediated clearance of damaged axons by activated glia is extremely rapid and specific, suggesting that it is mediated by upregulation of ligands for Draper on the surfaces of axons marked for degeneration (Awasaki et al., 2006; MacDonald et al., 2006). Thus, one way to reduce axonal degeneration after injury in mammals may be by preventing or changing surface expression of such "eat me" signals on transected axons.

If severed distal axons can be induced to survive and maintain their structure, what is the likelihood of re-establishing neuronal connectivity by fusion of the cut ends? Severed distal segments of nerve axons have been reported to survive for weeks in diverse organisms from different phyla (Bittner, 1991). Although transected mammalian nerves survive and maintain their structure for much shorter periods, cooling to 6–9°C retards Wallerian degeneration in both sciatic and spinal axons (Tsao et al., 1999; Marzullo et al., 2002). Sciatic nerve segments retained the ability to generate action potentials for up to 7 days when maintained at 6–9°C, while spinal segments retained this capacity for up to 48 h. Thus, even without intervention via mechanisms that mimic Wlds, it may be possible to lower temperature locally to maintain a clinical window of opportunity for reconnection of transected axons. Fusing axonal cut ends was first demonstrated in invertebrates such as earthworms and Aplysia (Lore et al., 1999; Bedi and Glanzman, 2001). In mammals, application of Ca^{2+}-free solutions of polyethylene glycol (PEG) to cut or crushed ends of rat myelinated sciatic or spinal axons in vitro demonstrated the possibility of rapid restoration of functional and morphological continuity (Lore et al., 1999). It should be noted that this approach is highly inefficient if the cut ends are not well aligned and very close to each other, therefore we are still a long way away from being able to fuse cut axons in the clinic. Nonetheless, the findings summarized above encourage additional mechanistic studies on how to maintain isolated nerve segments after injury, and on mechanisms and modalities of fusion of such segments in a surgical setting.

SUMMARY

A successful axonal response to injury requires retrograde signaling to induce changes in the cell body response, and mobilization of outgrowth programs while integrating growth-promoting and growth-inhibiting signals from the environment. CNS and PNS axons differ in their capacity for useful regeneration, most likely due to differences in intrinsic growth capacity coupled with differential composition of extracellular growth-modulating agents in central versus peripheral environments. Combinatorial manipulation of the environment together with re-activation of intrinsic growth programs may allow sufficient elongating growth of injured axons for some degree of functional recovery. An alternative and perhaps complementary (albeit more speculative) approach might be the combination of axonal maintenance mechanisms derived from studies of the Wlds phenomenon together with procedures to fuse adjacent segments of cut axons. Recent years have seen significant progress on understanding the different mechanistic aspects of the axonal response to injury, to a degree that we are finally beginning to appreciate the depths of our ignorance in this field. Since appreciation of ignorance is the first small step on the road to wisdom, one hopes for more enlightenment in the years to come.

ACKNOWLEDGEMENTS

I gratefully acknowledge research funding from the Dr. Miriam and Sheldon Adelson Medical Research Foundation (AMRF), the Christopher Reeve Foundation, and the International Institute for Research in Paraplegia (IFP).

REFERENCES

Ahmed FA, Ingoglia NA, and Sharma SC (2001). Axon resealing following transection takes longer in central axons than in peripheral axons: implications for axonal regeneration. *Exp Neurol* 167: 451–455.

Al-Majed AA, Tam SL, and Gordon T (2004). Electrical stimulation accelerates and enhances expression of regeneration-associated genes in regenerating rat femoral motoneurons. *Cell Mol Neurobiol* 24: 379–402.

Ambron RT and Walters ET (1996). Priming events and retrograde injury signals. A new perspective on the cellular and molecular biology of nerve regeneration. *Mol Neurobiol* 13: 61–79.

Ambron RT, Schmied R, Huang CC, and Smedman M (1992). A signal sequence mediates the retrograde transport of proteins from the axon periphery to the cell body and then into the nucleus. *J Neurosci* 12: 2813–2818.

Ambron RT, Zhang XP, Gunstream JD et al. (1996). Intrinsic injury signals enhance growth, survival, and excitability of Aplysia neurons. *J Neurosci* 16: 7469–7477.

Awasaki T, Tatsumi R, Takahashi K et al. (2006). Essential role of the apoptotic cell engulfment genes draper and ced-6 in programmed axon pruning during drosophila metamorphosis. *Neuron* 50: 855–867.

Bedi SS and Glanzman DL (2001). Axonal rejoining inhibits injury-induced long-term changes in Aplysia sensory neurons in vitro. *J Neurosci* 21: 9667–9677.

Beirowski B, Adalbert R, Wagner D et al. (2005). The progressive nature of Wallerian degeneration in wild-type and slow Wallerian degeneration (WldS) nerves. *BMC Neurosci* 6: 6.

Benowitz LI and Yin Y (2007). Combinatorial treatments for promoting axon regeneration in the CNS: Strategies for overcoming inhibitory signals and activating neurons' intrinsic growth state. *Dev Neurobiol* 67: 1148–1165.

Bittner GD (1991). Long-term survival of anucleate axons and its implications for nerve regeneration. *Trends Neurosci* 14: 188–193.

Bosse F, Hasenpusch-Theil K, Kury P et al. (2006). Gene expression profiling reveals that peripheral nerve regeneration is a consequence of both novel injury-dependent and reactivated developmental processes. *J Neurochem* 96: 1441–1457.

Bronfman FC, Escudero CA, Weis J et al. (2007). Endosomal transport of neurotrophins: roles in signaling and neurodegenerative diseases. *Dev Neurobiol* 67: 1183–1203.

Buki A and Povlishock JT (2006). All roads lead to disconnection?–Traumatic axonal injury revisited. *Acta Neurochir (Wien)* 148: 181–193; discussion 193–184.

Cavalli V, Kujala P, Klumperman J, and Goldstein LS (2005). Sunday driver links axonal transport to damage signaling. *J Cell Biol* 168: 775–787.

Chierzi S, Ratto GM, Verma P, and Fawcett JW (2005). The ability of axons to regenerate their growth cones depends on axonal type and age, and is regulated by calcium, cAMP and ERK. *Eur J Neurosci* 21: 2051–2062.

Coleman M (2005). Axon degeneration mechanisms: commonality amid diversity. *Nat Rev Neurosci* 6: 889–898.

Conforti L, Fang G, Beirowski B et al. (2007). NAD+ and axon degeneration revisited: Nmnat1 cannot substitute for WldS to delay Wallerian degeneration. *Cell Death Differ* 14: 116–127.

Costigan M, Befort K, Karchewski L et al. (2002). Replicate high-density rat genome oligonucleotide microarrays reveal hundreds of regulated genes in the dorsal root ganglion after peripheral nerve injury. *BMC Neurosci* 3: 16.

Dent EW, Tang F, and Kalil K (2003). Axon guidance by growth cones and branches: common cytoskeletal and signaling mechanisms. *Neuroscientist* 9: 343–353.

Erez H, Malkinson G, Prager-Khoutorsky M et al. (2007). Formation of microtubule-based traps controls the sorting and concentration of vesicles to restricted sites of regenerating neurons after axotomy. *J Cell Biol* 176: 497–507.

Fainzilber M and Twiss JL (2006). Tracking in the Wlds – the hunting of the SIRT and the luring of the Draper. *Neuron* 50: 819–821.

Filbin MT (2006). Recapitulate development to promote axonal regeneration: good or bad approach? *Philos Trans R Soc Lond B Biol Sci* 361: 1565–1574.

Fischer D, He Z, and Benowitz LI (2004a). Counteracting the Nogo receptor enhances optic nerve regeneration if retinal ganglion cells are in an active growth state. *J Neurosci* 24: 1646–1651.

Fischer D, Petkova V, Thanos S, and Benowitz LI (2004b). Switching mature retinal ganglion cells to a robust growth state in vivo: gene expression and synergy with RhoA inactivation. *J Neurosci* 24: 8726–8740.

Gitler D and Spira ME (1998). Real time imaging of calcium-induced localized proteolytic activity after axotomy and its relation to growth cone formation. *Neuron* 20: 1123–1135.

Goldberg JL (2004). Intrinsic neuronal regulation of axon and dendrite growth. *Curr Opin Neurobiol* 14: 551–557.

Hanz S and Fainzilber M (2006). Retrograde signaling in injured nerve – the axon reaction revisited. *J Neurochem* 99: 13–19.

Hanz S, Perlson E, Willis D et al. (2003). Axoplasmic importins enable retrograde injury signaling in lesioned nerve. *Neuron* 40: 1095–1104.

Harvey PJ, Grochmal J, Tetzlaff W et al. (2005). An investigation into the potential for activity-dependent regeneration of the rubrospinal tract after spinal cord injury. *Eur J Neurosci* 22: 3025–3035.

Hirata A, Masaki T, Motoyoshi K, and Kamakura K (2002). Intrathecal administration of nerve growth factor delays GAP 43 expression and early phase regeneration of adult rat peripheral nerve. *Brain Res* 944: 146–156.

Hoopfer ED, McLaughlin T, Watts RJ et al. (2006). Wlds protection distinguishes axon degeneration following injury from naturally occurring developmental pruning. *Neuron* 50: 883–895.

Howard MJ, David G, and Barrett JN (1999). Resealing of transected myelinated mammalian axons in vivo: evidence for involvement of calpain. *J Neurosci* 93: 807–815.

Howe CL and Mobley WC (2005). Long-distance retrograde neurotrophic signaling. *Curr Opin Neurobiol* 15: 40–48.

Iwata A, Stys PK, Wolf JA et al. (2004). Traumatic axonal injury induces proteolytic cleavage of the voltage-gated sodium channels modulated by tetrodotoxin and protease inhibitors. *J Neurosci* 24: 4605–4613.

Kenney AM and Kocsis JD (1998). Peripheral axotomy induces long-term c-Jun amino-terminal kinase-1 activation and activator protein-1 binding activity by c-Jun and junD in adult rat dorsal root ganglia In vivo. *J Neurosci* 18: 1318–1328.

Lee N, Neitzel KL, Devlin BK, and MacLennan AJ (2004). STAT3 phosphorylation in injured axons before sensory and motor neuron nuclei: potential role for STAT3 as a retrograde signaling transcription factor. *J Comp Neurol* 474: 535–545.

Lindwall C and Kanje M (2005). Retrograde axonal transport of JNK signaling molecules influence injury induced nuclear changes in p-c-Jun and ATF3 in adult rat sensory neurons. *Mol Cell Neurosci* 29: 269–282.

Lindwall C, Dahlin L, Lundborg G, and Kanje M (2004). Inhibition of c-Jun phosphorylation reduces axonal outgrowth of adult rat nodose ganglia and dorsal root ganglia sensory neurons. *Mol Cell Neurosci* 27: 267–279.

Lore AB, Hubbell JA, Bobb DS Jr et al. (1999). Rapid induction of functional and morphological continuity between severed ends of mammalian or earthworm myelinated axons. *J Neurosci* 19: 2442–2454.

Lu P, Yang H, Jones LL et al. (2004). Combinatorial therapy with neurotrophins and cAMP promotes axonal regeneration beyond sites of spinal cord injury. *J Neurosci* 24: 6402–6409.

Lu P, Jones LL, and Tuszynski MH (2007). Axon regeneration through scars and into sites of chronic spinal cord injury. *Exp Neurol* 203: 8–21.

Lykissas MG, Batistatou AK, Charalabopoulos KA, and Beris AE (2007). The role of neurotrophins in axonal growth, guidance, and regeneration. *Curr Neurovasc Res* 4: 143–151.

MacDonald JM, Beach MG, Porpiglia E et al. (2006). The Drosophila cell corpse engulfment receptor Draper mediates glial clearance of severed axons. *Neuron* 50: 869–881.

Mandolesi G, Madeddu F, Bozzi Y et al. (2004). Acute physiological response of mammalian central neurons to axotomy: ionic regulation and electrical activity. Faseb J 18: 1934–1936.

Marzullo TC, Britt JM, Stavisky RC, and Bittner GD (2002). Cooling enhances in vitro survival and fusion-repair of severed axons taken from the peripheral and central nervous systems of rats. *Neurosci Lett* 327: 9–12.

McKerracher L and Higuchi H (2006). Targeting Rho to stimulate repair after spinal cord injury. *J Neurotrauma* 23: 309–317.

Neumann S, Bradke F, Tessier-Lavigne M, and Basbaum AI (2002). Regeneration of sensory axons within the injured spinal cord induced by intraganglionic cAMP elevation. *Neuron* 34: 885–893.

Perlson E, Hanz S, Medzihradszky KF et al. (2004). From snails to sciatic nerve: retrograde injury signaling from axon to soma in lesioned neurons. *J Neurobiol* 58: 287–294.

Perlson E, Hanz S, Ben-Yaakov K et al. (2005). Vimentin-dependent spatial translocation of an activated MAP kinase in injured nerve. *Neuron* 45: 715–726.

Perlson E, Michaelevski I, Kowalsman N et al. (2006). Vimentin binding to phosphorylated Erk sterically hinders enzymatic dephosphorylation of the kinase. *J Mol Biol* 364: 938–944.

Perry VH, Lunn ER, Brown MC et al. (1990). Evidence that the Rate of Wallerian Degeneration is Controlled by a Single Autosomal Dominant Gene. *Eur J Neurosci* 2: 408–413.

Pfister KK, Shah PR, Hummerich H et al. (2006). Genetic analysis of the cytoplasmic dynein subunit families. *PLoS Genet* 2, e1.

Povelones M, Tran K, Thanos D, and Ambron RT (1997). An NF-kappaB-like transcription factor in axoplasm is rapidly inactivated after nerve injury in Aplysia. *J Neurosci* 17: 4915–4920.

Qiu J, Cai D, Dai H et al. (2002). Spinal axon regeneration induced by elevation of cyclic AMP. *Neuron* 34: 895–903.

Qiu J, Cafferty WB, McMahon SB, and Thompson SW (2005). Conditioning injury-induced spinal axon regeneration requires signal transducer and activator of transcription 3 activation. *J Neurosci* 25: 1645–1653.

Raivich G and Makwana M (2007). The making of successful axonal regeneration: genes, molecules and signal transduction pathways. *Brain Res Rev* 53: 287–311.

Raivich G, Hellweg R, and Kreutzberg GW (1991). NGF receptor-mediated reduction in axonal NGF uptake and retrograde transport following sciatic nerve injury and during regeneration. *Neuron* 7: 151–164.

Reynolds AJ, Hendry IA, and Bartlett SE (2001). Anterograde and retrograde transport of active extracellular signal-related kinase 1 (ERK1) in the ligated rat sciatic nerve. *J Neurosci* 105: 761–771.

Robles E, Huttenlocher A, and Gomez TM (2003). Filopodial calcium transients regulate growth cone motility and guidance through local activation of calpain. *Neuron* 38: 597–609.

Rossi F, Gianola S, and Corvetti L (2007). Regulation of intrinsic neuronal properties for axon growth and regeneration. *Prog Neurobiol* 81: 1–28.

Schmied R and Ambron RT (1997). A nuclear localization signal targets proteins to the retrograde transport system, thereby evading uptake into organelles in aplysia axons. *J Neurobiol* 33: 151–160.

Schwab ME (2002). Increasing plasticity and functional recovery of the lesioned spinal cord. *Prog Brain Res* 137: 351–359.

Shadiack AM, Sun Y, and Zigmond RE (2001). Nerve growth factor antiserum induces axotomy-like changes in neuropeptide expression in intact sympathetic and sensory neurons. *J Neurosci* 21: 363–371.

Smith DS and Skene JH (1997). A transcription-dependent switch controls competence of adult neurons for distinct modes of axon growth. *J Neurosci* 17: 646–658.

Spira ME, Oren R, Dormann A, and Gitler D (2003). Critical calpain-dependent ultrastructural alterations underlie the transformation of an axonal segment into a growth cone after axotomy of cultured Aplysia neurons. *J Comp Neurol* 457: 293–312.

Steinmetz MP, Horn KP, Tom VJ et al. (2005). Chronic enhancement of the intrinsic growth capacity of sensory neurons combined with the degradation of inhibitory proteoglycans allows functional regeneration of sensory axons through the dorsal root entry zone in the mammalian spinal cord. *J Neurosci* 25: 8066–8076.

Steward O, Zheng B, Banos K, and Yee KM (2007). Response to: Kim et al., "axon regeneration in young adult mice lacking Nogo-A/B." *Neuron* 38: 187–199. *Neuron* 54: 191–195.

Sung YJ, Povelones M, and Ambron RT (2001). RISK-1: a novel MAPK homologue in axoplasm that is activated and retrogradely transported after nerve injury. *J Neurobiol* 47: 67–79.

Sung YJ, Walters ET, and Ambron RT (2004). A neuronal isoform of protein kinase G couples mitogen-activated protein kinase nuclear import to axotomy-induced long-term hyperexcitability in Aplysia sensory neurons. *J Neurosci* 24: 7583–7595.

Sung YJ, Chiu DT, and Ambron RT (2006). Activation and retrograde transport of protein kinase G in rat nociceptive neurons after nerve injury and inflammation. *J Neurosci* 141: 697–709.

Tsao JW, George EB, and Griffin JW (1999). Temperature modulation reveals three distinct stages of Wallerian degeneration. *J Neurosci* 19: 4718–4726.

Verma P, Chierzi S, Codd AM et al. (2005). Axonal protein synthesis and degradation are necessary for efficient growth cone regeneration. *J Neurosci* 25: 331–342.

Voeltz GK, Prinz WA, Shibata Y et al. (2006). A class of membrane proteins shaping the tubular endoplasmic reticulum. *Cell* 124: 573–586.

Willis D, Li KW, Zheng JQ et al. (2005). Differential transport and local translation of cytoskeletal, injury-response, and neurodegeneration protein mRNAs in axons. *J Neurosci* 25: 778–791.

Willis DE and Twiss JL (2006). The evolving roles of axonally synthesized proteins in regeneration. *Curr Opin Neurobiol* 16: 111–118.

Wolf JA, Stys PK, Lusardi T et al. (2001). Traumatic axonal injury induces calcium influx modulated by tetrodotoxin-sensitive sodium channels. *J Neurosci* 21: 1923–1930.

Woolf CJ (2003). No Nogo: now where to go? *Neuron* 38: 153–156.

Yamashita T, Fujitani M, Yamagishi S et al. (2005). Multiple signals regulate axon regeneration through the nogo receptor complex. *Mol Neurobiol* 32: 105–111.

Zhai RG, Cao Y, Hiesinger PR et al.(2006). Drosophila NMNAT maintains neural integrity independent of its NAD synthesis activity. *PLoS Biol* 4: e416.

Zheng JQ, Kelly TK, Chang B et al.(2001). A functional role for intra-axonal protein synthesis during axonal regeneration from adult sensory neurons. *J Neurosci* 21: 9291–9303.

Zheng B, Atwal J, Ho C et al.(2005). Genetic deletion of the Nogo receptor does not reduce neurite inhibition in vitro or promote corticospinal tract regeneration in vivo. *Proc Natl Acad Sci U S A* 102: 1205–1210.

Zhou FQ and Snider WD (2006). Intracellular control of developmental and regenerative axon growth. *Philos Trans R Soc Lond B Biol Sci* 361: 1575–1592.

Ziv NE and Spira ME (1993). Spatiotemporal distribution of Ca^{2+} following axotomy and throughout the recovery process of cultured Aplysia neurons. *Eur J Neurosci* 5: 657–668.
Ziv NE and Spira ME (1995). Axotomy induces a transient and localized elevation of the free intracellular calcium concentration to the millimolar range. *J Neurophysiol* 74: 2625–2637.
Ziv NE and Spira ME (1998). Induction of growth cone formation by transient and localized increases of intracellular proteolytic activity. *J Cell Biol* 140: 223–232.
Zrouri H, Le Goascogne C, Li WW et al. (2004). The role of MAP kinases in rapid gene induction after lesioning of the rat sciatic nerve. *Eur J Neurosci* 20: 1811–1818.

3

Glial Cells, Inflammation, and CNS Trauma

Modulation of the Inflammatory Environment After Injury Can Lead to Long-Distance Regeneration Beyond the Glial Scar

Michael T. Fitch* and Jerry Silver†

*Department of Emergency Medicine, Wake Forest University School of Medicine, Medical Center Boulevard, Winston-Salem, NC 27157

†Department of Neurosciences, Case Western Reserve University School of Medicine, 10900 Euclid Avenue, Cleveland, OH 44106

Abstract

Injury to the central nervous system (CNS) in adult mammals leads to significant pathology associated with permanent disability. The reactivity of glial cells to injuries in the brain and spinal cord, including the importance of inflammatory influences, has been identified as one component of the failure of the nervous system to regenerate when healing occurs. This chapter will review selected aspects of reactive gliosis at the tissue, cellular, and molecular levels as it relates to oligodendrocyte, astrocyte, and microglial/macrophage responses to trauma and the abortive attempts of neuronal regeneration. The historical perspective and modern approaches detailed in this review will demonstrate that the field of glial cell biology has allowed us to go beyond purely mechanical considerations of the glial scar, and in doing so has provided new insights into the complex reactions and interactions of glial cells following injury that generate the generally nonpermissive nature

Copyright © 2008 Elsevier Inc.
All rights of reproduction in any form reserved.

of lesion sites in the adult CNS. Recent advances in the field have demonstrated that significant regeneration can occur when modifications to the inflammatory sequelae are made to create optimal conditions for axon growth.

INTRODUCTION

Glial cells of the central nervous system (CNS), as first described by Virchow (1846), were originally considered to be simply a mechanical framework to support neurons, as evidenced by their designation as "neuroglia," which literally translated means "nerve glue" (Weigert, 1895). In the years since their discovery, glial cells of the developing and mature CNS have been extensively studied and recognized as important functional components of the brain and spinal cord, in addition to their role in the structural arrangement of the CNS. Oligodendrocytes are recognized as the cells responsible for the myelination of axons within the CNS (Bunge, 1968; Wood and Bunge, 1984), while astrocytes have been demonstrated to participate in formation and maintenance of the blood–brain barrier (Wolburg and Risau, 1995), ion homeostasis (Walz, 1989), neurotransmitter transport and re-uptake (Schousboe and Westergaarde, 1995), production of extracellular matrix (Liesi et al., 1983; Bernstein et al., 1985; Liesi and Silver, 1988; Tomaselli et al., 1988; Grierson et al., 1990; McKeon et al., 1991; Ard et al., 1993; Smith-Thomas et al., 1994; McKeon et al., 1995; Canning et al., 1996; Tom et al., 2004a), and other functional roles. Microglial cells, widely believed to be related to cells of the peripheral macrophage and monocyte lineage (Ling and Wong, 1993; Theele and Streit, 1993), are the CNS resident phagocytic cells that maintain a resting phenotype during nonpathological conditions (Banati and Graeber, 1994; Gehrmann et al., 1995; Davalos et al., 2005).

In the adult peripheral nervous system (PNS), it has been known for many years that traumatic crush injury leads to a cellular response that frequently results in successful regeneration of injured axons (for review, see Guth, 1956). Unlike this robust regenerative response in the PNS, injury to the adult mammalian CNS leads, at best, to abortive regeneration or short range sprouting (Weidner et al., 2001). This occurs from or within the long myelinated axon tracts resulting in permanent disability with little or no functional regeneration of injured axons (Ramon y Cajal, 1928). For many years, the hypothesis was maintained that regeneration failure following injury was due primarily to a structural barrier to axon growth, the so-called "glial scar," composed chiefly of glial cells (primarily astrocytes) and connective tissue elements (Windle and Chambers, 1950; Windle et al., 1952; Clemente and Windle, 1954; Windle, 1956; Clemente, 1958).

The suggestion that the glial scar is an impenetrable mechanical barrier to regenerating axons was consistent with early histological observations and the classic idea that glial cells were merely structural components of the nervous system. However, in light of insights into the functional biology of glial cells in the past 30 years, the role of the glial scar in preventing CNS regeneration following traumatic injury has been re-evaluated (for reviews, see Reier et al., 1983; Fitch and Silver, 2001, 2007). Recent studies have taken the field of CNS regeneration beyond the physical ramifications of the glial scar, including the importance of inflammation and how inflammatory products can be modulated to create an environment that will support robust long-distance regeneration. This chapter is dedicated to a discussion of the cellular as well as molecular responses of oligodendrocytes, astrocytes, and microglial cells to traumatic injury and their proposed roles in the failure of functional regeneration of the adult mammalian CNS.

ARE ADULT CNS AXONS CAPABLE OF ROBUST REGENERATION?

Before addressing the issue of glial cell influences on CNS injury responses, the simplest question related to CNS regenerative failure must be considered – perhaps adult CNS neurons are not even capable of regenerating severed axons, unlike those in the PNS? (for review of peripheral nerve regeneration, see Lazar et al., 2004). This would be one possible explanation for the failure of axon regeneration in the adult CNS. Le Gros Clark (1942) suggested that adult CNS axons had a "feeble capacity" for axon growth as compared to PNS axons and even went so far as to claim that perhaps CNS neurons had no capacity to regenerate at all (Le Gros Clark, 1943). However, these suggestions were contradictory to the early observations that had been made by Ramon y Cajal (1928), which indicated that adult CNS axons could undergo limited growth and sprouting at the site of a lesion, but could not enter or cross the area of damage. Liu and Chambers (1958) demonstrated that axons of the spinal cord were capable of collateral sprouting, suggesting a plasticity and potential for growth in the adult CNS that had not been previously appreciated. Observations of limited growth by CNS axons on the proximal side of a lesion in the spinal cord (Guth et al., 1981, 1985, 1986) also suggested that adult axons have at least a limited intrinsic ability to grow. Analyses of cut axon endings using orthograde labeling suggests that severed axons are, in fact, capable of dynamic albeit short-distance sprouting immediately following axotomy (Li and Raisman, 1994), but eventually the sprouting growth cones become quiescent and can persist for many months or even years without leaving the site of injury (Li and Raisman, 1995). These dystrophic growth cones have been recently studied in greater detail when associated with inhibitory molecules in lesion sites (Tom et al., 2004b; Kerschensteiner et al., 2005) and it is now evident via time-lapse video analysis that these axon endings are in fact dynamic areas of growth with constant turnover of membrane components. However, the native environment that exists in the adult CNS after injury causes these classically dystrophic endballs to form and will not allow functional long-distance regeneration under ordinary circumstances.

While Tello (1911a,b) had reported growth of CNS axons within transplanted peripheral nerves, it was not until Aguayo and his colleagues demonstrated long-distance axon growth by intrinsic CNS neurons into grafts of peripheral nerve almost 70 years later that the robust intrinsic potential of certain adult CNS axons to regenerate in vivo became widely accepted (Richardson et al., 1980; David and Aguayo, 1981). These studies have led to the notion that the environment of the PNS is more conducive to axon elongation than the environment of the adult CNS, and that adult CNS axons are capable of long-distance axon regeneration under certain conditions. However, the failure of large numbers of regenerating axons to successfully leave peripheral nerve grafts and regenerate long distances back into the CNS in the absence of other manipulations highlights the transition between CNS and PNS environments as being a particularly important interface. A recent breakthrough in this technique (Houle et al., 2006) further demonstrates that with additional modification of the extracellular matrix, this difficult transition can be overcome to allow long-distance functional regeneration of adult axon fibers (see additional discussion of this work below).

Further evidence that the interface between the adult CNS environment and the PNS is nonpermissive for axon regeneration comes from studies of injured dorsal roots of the spinal cord. The cell bodies of the sensory dorsal root ganglion (DRG) neurons are situated in the PNS and extend bipolar axons both peripherally and centrally. Crush injury to the peripheral branch of this axon leads to successful regeneration in the PNS, while similar injury to the central branch does not. However, careful analysis of the central branch of the axon after injury suggests that this portion of the axon does, indeed, possess regenerative potential, as it can regenerate from the ganglion through the dorsal root right up to the surface of the spinal cord, where it stops or turns abruptly at the dorsal root entry zone (DREZ), the interface between PNS and CNS (Tower, 1931; Kimmel and Moyer, 1947; Perkins et al., 1980; Kliot et al., 1990; Golding et al., 1997). Thus, the axons of adult DRG neurons seem to be capable of regeneration, but regenerative failure occurs when these growing axons reach the interface between the cellular environments of the PNS and the CNS. Even axons from the ventral roots, known for their aggressive regenerative properties in the PNS, are unable to enter the CNS compartment when grafted onto injured dorsal roots (Carlstedt, 1983, 1985a,b). It is important to note that in very young animals (less than one week of age) regenerating dorsal root axons of the C-fiber system are capable of re-entering the CNS compartment (Carlstedt et al., 1987, 1988; Carlstedt, 1988), demonstrating in vivo that the interface between the CNS environment and the PNS environment in immature animals is somewhat permissive for axon navigation of this boundary, while this CNS/PNS interface in the adult is nonpermissive for axon regeneration into the CNS compartment without further modification. Recent investigations have shown that DRG neurons can indeed regenerate long distances into the CNS when modulation of inflammatory responses and extracellular matrix are coordinated to create the proper balance of factors to allow regenerating axons to traverse this critical CNS/PNS barrier (Steinmetz et al., 2005).

Adult DRG neurons have also been shown to have remarkable potential for long-distance regeneration in vivo when these cells are directly "micro" transplanted into adult white matter tracts of the brain and spinal cord (Davies et al., 1997, 1999) This special, minimally damaging transplant technique avoids trauma to CNS tissue and eliminates the need to modulate inflammatory factors altogether. These studies conclusively demonstrate the intrinsic ability of at least one subset of adult neurons to regenerate long distances within the normal CNS environment without further manipulation. However, when faced with the environment of a traumatic injury, even these rapidly regenerating axons are unable to penetrate in significant numbers past the area of blood–brain barrier breakdown and the resulting inflammatory milieu and inhibitory extracellular matrix (Davies et al., 1999), emphasizing that even adult neurons with remarkable regenerative potential are not enough to lead to functional repair alone.

It is not known whether the intrinsic growth ability for adult neuronal cells is present among all neuron cell types and whether it always equals the robust axon outgrowth capacity of transplanted DRG neurons or embryonic neurons during development. Interestingly, it does appear that mature DRG neurons are intrinsically different than CNS neurons when transplanted following the total neurotomy that is needed to harvest the cells in the first place. Mature CNS neurons, when completely neurotomized, tend to form only dendrites when explanted onto mature CNS tissue slices (Tom et al., 2004a) or when microtransplanted within white matter tracts in vivo (J. Silver, unpublished observations) and may not extend axons by default (Goldberg et al., 2002). Obviously, damaged CNS neurons can re-extend axons (e.g., into PNS grafts) when the axon hillock or beyond remains following axotomy, but they may do so at a slower rate than the DRG. The subset of adult neurons which elongate into PNS implants have been shown to dramatically increase expression of GAP-43, suggesting that the relative ability of neurons to upregulate GAP-43 may play an important role in determining which adult axons are most capable of robust axonal growth (Campbell et al., 1991; Vaudano et al., 1995), as upregulation of GAP-43 has been implicated as a necessary but not sufficient condition for neuron regeneration (Tetzlaff et al., 1994). Increases in GAP-43 are also demonstrated in successful regeneration of sensory axons through the DREZ following the addition of an inflammatory modulator and modification of the extracellular matrix (Steinmetz et al., 2005). In vitro studies have demonstrated relatively slow growth for adult neurons by showing that various neuronal cell types grow at significantly faster rates at early ages (embryonic or postnatal) than adult neurons when presented with similar substrates, including superior cervical ganglion neurons (Argiro et al., 1984), DRG neurons (Shewan et al., 1995), and retinal ganglion neurons (Bahr et al., 1995; Chen et al., 1995), although these studies contrast with the in vivo documentation of adult DRG axon growth at the remarkable rate of 1–2 mm per day (Davies et al., 1997). In addition, adult neurons may also respond differently to trophic factors, perhaps due to an intrinsic tendency to ramify. Thus, exogenous nerve growth factor (NGF) leads to increased branching of adult DRG neurons as opposed to the increased axon length seen with NGF-treated neonatal DRG cells (Yasuda et al., 1990). Such observations may help to explain why some adult axons will sprout and branch at the lesion site instead of extending axons linearly, which is more characteristic of embryonic neurite outgrowth. Therefore, improving the regeneration of adult axons may have to be done selectively for various neuronal subpopulations. Strategies to improve regeneration in the CNS may also need to consider utilizing methods to evoke a conditioning-like response within axons (Filbin, 2003) and may also need to consider incorporating molecules such as L1 that may play a role in increasing the rectilinear growth potential of different types of neurons (Brittis et al., 1995, Chen et al., 2005), as well as making a cellular environment generally more conducive to axon growth.

DO GLIAL CELLS CONTRIBUTE TO CNS REGENERATIVE FAILURE?

OLIGODENDROCYTES AND MYELIN

Oligodendrocytes, the cells responsible for the production of myelin in the CNS, have not been extensively studied in terms of their cellular responses to injury, and traditionally they are not thought to play a major reactive role in the glial response to trauma. While oligodendrocytes have been shown to proliferate following injury (Ludwin, 1984, 1985), this activity is found in the wound tract only transiently (Xie et al., 1995), and no changes in the expression of myelin proteins have been demonstrated (Ludwin, 1985). However, recent work suggests that a small amount of endogenous neurogenesis may replace oligodendrocytes after primate spinal cord injury and possibly remyelinate host axons (Yang et al., 2006). When considering CNS regenerative failure, much attention has been given to the hypothesis that the native cell membranes and myelin formed by mature oligodendrocytes throughout the CNS may be partially responsible for the failure of adult CNS axons to regenerate.

A number of studies have contributed evidence which suggests that the presence of myelin in the CNS may create an environment less conducive to axon growth (Caroni et al., 1988; Savio and Schwab, 1990; Bastmeyer et al., 1991; Stuermer et al., 1992; Schwab et al., 1993; Spillmann et al., 1997). For example, the chicken nervous system is capable of complete recovery following spinal cord transection prior to myelin formation, but after the developmental onset of myelination this capacity is lost (Keirstead et al., 1992). Treatments that delay the onset of myelination in both chicken and rat lead to greater regrowth of cut axons at developmental time periods during which regeneration is normally absent (Savio and Schwab, 1990; Keirstead et al., 1992), suggesting that the normal presence of myelin is at least associated with the time of the transition between a permissive and a nonpermissive environment for axon growth in the CNS. Some tissue culture studies have suggested that embryonic retinal neurons and certain peripheral neurons (such as early postnatal DRG or sympathetic ganglion neurons) avoid oligodendrocytes in vitro and have limited neurite outgrowth abilities on CNS myelin membranes when compared to myelin from the PNS (Caroni et al., 1988; Fawcett et al., 1989).

Isolation of protein fractions from CNS myelin suggested that the nonpermissive nature of this substrate was possibly due to a specific molecule (Caroni and Schwab, 1988b), and further work has characterized a novel membrane protein of spinal cord myelin named Nogo-A (Chen et al., 2000). An antibody to this protein has been used in an attempt to neutralize the nonpermissive nature of CNS white matter in a variety of studies in vitro (Caroni and Schwab, 1988a; Savio and Schwab, 1989; Bandtlow et al., 1990; Spillmann et al., 1997), demonstrating that the inhibitory nature of CNS myelin can be modified to be more permissive to axonal outgrowth. Antibody experimentation has also been utilized in vivo in several studies and results in histological and behavioral recovery from injury (Schnell and Schwab, 1990; Bregman et al., 1995; Freund et al., 2006). However, even a recent study in primates suggesting possible benefit from antibody treatments is difficult to interpret due to limited numbers of animals (12), problems with reproducibility of injuries (one-third of animals with incomplete lesions), nonsignificant outcomes when all animals are included in analysis, and an injury model with only minor deficits that has unclear relevance to human spinal cord injuries (Freund et al., 2006).

Myelin associated glycoprotein (MAG) has also been suggested as an identified component of myelin that inhibits neurite outgrowth (McKerracher et al., 1994), but studies using MAG-deficient mice have demonstrated that the absence of MAG alone does not abolish inhibitory aspects of oligodendrocytes and myelin both in vitro and in vivo (Bartsch et al., 1995; Bartsch, 1996). However, other evidence suggests that MAG may in fact play some role in regenerative failure (Tang et al., 1997). Oligodendrocyte-myelin glycoprotein (OMgp) has also been identified as a nonpermissive substrate for in vitro axon growth that may induce growth cone collapse (Kottis et al., 2002; Vourc'h and Andres, 2004). Until further studies identify MAG, OMgp, or other specific myelin components as instrumental in preventing axon regeneration in vivo, the investigations of myelin as an inhibitory component of the CNS suggest that at least Nogo-A may contribute at some level to the regenerative failure seen in the adult CNS after injury and that modulation of these inhibitors may be therapeutic. However, myelin factors alone are not likely to explain the failure of regeneration after injury since several studies have shown robust regeneration of adult axons within white matter tracts without any modulation of these purported myelin inhibitors (Davies et al., 1997, 1999; Steinmetz et al., 2005; Houle et al., 2006).

Although the histological and functional recovery demonstrated using the antibodies to Nogo-A in vivo suggests the importance of modifying myelin components following a CNS lesion, the specificity of this treatment as a therapeutic agent to allow axon regeneration remains to be further elucidated, particularly in light of studies demonstrating local neurite sprouting in response to such antibody treatment without long-distance regeneration through the lesion site (Guest et al., 1997; Thallmair et al., 1998; Z'Graggen et al., 1998). Many of the experiments using this antibody in vivo utilized control antibodies raised against horseradish peroxidase (Schnell and Schwab, 1990; Bregman et al., 1995; Guest et al., 1997; Thallmair et al., 1998; Z'Graggen et al., 1998), an antigen not present within normal myelin structure where the IN-1 (Nogo-A) antigens are located (Rubin et al., 1994). This raises the possibility that antibodies that recognize other myelin proteins may have similar beneficial effects as those demonstrated in these studies, not as a direct effect of blocking a specific inhibitory component of myelin, but as an indirect effect by causing a physical disruption to the myelin structure or stimulating an immunologic reaction. In fact, experiments by other investigators have demonstrated regeneration of axons and electrophysiological recovery following disruption of the myelin structure via in vivo treatments with Gal-C or O4 antibodies (against antigens found in normal myelin) combined with complement (Keirstead et al., 1995). Such results demonstrate that less-specific disruption of myelin is beneficial to axon regeneration, and that a component of the normal inflammatory response can mediate such alterations in myelin structure. Importantly, studies using the antibodies to myelin components as therapeutic agents have not addressed the possibility that antibodies that bind to oligodendrocytes and myelin membranes may, in turn, create a persistent low-grade inflammatory reaction that is secondary to antibody binding or myelin destruction. Such limited inflammatory responses have previously been associated with beneficial effects on axon regeneration in vivo (Windle and Chambers, 1950; Kliot et al., 1990; Siegal et al., 1990), perhaps through the direct effects of secreted inflammatory products on axon growth or indirectly via cytokine modulation of reactive astrocytes.

While myelin-associated inhibitors have been demonstrated as one component of the CNS that may contribute to the failure of regeneration, the relative potency of their actions remains to be determined (for a detailed discussion, see Fitch and Silver, 2000). A number of studies have questioned the role of myelin and myelin components as the sole inhibitory influences for axon growth. Tissue culture experiments have shown that while DRG neurons are inhibited by oligodendrocytes (Kobayashi et al., 1995), retinal ganglion cell axons are not altered by contact with these myelinating cells (Ard et al., 1991; Kobayashi et al., 1995), observations which contrast with the initial experiments that purported to establish the "potently" inhibitory properties of oligodendrocytes (Caroni et al., 1988). In addition, unmyelinated optic nerves from late stage embryos have been shown to be nonpermissive for neonatal and adult axon growth in vitro, despite the lack of myelin components (Shewan et al., 1993), while embryonic DRG neurons can grow on both unmyelinated neonatal and fully myelinated adult optic nerves (Shewan et al., 1995), suggesting that issues of neuronal age and subtype may be critical to consider before proposing global inhibitory effects for myelin substrates. The Bowman-Wyse mutant rat has areas of the optic nerve devoid of myelin and oligodendrocytes that still demonstrate a lack of regenerative ability in vivo (Berry et al., 1992), suggesting that the absence of myelin and oligodendrocytes is not sufficient to allow regeneration to occur. Embryonic neurons, transplanted gently into the adult nervous system, have the ability to extend long axons through the heavily myelinated environment of an adult white matter axon tract in vivo (Li and Raisman, 1993). A study by Berry et al. (1996) suggests that myelin components do not necessarily need to be neutralized to get regeneration in the acutely lesioned optic nerve, as trophic factors administered to the retinal ganglion cell bodies via a crude PNS graft can partially overcome any inhibitory factors that may be acting on the severed axons distally in the injured optic nerve. The most convincing evidence to date that the presence of myelin in adult CNS white matter tracts is not sufficient to inhibit the regeneration of adult CNS axons is the series of in vivo experiments that demonstrate robust regeneration of transplanted DRG cell axons across the corpus callosum into the contralateral hemisphere (Davies et al., 1997) and follow-up experiments demonstrating similar robust regeneration within white matter tracts of the spinal cord (Davies et al., 1999). These studies illustrate that the nonpermissive nature of myelin is not the universal factor which leads to regenerative failure in the CNS. It remains to be determined under what circumstances myelin inhibitors may play a role as a component of regenerative failure, and where factors other than myelin work to establish a truly inhibitory environment for regeneration of the adult CNS.

ASTROCYTES AND THE GLIAL SCAR

Astrogliosis

The astrocytic cell responses to injury have been extensively studied in a variety of experimental models, and the terms "gliosis" and "astrogliosis" are often used to describe the astrocyte reactions to injury. Astrocyte cellular hypertrophy, hyperplasia, and increased production of intermediate filaments characterize astrocyte gliosis, and cells responding in these ways to injury are often referred to as "reactive astrocytes." Astrocytes are easily identified by immunocytochemical methods directed toward the astrocyte specific glial fibrillary acidic protein (GFAP) (Bignami et al., 1972; Bignami and Dahl, 1974; Eng, 1985), and astrocyte hypertrophy and increased GFAP following injury have been demonstrated using these techniques to label reactive astrocytes (Bignami and Dahl, 1976). The issue of astrocyte cell division following CNS injury is less clearly recognized, as various studies have determined that some reactive astrocytes divide after injury but the majority of them do not (Norton et al., 1992). A number of investigators have suggested that there are relatively few proliferating astrocytes (ranging from 1–2% to 5–6%), that they are restricted to the immediate area of the wound, and that the apparent increase in numbers of reactive astrocytes is not primarily due to proliferation, but rather due to migration of cells and the enhanced ability to visualize them with GFAP antibodies (Adrian and Williams, 1973; Murray and Walker, 1973; Latov et al., 1979; Ludwin, 1985; Miyake et al., 1988; Takamiya et al., 1988; Murray et al., 1990; Hatten et al., 1991; Miyake et al., 1992; Amat et al., 1996). Thus, astrocytic hypertrophy and increased expression of GFAP are widely recognized as characteristics of reactive astrocytes and are often used as markers for gliosis. The previously described cellular hyperplasia as a component of gliosis is predominantly due to increases in inflammatory cells within the lesion.

Astrocytes as a Mechanical Barrier

The idea that the astrocytes that comprise the glial scar serve only as a mechanical barrier to regeneration has been shown to be incomplete as an explanation as to how the glial scar functions as a barrier to growing axons (for review, see Reier et al., 1983). While this hypothesis was originally widely supported (Windle and Chambers, 1950; Windle et al., 1952; Clemente and Windle, 1954; Windle, 1956; Clemente, 1958), other work suggests that the astrocytic scar that forms after traumatic injury does not prevent axon growth simply via a mechanical mechanism. Glial scarring appears to be a critical feature of wound healing in the CNS, since elimination of the mitotically active contingent of reactive astrocytes leads to increase in the size of the wound and an exaggerated leakiness of the compromised blood–brain barrier (Faulkner et al., 2004; Myer et al., 2006). Even in the absence of a large scar, all regeneration fails. Spinal cord injury in hibernating squirrels does not lead to formation of a dense glial scar, but regeneration still does not occur in the absence of a histologically apparent barrier (Guth et al., 1986), demonstrating that the lack of a histologically dense and mechanically obstructive glial scar does not in itself lead to successful regeneration. Instead, it is now believed that the molecular nature of glial scar, including the production of inhibitory molecules by astrocytes, is primarily responsible for the failure of axon regeneration (Fitch and Silver, 2000). Current theories of how astrocytes may contribute to the lack of CNS regeneration include the following: (1) failure to provide a proper cellular substrate for axon growth, (2) failure to produce molecules that support axons, and (3) production of molecules that actively inhibit axon elongation.

Astrocytes as Substrates for Axon Elongation

Astrocytes have been implicated as negative components of the mature mammalian CNS response to injury that may be responsible in part for the failure of regeneration, and much attention has been given to the role of astrocytes as substrates for neuron outgrowth. Glial cells of the astrocyte lineage play an integral role during development of the nervous system as a substrate for neuronal migration and axon elongation in vivo (Silver et al., 1993; Rakic, 1995). Neonatal astrocytes have been shown to be a supportive substrate for axon growth in vitro (Fallon, 1985a,b; Rudge et al., 1989; Ard et al., 1991; Bahr et al., 1995), while reactive astrocytes are a nonpermissive substrate for axon growth in vitro (Rudge et al., 1989; Smith et al., 1990; Geisert and Stewart, 1991; Bahr et al., 1995; Le Roux and Reh, 1996), suggesting that the reactive astrocytes present following a traumatic injury to the CNS may contribute to an inhibitory cellular environment. However, it is interesting to note that although axon elongation is severely limited, dendritic growth is much less altered by reactive astrocytes, at least in two dimensions (Le Roux and Reh, 1996). This observation suggests that perhaps some of the limited neurite growth and abortive sprouting seen in the vicinity of wounds in grey matter are dendritic processes instead of short axonal projections. The lack of axonal outgrowth on adult reactive astrocytes in vitro suggests that these cells may contribute to CNS regenerative failure either by actively inhibiting or by not providing an appropriately supportive environment.

While mature reactive astrocytes may not be favorable substrates for axon elongation, the presence of a cellular terrain for axons to grow on is certainly essential (Guth et al., 1981, 1985), especially considering that axons will not traverse a purely fluid environment (Harrison, 1910, 1914). A problem that has plagued the field of CNS injury research for many years is the progressive necrosis and development of cavities or cysts as secondary events following trauma to the CNS. Such cavitation can develop from a small initial lesion that progresses to a large cavity extending far rostral and caudal to the original area of injury (Balentine, 1978). Various investigators have hypothesized that cavitation and central necrosis is related to ischemic injury (Balentine, 1978), hemorrhage (Ducker et al., 1971; Wallace et al., 1987), neuronal lysozyme activity (Kao et al., 1977a,b), leakage of serum proteins across the blood–brain barrier (Fitch and Silver, 1997a), and macrophage infiltration and inflammation (Blight, 1994; Fujiki et al., 1996; Szczepanik et al., 1996; Zhang et al., 1996; Fitch and Silver, 1997a). The underlying process controlling progressive axon damage and astrocyte cavity formation are not completely understood, but a series of studies has implicated inflammatory events as important in the cascade of events leading to cavitation (Fitch et al., 1999). Video microscopy allowed direct visualization of a novel mechanism for cavity formation following CNS injury that occurs via astrocyte migratory behaviors. This physical process of cavitation is a direct result of astrocyte movement away from inflammatory infiltrates and can lead to astrocyte abandonment of neuronal processes, dynamic stretching of neurites, and secondary injury (Fitch et al., 1999). This kind of astrocyte migration in injury models leads to a large cavity devoid of astrocytes (Mathewson and Berry, 1985; Fujiki et al., 1996; Zhang et al., 1996; Fitch and Silver, 1997b; Zhang et al., 1997; Fitch et al., 1999), which is a significant impediment to regeneration, as axons have no cellular substrate on which to elongate.

Astrocyte Production of Axon Supportive Molecules

Another possible way in which astrocytes could play a role in the failure of regeneration is to not supply appropriate factors that are supportive for axon growth or to produce adhesion molecules inappropriate for most axons. The developing nervous system contains a number of molecules produced by astrocytes or their precursors that encourage axon growth (Tomaselli et al., 1988; Smith et al., 1990), but until recently it was believed that adult astrocytes did not produce the same types of supportive molecules. It is now recognized that adult astrocytes can produce laminin in vitro but it remains controversial whether they do so in vivo (Liesi et al., 1983, 1984; Bernstein et al., 1985; Liesi and Silver, 1988; Frisen et al., 1995; Grimpe et al., 2002). Laminin has been described as a component of glial scars in vivo (Bernstein et al., 1985; McKeon et al., 1991; Frisen et al., 1995; McKeon et al., 1995) but most of this critical adhesion molecule is shielded from axons because it is located within basal laminae of blood vessels. In addition, astrocyte-associated fibronectin was recently shown to be present robustly within white matter tracts of the adult CNS and appears necessary for the regeneration of lengthy axons from adult DRG neurons (Tom et al., 2004a). Many other molecules thought to promote axon regeneration have been shown to be produced by reactive astrocytes at least in vitro (for review see Eddleston and Mucke, 1993 and Grimpe and Silver, 2002). However, questions remain that may prove to be important in determining the regenerative response of injured axons in relation to other molecules of the CNS. Thus, while astrocytes certainly do produce axon supportive molecules in the vicinity of CNS lesions, it is not known whether sufficient quantities, correct temporal or spatial sequences, or combinations of these axon growth supportive molecules are produced. It is also not well appreciated whether axotomized neurons re-express the appropriate receptor molecules that would enable regeneration to occur (Condic, 2001). It is likely that interventions designed to increase concentrations or combinations of trophic and/or tropic molecules may be beneficial at certain time points following a traumatic injury, and active investigations into these issues are ongoing.

For example, some evidence indicates that growing adult axons can associate with reactive astrocytes after injury when supplied with exogenous trophic molecules such as NGF, although the temporal sequence of these cellular events has not been completely determined. Implantation of nitrocellulose filters containing NGF along with fetal tissue into the lesioned spinal cord leads to increases in axon growth from re-implanted cut dorsal roots, perhaps due to effects of the NGF on the glial cells which organize around the implant (Houle, 1992). Fibroblasts that are genetically engineered to secrete NGF have been implanted into the sites of CNS damage, and this treatment encourages large numbers of axons to regenerate into the usually nonpermissive central region of injury (Kawaja and Gage, 1991; Kawaja et al., 1992; Tuszynski et al., 1994, 1996, 1997, 2002;

Jones et al., 2003b). However, while these approaches have shown the ability to stimulate sprouting into the lesion site, the difficulty appears to be in encouraging the axons to leave the area of trophic support and regenerate out of the immediate site of injury toward their proper functional connections. This highlights a major dilemma inherent in any repair strategy that uses local application of trophic molecules at the cut ends of axons to stimulate regeneration. Unfortunately, the axons that are stimulated to grow toward the site of factor release are possibly unable to leave the "trophic oasis" that is present at the source of molecules and, thus, simply remain indefinitely in the immediate vicinity of the exogenous factors. A successful modulation of trophic factors must not only encourage axons to grow following injury, but must also provide a stimulus for elongation through the lesion site and back into the CNS parenchyma. Such a combinatorial therapy has shown some success when combining NT-3 treatment with cAMP stimulation of neuronal cell bodies leading to regeneration beyond spinal cord injury sites (Lu et al., 2004). Alternatively, delivery of such potential therapeutic agents could encourage growth of axons without attracting regeneration directly into the cellular graft, as demonstrated with grafts of NT-3-secreting cells leading to partial functional recovery and growth of corticospinal axons in adjacent grey matter (Grill et al., 1997). Perhaps the addition of trophic factors combined with other therapeutic agents may provide another therapeutic avenue, as demonstrated by Schnell et al. (1994), using NT-3 combined with an antibody to neutralize myelin inhibitors.

Another intriguing issue concerning the availability of trophic factors in the area of a CNS wound has been suggested by Frisen et al. (1993) and Fryer et al. (1997). A truncated form of the neurotrophin TrkB receptor that lacks the catalytic domain has been shown to be expressed by astrocytes and oligodendrocytes and is upregulated in the glial scar following injury (Frisen et al., 1993; Fryer et al., 1997). The hypothesis has been suggested that neurotrophins bound to the surface of glial cells via the truncated TrkB receptors may be used as a growth-promoting substrate for sprouting axons at the site of a lesion (Frisen et al., 1993). However, this theory fails to explain why axon regrowth does not progress beyond minimal sprouting. Perhaps a more satisfying hypothesis is supported by in vitro experiments, which demonstrate that cells expressing the truncated TrkB receptor fail to support neurite outgrowth from BDNF responsive neurons even in the presence of exogenous BDNF (Fryer et al., 1997), suggesting that the truncated receptors may be binding and internalizing neurotrophins. Such a situation could, in a sense, form a "molecular sponge", which acts to soak up and remove trophic factors that may be required by growing axons. Thus, factors bound to truncated TrkB receptors (or other similar "molecular sponges") may be unavailable for growing axons, which could help to explain the lack of functional regeneration even in wounds where abundant trophic molecules have been demonstrated. Astrocytes in the immediate vicinity of an injury may be inhibitory to axon regeneration in part by virtue of removing functional access to essential growth factors in the lesion area.

Production of Inhibitory Molecules by Astrocytes

A number of studies have suggested that astrocyte production of axon inhibitory molecules may explain several aspects of CNS regenerative failure (for reviews, see Fitch and Silver, 1997b; Silver and Miller, 2004). A particular emphasis has been placed on extracellular matrix molecules produced by reactive astrocytes that are upregulated following injury, a notable example being tenascin (Laywell and Steindler, 1991; McKeon et al., 1991; Laywell et al., 1992; Brodkey et al., 1995; Lips et al., 1995; McKeon et al., 1995; Zhang et al., 1995). Some forms of tenascin have been implicated as negative influences of axon growth on astrocytes (Grierson et al., 1990; Ard et al., 1993; Smith-Thomas et al., 1994; Chiquet-Ehrismann et al., 1995; Gates et al., 1997), and tenascin has been demonstrated as a component of glial scars extracted from adult brain that are poorly supportive of axon growth (McKeon et al., 1991, 1995). Other growth inhibitory molecules such as Semaphorin 3 (Pasterkamp et al., 2001), ephrin-B2 (Bundeson et al., 2003), and slit proteins (Hagino et al., 2003) are also candidates for playing roles in regenerative failure that have not yet been fully characterized.

Proteoglycans are an important class of molecules produced by reactive astrocytes that have been shown to play a pivotal role in the modulation of axon growth and regeneration. Proteoglycans are molecules consisting of a protein core with attached sugar moieties called glycosaminoglycans (GAGs) and are characterized by their GAG compositions as chondroitin sulfate, heparan sulfate, keratan sulfate, and dermatan sulfate (for a review of nervous tissue proteoglycans, see Margolis and Margolis, 1993). The upregulation of such proteoglycans is found in many tissues throughout the body in pathological conditions. Changes in proteoglycans are found in regenerating skeletal muscle (Carrino et al., 1988), arterial injury (Richardson and Hatton, 1993; Nikkari et al., 1994; Jain et al., 1996), atherogenesis (Srinivasan et al., 1995), and corneal injury (Brown et al., 1995). The nervous system is no exception, as increases in proteoglycans have also been demonstrated in vivo following trauma to the adult CNS. For example, the NG2 proteoglycan is increased transiently after injury to the CNS (Levine, 1994), phosphacan is increased in glial scars (McKeon et al., 1995), and neurocan, brevican, versican, and phosphacan are all produced in vivo within injured spinal cord (Jones et al., 2003a,b). Chondroitin sulfate proteoglycans are increased and have been shown to persist in the extracellular matrix of the CNS following injury, including the spinal cord following dorsal root injury (Pindzola et al., 1993), in the fornix following transection (Lips et al., 1995), in the brain following stab wound (Fitch and Silver, 1997a), in explants of wounded striatum (Gates et al., 1997), and in the spinal cord following injury (Fitch and Silver, 1997a; Jones et al., 2003a,b). The presence of these putative inhibitory proteoglycans in vivo following injury suggests a role for these molecules in contributing to the nonpermissive environment encountered in the CNS. Along these lines, studies have shown that changes in the ratios of chondroitin sulfate and heparan sulfate proteoglycans in the developing chicken spinal cord occur precisely at the transition between the permissive and the nonpermissive environments for spinal cord regeneration (Dow et al., 1994). Proteoglycans and other inhibitory molecules have been implicated in the failure of the adult CNS environment to support robust axon growth following traumatic injury, and several experiments have demonstrated the direct association of these molecules in reactive glial extracellular matrix with the failure of adult axon regeneration in otherwise robust growth environments (Davies et al., 1997, 1999; Grimpe and Silver, 2004).

In support of these ideas, astrocytes in vitro have been demonstrated to produce proteoglycans and associated GAGs (Norling et al., 1984; Gallo et al., 1987; Gallo and Bertolotto, 1990; Johnson-Green et al., 1991; Geisert et al., 1992; Smith-Thomas et al., 1994; Grimpe and Silver, 2004), and proteoglycans associated with reactive astrocytes have been demonstrated to inhibit neurite outgrowth (Snow et al., 1990, 1991; McKeon et al., 1991, 1995; Canning et al., 1993, 1996; Dou and Levine, 1994; Tom et al., 2004b). Such proteoglycans have been shown to inhibit neurite outgrowth as a result of their GAG chains in vitro (Snow et al., 1990; Cole and McCabe, 1991; Fichard et al., 1991; Snow et al., 1991; Bovolenta et al., 1993; Grimpe and Silver, 2004; Tom et al., 2004b) and sometimes as a function of their core proteins (Oohira et al., 1991; Geisert and Bidanset, 1993; Dou and Levine, 1994). It has also been suggested that in certain situations proteoglycans, especially their GAG moieties, when in solution, may positively influence neuronal survival, attachment, and /or axon growth as well (Iijima et al., 1991; Maeda et al., 1995; Challacombe and Elam, 1997; Gates et al., 1997; Kappler et al., 1997). However, an intriguing observation is that the increased neurite outgrowth seen on certain proteoglycans is often dendritic rather than axonal (Maeda and Noda, 1996), which is consistent with observations that neurite outgrowth demonstrated on reactive astrocyte substrates is also largely dendritic (Le Roux and Reh, 1996). The growth of axons on glial scar tissue is inhibited by proteoglycans present in the extracellular matrix, and this inhibition can be partially reversed by enzymatic treatments to remove or prevent the production of specific sugar epitopes from the proteoglycan molecules (McKeon et al., 1991, 1995; Bradbury et al., 2002; Grimpe and Silver, 2004; Steinmetz et al., 2005). Furthermore, neutralizing antibodies directed against a heparan/chondroitin sulfate proteoglycan expressed after brain injury block the inhibition of neurite outgrowth and growth cone collapse activity in vitro (Bovolenta ct al., 1997). While such studies provide evidence for a direct effect of proteoglycans on neurite extension, another report suggests that in some situations chondroitin sulfate may regulate and organize other extracellular matrix-associated molecules, perhaps by directly binding to various undefined growth-promoting or inhibiting factors (Emerling and Lander, 1996). Such findings indicate that proteoglycans present in glial scars may play direct and/or indirect roles in the lack of regeneration of the injured adult CNS, and that modification of the inhibitory effects of post-injury CNS scar tissue may be possible.

Several in vivo experiments have demonstrated that, in fact, this is possible and remarkable long-distance regeneration of adult axons though CNS white matter tracts can occur after enzymatic treatment to digest proteoglycan side chains. In one of these experiments (Steinmetz et al., 2005), the DREZ was utilized as a model for regenerative failure at an important transition between the PNS and the CNS. Many authors have demonstrated that while DRG axons can regenerate in the PNS environment, they abruptly stop at the interface of the DREZ (Tower, 1931; Kimmel and Moyer, 1947; Perkins et al., 1980; Kliot et al., 1990, Golding et al., 1997). This powerful inhibitory interface was only overcome with a combination strategy that utilized zymosan, a specific inflammatory activator of macrophages, and enzymatic digestion of inhibitory proteoglycans with chondroitinase ABC (ChABC). While neither of these treatments alone were successful in vitro or in vivo, the combination led to robust regeneration of nerve fibers in both in vitro assays and in vivo with accompanying electrophysiologically proven synapse formation in the CNS. This study highlights the importance of creating a specific inflammatory environment to condition neurons for regeneration and the need to modify the extracellular matrix to remove inhibitory molecules that prevent successful axon regrowth.

This modification of the inhibitory extracellular matrix is again featured in a second study with robust functional regeneration using a different combination strategy (Houle et al., 2006). In this recent work, investigators for the first time were able to overcome another historically important barrier between the PNS and the CNS environments when using peripheral nerve bridges as a surgical technique to enhance CNS regeneration. It has been well established that CNS axons will grow readily in a transplanted segment of PNS nerve; however, few of the regenerating fibers will cross the PNS/CNS barrier to re-enter the relatively inhospitable CNS environment (Richardson et al., 1980; David and Aguayo, 1981). Combining the use of PNS bridges with extracellular matrix modification with ChABC allowed regenerating axons to bypass and hemisection lesion of the spinal cord and undergo long-distance regeneration with subsequent increases in functional tasks. Functional recovery was demonstrated by return of forelimb swing during locomotion, enhanced ability to use forelimbs, and improved balance and weight bearing on a horizontal rope. These improvements were diminished after transection of the PNS bridge – further demonstrating that the regenerated fibers were responsible for the remarkable recovery of function. Modification of this inhibitory extracellular matrix that is created by astrocytes in association with inflammatory infiltrates after CNS injury appears to be a key step in creating a permissive environment for functional regeneration of severed axons.

THE FUNCTION OF INHIBITORY MOLECULES

The function of increased levels of proteoglycans or other inhibitory molecules following injury to the adult CNS remains unclear. Normal environmental factors of the adult CNS may favor the inhibition of axon growth in an attempt by the body to maintain normal synaptic connections (Hockfield et al., 1990), and after injury the CNS cellular environment may become even more inhibitory in an effort to prevent the aberrant growth of axons and the formation of inappropriate connections. Alternatively, the upregulation of proteoglycans may be part of the protective CNS response to injury much as it is for other body tissues, and the functional aspects of the increases in these molecules may be unrelated to axon growth and regeneration. In other words, the negative effects of proteoglycan upregulation or other inhibitors on axon growth may simply be an unfortunate side effect of the normal wound-healing response by an injured tissue. This concept has been supported by recent work demonstrating that excessive degradation of proteoglycans with chondrotinase in the vicinity of a large CNS lesion can actually exacerbate the wound (Tom et al., 2006).

Thus, while it is clear that one effect of proteoglycan upregulation following injury may be the inhibition of functional regeneration, it is now becoming a bit more clear what the intended function for these molecules may be in the CNS response to injury. Proteoglycans have been suggested to play a role in modulating growth factors (Ruoslahti and Yamaguchi, 1991), a function which could help regeneration by making appropriate growth factors available, or alternatively could hinder axon growth by binding and functionally removing important growth signals from the injury site. Further investigations are required to determine whether modulation of growth factors by proteoglycans has any effect on the regenerative responses of CNS axons. Proteoglycans have also been implicated as modulators of cell adhesion or migration (Grumet et al., 1993), and have been shown to lead to the increased migration of astrocytes in response to wound is equal to associated factors (Faber-Elman et al., 1996), another role that could certainly exert a positive influence on the CNS repair response by directing the migration of astrocytes into areas of damage to stabilize the tissue structure. Unfortunately, as discussed above in the section on reactive astrocytes, such cell migration often occurs in the opposite direction, as the necrotic cyst characteristic of many CNS wounds develops into a large cavity devoid of astrocytes (Mathewson and Berry, 1985; Fujiki et al., 1996; Zhang et al., 1996, 1997; Fitch and Silver, 1997a; Fitch et al., 1999).

Proteoglycans and other astrocyte extracellular matrix products have been implicated in the formation of these cystic cavities in the CNS. Astrocytes normally produce a basal lamina (which contains proteoglycans) at the pial surface of the CNS, and the astrocyte is polarized with respect to the production of this basal lamina that is produced on one side of the cell only (Kusaka et al., 1985). Following traumatic injury to the CNS, astrocytes produce ectopic basal lamina components at the borders of the cut edges of the injured tissue, again in a polarized fashion on the side facing the lesion cavity (Lawrence et al., 1984; Bernstein et al., 1985). Chondroitin sulfate proteoglycans have been described at the interface between developing cavities and the surrounding viable tissue (MacLaren, 1996; Fitch and Silver, 1997a; Fitch et al., 1999), perhaps as a component of an ectopic basal lamina secreted by reactive astrocytes. The function of extracellular matrix products surrounding a necrotic cavity of the

CNS is open for speculation, and it is an intriguing possibility that components such as proteoglycans may play a role in "walling off" the injured tissue in an effort to protect the surrounding viable cellular environment from further damage. In fact, proteoglycans have been demonstrated to inhibit phagocytosis and destruction of ß-amyloid protein by macrophages (Shaffer et al., 1995). Thus, it is possible that the CNS uses proteoglycans as a molecular protectant of tissue destruction by degradative enzymes or secondary tissue damage by inflammatory cells following a traumatic injury (Tom et al., 2006). Certainly the association between proteoglycan upregulation and intense inflammatory stimuli supports this concept (Fitch et al., 1999), as discussed in further detail in the next section.

WHAT CAUSES ASTROCYTE GLIOSIS AND INCREASES IN INHIBITORY MOLECULES?

TRIGGERS OF ASTROCYTE GLIOSIS

While astrocyte gliosis in response to injury to the CNS has been studied for many years, the specific causes and mechanisms leading to astrocyte hypertrophy, hyperplasia, and increased production of GFAP remain unknown. Several investigators have suggested that astrocyte gliosis can be triggered by the degeneration of severed axon tracts, perhaps secondary to Wallerian degeneration or retrograde axon changes, based on observations of the spread of GFAP immunoreactivity along white matter tracts far from the site of trauma to the spinal cord or dorsal roots (Barrett et al., 1981, 1984; Murray et al., 1990), or the delayed increases in GFAP mRNA in mutant animals with delayed Wallerian degeneration (Steward and Trimmer, 1997). Lesions to one side of the brain leads to gliosis throughout the ipsilateral hemisphere (Berry et al., 1983; Mathewson and Berry, 1985), and sometimes gliosis is seen to extend to areas of the contralateral hemisphere as well (Amaducci et al., 1981; Ludwin, 1985; Schiffer et al., 1986; Takamiya et al., 1988; Xie et al., 1995; Fitch and Silver, 1997a), suggesting that axons that span the brain hemispheres may contribute to the glial responses on the side contralateral to the injury. Various cytokines and other molecules have been implicated as possible triggers of astrocyte gliosis, including IL-1 (Giulian and Lachman, 1985; Giulian et al., 1988; Rostworowski et al., 1997), IL-6 (Chiang et al., 1994; Klein et al., 1997), thrombin (Nishino et al., 1993, 1994), ciliary neurotrophic factor (CNTF) (Kahn et al., 1995, 1997; Winter et al., 1995), endothelin-1 (Hama et al., 1997), and TNF-alpha (Rostworowski et al., 1997). Such factors that potentially play a role in astrocyte gliosis may derive from the injured astrocytes themselves, endogenous serum factors, activated microglial cells, or the invading inflammatory cells from the periphery.

While it is certainly important to identify molecules that may play a role in inducing astrocyte gliosis, it is becoming increasingly apparent that using GFAP upregulation and cellular hypertrophy as markers for functionally "bad" gliosis (as most studies have done) may be of limited benefit. Observations by several investigators suggest that astrocyte gliosis may be heterogeneous (Miller et al., 1986; Hill et al., 1996), particularly in its association with the production of inhibitory molecules (McKeon et al., 1991, 1995; Fitch and Silver, 1997a; Fitch et al., 1999). While some reactive astrocytes, particularly those near the wound epicenter, have been associated with the production of boundary molecules, it is certainly apparent that not all astrocytes with the morphological characteristics of reactive astrocytes (i.e., increased GFAP) are present in areas with increased levels of extracellular matrix molecules (McKeon et al., 1991, 1995; Pindzola et al., 1993; Levine, 1994; Davies et al., 1997; Fitch and Silver, 1997a; Fitch et al., 1999). These observations suggest that perhaps not all astrocytes that react to injury play a role in the failure of CNS regeneration, and that only those astrocytes associated with inhibitory molecules are detrimental to axon growth while those further away from the lesion may be more conducive to neurite sprouting, functional plasticity, and long-distance regeneration.

TRIGGERS FOR INHIBITORY MOLECULES

Therefore, the pivotal question that remains is, what molecular triggers are responsible for the production of astroglial inhibitory extracellular matrix? A series of studies have identified ß-amyloid protein as one trigger of reactive astrogliosis that leads to increases in the production of inhibitory molecules (Canning et al., 1993, 1996; Hoke et al., 1994; Hoke and Silver, 1996). Since it is unlikely that ß-amyloid is a trigger for the production of inhibitory molecules following traumatic injury to the CNS, it is important to begin considering what other factors may play such a role. The role of degenerating axons as a trigger for gliosis is unclear at least in vitro, as one study claims that dying axons are not sufficient to signal matrix production by astrocytes (Ard et al., 1993), while another study indicates that degenerating axons can lead to increases in extracellular matrix production by astrocytes (Guenard et al., 1996). However, observations in vivo illustrate that the degeneration of injured axons is not sufficient to induce the upregulation of inhibitory molecules outside of the immediate vicinity of a CNS wound (Fitch and Silver, 1997a). It should be noted, however, that degenerating axons could indirectly lead to slow increases in inhibitory molecules, as the presence of dying axons has been demonstrated to lead to inflammation within the CNS (Zhang et al., 1996), and such inflammatory cells may in turn trigger the upregulation of extracellular matrix molecules. The importance of inflammation has been highlighted further using lesion models of the brain and spinal cord, where such in vivo experimentation suggests that the increases in one family of inhibitory molecules (chondroitin sulfate proteoglycans) are associated with a breakdown of the blood–brain barrier and infiltrating macrophages present within the lesion site (Fitch and Silver, 1997a; Fitch et al., 1999). As detailed in the next section, introduction of pro-inflammatory molecules to specifically activate microglial cells and macrophages is sufficient to trigger intense astrocyte gliosis and upregulation of proteoglycan production (Fitch et al., 1999). In other studies, cytokines associated with inflammatory infiltrates have been shown to modulate extracellular matrix production by astrocytes in vitro (DiProspero et al., 1997). These hypotheses are in agreement with previous studies concerning general triggers of astrocyte gliosis (Giulian et al., 1988; Nishino et al., 1993, 1994), but work remains to be done to confirm whether all of these known triggers of widespread gliosis also induce inhibitory molecules in the discrete region surrounding the wound.

WHAT IS THE ROLE OF INFLAMMATION IN CNS INJURY?

MACROPHAGES AND MICROGLIA

The inflammatory response in the CNS following injury is composed primarily of two components: activation of intrinsic microglial cells (see Kim and de Vellis, 2005) and recruitment of bone marrow-derived inflammatory cells from the peripheral bloodstream (for review, see Perry et al., 1993). Chemical injuries to the brain appear to lead to a predominantly microglial cell inflammatory response, while direct stab wounds and injections are composed mostly of peripheral monocytes (Murabe et al., 1982; Riva-Depaty et al., 1994). However, it is generally accepted that both microglia and peripherally derived macrophages respond to injury in various proportions depending on the type and severity of the lesion (Giulian et al., 1989). Some authors have suggested that the limited and delayed recruitment of macrophages into a CNS lesion may explain the lack of efficient myelin clearance seen after such injury (Perry et al., 1987; George and Griffin, 1994), and the persistence of myelin has been implicated as one component that may explain some aspects of CNS regenerative failure as discussed previously in this chapter (for review, see Schwab et al., 1993).

The inflammatory response to injury is thought by some investigators to contribute to secondary tissue damage within the CNS (Blight, 1994; Fitch et al., 1999). Microglial cytokines have been suggested as possible sources of nervous system impairment following injury (Giulian et al., 1989), and neutrophilic leukocytes may augment necrosis and inflammation following a CNS wound (Means and Anderson, 1983). Microglial cells are capable of releasing cytotoxic factors that can kill neurons (Banati et al., 1993; Giulian, 1993) and inhibit neurogenesis (Ekdahl et al., 2003), and have been suggested to play a role in disconnecting existing neuronal connections and destroying neurons surrounding areas of injury (Giulian et al., 1994a,b). Many authors have advocated the use of therapeutic agents to modify the secretory activity of microglia and macrophages as a way to limit secondary damage to the CNS (Giulian and Lachman, 1985; Giulian et al., 1989; Banati et al., 1993; Guth et al., 1994a,b; Zhang et al., 1997; He et al., 2004), and in vitro studies have shown that preventing inflammatory activation with peroxisome proliferator-activated receptor (PPAR)-gamma agonists, a class of anti-inflammatory agents, may provide a novel therapy for preventing progressive cavitation by limiting inflammation and its subsequent secondary damage after a CNS injury (Fitch et al., 1999).

Work in our laboratory has demonstrated that inflammatory processes are sufficient to initiate a cascade of secondary tissue damage, progressive cavitation, and glial scarring in the CNS (Fitch et al., 1999). In vivo microinjection of zymosan, a specific phagocytic activator of the macrophage mannose receptor and the beta-glucan site of the CR3 integrin receptor, leads to a large inflammatory infiltration of activated macrophages and microglial cells. This inflammation leads to a statistically significant increase in astrocyte cavity size, while control injections of latex microspheres do not induce this inflammation-induced cavity formation. This supports the idea that a specific activation of macrophages is required to cause sufficient secondary damage to lead to progressive cavitation in vivo.

Perhaps a bit paradoxically, the inflammatory response within the CNS has also been suggested to have positive effects on the healing of nervous system wounds (Lotan and Schwartz, 1994; Klusman and Schwab, 1997). Experiments that demonstrated regeneration of PNS axons into the CNS environment noted the presence of a mild inflammatory reaction at the site of axon entry into the CNS, suggesting a positive role for regeneration (Le Gros Clark, 1943; Kliot et al., 1990; Siegal et al., 1990), and the use of certain pro-inflammatory agents was demonstrated to have positive effects on axon growth (Windle and Chambers, 1950; Windle et al., 1952; Clemente and Windle, 1954; Windle, 1956; Clemente, 1958; Guth et al., 1994a,b). Macrophages secrete factors that can promote axon growth, such as NGF, NT-3 (Elkabes et al., 1996), thrombospondin (Chamak et al., 1994), and IL-1 (Giulian et al., 1994a,b), suggesting that appropriate secretory activity by inflammatory cells may indeed promote regeneration of axons. Transplantation of inflammatory cells may be beneficial for some axonal sprouting or limited regenerative responses (Lazarov-Spiegler et al., 1996; Prewitt et al., 1997; Rabchevsky and Streit 1997), and use of zymosan to stimulate inflammation in the vitreous chamber adjacent to the retinal ganglion cells can drive regenerating optic fibers past a crush lesion of the optic nerve (Leon et al., 2000). In terms of the resident functions of brain macrophages, Banati and Graeber (1994) describes microglial cells as "sensors of pathology" and maintains that there is little evidence that activation of microglia alone is harmful for the CNS. In fact, one recent study suggests that activated microglial cells may actually support axon regeneration via the production of growth-promoting molecules (Rabchevsky and Streit, 1997). A recent study from our laboratory found that zymosan-induced inflammation was essential for the successful regeneration of DRG axons past the DREZ when combined with modification of the extracellular matrix (Steinmetz et al., 2005), and this highlights that positive effects of inflammation can be harnessed outside of the CNS compartment.

These apparently conflicting results for the effects of inflammation in the CNS after injury highlight the importance of the various states of activation exhibited by inflammatory cells that respond to trauma. While activation of the macrophage mannose receptor and the beta-glucan site of the CR3 integrin receptor together have been implicated as one way to activate macrophages to a state that leads to detrimental astrocyte reactions within the parenchyma of the brain (Fitch et al., 1999), it appears that a related activation state elicited outside of the CNS within the DRG neurons themselves allows their axons to regenerate into the spinal cord when combined with enzymatic digestion of proteoglycans (Steinmetz et al., 2005). These findings exemplify the apparently contradictory results for the role of inflammation in the creation of cellular environments or cellular activation states that are inhibitory or permissive for regeneration. This suggests that some undefined aspects of inflammation may be useful to promote growth, but that the presence of intense inflammation can surpass the growth promoting aspects by the production of molecules that inhibit the full regenerative potential of growing axons in the absence of external modifications to the extracellular matrix. These differences may reflect subpopulations of macrophages secreting different pro- and anti-inflammatory cytokines (Calvo et al., 2005), the simultaneous production of a balance of permissive and inhibitory molecules (Jones et al., 2003b) or whether the inflammation occurs centrally or peripherally. This dichotomy highlights an important issue for further experimental study, as identification of factors present at the site of injury that activate macrophages to secrete growth-promoting versus growth-inhibiting factors will allow future experimentation to modulate this critical balance.

MACROPHAGES, MICROGLIA, AND INHIBITORY MOLECULES

The relationship between inflammatory macrophages, activated microglia, and proteoglycan upregulation following trauma to the CNS has been discussed previously in this chapter. In addition to their potential role as a trigger for astrocyte production of extracellular matrix molecules, microglia have been suggested as a possible source of proteoglycans with neurite inhibitory properties (Bovolenta et al., 1993). In fact, peripheral macrophages have been shown to produce cell surface and secreted proteoglycans and their associated GAGs in tissue culture experiments (Kolset et al., 1986, 1988; Uhlin-Hansen and Kolset, 1987, 1988; Kolset and Larsen, 1988; Uhlin-Hansen et al., 1989, 1993; Petricevich and Michelacci, 1990; Owens and Wagner, 1992; Yeaman and Rapraeger, 1993a, b; Haidl and Jefferies, 1996; Kolset et al., 1996). However, the majority of proteoglycan production by macrophages appears to be secreted into the fluid media compartment in vitro, unlike the culture experiments in which astrocytes deposited quantities of proteoglycans onto culture substrates (McKeon et al., 1991; Canning et al., 1993, 1996; Hoke et al., 1994; Hoke and Silver, 1996). In addition to production, macrophages are also active in degrading and inducing other cells to degrade proteoglycan molecules (Laub et al., 1982). The proteoglycans secreted by macrophages have been suggested to protect themselves from their own degradative enzymes (Kolset and Larsen, 1988), or perhaps to regulate the inflammatory response itself

(Kolset et al., 1996). The possible roles of macrophage-derived proteoglycans in the CNS injury response remain to be elucidated.

CAN GLIAL RESPONSES BE MODIFIED TO ENHANCE REGENERATION?

This chapter has been dedicated to a discussion of the glial cell responses to CNS injury and their possible roles in contributing to regenerative failure. Oligodendrocytes and myelin, astrocytes and extracellular matrix, and microglia/macrophages and inflammation have all been examined as potential players in the abortive regenerative responses seen in adult CNS lesions. Many potential repair strategies have been attempted in the past, and recent studies have shown encouraging results suggesting that the use of a combination of approaches will be beneficial in improving the CNS regenerative response. Combinations of therapeutic agents designed to modify inflammatory responses (Bracken et al., 1990, 1992, 1997; Guth et al., 1994a,b Taoka et al., 1997), peripheral nerve bridges supplemented with acidic fibroblast growth factor combined with expert surgical technique (Cheng et al., 1996), the use of inflammatory modulators combined with enzymatic digestion of proteoglycans (Steinmetz et al., 2005), and the use of peripheral nerve bridges combined with proteoglycan digestion (Houle et al., 2006) are examples of successful approaches to these problems that supply insight into the mechanisms important for repair after CNS trauma. A common thread that is emerging in recent work is that the combination of multiple strategies to enhance regeneration is important to eliminate the inhibitory components of the extracellular matrix that are produced in association with inflammatory infiltration of traumatic injuries. Limiting the amount of secondary damage done by inflammation to reduce cavitation, encouraging the production of molecules supportive of regeneration, and decreasing factors inhibiting axon growth will tip the delicate balance of growth-promoting and growth-inhibiting factors to a net environment that supports functional regrowth after CNS injury.

A combination of the factors discussed in this chapter work together to create a local CNS environment that overall is not completely conducive to axonal regeneration after regeneration, and the challenge that remains is to find practical ways to assist injured axons at the site of a lesion to grow beyond the glial scar in both the brain and the spinal cord. Further research will elucidate repair strategies that may include modification of inflammatory and glial cell responses in the immediate vicinity of trauma along with measures to enhance the regenerative response by adult neurons. The many studies discussed in this chapter demonstrate that there is no single easy answer to this difficult clinical problem, and the success of functional regeneration will require a combination strategy and an interdisciplinary approach to solve the problem of CNS injury and the resulting disability.

REFERENCES

Adrian EJ and Williams MG (1973). Cell proliferation in injured spinal cord. An electron microscopic study. *J Comp Neurol* 151: 1–24.

Amaducci L, Forno KI, and Eng LF (1981). Glial fibrillary acidic protein in cryogenic lesions of the rat brain. *Neurosci Lett* 21: 27–32.

Amat JA, Ishiguro H, Nakamura K, and Norton WT (1996). Phenotypic diversity and kinetics of proliferating microglia and astrocytes following cortical stab wounds. *Glia* 16: 368–382.

Ard MD, Bunge MB, Wood PM et al. (1991). Retinal neurite growth on astrocytes is not modified by extracellular matrix, anti-L1 antibody, or oligodendrocytes. *Glia* 4: 70–82.

Ard MD, Schachner M, Rapp JT, and Faissner A (1993). Growth and degeneration of axons on astrocyte surfaces: effects on extracellular matrix and on later axonal growth. *Glia* 9, 248–259.

Argiro V, Bunge MB, and Johnson MI (1984). Correlation between growth form and movement and their dependence on neuronal age. *J Neurosci* 4: 3051–3062.

Bahr M, Przyrembel C, and Bastmeyer M (1995). Astrocytes from adult rat optic nerves are non-permissive for regenerating retinal ganglion cell axons. *Exp Neurol* 131: 211–220.

Balentine JD (1978). Pathology of experimental spinal cord trauma. I. the necrotic lesion as a function of vascular injury. *Lab Invest* 39: 236–253.

Banati RB and Graeber MB (1994). Surveillance, intervention and cytotoxicity: is there a protective role of microglia? *Dev Neurosci* 16: 114–127.

Banati RB, Gehrmann J, Schubert P, and Kreutzberg GW (1993). Cytotoxicity of microglia. *Glia* 7: 111–118.

Bandtlow C, Zachleder T, and Schwab ME (1990). Oligodendrocytes arrest neurite growth by contact inhibition. *J Neurosci* 10: 3837–3848.

Barrett CP, Guth L, Donati EJ, and Krikorian JG (1981). Astroglial reaction in the gray matter lumbar segments after midthoracic transection of the adult rat spinal cord. *Exp Neurol* 73: 365–377.

Barrett CP, Donati EJ, and Guth L (1984). Differences between adult and neonatal rats in their astroglial response to spinal injury. *Exp Neurol* 84: 374–385.

Bartsch U (1996). Myelination and axonal regeneration in the central nervous system of mice deficient in the myelin-associated glycoprotein. *J Neurocytol* 25: 303–313.

Bartsch U, Bandtlow CE, Schnell L et al. (1995). Lack of evidence that myelin-associated glyco-protein is a major inhibitor of axonal regeneration in the CNS. *Neuron* 15: 1375–1381.

Bastmeyer M, Beckmann M, Schwab ME, and Stuermer CA (1991). Growth of regenerating goldfish axons is inhibited by rat oligodendrocytes and CNS myelin but not by goldfish optic nerve tract oligodendrocytelike cells and fish CNS myelin. *J Neurosci* 11: 626-40.

Bernstein JJ, Getz R, Jefferson M, and Kelemen M (1985). Astrocytes secrete basal lamina after hemisection of rat spinal cord. *Brain Res* 327: 135–1341.

Berry M, Maxwell WL, Logan A et al. (1983). Deposition of scar tissue in the central nervous system. *Acta Neurochir Suppl Wien* 32: 31–53.

Berry M, Hall S, Rees L et al. (1992). Regeneration of axons in the optic nerve of the adult Bowman-Wyse (BW) mutant rat. *J Neurocytol* 21: 426–448.

Berry M, Carlile J, and Hunter A (1996). Peripheral nerve explants grafted into the vitreous body of the eye promote the regeneration of retinal ganglion cell axons severed in the optic nerve. *J Neurocytol* 25: 147–170.

Bignami A and Dahl D (1974). Astrocyte-specific protein and neuroglial differentiation. An immunofluorescence study with antibodies to the glial fibrillary acidic protein. *J Comp Neurol* 153: 27–38.

Bignami A and Dahl D (1976). The astroglial response to stabbing. Immunofluorescence studies with antibodies to astrocyte-specific protein (GFA) in mammalian and submammalian vertebrates. *Neuropathol Appl Neurobiol* 2: 99–110.

Bignami A, Eng LF, Dahl D, and Uyeda CT (1972). Localization of the glial fibrillary acidic protein in astrocytes by immunofluorescence. *Brain Res* 43: 429–435.

Blight AR (1994). Effects of silica on the outcome from experimental spinal cord injury: implication of macrophages in secondary tissue damage. *J Neurosci* 60: 263–273.

Bovolenta P, Wandosell F, and Nieto SM (1993). Neurite outgrowth inhibitors associated with glial cells and glial cell lines. *Neuroreport* 5: 345–348.

Bovolenta P, Fernaud-Espinosa I, Mendez-Otero R, and Nieto-Sampedro M (1997). Neurite outgrowth inhibitor of gliotic brain tissue. Mode of action and cellular localization, studied with specific monoclonal antibodies. *Eur J Neurosci* 9(5): 977–989.

Bracken MB, Shepard MJ, Collins WF et al. (1990). A randomized, controlled trial of methylprednisolone or naloxone in the treatment of acute spinal-cord injury. Results of the Second National Acute Spinal Cord Injury Study. *N Engl J Med* 322: 1405–1411.

Bracken MB, Shepard MJ, Collins WJ et al. (1992). Methylprednisolone or naloxone treatment after acute spinal cord injury: 1-year follow-up data. Results of the second National Acute Spinal Cord Injury Study. *J Neurosurg* 76: 23–31.

Bracken MB, Shepard MJ, Holford TR et al. (1997). Administration of methylprednisolone for 24 or 48 hours or tirilazad mesylate for 48 hours in the treatment of acute spinal cord injury. Results of the Third National Acute Spinal Cord Injury Randomized Controlled Trial. National Acute Spinal Cord Injury Study. *JAMA* 277: 1597–1604.

Bradbury EJ, Moon LD, Popat RJ et al. (2002). Chondroitinase ABC promotes functional recovery after spinal cord injury. *Nature* 416(6881): 636–640.

Bregman BS, Kunkel BE, Schnell L et al. (1995). Recovery from spinal cord injury mediated by antibodies to neurite growth inhibitors. *Nature* 378: 498–501.

Brittis PA, Lemmon V, Rutishauser U, and Silver J (1995). Unique changes of ganglion cell growth cone behavior following cell adhesion molecule perturbations: a time lapse study of the living retina. *Mol Cell Neurosci* 6: 433–449.

Brodkey JA, Laywell ED, O'Brien TF et al. (1995). Focal brain injury and upregulation of a developmentally regulated extracellular matrix protein. *J Neurosurg* 82: 106–112.

Brown CT, Applebaum E, Banwatt R, and Trinkaus RV (1995). Synthesis of stromal glycosaminoglycans in response to injury. *J Cell Biochem* 59: 57–68.

Bundeson LQ, Scheel TA, Bregman BS, and Kromer LF (2003). Ephrin-B2 and EphB2 regulation of astrocyte-meningeal fibroblast interactions in response to spinal cord lesions in adult rats. *J Neurosci* 23: 7789–7800.

Bunge RP (1968). Glial cells and the central myelin sheath. *Phys Reviews* 48: 197.

Calvo CF, Amigou E, Desaymard C, and Glowinski J (2005). A pro- and anti-inflammatory cytokine are synthetised in distinct brain macrophage cells during innate activation. *J Neuroimmunol* 170: 21–30.

Campbell G, Anderson PN, Turmaine M, and Lieberman AR (1991). GAP-43 in the axons of mammalian CNS neurons regenerating into peripheral nerve grafts. *Exp Brain Res* 87: 67–74.

Canning DR, McKeon RJ, DeWitt DA et al. (1993). Beta-Amyloid of Alzheimer's disease induces reactive gliosis that inhibits axonal outgrowth. *Exp Neurol* 124: 289–298.

Canning DR, Hoke A, Malemud CJ, and Silver J (1996). A potent inhibitor of neurite outgrowth that predominates in the extracellular matrix of reactive astrocytes. *Int J Devl Neurosci* 14: 153–175.

Carlstedt T (1983). Regrowth of anastomosed ventral root nerve fibers in the dorsal root of rats. *Brain Res* 272: 162–165.

Carlstedt T (1985a). Regenerating axons form nerve terminals at astrocytes. *Brain Res* 347: 188–191.

Carlstedt T (1985b). Regrowth of cholinergic and catecholaminergic neurons along a peripheral and central nervous pathway. *J Neurosci* 15: 507–518.

Carlstedt T (1988). Reinnervation of the mammalian spinal cord after neonatal dorsal root crush. *J Neurocytol* 17: 335–350.

Carlstedt T, Dalsgaard CJ, and Molander C (1987). Regrowth of lesioned dorsal root nerve fibers into the spinal cord of neonatal rats. *Neurosci Lett* 74: 14–18.

Carlstedt T, Cullheim S, Risling M, and Ulfhake B. (1988). Mammalian root-spinal cord regeneration. *Prog Brain Res* 78: 225–229.

Caroni P and Schwab ME (1988a). Antibody against myelin-associated inhibitor of neurite growth neutralizes nonpermissive substrate properties of CNS white matter. *Neuron* 1: 85–96.
Caroni P and Schwab ME (1988b). Two membrane protein fractions from rat central myelin with inhibitory properties for neurite growth and fibroblast spreading. *J Cell Biol* 106: 1281–1288.
Caroni P, Savio T, and Schwab ME (1988). Central nervous system regeneration: oligodendrocytes and myelin as non-permissive substrates for neurite growth. *Prog Brain Res* 78: 363–370.
Carrino DA, Oron U, Pechak DG, and Caplan AI (1988). Reinitiation of chondroitin sulphate proteoglycan synthesis in regenerating skeletal muscle. *Development* 103: 641–656.
Challacombe JF and Elam JS (1997). Chondroitin 4-sulfate stimulates regeneration of goldfish retinal axons. *Exp Neurol* 143: 10–17.
Chamak B, Morandi V, and Mallat M (1994). Brain macrophages stimulate neurite growth and regeneration by secreting thrombospondin. *J Neurosci Res* 38: 221–233.
Chen DF, Jhaveri S, and Schneider GE (1995). Intrinsic changes in developing retinal neurons result in regenerative failure of their axons. *Proc Natl Acad Sci U S A* 92: 7287–7291.
Chen MS, Huber AB, Van Der Haar ME et al. (2000). Nogo-A is a myelin-associated neurite outgrowth inhibitor and an antigen for monoclonal antibody IN-1. *Nature* 403: 434–439.
Chen J, Bernreuther C, Dihne M, and Schachner M (2005). Cell adhesion molecule l1-transfected embryonic stem cells with enhanced survival support regrowth of corticospinal tract axons in mice after spinal cord injury. *J Neurotrauma* 22(8): 896–906.
Cheng H, Cao Y, and Olson L (1996). Spinal cord repair in adult paraplegic rats: partial restoration of hind limb function. *Science* 273: 510–513.
Chiang CS, Stalder A, Samimi A, and Campbell IL (1994). Reactive gliosis as a consequence of interleukin-6 expression in the brain: studies in transgenic mice. *Dev Neurosci* 16: 212–221.
Chiquet-Ehrismann R, Hagios C, and Schenk S (1995). The complexity in regulating the expression of tenascins. *Bioessays* 17: 873–878.
Clemente CD (1958). The regeneration of peripheral nerves inserted into the cerebral cortex and the healing of cerebral lesions. *J Comp Neurol* 109: 123–143.
Clemente CD and Windle WF (1954). Regeneration of severed nerve fibers in the spinal cord of the adult cat. *J Comp Neurol* 101: 691–731.
Cole GJ and McCabe CF (1991). Identification of a developmentally regulated keratan sulfate proteoglycan that inhibits cell adhesion and neurite outgrowth. *Neuron* 7: 1007–1118.
Condic M (2001). Adult neuronal regeneration induced by transgenic integrin expression. *J Neurosci* 21(13): 4782–4788.
Davalos D, Grutzendler J, Yang G et al. (2005). *Nat Neurosci* 8(6): 752–758.
David S and Aguayo AJ (1981). Axonal elongation into peripheral nervous system "bridges" after central nervous system injury in adult rats. *Science* 214: 931–933.
Davies SJA, Fitch MT, Memberg SP et al. (1997). Regeneration of adult axons in white matter tracts of the central nervous system. *Nature* 390: 680–683.
Davies SJA, Goucher DR, Doller C, and Silver J (1999). Robust regeneration of adult sensory axons in degenerating white matter of the adult rat spinal cord. *J Neurosci* 19: 5810–5822.
DiProspero NA, Meiners S, and Geller HM (1997). Inflammatory cytokines interact to modulate extracellular matrix and astrocytic support of neurite outgrowth. *Exp Neurol* 148: 628–639.
Dou C-L and Levine JM (1994). Inhibition of neurite growth by the NG2 chondroitin sulfate proteoglycan. *J Neurosci* 14: 7616–7628.
Dow KE, Ethell DW, Steeves JD, and Riopelle RJ (1994). Molecular correlates of spinal cord repair in the embryonic chick: heparan sulfate and chondroitin sulfate proteoglycans. *Exp Neurol* 128: 233–238.
Ducker TB, Kindt GW, and Kempe LG (1971). Pathological findings in acute experimental spinal cord trauma. *J Neurosurg* 35: 700–707.
Eddleston M and Mucke L (1993). Molecular profile of reactive astrocytes–implications for their role in neurologic disease. *J Neurosci* 54: 15–36.
Ekdahl CT, Claasen JH, Bonde S et al. (2003). Inflammation is detrimental for neurogenesis in adult brain. *PNAS* 100(23): 13632–13637.

Elkabes S, DiCicco BE, and Black IB (1996). Brain microglia/macrophages express neuro-trophins that selectively regulate microglial proliferation and function. *J Neurosci* 16: 2508–2521.

Emerling DE and Lander AD (1996). Inhibitors and promoters of thalamic neuron adhesion and outgrowth in embryonic neocortex: functional association with chondroitin sulfate. *Neuron* 17: 1089–1100.

Eng LF (1985). Glial fibrillary acidic protein (GFAP): the major protein of glial intermediate filaments in differentiated astrocytes. *J Neuroimmunol* 8: 203–214.

Faber-Elman A, Solomon A, Abraham JA et al. (1996). Involvement of wound-associated factors in rat brain astrocyte migratory response to axonal injury: in vitro simulation. *J Clin Invest* 97: 162–171.

Fallon JR (1985a). Neurite guidance by non-neuronal cells in culture: Preferential outgrowth of peripheral neurites on glial as compared to nonglial cell surfaces. *J Neurosci* 5: 3169–3177.

Fallon JR (1985b). Preferential outgrowth of central nervous system neurites on astrocytes and schwann cells as compared with nonglial cells in vitro. *J Cell Biol* 100: 198–207.

Faulkner JR, Herrmann JE, Woo MJ et al. (2004). Reactive astrocytes protect tissue and preserve function after spinal cord injury. *J Neurosci* 24(9): 2143–2155.

Fawcett JW, Rokos J, and Bakst I (1989). Oligodendrocytes repel axons and cause axonal growth cone collapse. *J Cell Sci* 92: 93–100.

Fichard A, Verna JM, Olivares J, and Saxod R (1991). Involvement of a chondroitin sulfate proteoglycan in the avoidance of chick epidermis by dorsal root ganglia fibers: a study using beta-D-xyloside. *Dev Biol* 148: 1–9.

Filbin MT (2003). Myelin-associated inhibitors of axonal regeneration in the adult mammalian CNS. *Nat Rev Neurosci* 4(9): 703–713.

Fitch MT and Silver J (1997a). Activated macrophages and the blood brain barrier: Inflammation after CNS injury leads to increases in putative inhibitory molecules. *Exp Neurol* 148: 587–603.

Fitch MT and Silver J (1997b). Glial cell extracellular matrix: boundaries for axon growth in development and regeneration. *Cell Tissue Res* 290: 379–384.

Fitch MT and Silver J (2000). Inflammation and the glial scar: factors at the site of injury that influence regeneration in the central nervous system. In NR Saunders and KM Dziegielewska, eds. *Degeneration and Regeneration in the Nervous System.* Harwood Academic Publishers. pp. 77–92.

Fitch MT and Silver J (2001). Astrocytes are dynamic participants in central nervous system development and injury responses. In KR Jessen and WD Richardson, eds. *Glial Cell Development: Basic Principles and Clinical Relevance.* Second Edition. Oxford University Press. pp. 263–269.

Fitch MT and Silver J (2007). CNS Injury, Glial Scars, and Inflammation: Inhibitory extracellular matrices and regeneration failure. *Exp Neurol.* May 31 [Epub ahead of print].

Fitch MT, Doller C, Combs CK et al. (1999). Cellular and molecular mechanisms of glial scarring and progressive cavitation: *in vivo* and *in vitro* analysis of inflammation-induced secondary injury after CNS trauma. *J Neurosci* 19: 8182–8198.

Freund P, Schmidlin E, Wannier T et al. (2006). Nogo-A-specific antibody treatment enhances sprouting and functional recovery after cervical lesion in adult primates. *Nat Med* 12(7): 790–792.

Frisen J, Verge VM, Fried K et al. (1993). Characterization of glial trkB receptors: differential response to injury in the central and peripheral nervous systems. *Proc Natl Acad Sci U S A* 90: 4971–4975.

Frisen J, Haegerstrand A, Risling M et al. (1995). Spinal axons in central nervous system scar tissue are closely related to laminin-immunoreactive astrocytes. *J Neurosci* 65: 293–304.

Fryer RH, Kaplan DR, and Kromer LF (1997). Truncated TrkB receptors on non-neuronal cells inhibit BDNF induced neurite outgrowth *in vitro*. *Exp Neurol* 148: 616–627.

Fujiki M, Zhang Z, Guth L, and Steward O (1996). Genetic influences on cellular reactions to spinal cord injury: activation of macrophages/microglia and astrocytes is delayed in mice carrying a mutation (Wld) that causes delayed wallerian degeneration. *J Comp Neurol* 371: 469–484.

Gallo V and Bertolotto A (1990). Extracellular matrix of cultured glial cells: selective expression of chondroitin sulfate by type-2 astrocytes and progenitors. *Exp Cell Res* 187: 211–223.

Gallo V, Bertolotto A, and Levi G (1987). The proteoglycan chondroitin sulfate is present in a subpopulation of cultured astrocytes and in their precursors. *Dev Biol* 123: 282–285.

Gates MA, Fillmore H, and Steindler DA (1997). Chondroitin sulfate proteoglycan and tenascin in the wounded adult mouse neostriatum *in vitro*: Dopamine neuron attachment and process outgrowth. *J Neurosci* 16: 8005–8018.

Gehrmann J, Matsumoto Y, and Kreutzberg GW (1995). Microglia: intrinsic immuneffector cell of the brain. *Brain Res Reviews* 20: 269–287.

Geisert EE and Stewart AM (1991). Changing interactions between astrocytes and neurons during CNS maturation. *Dev Biol* 143: 335–345.

Geisert EJ and Bidanset DJ (1993). A central nervous system keratan sulfate proteoglycan: localization to boundaries in the neonatal rat brain. *Brain Res Dev Brain Res* 75: 163–173.

Geisert EJ, Williams RC, and Bidanset DJ (1992). A CNS specific proteoglycan associated with astrocytes in rat optic nerve. *Brain Res* 571: 165–168.

George R and Griffin JW (1994). Delayed macrophage responses and myelin clearance during Wallerian degeneration in the central nervous system: the dorsal radiculotomy model. *Exp Neurol* 129: 225–236.

Giulian D (1993). Reactive glia as rivals in regulating neuronal survival. *Glia* 7: 102–110.

Giulian D and Lachman, LB (1985). Interleukin-1 stimulation of astroglial proliferation after brain injury. *Science* 228: 497–499.

Giulian D, Woodward J, Young DG et al. (1988). Interleukin-1 injected into mammalian brain stimulates astrogliosis and neovascularization. *J Neurosci* 8: 2485–2490.

Giulian D, Chen J, Ingeman JE et al. (1989). The role of mononuclear phagocytes in wound healing after traumatic injury to adult mammalian brain. *J Neurosci* 9: 4416–4429.

Giulian D, Li J, Leara B, and Keenen C (1994a). Phagocytic microglia release cytokines and cytotoxins that regulate the survival of astrocytes and neurons in culture. *Neurochem Int* 25: 227–233.

Giulian D, Li J, Li X et al. (1994b). The impact of microglia-derived cytokines upon gliosis in the CNS. *Dev Neurosci* 16: 128–136.

Goldberg JL, Espinosa JS, Xu Y et al. (2002). Retinal ganglion cells do not extend axons by default: promotion by neurotrophic signaling and electrical activity. *Neuron* 33(5): 689–702.

Golding J, Shewan D, and Cohen J (1997). Maturation of the mammalian dorsal root entry zone–from entry to no entry. *Trends Neurosci* (7): 303–308.

Grierson JP, Petroski RE, Ling DS, and Geller HM (1990). Astrocyte topography and tenascin cytotactin expression: correlation with the ability to support neuritic outgrowth. *Brain Res Dev Brain Res* 55: 11–19.

Grill R, Murai K, Blesch A et al. (1997). Cellular delivery of neurotrophin-3 promotes corticospinal axonal growth and partial functional recovery after spinal cord injury. *J Neurosci* 17: 5560–5572.

Grimpe B and Silver J (2002). The extracellular matrix in axon regeneration. *Prog Brain Res* 137: 333–349.

Grimpe B and Silver J (2004). A novel DNA enzyme reduces glycosaminoglycan chains in the glial scar and allows microtransplanted dorsal root ganglia axons to regenerate beyond lesions in the spinal cord. *J Neurosci* 24: 1393–1397.

Grimpe B, Dong S, Doller C et al. (2002). The critical role of basement membrane-independent laminin gamma 1 chain during axon regeneration in the CNS. *J Neurosci* 22(8): 3144–3160.

Grumet M, Flaccus A, and Margolis RU (1993). Functional characterization of chondroitin sulfate proteoglycans of brain: interactions with neurons and neural cell adhesion molecules. *J Cell Biol* 120: 815–824.

Guenard V, Frisch G, and Wood PM (1996). Effects of axonal injury on astrocyte proliferation and morphology in vitro: implications for astrogliosis. *Exp Neurol* 137: 175–190.

Guest JD, Hesse D, Schnell L et al. (1997). Influence of IN-1 antibody and acidic FGF-fibrin glue on the response of injured corticospinal tract axons to human schwann cell grafts. *J Neurosci Res* 50: 888–905.

Guth L (1956). Regeneration in the mammalian peripheral nervous system. *Physiological Reviews* 36: 441–478.

Guth L, Barrett CP, Donati EJ et al. (1981). Histopathological reactions an axonal regeneration in the transected spinal cord of hibernating squirrels. *J Comp Neurol* 203: 297–308.

Guth L, Barrett CP, Donati EJ et al. (1985). Essentiality of a specific cellular terrain for growth of axons into a spinal cord lesion. *Exp Neurol* 88: 1–12.

Guth L, Barrett CP, and Donati EJ (1986). Histological factors influencing the growth of axons into lesions of the mammalian spinal cord. *Exp Brain Res* Suppl. 13: 271–282.

Guth L, Zhang Z, DiProspero NA et al. (1994a). Spinal cord injury in the rat: treatment with bacterial lipopolysaccharide and indomethacin enhances cellular repair and locomotor function. *Exp Neurol* 126: 76–87.

Guth L, Zhang Z, and Roberts E (1994b). Key role for pregnenolone in combination therapy that promotes recovery after spinal cord injury. *Proc Natl Acad Sci U S A* 91: 12308–12312.

Hagino S, Iseki K, Mori T et al. (2003). Slit and glypican-1 mRNAs are coexpressed in the reactive astrocytes of the injured adult brain. *Glia* 42: 130–138.

Haidl ID and Jefferies WA (1996). The macrophage cell surface glycoprotein F4/80 is a highly glycosylated proteoglycan. *Eur J Immunol* 26: 1139–1146.

Hama H, Kasuya Y, Sakurai T et al. (1997). Role of endothelin-1 in astrocyte responses after acute brain damage. *J Neurosci Res* 47: 590–602.

Harrison RG (1910). The outgrowth of the nervbe fiber as a mode of protoplasmic movement. *J Exp Zool* 9: 787–848.

Harrison RG (1914). The reaction of embryonic cells to sold structures. *J Exp Zool* 17: 521–544.

Hatten ME, Liem RK, Shelanski ML, and Mason CA (1991). Astroglia in CNS injury. *Glia* 4: 233–243.

He J, Evans CO, Hoffman SW et al. (2004). Progesterone and allopregnanolone reduce inflammatory cytokines after traumatic brain injury. *Exp Neurol* 189(2): 404–412.

Hill SJ, Barbarese E, and McIntosh TK (1996). Regional heterogeneity in the response of astrocytes following traumatic brain injury in the adult rat. *J Neuropathol Exp Neurol* 55: 1221–1229.

Hockfield S, Kalb RG, Zaremba S, and Fryer HJ (1990). Expression of neural proteoglycan correlates with the acquisition of mature neuronal properties in the mammalian brain. *Cold Spring Harbor Symp Quant Biol* 55: 505–514.

Hoke A and Silver J (1996). Proteoglycans and other repulsive molecules in glial boundaries during development and regeneration of the nervous system. *Prog Br Res* 108: 149–163.

Hoke A, Canning DR, Malemud CJ and Silver J (1994). Regional differences in reactive gliosis induced by substrate-bound beta-amyloid. *Exp Neurol* 130: 56–66.

Houle JD (1992). Regeneration of dorsal root axons is related to specific non-neuronal cells lining NGF-treated intraspinal nitrocellulose implants. *Exp Neurol* 118: 133–142.

Houle JD, Tom VJ, Mayes D et al. (2006). Combining an autologous peripheral nervous system "bridge" and matrix modification by chondroitinase allows robust, functional regeneration beyond a hemisection lesion of the adult rat spinal cord. *J Neurosci* 26(28): 7405–7415.

Iijima N, Oohira A, Mori T et al. (1991). Core protein of chondroitin sulfate proteoglycan promotes neurite outgrowth from cultured neocortical neurons. *J Neurochem* 56: 706–708.

Jain M, He Q, Lee WS et al. (1996). Role of CD44 in the reaction of vascular smooth muscle cells to arterial wall injury. *J Clin Invest* 97: 596–603.

Johnson-Green P, Dow KE, and Riopelle RJ (1991). Characterization of glycosaminoglycans produced by primary astrocytes in vitro. *Glia* 4: 314–321.

Jones LL, Margolis RU, and Tuszynski MH (2003a). The chondroitin sulfate proteoglycans neurocan, brevican, phosphacan, and versican are differentially regulated following spinal cord injury. *Exp Neurol* 182(2): 399–411.

Jones LL, Sajed D, and Tuszynski MH (2003b). Axonal regeneration through regions of chondroitin sulfate proteoglycan deposition after spinal cord injury: A balance of permissiveness and inhibition. *J Neurosci* 23(28): 9276–9288.

Kahn MA, Ellison JA, Speight GJ, and de Vellis J (1995). CNTF regulation of astrogliosis and the activation of microglia in the developing rat central nervous system. *Brain Res* 685: 55–67.

Kahn MA, Huang CJ, Caruso A et al. (1997). Ciliary neurotrophic factor activates JAK/Stat signal transduction cascade and induces transcriptional expression of glial fibrillary acidic protein in glial cells. *J Neurochem* 68: 1413–1423.

Kao CC, Chang LW, and Bloodworth JJ (1977a). Axonal regeneration across transected mammalian spinal cords: an electron microscopic study of delayed microsurgical nerve grafting. *Exp Neurol* 54: 591–615.

Kao CC, Chang LW, and Bloodworth JM (1977b). The mechanism of spinal cord cavitation following spinal cord transection. Part 3: Delayed grafting with and without spinal cord retransection. *J Neurosurg* 46: 757–766.

Kappler J, Junghans U, Koops A et al. (1997). Chondroitin/dermatan sulphate promotes the survival of neurons from rat embryonic neocortex. *Eur J Neurobiol* 9: 306–318.

Kawaja MD and Gage FH (1991). Reactive astrocytes are substrates for the growth of adult CNS axons in the presence of elevated levels of nerve growth factor. *Neuron* 7: 1019–1030.

Kawaja MD, Rosenberg MB, Yoshida K, and Gage FH (1992). Somatic gene transfer of nerve growth factor promotes the survival of axotomized septal neurons and the regeneration of their axons in adult rats. *J Neurosci* 12: 2849–2864.

Keirstead HS, Hasan SJ, Muir GD, and Steeves JD (1992). Suppression of the onset of myelination extends the permissive period for the functional repair of embryonic spinal cord. *Proc Natl Acad Sci U S A* 89: 11664–11668.

Keirstead HS, Dyer JK, Sholomenko GN et al. (1995). Axonal regeneration and physiological activity following transection and immunological disruption of myelin within the hatchling chick spinal cord. *J Neurosci* 15: 6963–6974.

Kerschensteiner M, Schwab ME, Lichtman JW, and Misgeld T (2005). In vivo imaging of axonal degeneration and regeneration in the injured spinal cord. *Nat Med* 11(5): 572–577.

Kim SU and de Vellis J (2005). Microglia in health and disease. *J Neurosci Res* 81: 302–313.

Kimmel DL and Moyer EK (1947). Dorsal roots following anastomosis of the central stumps. *J Comp Neurol* 87: 289–321.

Klein MA, Moller JC, Jones LL et al. (1997). Impaired neuroglial activation in interleukin-6 deficient mice. *Glia* 19: 227–233.

Kliot M, Smith GM, Siegal JD, and Silver J (1990). Astrocyte-polymer implants promote regeneration of dorsal root fibers into the adult mammalian spinal cord. *Exp Neurol* 109: 57–69.

Klusman I and Schwab ME (1997). Effects of pro-inflammatory cytokines in experimental spinal cord injury. *Brain Res* 762: 173–184.

Kobayashi H, Watanabe E, and Murikami F (1995). Growth cones of dorsal root ganglion but not retina collapse and avoid oligodendrocytes in culture. *Dev Biol* 168, 383–394.

Kolset SO and Larsen T (1988). Sulfur-containing macromolecules in cultured monocyte-like cells. *Acta Histochem* 84: 67–75.

Kolset SO, Ehlorsson J, Kjellen L, and Lindahl U (1986). Effect of benzyl beta-D-xyloside on the biosynthesis of chondroitin sulphate proteoglycan in cultured human monocytes. *Biochem J* 238: 209–216.

Kolset SO, Ivhed I, Overvatn A, and Nilsson K (1988). Differentiation-associated changes in the expression of chondroitin sulfate proteoglycan in induced U-937 cells. *Cancer Res* 48: 6103–6108.

Kolset SO, Mann DM, Uhlin HL et al. (1996). Serglycin-binding proteins in activated macrophages and platelets. *J Leukoc Biol* 59: 545–554.

Kottis V, Thibault P, Mikol D et al. (2002). Oligodendrocyte-myelin glycoprotein (OMgp) is an inhibitor of neurite outgrowth. *J Neurochem* 82(6): 1566–1569.

Kusaka H, Hirano A, Bornstein MB, and Raine CS (1985). Basal lamina formation by astrocytes in organotypic cultures of mouse spinal cord tissue. *J Neuropathol Exp Neurol* 44: 295–303.

Latov N, Nilaver G, Zimmerman EA et al. (1979). Fibrillary astrocytes proliferate in response to brain injury: a study combining immunoperoxidase technique for glial fibrillary acidic protein and radioautography of tritiated thymidine. *Dev Biol* 72: 381–384.

Laub R, Huybrechts GG, Peeters JC, and Vaes G (1982). Degradation of collagen and proteoglycan by macrophages and fibroblasts. Individual potentialities of each cell type and cooperative effects through the activation of fibroblasts by macrophages. *Biochim Biophys Acta* 721: 425–433.

Lawrence JM, Huang SK, and Raisman G (1984). Vascular and astrocytic reactions during establishment of hippocampal transplants in adult host brain. *J Neurosci* 12: 745–760.

Laywell ED and Steindler DA (1991). Boundaries and wounds, glia and glycoconjugates. Cellular and molecular analyses of developmental partitions and adult brain lesions. *Ann N Y Acad Sci* 633: 122–141.

Laywell ED, Dorries U, Bartsch U et al. (1992). Enhanced expression of the developmentally regulated extracellular matrix molecule tenascin following adult brain injury. *Proc Natl Acad Sci U S A* 89: 2634–2638.

Lazar DA, Fitch MT, Silver J, and Kliot M (2004). Cellular and molecular mechanisms mediating injury and recovery in the nervous system. In H Richard Winn, ed. *Youmans Neurological Surgery*. Fifth Edition. WB Saunders. pp. 195–213.

Lazarov-Spiegler O, Solomon AS, Zeev-Brann AB et al. (1996). Transplantation of activated macrophages overcomes central nervous system regrowth failure. *FASEB J* 10: 1296–1302.

Le Gros Clark WE (1942). The problem of neuronal regeneration in the central nervous system. I. The influence of spinal ganglia and nerve fragments grafted in the brain. *J Anatomy* 77: 20–45.

Le Gros Clark WE (1943). The problem of neuronal regeneration in the central nervous system. II. The insertion of peripheral nerve stumps into the brain. *J Anatomy* 77: 251–259.

Leon S, Yin Y, Nguyen J et al. (2000). Lens injury stimulates axon regeneration in the mature rat optic nerve. *J Neurosci* 20(12): 4615–4626.

Le Roux PD and Reh TA (1996). Reactive astroglia support primary dendritic but not axonal outgrowth from mouse cortical neurons in vitro. *Exp Neurol* 137: 49–65.

Levine JM (1994). Increased expression of the NG2 chondroitin-sulfate proteoglycan after brain injury. *J Neurosci* 14: 4716–4730.

Li Y and Raisman G (1993). Long axon growth from embryonic neurons transplanted into myelinated tracts of the adult rat spinal cord. *Brain Res* 629: 115–127.

Li Y and Raisman G (1994). Schwann cells induce sprouting in motor and sensory axons in the adult rat spinal cord. *J Neurosci* 14: 4050–4063.

Li Y and Raisman G (1995). Sprouts from cut corticospinal axons persist in the presence of astrocytic scarring in long-term lesions of the adult rat spinal cord. *Exp Neurol* 134: 102–111.

Liesi P and Silver J (1988). Is astrocyte laminin involved in axon guidance in the mammalian CNS? *Dev Biol* 130: 774–85.

Liesi P, Dahl D, and Veheri A (1983). Laminin is produced by early rat astrocytes in primary culture. *J Cell Biol* 96: 920–924.

Liesi P, Kaakkola S, Dahl D, and Vaheri A (1984). Laminin is induced in astrocytes of adult brain by injury. *EMBO J* 3: 683–686.

Ling EA and Wong WC (1993). The origin and nature of ramified and amoeboid microglia: a historical review and current concepts. *Glia* 7: 9–18.

Lips K, Stichel CC, and Muller HW (1995). Restricted appearance of tenascin and chondroitin sulphate proteoglycans after transection and sprouting of adult rat postcommissural fornix. *J Neurocytol* 24: 449–464.

Liu C-N and Chambers WW (1958). Intraspinal sprouting of dorsal root axons. *Arch Neurol Psychiat* 79: 46–61.

Lotan M and Schwartz M (1994). Cross talk between the immune system and the nervous system in response to injury: implications for regeneration. *FASEB J* 8: 1026–1033.

Lu P, Yang H, Jones LL et al. (2004). Combinatorial therapy with neurotrophins and cAMP promotes axonal regeneration beyond sites of spinal cord injury. *J Neurosci* 24(28): 6402–6409.

Ludwin SK (1984). Proliferation of mature oligodendrocytes after trauma to the central nervous system. *Nature* 308: 274–275.

Ludwin SK (1985). Reaction of oligodendrocytes and astrocytes to trauma and implantation. A combined autoradiographic and immunohistochemical study. *Lab Invest* 52: 20–30.

MacLaren RE (1996). Development and role of retinal glia in regeneration of ganglion cells following retinal injury. *Br J Ophthalmol* 80: 458–464.

Maeda N and Noda M (1996). 6B4 proteoglycan/phosphacan is a repulsive substratum but promotes morphological differentiation of cortical neurons. *Development* 122: 647–658.

Maeda N, Hamanaka H, Oohira A, and Noda M (1995). Purification, characterization and developmental expression of a brain-specific chondroitin sulfate proteoglycan, 6B4 proteoglycan/phosphacan. *J Neurosci* 67: 23–35.

Margolis RK and Margolis RU (1993). Nervous tissue proteoglycans. *Experientia* 49: 429–446.

Mathewson AJ and Berry M (1985). Observations on the astrocyte response to a cerebral stab wound in adult rats. *Brain Res* 327: 61–69.

McKeon RJ, Schreiber RC, Rudge JS, and Silver J (1991). Reduction of neurite outgrowth in a model of glial scarring following CNS injury is correlated with the expression of inhibitory molecules on reactive astrocytes. *J Neurosci* 11: 3398–3411.

McKeon RJ, Hoke A, and Silver J (1995). Injury-induced proteoglycans inhibit the potential for laminin-mediated axon growth on astrocytic scars. *Exp Neurol* 136: 32–43.

McKerracher L, David S, Jackson DL et al. (1994). Identification of myelin-associated glycoprotein as a major myelin-derived inhibitor of neurite growth. *Neuron* 13: 805–811.

Means ED and Anderson DK (1983). Neuronophagia by leukocytes in experimental spinal cord injury. *J Neuropathol Exp Neurol* 42: 707–719.

Miller RH, Abney ER, David S et al. (1986). Is reactive gliosis a property of a distinct subpopulation of astrocytes? *J Neurosci* 6: 22–29.

Miyake, T, Hattori T, Fukuda M et al. (1988). Quantitative studies on proliferative changes of reactive astrocytes in mouse cerebral cortex. *Brain Res* 451: 133–138.

Miyake T, Okada M, and Kitamura T (1992). Reactive proliferation of astrocytes studied by immunohistochemistry for proliferating cell nuclear antigen. *Brain Res* 590, 300–302.

Murabe Y, Ibata Y, and Sano Y (1982). Morphological studies on neuroglia III. Macrophage response and "microgliocytosis" in kainic acid-induced lesions. *Cell Tissue Res* 218: 75–86.

Murray HM and Walker BE (1973). Comparative study of astrocytes and mononuclear leukocytes reacting to brain trauma in mice. *Exp Neurol* 41: 290–302.

Murray M, Wang SD, Goldberger ME, and Levitt P (1990). Modification of astrocytes in the spinal cord following dorsal root or peripheral nerve lesions. *Exp Neurol* 110: 248–257.

Myer DJ, Gurkoff GG, Lee SM et al. (2006). Essential protective roles of reactive astrocytes in traumatic brain injury. *Brain* 129(Pt 10): 2761–2772.

Nikkari ST, Jarvelainen HT, Wight TN et al. (1994). Smooth muscle cell expression of extracellular matrix genes after arterial injury. *Am J Pathol* 144: 1348–1356.

Nishino A, Suzuki M, Ohtani H et al. (1993). Thrombin may contribute to the pathophysiology of central nervous system injury. *J Neurotrauma* 10: 167–179.

Nishino A, Suzuki M, Yoshimoto T et al. (1994). A novel aspect of thrombin in the tissue reaction following central nervous system injury. *Acta Neurochir Suppl Wien* 60: 86–88.

Norling B, Glimelius B, and Wasteson A (1984). A chondroitin sulphate proteoglycan from human cultured glial and glioma cells. *Biochem J* 221: 845–853.

Norton WT, Aquino DA, Hozumi I et al. (1992). Quantitative aspects of reactive gliosis: a review. *Neurochem Res* 17: 877–885.

Oohira A, Matsui F, and Katoh SR (1991). Inhibitory effects of brain chondroitin sulfate proteoglycans on neurite outgrowth from PC12D cells. *J Neurosci* 11: 822–827.

Owens RT and Wagner WD (1992). Chondroitin sulfate proteoglycan and heparan sulfate proteoglycan production by cultured pigeon peritoneal macrophages. *J Leukoc Biol* 51: 626–633.

Pasterkamp RJ, Anderson PN, and Verhaagen J (2001). Peripheral nerve injury fails to induce growth of lesioned ascending dorsal column axons into spinal cord scar tissue expressing the axon repellent Semaphorin3A. *Eur J Neurosci* 13: 467–471.

Perkins S, Carlstedt T, Mizuno K, and Aguayo AJ (1980). Failure of regenerating dorsal root axons to regrow into the spinal cord. *Canad J Neurol Sci* 7: 323.

Perry VH, Brown MC, and Gordon S (1987). The macrophage response to central and peripheral nerve injury. A possible role for macrophages in regeneration. *J Exp Med* 165: 1218–1223.

Perry VH, Andersson PB, and Gordon S (1993). Macrophages and inflammation in the central nervous system. *Trends Neurosci* 16, 268–273.

Petricevich VL and Michelacci YM (1990). Proteoglycans synthesized in vitro by nude and normal mouse peritoneal macrophages. *Biochimi Biophys Acta* 1053: 135–143.

Pindzola RR, Doller C, and Silver J (1993). Putative inhibitory extracellular matrix molecules at the dorsal root entry zone of the spinal cord during development and after root and sciatic nerve lesions. *Dev Biol* 156: 34–48.

Prewitt CM, Niesman IR, Kane CJ, and Houle JD (1997). Activated macrophage/microglial cells can promote the regeneration of sensory axons into the injured spinal cord. *Exp Neurol* 148: 433–443.

Rabchevsky AG and Streit WJ (1997). Grafting of cultured microglial cells into the lesioned spinal cord of adult rats enhances neurite outgrowth. *J Neurosci Res* 47: 34–48.

Rakic P (1995). Radial glial cells: scaffolding for brain construction. In H Kettenmann and BR Ransom, eds. *Neuroglia*. Oxford University Press, New York. pp. 746–762.

Ramon y Cajal S. (1928). Degeneration and regeneration of the nervous system. May RM, trans., Oxford Univeristy Press, London.

Reier PJ, Stensaas LJ, and Guth L (1983). The Astrocytic Scar As an Impediment to Regeneration in the Central Nervous System. In RP Bunge, CC Kao, and PJ Reier, eds. *Spinal Cord Reconstruction*. Raven Press, New York. pp. 163–195.

Richardson M and Hatton MW (1993). Transient morphological and biochemical alterations of arterial proteoglycan during early wound healing. *Exp Mol Pathol* 58: 77–95.

Richardson PM, McGuinness UM, and Aguayo AJ (1980). Axons from CNS neurons regenerate into PNS grafts. *Nature* 284: 264–265.

Riva-Depaty I, Fardeau C, Mariani J et al. (1994). Contribution of peripheral macrophages and microglia to the cellular reaction after mechanical or neurotoxin-induced lesions of the rat brain. *Exp Neurol* 128: 77–87.

Rostworowski M, Balsingam V, Chabot S et al. (1997). Astrogliosis in the neonatal and adult murine brain post-trauma: elevation of inflammatory cytokines and the lack of requirement for endogenous interferon-gamma. *J Neurosci* 17(10): 3664–3674.

Rubin BP, Dusart I, and Schwab ME (1994). A monoclonal antibody (IN-1) which neutralizes neurite growth inhibitory proteins in the rat CNS recognizes antigens localized in CNS myelin. *J Neurocytol* 23: 209–217.

Rudge JS, Smith GM, and Silver J (1989). An in vitro model of wound healing in the CNS: analysis of cell reaction and interaction at different ages. *Exp Neurol* 103: 1–16.

Ruoslahti E and Yamaguchi Y (1991). Proteoglycans as modulators of growth factor activities. *Cell* 64: 867–869.

Savio T and Schwab ME (1989). Rat CNS white matter, but not gray matter, is nonpermissive for neuronal cell adhesion and fiber outgrowth. *J Neurosci* 9: 1126–1133.

Savio T and Schwab, ME (1990). Lesioned corticospinal tract axons regenerate in myelin-free rat spinal cord. *Proc Natl Acad Sci U S A* 87: 4130–4133.

Schiffer D, Giordana MT, Migheli A et al. (1986). Glial fibrillary acidic protein and vimentin in the experimental glial reaction of the rat brain. *Brain Res* 374: 110–118.

Schnell L and Schwab ME (1990). Axonal regeneration in the rat spinal cord produced by an antibody against myelin-associated neurite growth inhibitors. *Nature* 343, 269–272.

Schnell L, Schneider R, Kolbeck R et al. (1994). Neurotrophin-3 enhances sprouting of corticospinal tract during development and after adult spinal cord lesion. *Nature* 367: 170–173.

Schousboe A and Westergaarde N (1995). Transport of neuroactive amino acids in astrocytes. In H Kettenmann and BR Ransom, eds. *Neuroglia* Oxford University Press, New York. pp. 246–258.
Schwab ME, Kapfhammer JP, and Bandtlow CE (1993). Inhibitors of neurite growth. *Annu Rev Neurosci* 16: 565–595.
Shaffer LM, Dority MD, Gupta BR et al. (1995). Amyloid beta protein (A beta) removal by neuroglial cells in culture. *Neurobiol Aging* 16: 737–745.
Shewan D, Berry M, Bedi K, and Cohen J (1993). Embryonic optic nerve tissue fails to support neurite outgrowth by central and peripheral neurons in vitro. *Eur J Neurosci* 5: 809–817.
Shewan D, Berry M, and Cohen, J (1995). Extensive regeneration in vitro by early embryonic neurons on immature and adult CNS tissue. *J Neurosci* 15: 2057–2062.
Siegal JD, Kliot M, Smith GM, and Silver, J (1990). A comparison of the regeneration potential of dorsal root fibers into gray or white matter of the adult rat spinal cord. *Exp Neurol* 109: 90–97.
Silver J and Miller JH (2004). Regeneration beyond the glial scar. *Nat Rev Neurosci* 5(2): 146–156.
Silver J, Edwards MA, and Levitt P (1993). Immunocytochemical demonstration of early appearing astroglial structures that form boundaries and pathways along axon tracts in the fetal brain. *J Comp Neurol* 328: 415–436.
Smith GM, Rutishauser U, Silver J, and Miller RH (1990). Maturation of astrocytes in vitro alters the extent and molecular basis of neurite outgrowth. *Dev Biol* 138: 377–390.
Smith-Thomas L, Fok SJ, Stevens J et al. (1994). An inhibitor of neurite outgrowth produced by astrocytes. *J Cell Sci* 107: 1687–1695.
Snow DM, Lemmon V, Carrino DA et al. (1990). Sulfated proteoglycans in astroglial barriers inhibit neurite outgrowth in vitro. *Exp Neurol* 109: 111–130.
Snow DM, Watanabe M, Letourneau PC, and Silver J. (1991). A chondroitin sulfate proteoglycan may influence the direction of retinal ganglion cell outgrowth. *Development* 113: 1473–1485.
Spillmann AA, Amberger VR, and Schwab ME (1997). High molecular weight protein of human central nervous system myelin inhibits neurite outgrowth: an effect which can be neutralized by the monoclonal antibody IN-1. *Eur J Neurosci* 9: 549–555.
Srinivasan SR, Xu JH, Vijayagopal P et al. (1995). Low-density lipoprotein binding affinity of arterial chondroitin sulfate proteoglycan variants modulates cholesteryl ester accumulation in macrophages. *Biochim Biophys Acta* 1272: 61–67.
Steinmetz MP, Horn KP, Tom VJ et al. (2005). Chronic enhancement of the intrinsic growth capacity of sensory neurons combined with the degradation of inhibitory proteoglycans allows functional regeneration of sensory axons through the dorsal root entry zone in the mammalian spinal cord. *J Neurosci* 25(35): 8066–8076.
Steward O and Trimmer PA (1997). Genetic influences on cellular reactions to CNS injury: the reactive response of astrocytes in denervated neuropil regions in mice carrying a mutation (Wld(S)) that causes delayed Wallerian degeneration. *J Comp Neurol* 380: 70–81.
Stuermer CAO, Bastmeyer M, Bahr M et al. (1992). Trying to understand axonal regeneration in the CNS of fish. *J Neurobiol* 23: 537–550.
Szczepanik AM, Fishkin RJ, Rush DK, and Wilmot CA (1996). Effects of chronic intrahippocampal infusion of lipopolysaccharide in the rat. *J Neurosci* 70: 57–65.
Tang S, Woodhall RW, Shen YJ et al. (1997). Soluble myelin-associated glycoprotein (MAG) found in vivo inhibits axonal regeneration. *Mol Cell Neurosci* 9: 333–346.
Takamiya Y, Kohsaka S, Toya S et al. (1988). Immunohistochemical studies on the proliferation of reactive astrocytes and the expression of cytoskeletal proteins following brain injury in rats. *Brain Res* 466: 201–210.
Taoka Y, Okajima K, Uchiba M et al. (1997). Reduction of spinal cord injury by administration of iloprost, a stable prostacyclin analog. *J Neurosurg* 86: 1007–1011.
Tello F (1911a). La influencia del neurotropismo en la regeneracion de los centros nerviosos. *Trab. Lab. Invest. Univ. Madrid* 9: 123–159.
Tello F (1911b). Un experimento sobre las influencia del neurotropismo en la regeneracion de la corteza cerebral. *Rev Clin Madrid* 5: 292–294.

Tetzlaff W, Kobayashi NR, Giehl KMG et al. (1994). Response of rubrospinal and corticospinal neurons to injury and neurotrophins. *Prog Brain Res* 103: 271–286.

Thallmair M, Metz GAS, Z'Graggen WJ et al. (1998). Neurite growth inhibitorys restrict plasticity and functional recovery following corticospinal tract lesions. *Nat Neurosci* 1: 124–131.

Theele DP and Streit WJ (1993). A chronicle of microglial ontogeny. *Glia* 7: 5–8.

Tom VJ, Doller CM, Malouf AT, and Silver J (2004a). Astrocyte-associated fibronectin is critical for axonal regeneration in adult white matter. *J Neurosci* 24(42): 9282–9290.

Tom VJ, Steinmetz MP, Miller JH et al. (2004b). Studies on the development and behavior of the dystrophic growth cone, the hallmark of regeneration failure, in an in vitro model of the glial scar and after spinal cord injury. *J Neurosci* 24(29): 6531–6539.

Tom VJ, Coleman C, Shumsky JS, and Houle, JD (2006). Digestion of chondroitin sulfate proteoglycans with high concentrations of chondroitinase ABC after spinal cord injury augments tissue damage and increases functional deficits. *Soc Neurosci Abst* 383: 18/NN71.

Tomaselli KJ, Neugebauer KM, Bixby JL et al. (1988). N-cadherin and integrins: two receptor systems that mediate neuronal process outgrowth on astrocyte surfaces. *Neuron* 1: 33–43.

Tower S (1931). A search for trophic influence of posterior spinal roots on skeletal muscle, with a note on the nerve fibers found in the proximal stumps of the roots after excision of the root ganglia. *Brain* 54: 99–110.

Tuszynski MH, Peterson DA, Ray J et al. (1994). Fibroblasts genetically modified to produce nerve growth factor induce robust neuritic ingrowth after grafting to the spinal cord. *Exp Neurol* 126: 1–14.

Tuszynski MH, Gabriel K, Gage FH et al. (1996). Nerve growth factor delivery by gene transfer induces differential outgrowth of sensory, motor, and noradrenergic neurites after adult spinal cord injury. *Exp Neurol* 137: 157–173.

Tuszynski MH, Murai K, Blesch A et al. (1997). Functional characterization of NGF-secreting cell grafts to the acutely injured spinal cord. *Cell Transplant* 6: 361–368.

Tuszynski MH, Grill R, Jones LL et al. (2002). Spontaneous and augmented growth of axons in the primate spinal cord: effects of local injury and nerve growth factor-secreting cell grafts. *J Comp Neurol* 449(1): 88–101.

Uhlin-Hansen L and Kolset SO (1987). Proteoglycan biosynthesis in relation to differentiation of cord blood monocytes in vitro. *Cell Differ* 21: 189–197.

Uhlin-Hansen L and Kolset SO (1988). Cell density-dependent expression of chondroitin sulfate proteoglycan in cultured human monocytes. *J Biol Chem* 263: 2526–2531.

Uhlin-Hansen L, Eskeland T, and Kolset SO (1989). Modulation of the expression of chondroitin sulfate proteoglycan in stimulated human monocytes. *J Biol Chem* 264: 14916–14922.

Uhlin-Hansen L, Wik T, Kjellen L et al. (1993). Proteoglycan metabolism in normal and inflammatory human macrophages. *Blood* 82: 2880–2889.

Vaudano E, Campbell G, Anderson PN et al. (1995). The effects of a lesion or a peripheral nerve graft on GAP-43 upregulation in the adult rat brain: an in situ hybridization and immunocytochemical study. *J Neurosci* 15: 3594–3611.

Virchow R (1846). Ueber das granulierte Ansehen der Wandungen der Gerhirnventrikel. *Allg Z Psychiatr Psych.-Gerichtl Med* 3: 424–450.

Vourc'h P and Andres C (2004). Oligodendrocyte myelin glycoprotein (OMgp): evolution, structure and function. *Brain Res Rev* 45: 115–124.

Wallace MC, Tator CH, and Lewis AJ (1987). Chronic regenerative changes in the spinal cord after cord compression injury in rats. *Surg Neurol* 27, 209–219.

Walz W (1989). Role of glial cells in the regulation of the brain ion microenvironment. *Prog Neurobiol* 33: 309–333.

Weidner N, Ner A, and Tuszynski MH (2001). Spontaneous corticospinal axonal plasticity and functional recovery after adult central nervous system injury. *Proc Natl Acad Sci U S A.* 98(6): 3513–3518.

Weigert C (1895). Beitrage zur Kenntnis der normalen menshlichen Neuroglia. Moritz Diesterweg, Frankfurt aum Maim, Weisbrod.

Windle WF (1956). Regeneration of axons in the vertebrate central nervous system. *Phys Rev* 36: 427–439.

Windle WF and Chambers WW (1950). Regeneration in the spinal cord of the cat and dog. *J Comp Neurol* 93: 241–258.

Windle WF, Clemente CD, and Chambers WW (1952). Inhibition of formation of a glial barrier as a means of permitting a peripheral nerve to grow into the brain. *J Comp Neurol* 96: 359–369.

Winter CG, Saotome Y, Levison SW, and Hirsh D (1995). A role for ciliary neurotrophic factor as an inducer of reactive gliosis, the glial response to central nervous system injury. *Proc Natl Acad Sci U S A* 92: 5865–5869.

Wolburg H and Risau W (1995). Formation of the blood-brain-barrier. In H Kettenmann and BR Ransom, eds. *Neuroglia.* Oxford University Press, New York. pp. 763–776.

Wood P and Bunge RP (1984). The biology of the oligodendrocyte. In WT Norton, eds. *Oligodendroglia.* Plenum Press, New York. pp. 1–46.

Xie D, Schultz RL, and Whitter EF (1995). The oligodendroglial reaction to brain stab wounds: an immunohistochemical study. *J Neurocytol* 24: 435–448.

Yang H, Lu P, McKay HM et al. (2006). Endogenous neurogenesis replaces oligodendrocytes and astrocytes after primate spinal cord injury. *J Neurosci* 26(8): 2157–2166.

Yasuda T, Sobue G, Ito T et al. (1990). Nerve growth factor enhances neurite arborization of adult sensory neurons; a study in single-cell culture. *Brain Res* 524: 54–63.

Yeaman C and Rapraeger AC (1993a). Membrane-anchored proteoglycans of mouse macrophages: P388D1 cells express a syndecan-4-like heparan sulfate proteoglycan and a distinct chondroitin sulfate form. *J Cell Physiol* 157: 413–425.

Yeaman C and Rapraeger AC (1993b). Post-transcriptional regulation of syndecan-1 expression by cAMP in peritoneal macrophages. *J Cell Biol* 122: 941–950.

Z'Graggen WJ, Metz GAS, Kartje GL et al. (1998). Functional recovery and enhanced corticofugal plasticity after unilateral pyramidal tract lesion and blockage of myelin-associated neurite growth inhibitors in adult rats. *J Neurosci* 18: 4744–4757.

Zhang Y, Anderson PN, Campbell G et al. (1995). Tenascin-C expression by neurons and glial cells in the rat spinal cord: changes during postnatal development and after dorsal root or sciatic nerve injury. *J Neurocytol* 24: 585–601.

Zhang Z, Fujiki M, Guth L, and Steward O (1996). Genetic influences on cellular reactions to spinal cord injury: A wound-healing response present in normal mice is impaired in mice carrying a mutation (Wld) that causes delayed wallerian degeneration. *J Comp Neurol* 371: 485–495.

Zhang Z, Krebs CJ, and Guth L (1997). Experimental analysis of progressive necrosis after spinal cord trauma in the rat: etiological role of the inflammatory response. *Exp Neurol* 143: 141–152.

PART II

Stem Cells and Trophic Factors in CNS Repair

4

NEUROTROPHIC FACTORS

EUGENE M. JOHNSON, JR*
AND MARK H. TUSZYNSKI†,‡

*Department of Neurology and of Molecular Biology and Pharmacology, Washington University School of Medicine, St. Louis, MO 63110

†Department of Neurosciences-0626, University of California-San Diego, La Jolla, CA 92093

‡Veterans Affairs Medical Center, San Diego, CA 92161

ABSTRACT

The nervous system growth factors, or neurotrophic factors, comprise several families of proteins that influence neuronal survival, neuritic outgrowth, and cell function from embryonic development through old age. Given their extensive influences on neurons, neurotrophic factors represent candidate mechanisms for treating neurodegenerative and axonal disorders. This chapter will review biochemical, physiological, and systems-level effects of several growth factor families.

Keywords: neurotrophic factors, nerve growth factor, brain-derived neurotrophic factor, ciliary neurotrophic factor, Alzheimer's disease, gene therapy

INTRODUCTION

Neurotrophic factors are usually small proteins so named because they exert survival and/or function-enhancing effects on neuronal populations. The first described factor was nerve growth factor (NGF), discovered and initially characterized by Levi-Montalcini, Hamburger, Cohen, and their colleagues at Washington University St. Louis a half-century ago (Cohen et al., 1954; Levi-Montalcini, 1987). Since that time, many molecules have been described that exert such effects physiologically and pharmacologically. NGF, as were most subsequently so-described factors, was discovered and originally studied in the context of its role in the development of the nervous system, especially for its ability to suppress naturally occurring neuronal death. For example, NGF

Copyright © 2008 Elsevier Inc.
All rights of reproduction in any form reserved.

deprivation in vivo, originally produced by antibodies to the factor, accentuates markedly the naturally occurring neuronal death in developing sympathetic (Levi-Montalcini and Angeletti, 1963) and sensory (Johnson et al., 1980) neurons. Conversely, exogenous administration of supraphysiological levels of NGF suppresses the death of these neurons that occurs during normal development or that results from axotomy (reviewed in Levi-Montalcini, 1987). NGF is produced by the targets of these neurons, binds to receptors for the factor on nerve terminals, and is retrogradely transported to the cell body in which gene expression presumably becomes altered. Thus, in the classic "neurotrophic hypothesis", neurotrophic factors were described as proteins elaborated by targets that acted to inhibit neuronal death, a process we now know is caused by apoptosis. This definition of neurotrophic factor was developed largely by studying neurons in, or projecting to, the periphery, where clear separation of neuronal soma from target is experimentally simple. However, this definition is much too limiting and specific, based on our current knowledge of these molecules.

We appreciate now that neurotrophic factors exert a much broader range of *physiological* functions on neurons, and may do so by means other than acting in a target-derived manner. Molecules exerting "trophic" effects on neurons may be made in targets, in close proximity to the neuronal soma, or in tracts or nerves along which neurons may project. Neurotrophic factors, in addition to suppressing apoptosis, also maintain or stimulate neuronal metabolism (glucose uptake, protein synthesis, RNA synthesis, etc.). Some neurotrophic factors also serve a neurotro*pic* function in guiding neuronal migration and axonal projection during development. In addition to these actions that occur over rather protracted timeframes (hours to days), neurotrophic factors can demonstrate acute effects in modulating several nervous system functions including synaptic efficacy and neurotransmitter turnover (see Levi-Montalcini, 1987; Thoenen, 1995; Reichardt, 2006).

Many neurotrophic factors have critical roles in non-neuronal cells. How we think about, or especially *name* a factor, may be an historical accident of the context in which the actions of a factor were first observed, their initial source, or of the bioassay that was used to monitor activity in the original isolation of the molecule. Many of the neurotrophic factors we will discuss below were originally described and largely studied outside the nervous system (e.g., fibroblast growth factor). Conversely, some factors originally described based upon trophic or survival-promoting actions on neurons have profound roles on non-neuronal cells (e.g., glial cell line-derived neurotrophic factor, or GDNF, on developing kidney and on spermatogenesis). Lastly, neurotrophic factors have a critical function outside of the context of neuronal development. Neurons never outgrow their need for trophic support, despite their progressive loss of acute dependency for trophic factors to suppress apoptosis after embryonic development. Neurotrophic factors in the *adult* animal are required to maintain the morphology, function, and connectivity of mature neurons. This was first shown with NGF, wherein administration or generation of NGF antibodies in adult animals led to a profound atrophy and functional impairment of sympathetic neurons (Otten et al., 1979; Gorin and Johnson, 1980). In some cases the neurotrophic factor acting to provide trophic support in the adult is not the same factor that was critical to its survival during development (e.g., Molliver et al., 1997).

Of particular interest for the purposes of this chapter are the *therapeutic,* apart from the *physiological,* actions discussed above. Studies performed over more than 30 years demonstrate that neurotrophic factors prevent the death of developing or mature neurons in response to an enormous variety of mechanical, chemical, metabolic, and immunologic insults that kill neurons by apoptotic or non-apoptotic mechanisms (see below for many examples). In addition to preventing frank death of neurons, they can "rescue" atrophic, hypofunctional neurons. Neurotrophic factors can increase the size, projections, metabolic activity, and neurotransmitter synthesis of neurons injured by the above insults and in the context of neurodegenerative disease. This profile of pharmacologic action has made neurotrophic factors important therapeutic candidates. They are perhaps unique in that their biological properties offer the hope of exerting trophic effects to enhance the function of atrophic, hypofunctional neurons that may translate into *symptomatic improvement* over a relatively short timeframe, while at the same time promoting survival by arresting the pathogenic mechanism(s), thus halting or slowing the death of neurons, i.e., *neuroprotection.*

Neurotrophic factors offer not only unique opportunities, but also unique challenges as potential therapeutic agents. Selecting a specific factor or defining a potentially treatable disease or condition involves many considerations:

Which factor for which condition? Since most factors will act upon a limited set of neuronal populations, those conditions for which the major symptoms of the disease can be ascribed to a specific neuronal population are more tractable than conditions that involve many neuronal types that presumably do not all respond to the same factor, and/or the relationship between neuronal type and the most debilitating aspect of the condition is unclear. If a target neuronal population is defined, one can match the known actions of trophic factors to that population.

How to deliver the molecule to the target neuronal population(s) with minimal exposure of non-targeted yet responsive neurons? Several issues affect this decision. Since neurotrophic factors are proteins, administration and distribution of the molecules present several problems. The ability of these molecules to pass the blood–brain-barrier (BBB) is very limited. If the condition under consideration is in the periphery (e.g., diabetic sensory or autonomic neuropathy), this is an advantage in that direct CNS actions of a peripherally administered factor are unlikely. If the target neuronal population is within the CNS (e.g., nigral-striatal dopamine neurons in Parkinson's disease), then means must be found to circumvent the BBB. An upside to this limited distribution is that if one can deliver the factor to the target neuronal population in the CNS, direct action of the factor on peripheral neurons is of little concern. If CNS delivery is required, several issues emerge. The easiest delivery into the CNS is to administer the factor into the CSF. However, this is problematic for at least two major reasons. First, the factor is not localized but rather potentially exposes the whole CNS to the drug with the attendant increase in the likelihood of undesired effects. Second, many of the factors bind tightly to extracellular matrix such that diffusion into the parenchyma of the brain is limited. Not surprisingly, administration of factors into the CSF of patients in the few trials attempted have not produced positive results on the desired populations, but have been associated with adverse effects because of off-target actions (examples discussed below).

The alternative is to administer the factor into the parenchyma of the brain. Although more difficult, invasive and technically more complex, intraparenchymal administration has the potential advantage of providing more restricted distribution of the factor to the cell bodies or processes of the affected neuronal population(s), thus reducing side effects associated with exposure to distant structures. Two general approaches have been used: direct administration of protein and administration of the factor via in vivo or ex vivo gene therapy. In the former, a solution of the protein is injected or infused; in a chronic neurodegenerative disease this would almost certainly involve an indwelling cannula attached to a pump that can deliver the protein by constant or episodic infusion. This approach has the advantage that the dose can be controlled and altered, and the factor infusion can be stopped in the event of adverse event. The disadvantages are that this strategy requires the continued presence of indwelling hardware, periodic refilling of the pump reservoir, and the increased potential of an immune response to recombinant protein. Gene therapy approaches have the advantages that the protein is expressed after production in a human cell (in the patient's own cells in the case of in vivo gene therapy) and, thus, is less likely to be immunogenic, and that after injection the factor will likely be produced indefinitely. No indwelling hardware is required and no need to replenish the factor. The disadvantage of gene therapy is that the dose of factor to which the brain will be exposed is difficult to control, there is no opportunity to adjust dose in response to clinical response or side effect, and one caanot cease neurotrophic factor administration in the event of an adverse event.

Other considerations: As noted above, many of the proteins with neurotrophic factor-like activities were discovered in a non-neurotrophic context and , indeed, were most commonly thought of in different contexts (fibroblast growth factors, hematopoietic cytokines). Thus, these factors may exert effects on non-neuronal cells that can produce potentially undesirable effects (e.g., angiogenesis) from the perspective of a potential neurotherapeutic agent. Another complicating issue is the adequacy of existing animal models of the human conditions/diseases toward which NTF activity represents a therapeutic potential. This can be seen in acute settings such as stroke, where a myriad of agents, traditional neurotrophic factors and otherwise have shown effectiveness, only to fail in clinical trial. Animal models of neurodegenerative diseases usually fail to reproduce the chronic progressive nature of these diseases and/or fail to reproduce the full spectrum of neuropathological changes seen in the human disease. Lastly, other real-world issues may affect the choice of which factor to choose such as ease of production, stability of protein, and even intellectual property considerations that may impact the likelihood that a particular factor will attract the investment needed to develop the factor into a marketable product that can be widely used.

Currently, more than three dozen nervous system growth factors have been identified. These growth factors possess a wide range of structures, receptor signaling mechanisms, responsive neurons, and in vivo biological effects. The actions of a specific growth factor are determined by the location and extent of expression of the growth factor itself, by the expression of an appropriate receptor set for the factor on a target cell, and by the cascade of intracellular pathways that receptor binding activates. Thus, the spatial and temporal balance between neurotrophic factor and receptor expression determines the spectrum of biological activity. Based on these properties, different growth factors may be useful for treating of a variety of neurological diseases.

The known neurotrophic factors can be divided into several families based on shared coding sequences and structures. Below we discuss the most extensively studied of these families of neurotrophic molecules.

THE CLASSIC NEUROTROPHIN FAMILY

The first neurotrophic factor family identified was the "classic" neurotrophin family consisting of NGF (Levi-Montalcini and Hamburger, 1953; Thoenen and Barde, 1980; Levi-Montalcini, 1987), brain-derived neurotrophic factor (BDNF) (Barde et al., 1982), neurotrophin-3 (NT-3) (Maisonpierre et al., 1990; Rosenthal et al., 1990), neurotrophin-4/5 (NT-4/5) (Berkmeier et al., 1991; Ip et al., 1992a), and the non-mammalian growth factors neurotrophin-6 (NT-6) (Gotz et al., 1994) and neurotrophin-7 (NT-7) (Nilsson et al., 1998). NGF was discovered serendipitously 50 years ago in the course of examining properties of a sarcoma cell line in vitro (Levi-Montalcini and Hamburger, 1951; Levi-Montalcini et al., 1954). A role for NGF as an essential survival factor for peripheral sensory and sympathetic neurons during development of the nervous system was soon described (Levi-Montalcini and Angeletti, 1963; Levi-Montalcini 1987). However, the presence and importance of NGF in the adult nervous system was not appreciated until nearly 35 years later. Beginning in 1979, it became clear that NGF is transported (Schwab et al., 1979; Seiler and Schwab, 1984) and produced (Johnson et al., 1971; Korsching et al., 1985) in the adult CNS. Several years later, the potent neuroprotective effects of NGF on injured adult neurons of the cholinergic basal forebrain were described (Hefti, 1986a,b; Kromer, 1987; Buzsaki, 1988), revolutionizing the previous perception that the adult CNS was an inflexible, non-plastic structure. These findings helped usher in the modern era of research directed at enhancing CNS plasticity, preventing neuronal loss, and promoting CNS regeneration.

In 1983, NGF was cloned and sequenced (Scott et al., 1983). In humans, the NGF gene is located on the short arm of chromosome 1. Initial studies regarding the biochemistry of NGF were performed on protein isolated from the adult male mouse submandibular gland because it contains the highest natural levels of NGF in mammals (Varon et al., 1967a,b; Petrides and Shooter, 1986). NGF produced in the submandibular gland exists as a protein pentamer consisting of two alpha, one beta, and two gamma subunits with a total molecular weight of 130 kDa and a sedimentation rate of 7S (Varon et al., 1967a,b). The biological activity of the NGF molecule is contained within the β-subunit, a 118 amino acid polypeptide with a molecular weight of 12 250 Da and a sedimentation rate of 2.5S (Angeletti et al., 1973). NGF is stable in the pH range of 5–8 (Thoenen and Barde, 1980). The gamma subunit of the NGF pentamer is a peptidase involved in cleavage of the prepro molecule to release the active β-fragment, presenting NGF as a diffusible molecule in the extracellular space. Because NGF is a basic molecule with an isoelectric point greater than 10, NGF can also be bound to charged molecules on the extracellular aspect of the cell membrane after secretion from the cell (Blochl and Thoenen, 1996). Neurotrophin precursors are cleaved into mature, active peptides by specific prohormone convertases (Seidah et al., 1996a,b), or released from cells retaining the "pro" peptide segments (Lee et al., 2001)

The second neurotrophic factor identified, BDNF, was isolated in 1982 from pooled extracts of porcine brain (Barde et al., 1982). The BDNF gene, subsequently cloned and sequenced (Leirock et al., 1989), consists of a 120-amino acid protein with a molecular weight of 12 300 Da. Like NGF, the BDNF gene codes for a large prepro molecule with a secretory signal peptide that presents BDNF as an extracellular factor.

The finding that NGF and BDNF share approximately 50% homology for base pair codons suggested that a family of related proteins might exist possessing similar neurotrophic properties. This led to a search for additional molecules. Using hybridization probes to homologous NGF and BDNF coding sequences, NT-3 (Hohn et al., 1990; Maisonpierre et al., 1990; Rosenthal et al., 1990), NT-4 (Berkmeier et al., 1991; Ip et al., 1992a), NT-6 (Gotz et al., 1994), and NT-7 (Nilsson et al., 1998) were identified in the 1990s. A molecule tentatively identified as NT-5 was shown to be identical to NT-4 (Berkmeier et al., 1991). NT-6 and NT-7 are only present in non-mammalian species, however. Like NGF and BDNF, NT-3 and NT-4 are naturally expressed in the nervous system during development and in adulthood. The regional patterns of expression of neurotrophins in the adult brain exhibit both specificity and overlap, suggesting that each neurotrophic factor has a distinct role in modulating neuronal function. Thus, each neurotrophin may possess potential for treating degeneration of specific neuronal populations, and combinations of trophic factors could act synergistically to promote the survival of single or multiple neuronal populations in nervous system disease (Mitsumoto et al., 1994a,b).

In 2001, Hempstead and colleagues reported the surprising finding that neurotrophins are also released from neurons in a "pro" form, in which the leader sequences of amino acids that allow growth factor packaging and secretion from cells are not fully cleaved on growth factor release (Lee et al., 2001). In their pro-neurotrophin state, growth factors exhibit preferential binding to p75 as a high-affinity ligand but have minimal affinity for binding to trk. Activation of p75 signaling can, in turn, lead to upregulation of apoptotic pathways in cells (Ibanez, 2002; Fahnestock et al., 2004; Harrington et al., 2004). Thus, neurotrophins can, depending on the form of their release from cells and the repertoire of receptors that are expressed in a given region at a given time, act as either a pro-survival or a death signal (Ibanez, 2002; Fahnestock et al., 2004; Lu et al., 2005). Modulation of the pro-state can also exert a modifying role in synaptic plasticity (Lu, 2003).

The neurotrophins exert their actions on subsets of neurons by binding to specific transmembrane receptors. Two distinct classes of neurotrophin receptors have been identified: the p75 receptor, to which all neurotrophins can bind, and a tyrosine kinase-linked (trk) family of receptors that bind different neurotrophins with unique and specific affinities (Kaplan and Miller, 2000; Reichardt, 2006).

p75, a glycosylated transmembrane protein, is a member of the tumor necrosis factor-alpha (TNF-alpha) receptor family (Carter, et al., 1996). p75 appears to serve two roles: it modulates neurotrophin interactions with trk receptors, and it influences cellular function directly via a p75 intracellular signaling domain (Chao, 1994; Rao et al., 1995; Kaplan and Miller, 2000). p75 is capable of associating with trk receptors to accelerate neurotrophin binding, leading to a high-affinity receptor state (Hempstead, et al., 1991; Mahadeo et al., 1994). This may occur by initial NGF homodimeric binding to p75, which prevents association with a second p75 receptor and allows subsequent binding with trk (He and Garcia, 2004), or by convergence of separate intracellular signaling through separate trk and p75 signaling pathways (Wehrman et al., 2007). Thus, neurons expressing both p75 and the neurotrophin-specific trk receptors can respond to lower concentrations of neurotrophins than neurons expressing trk alone (Chao, 1994). Independent of trk interactions, p75 signaling via its own set of intracellular cascades regulates *apoptosis*, or programmed cell death (Rabizadeh et al., 1993; Kaplan and Miller, 2000). Upon neurotrophin withdrawal or in the presence of pro-neurotrophins, transmembrane p75 receptors dimerize and activate the Jun amino-terminal kinase (*Jnk*) pathway, which leads through intermediary proteins to activation of p53, Bax, and caspaces to induce cell death (Kaplan and Miller, 2000). p75 activation also regulates cell cycle function and can influence actin assembly and neurite growth via RhoA (Kaplan and Miller, 2000). Indeed, a mechanism contributing to the limited ability of CNS axons to regenerate after injury is activation of p75 by signaling via p75, as described in Chapters 15 and 16. In addition, p75 modulates sphingomyelin metabolism via activation of NFkB (Dobrowsky et al., 1994).

Most "positive" effects of neurotrophins on neurons appear to be mediated by trk receptors. Biological responses mediated by trk receptors include neuronal survival, neuronal differentiation, axon outgrowth, synaptic plasticity, and neurotransmitter expression. Three forms of the trk receptor have been characterized and are designated trkA, trkB, and trkC (Kaplan et al., 1991; Cordon-Cardo et al., 1991; Lamballe et al., 1991; Kaplan and Miller, 2000; Reichardt, 2006). Each trk receptor binds specific ligands: NGF binds to *trkA*, BDNF and NT-4/5 bind to *trkB*, and NT-3 binds primarily to *trkC* (see Kaplan and Miller, 2000; Reichardt, 2006 for review). Trk receptors are transmembrane glycoproteins with intracellular tyrosine receptor kinase domains. Binding of a neurotrophin induces dimerization and autophosphorylation of trk, which then associates with and activates related proteins including Akt and phosphatidylinositol 3-kinase (PI3-K) (Kaplan and Miller, 2000). Other kinases and second messengers subsequently become activated, eliciting multiple biological responses within the cell. Thus, neurons exhibit responses to a given neurotrophic factor by signaling through distinct and specific receptors. The various biological effects of the neurotrophins in a given cell may be specified by the site of growth factor availability (e.g., axon versus cell soma), and trafficking of growth factor-receptor complexes within cells (e.g., the signaling endosome (Sofroniew et al., 2001)) to activate remote versus local signaling.

The biological effects of the neurotrophins in the developing and adult nervous system have been described extensively (Levi-Montalcini, 1987; Sofroniew et al., 2001; Tuszynski, 2002; Kordower, 2003; Reichardt, 2006 for reviews). Although originally thought to function primarily as cell survival factors, neurotrophic factors, in general, and the neurotrophin family, in particular, have now been shown capable of modulating several aspects of nervous system development and function, including cellular differentiation, axon growth, axon target finding, synaptic efficacy, synaptic plasticity, and neurotransmission/neurotransmitter turnover.

NEUROTROPHIN ROLES IN THE PNS

In the PNS, neurotrophic factors are survival factors for several classes of neurons. The availability of limited quantities of NGF in a target-derived manner is essential for the survival of sympathetic neurons of the paravertebral ganglia, and for 70–80% of sensory neurons (nociceptors) in the dorsal root ganglia (DRG) (Gorin and Johnson, 1980; Johnson et al., 1980; Pearson et al., 1983; Levi-Montalcini, 1987; Lewin and Mendell, 1993; Crowley et al., 1994; Smeyne et al., 1994; Silos-Santiago et al., 1995; Fagan et al., 1996). Specific subpopulations of other PNS neurons exhibit dependence on either BDNF (Ernfors et al., 1994a; Jones et al., 1994; Liu et al., 1995a), NT-3 (Ernfors et al., 1994b; Farinas et al., 1994), NT-4/5 (Liu et al., 1995a), or the cytokine GDNF (see below) during development.

Once a neuron has successfully survived the period of developmental cell death, it may subsequently downregulate expression of a trophic factor receptor and fail to respond to the trophic factor even if it is exogenously delivered. Cells may also upregulate expression of a different neurotrophic factor receptor to modulate another aspect of cell function. For example, a subpopulation of dorsal-root-ganglion nociceptive sensory neurons requires NGF for survival during development and expresses trkA prior to embryonic day 15 (E15). After E15, however, these neurons downregulate trkA expression and upregulate expression of the GDNF receptor RET, a switch in neurotrophin receptor expression that temporally correlates with a switch in the phenotype of the neuron (Molliver et al., 1997). Other nociceptive axons continue to express trkA throughout life although they do not depend upon NGF for survival; rather, NGF serves a neuroregulatory role in these neurons by altering neuropeptide/neurotransmitter levels, including substance P and calcitonin gene-related peptide (CGRP) levels (Otten, 1984; Lindsay and Harmar, 1989). In contrast to sensory neurons, sympathetic neurons continue to require NGF for survival into adulthood (Gorin and Johnson, 1980; Levi-Montalcini, 1987). Thus, three different classes of neurons that require NGF as a survival factor during development diverge in the nature of their subsequent NGF dependence: one class permanently downregulates NGF receptor expression and responds to a different growth factor, GDNF; a second class switches from NGF dependence for survival to NGF modulation of transmitter expression and function; and, a third class retains NGF dependence for survival through adulthood.

In addition to regulating neuronal survival and function, neurotrophins modulate neuronal differentiation and synapse function during PNS development. Thus, the administration of anti-NGF immunoglobulins during development leads to a loss of nociceptive neurons in the DRG and their replacement with D-hair afferent neurons (Ritter et al., 1991; Lewin et al., 1992), indicating that NGF modulates neuronal differentiation. NGF also modulates neuronal morphology, evidenced by an increase in the number of dendrites on sympathetic neurons treated with NGF (Ruit et al., 1990). Neurotrophin effects at the synaptic level are exhibited by BDNF and NT-3, which potentiate synapse function at the neuromuscular junction (Lohof et al., 1993). Thus, neurotrophins exert effects on cell survival, differentiation, synapse function, and neurotransmitter regulation in the PNS. In many cases these functions continue into adulthood: sensory neurons in the adult bind and retrogradely transport NGF, BDNF, NT-3, and NT-4/5 (DiStefano et al., 1992). Schwann cells, a source of NGF, BDNF, NT-4/5, GDNF, and CNTF, produce augmented amounts of neurotrophic factors after peripheral nerve injury and likely contribute to both the survival and the regrowth of injured neurons in the adult (Meyer et al., 1992; Funakoshi et al., 1993).

NEUROTROPHIN ROLES IN THE CNS

Neurotrophins modulate several neuronal systems in the CNS. During development, BDNF is required for the survival of vestibular neurons and NT-3 is required for the survival of most cochlear neurons (Ernfors et al., 1995). In the brain proper, however, less direct evidence exists that neurotrophins are required for neuronal survival during development. For example, despite the demonstration that NGF is capable of completely preventing the death of basal forebrain cholinergic neurons after lesions in the adult, mice with targeted homozygous (-/-) (Crowley et al., 1994) or heterozygous (+/-) (Chen et al., 1997) mutations of NGF or of trkA (Silos-Santiago et al., 1995; Fagan et al., 1996) do not exhibit a loss of basal forebrain cholinergic neurons during development. Rather, NGF influences the location and density, but not the number, of basal forebrain cholinergic inputs to the hippocampus (Chen, et al., 1997). Additional evidence also supports a role for NGF in modulating neuronal size, dendritic structure, and complexity (Purves, 1988), and synapse number and size in the brain (Garofalo et al., 1992).

BDNF-null mutant mice and NT-3-null mutant mice also exhibit relatively modest reductions in neuronal numbers in the CNS. Little if any reduction in numbers of cortical and hippocampal neurons are observed in these null mutants, despite the abundant developmental expression of trkB and trkC receptors in cortical and hippocampal regions (Ernfors et al., 1994a,b). In the spinal cord, BDNF-null mutants exhibit no loss of motor neurons (Ernfors et al., 1994a; Jones et al., 1994; Conover et al., 1995; Liu et al., 1995a,b), despite the fact that experiments in neonatal animals demonstrate that BDNF can prevent injury-induced death of motor neurons (Yan et al., 1991; Sendtner et al., 1992a,b; Clatterbuck et al., 1994). NT-3-null mutant mice exhibit a modest reduction in motor neuron numbers, accounted for by a reduction in the number of muscle spindle afferent Ia motor neurons (Klein et al., 1994; Kucera et al., 1995a,b). trkB-null mutant mice exhibit reductions in numbers of motor neurons in the facial nucleus and spinal cord (Klein et al., 1993), although these changes are inconsistently noted in different strains of $trkB^{-/-}$ mice. trkC-null mutants support findings from NT-3-null mutant mice by exhibiting fewer muscle spindle afferent motor neurons, therefore indicating that these neurons require NT-3 for survival during development (Klein et al., 1994; Kucera et al., 1995a,b).

Evidence also exists for neurotrophin-mediated modulation of target finding and synapse formation during CNS development (Bonhoeffer, 1996; Thoenen, 1995; McAllister, 1999; Lu, 2003). In the developing visual system, injections of NGF antibodies extend the developmental period over which ocular dominance column plasticity can be elicited (Domenici et al., 1995). Further, injections of BDNF, NT-4/5, or antagonists to trkB receptors disrupt normal formation of ocular dominance columns in the visual cortex (Cabelli et al., 1995; Cabelli et al., 1997). Injections of NGF, BDNF, NT-3, or NT-4 also alter dendritic pattern formation in the visual cortex (McAllister et al., 1995); indeed, formation of normal dendrites requires both neurotrophins and neuronal activity in the developing visual cortex (McAllister et al., 1996). Supporting the importance of neurotrophins in the CNS, mice with targeted mutations of BDNF (-/-) exhibit abnormal dendritic branching and pattern formation in the cerebellum (Schwartz et al., 1997). BDNF signaling is also required for migration of cerebellar precursor neurons (Zhou et al., 2007). Hippocampal development requires BDNF both embryonically (Luikart et al., 2005) and after birth (Itami et al., 2007). Cortical dendritic development is BDNF dependent (Jin et al., 2003), particularly for developing glutamatergic synapses (Gomes et al., 2006). Developmental sensory plasticity is also BDNF dependent (Katz and Meiri, 2006). Among the developmental functions of NT-3 is a role in patterning of thalamocortical projections (Ma et al., 2002). Thus, neurotrophins modulate neuronal differentiation, migration, axon target finding, and synapse function in the developing CNS.

Several studies demonstrate continuing and essential roles for neurotrophins in the adult CNS. These functions range from maintaining neuronal gene expression and phenotype to modulating synaptic function and neurotransmitter availability. NGF removal in adult rats induces retrograde atrophy of basal forebrain cholinergic neurons (Sofroniew et al., 1990). BDNF-null mutant mice exhibit impaired hippocampal long-term potentiation (LTP) (Korte et al., 1995) that can be restored by application of exogenous BDNF. Conversely, the induction of LTP in normal mice alters BDNF, NT-3, and trk expression (Patterson et al., 1992; Castren et al., 1993; Nawa et al., 1995), supporting an interrelation between neurotrophin production and synaptic mechanisms. Indeed, release of neurotrophins from neurons may be a function of activity in ensembles of neurons (Lewin and Barde, 1996). Neurotrophin levels in the brain appear to be regulated by events that lead to synaptic plasticity; neurotrophins in turn can affect the strength of synaptic connections. If neurotrophins contribute substantially to short-term neuronal signaling and synaptic plasticity, then mechanisms would need to exist for the rapid and efficient removal of neurotrophins from the extracellular space, as suggested by Lewin and Barde (Lewin and Barde, 1996). This function may be served in the adult CNS by splice variants of the trk receptors, which are truncated forms of full-length receptors (Barbacid, 1994). Truncated trk receptors lack the tyrosine kinase domains that normally mediate signal transduction after neurotrophin binding and could eliminate neurotrophins by binding and internalizing them without stimulating signal transduction. This elimination of neurotrophins would be analogous to uptake mechanisms that rapidly remove neurotransmitters from the extracellular space (Lewin and Barde, 1996).

Expanding beyond the historical concept that BDNF acts as a retrograde, target-derived growth factor, BDNF is also anterogradely transported down axons in many regions of the adult CNS (Altar et al., 1997; Conner et al., 1997). Further, BDNF mRNA is targeted specifically to the dendritic compartments of neurons in some neural systems, including the forebrain and hippocampus (Fryer et al., 1996; Tongiorgi et al., 1997). Indeed, BDNF compartmentalization and availability at axonal projection sites appear to modulate the nature and strength of projecting systems in an anterograde manner. In vitro preparations indicate that BDNF is present in dendrites of hippocampal neurons and is released upon neuronal depolarization, suggesting that local BDNF availability can modulate synaptic function.

Neurotrophins modulate synaptic efficacy over very short times of seconds and minutes, suggesting the existence of non-trk/non-p75 signaling mechanisms (Lewin and Barde, 1996). For example, NT-3 and BDNF increase spontaneous and evoked synaptic activity over minutes or less (Lohof et al., 1993; Knipper et al., 1994; Leßmann et al., 1994; Kang and Schuman, 1995). These effects occur at the presynaptic level and do not require kinase activation. Tissue plasminogen activator-mediated cleavage of pro-BDNF to mature BDNF leads to long-term synaptic potentiation in the hippocampus (Pang et al., 2004). Within seconds, BDNF and NT-3 can increase intracellular calcium in cultured hippocampal neurons (Berninger et al., 1993), which may also lead to increased transmitter release. NT-3 increases activity of hippocampal neurons in culture, possibly by decreasing GABAergic transmission (Kin et al., 1994). NGF also modulates hippocampal plasticity over longer times, likely acting as a direct modulator of acetylcholine release from septal projections in the hippocampus (Conner et al., 2007).

The diverse effects of neurotrophins in modulating neural mechanisms in the adult brain suggest that they may act as essential substrates in adult behavioral plasticity and learning. Indeed, evidence in favor of this hypothesis has been forthcoming in models of amygdala-dependent learning (Rattiner et al., 2004), social learning (Berton et al., 2006), and hippocampal-dependent memory (Heldt et al., 2007). Further, a polymorphism in the human BDNF gene has been identified in which a substitution of valine for methionine occurs in the 5′ pro-region of the human BDNF protein (Val66Met); this polymorphism is associated with impaired episodic memory and hippocampal activation (Egan et al., 2003) and diminished hippocampal volume (Bueller et al., 2006). The Val66Met mutation is also a risk factor for a number of human psychiatric disorders, ranging from depression to psychosis (Hall et al., 2003; Bath and Lee, 2006).

Thus, the family of "classic neurotrophins" modulates a number of functions in the intact CNS and PNS. In the PNS, neurotrophins act as survival, differentiation, and synaptic plasticity factors. In the CNS, neurotrophin effects on cell survival are less clear, but other neuronal functions are extensively modulated by neurotrophins, including axonal and/or dendritic target finding, neuritic complexity, neuronal phenotype, neurotransmitter expression, and modulation of synaptic function.

CELLULAR TARGETS AND EFFECTS OF THE CLASSIC NEUROTROPHINS IN THE INJURED NERVOUS SYSTEM

After injury, several neurotrophins protect injured or degenerating adult nervous system neurons and axons. These survival and growth-promoting capabilities of the neurotrophins are described extensively in subsequent chapters that review specific animal models of human diseases and are briefly summarized below.

NGF influences the survival and/or growth of several neuronal populations in the injured adult nervous system. NGF prevents the degeneration of basal forebrain cholinergic neurons after injury or as a result of normal aging in rats and primates (Hefti, 1986; Fischer et al., 1987; Kromer, 1987; Rosenberg et al., 1988; Nebes and Brady, 1989; Barnett et al., 1990; Tuszynski et al., 1991; Holtzman et al., 1993; Emerich et al., 1994; Smith, 1999; Conner, 2001; Cooper et al., 2001). For these reasons, NGF has been tested as a therapeutic agent for treating the cholinergic component of neuronal loss in Alzheimer's disease (see Chapter 8) (Jonhagen et al., 1998; Tuszynski et al., 2005). The mode of NGF administration into the CNS is an important practical issue in clinical trials, as broad NGF distribution within the CNS elicits serious adverse effects, including Schwann cell hyperplasia, pain, and weight loss (Williams, 1991; Winkler et al., 1997) (Crutcher, 1987). To be adequately tested in the CNS clinically, growth factors must be delivered intraparenchymally and they must be accurately targeted in such a manner that NGF diffusion is limited to targeted, degenerating neurons without extensive diffusion from the delivery site. In other models, NGF also reduces cholinergic and non-cholinergic neuronal degeneration in the striatum of rodents with quinolinic acid lesions (Kordower et al., 1996), suggesting its potential use in Huntington's disease. In the spinal cord, cells genetically modified to produce and secrete NGF induce extensive growth of supraspinal coerulospinal axons and the central processes of nociceptive sensory axons in acute and chronic spinal cord injury (Tuszynski et al., 1994; Tuszynski et al., 1996) (see Chapter 13). In the PNS, NGF ameliorates several parameters of nerve dysfunction in rodent models of diabetic neuropathy (Apfel et al., 1994; Apfel, 1997) and taxol-induced neuropathy (Apfel et al., 1991). Based on these findings, Phase III trials of NGF in diabetic neuropathy were conducted (Apfel, 2002). Efficacy on primary outcome measures was not achieved, because of inadequate NGF dosing after systemic administration, a high placebo-responder rate, or a failure of NGF to ameliorate the disease mechanism underlying diabetes-related axonal degeneration.

The BDNF influences several types of neurons in vitro, including sensory neurons, basal forebrain cholinergic neurons, GABAergic neurons of the ventral mesencephalon, as well as retinal ganglion cells (see Ip et al., 1993; Hyman et al., 1994; Lewin and Barde, 1996; Hefti, 1997). Messenger RNA (mRNA) for BDNF is diversely expressed in the hippocampus, cortex, hypothalamus, brainstem, and cerebellum (Schecterson and Bothwell, 1992; Lewin and Barde, 1996). TrkB mRNA is also diversely expressed in the hippocampus, cortex, basal forebrain, ventral mesencephalon, and spinal cord. When delivered in vivo to the injured nervous system, BDNF rescues basal forebrain cholinergic and GABAergic neurons (Widmer et al., 1993), although it is less potent in rescuing cholinergic neurons of this system than NGF. BDNF also prevents the degeneration of brainstem neurons after axotomy in spinal cord injury models (Kobayashi et al., 1997). BDNF prevents the death of injured primary motor neurons in neonatal animals (Sendtner et al., 1992a; Yan et al., 1992; Clatterbuck et al., 1994), and prevents the downregulation of the cholinergic neuronal phenotype after motor neuron injury in the adult (Chiu et al., 1994; Yan et al., 1994; Friedman et al., 1995; Xu et al., 1995; Tuszynski et al., 1996). In addition, BDNF reduces motor neuron degeneration and improves function in models of mouse mutant neuronopathy, a correlational model of the human disease, amyotrophic lateral sclerosis (ALS or Lou Gherig's disease) (Ikeda et al., 1995). Notably, BDNF rescues neurons in this motor-neuron disease model if administered to adult (Mitsumoto et al., 1994b; Ikeda et al., 1995) but not neonatal (Blondet et al., 1997) animals. Based on the effects of BDNF on motor-neuron degeneration in animal models, BDNF underwent clinical trials for ALS in humans (1999). In this trial, BDNF was injected subcutaneously, which likely severely limited quantities of BDNF actually reaching degenerating motor neurons across the BBB. BDNF is retrogradely transported to spinal motor neurons when injected into the neuromuscular junction or intravascularly (e.g., Yan et al., 1994), but it is unclear that pharmacological levels of BDNF can be attained in the spinal cord by this route of administration. Perhaps not surprisingly, the outcome of this clinical trial was negative (1999). Further preliminary studies examining the effects of intrathecal delivery of BDNF in ALS have been reported (Ochs et al., 2000), but access of the growth factor to degenerating motor neurons with this delivery technique remains potentially problematic. Chemically modifying the BDNF molecule through pegylation can improve its parenchymal penetration after intrathecal infusion (Ankeny et al., 2001), but improved penetration could also extend its adverse event profile by stimulating non-degenerating neurons and causing untoward effects. BDNF also influences the survival and function of a subset of peripheral sensory axons (Hofer and Barde, 1988; Acheson et al., 1995; Conover et al., 1995), leading to human clinical trails in diabetic neuropathy (Wellmer et al., 2001). Once again, findings of the trial were negative because of either inadequate dosing or the futility of this method of treating diabetic neuropathy. BDNF also promotes the growth of injured rubrospinal and reticulospinal axons in experimental spinal cord injury (see Chapter 13). Finally, BDNF promotes the survival of vestibular neurons and a subset of auditory neurons, suggesting a potential future use in treating disorders of the middle ear (Ernfors et al., 1995).

In vitro, NT-3 influences the survival of neurons of the dorsal root ganglion (Ernfors et al., 1994b; Kucera et al., 1995a,b), sympathetic ganglia, hippocampus, ventral mesencephalon (dopaminergic and GABAergic neurons), and a subpopulation of spinal cord motor neurons (muscle spindle afferents) (Kucera, 1995a,b) reviewed in Lewin and Barde, 1996. NT3 mRNA is present in muscle, hippocampus, neocortex, cerebellum, olfactory bulb, diencephalon, midbrain, and spinal cord. mRNA for trkC is present in the hippocampus, neocortex, cerebellum, ventral mesencephalon, and motoneurons (Lamballe et al., 1991, 1994; Lewin and Barde, 1996). After injury in the adult, NT-3 can augment the function of large caliber, proprioceptive peripheral sensory axons (Gao et al., 1995; Helgren et al., 1997); on this basis, NT-3 underwent preliminary clinical trials as a treatment for the neuropathy associated with Charcot-Marie-Tooth disease (Sahenk et al., 2005). Preliminarily findings from the early clinical trial supported extension of studies to additional patients. NT-3 also promotes the growth of corticospinal motor axons and partially ameliorates motor deficits after spinal cord injury (see Chapter 13) (Schnell et al., 1994; Grill et al., 1997) and in rat models (Grill et al., 1997), and promotes regeneration of dorsal root sensory axons after dorsal rhizotomy (Ramer et al., 2000) and spinal cord injury (Lu et al., 2004). In addition, NT-3 is a survival factor for the majority of cochlear neurons (Cabelli et al., 1995), indicating potential uses for hearing disorders. NT-3 further modulates functional features of oligodendrocytes (Cohen et al., 1996), raising the possibility of treatment for demyelinating conditions. Finally, BDNF and NT-3 prevent the death of cortical neurons after axotomy (Giehl and Tetzlaff, 1996).

The influences of NT-4 on neuronal populations show overlap with BDNF, attributable to the fact that the two neurotrophins share signal transduction through the trkB receptor. In vitro, NT-4 promotes the survival of neurons of the DRG, sympathetic ganglia, nodose ganglion, hippocampus, basal forebrain, ventral mesencephalon (dopamine and GABAergic neurons), motor neurons, and retinal ganglion cells (see Lewin and Barde, 1996). mRNA for NT-4 is present in muscle, the pons, medulla, hypothalamus, thalamus, and cerebellum. In the adult, NT-4 ameliorates motor-neuron dysfunction (Friedman et al., 1995; Tuszynski et al., 1996), reduces brainstem neuronal loss after spinal cord injury, and promotes axonal growth of supraspinal brainstem projections to the spinal cord (Kobayashi et al., 1997). NT-4/5 has not undergone clinical testing.

Thus, the first and most thoroughly studied family of growth factors illustrates fundamental features of neurotrophic factor biology that are common to the other families as well. Neurotrophins are diversely expressed during development and in adulthood in specific distributions that may overlap with other trophic factors. Most trophic factor actions are paracrine in nature, although autocrine effects are also observed (Acheson et al., 1995). Signaling occurs through specific receptors that modulate the action of receptor-linked tyrosine kinases or other intracellular transduction proteins. Both neurons and glia are influenced by trophic mechanisms. A number of cellular functions are influenced by growth factors, including survival, differentiation, process outgrowth, target finding, maintenance of the cellular phenotype, modulation of transmitter levels, and synaptic plasticity.

"GDNF FAMILY OF LIGANDS" NEUROTROPHIC FACTORS

The GDNF Family of Ligands (GFL) neurotrophic factors are critical to the development and maintenance of several neuronal types in the CNS and the PNS. They have efficacy in many animal models of neuronal injury and disease and have progressed to human clinical trials in one human neurodegenerative disease.

The GFL neurotrophic factors consists of four members. The original member, GDNF, was isolated from conditioned medium from a glial cell line based upon its ability to maintain the survival of cultured mesencephalic dopaminergic neurons (Lin et al., 1993). The second member, neurturin, was purified from the conditioned medium of Chinese hamster ovary (CHO) cells based up its ability to maintain the survival of dissociated sympathetic neurons (Kotzbauer et al., 1996). With the discovery of this second member of the family, the islands of conserved sequence between the two factors were discerned, which enabled the isolation of the cDNA for the third member, persephin, by a degenerate PCR strategy (Milbrandt et al., 1998). Subsequently, an EST for the last member, artemin, appeared that facilitated its rapid identification (Baloh et al., 1998). The GFL neurotrophic factors are heparin-binding members of the TGF-β superfamily. Indeed, these represent the most distant members of this superfamily, showing about 20% sequence identity with the TGF-βs, most of which are accounted for by the canonical cysteines involved in the cystine knot structure and the formation of the disulfide-linked homodimers of the members of the superfamily. The factors are synthesized as prepro-sequences that are cleaved during processing. Interestingly, most of the members have more than one RXXR cleavage site making possible multiple mature forms from the pro-sequence. The biological significance of these potential multiple forms is not understood.

The receptor system for these factors is unique, and distinct from that of other members of the TFF-β superfamily (for detailed review, see Baloh et al., 2000; Bespalov and Saarma, 2007). The signaling of all of the factors is mediated by the transmembrane tyrosine kinase, Ret, originally described as a proto-oncogene (reviewed in Takahashi, 2001), Ret remained an orphan receptor for many years until the identification of GDNF as a ligand. However, in contrast to the neurotrophin family of receptors and the Trk receptors, the GFL do not directly bind Ret. A family of GPI-anchored proteins, the GFRαs, function as binding subunits. These lipid-raft associated proteins serve to bind the ligands, recruit Ret to lipids rafts, and produce the formation of dimers that induce autophosphoylation of Ret and subsequent activation of multiple signal transduction pathways critical to GFL action, such as the PI 3-kinase, MAP kinase, and src-family kinase(s) (Tansey et al., 2000). The four members of the GFRα family are preferentially associated with a specific ligand: GFRα1 – GDNF, GFRα2 – neurturin, GFRα3 – artemin, and GFRα4 – persephin. Although the specific pairing of the factors to the specific ligands appears very strict in physiological settings (see below), in primary cultures, or in engineered cells, significant "cross talk" occurs. Most importantly is cross talk between neurturin on GFRα1 and, to lesser degrees, GDNF on GFRα2 and artemin on GFRα1.

The physiological roles of these factors and their receptors have been deduced largely by examination of knockout mice and, more recently, via conditional knockout mice. The most robust phenotype observed in knockout mice is seen in the GDNF, GFRα1, and Ret knockouts wherein the animals are born lacking kidneys (therefore, perinatal lethal) and lacking an enteric nervous system. Indeed, the dramatic and concordant phenotypes of the Ret knockout and the GDNF knockout provided the critical insight into the ligand(s) for the previous orphan receptor, c-Ret (Schuchardt et al., 1994; Moore et al., 1996; Pichel et al., 1996; Sanchez et al., 1996; Enomoto et al., 1998). Examination of the phenotypes of the knockouts of all of the four factors and of the four GFLα-receptors (GFRα1–4) has demonstrated that physiological pairing of the ligands and co-receptors are quite strict; i.e., the phenotypes of the GDNF and GFRα1 knockouts are virtually identical, the phenotypes of the neurturin and GFRα2 are virtually identical, etc. (reviewed in Baloh et al., 2000; Airaksinen and Saarma, 2002). For example, GDNF has a critical role in the development of the enteric and parasympathetic nervous systems, being required for the proliferation and migration of precursors during early development. In contrast, neurturin is required for the *maintenance* of enteric and parasympathetic systems; both are atrophied in neurturin or GFRα2 knockouts. GDNF, neurturin, and artemin all act on subpopulations of neurons that express appropriate GFRαs, although the phenotypes of the knockouts are relatively subtle, especially when compared to the dramatic effects of deletion of the neurotrophins NGF, BDNF, or NT-3 on the sensory nervous system. Similarly, modest effects are observed on motor neuron development. Non-neuronal tissues are also affected by the deletion of GDNF. As noted previously, loss of GDNF, or Ret or GFRα1 produces renal agenesis. In addition, GDNF/Ret signaling is required for spermatogenesis (Meng et al., 2000; Jain et al., 2004).

Although null animals in which the genes for neurturin, artemin, and persephin have been deleted survive into adulthood and reproduce, the perinatal lethality associated of the GDNF, Ret, or GFRα1 knockouts has precluded an analysis of the role of GDNF/Ret signaling in the adult animal. This gap in our understanding of the physiological role of GDNF has recently been lessened by the generation of dominant-negative transgenic mice that survive for weeks after birth and by animals engineered to have floxed Ret alleles necessary to make conditional knockouts. The dominant-negative animals have clarified to some extent the role of GDNF in the spermatogenesis and in kidney maintenance (Jain et al., 2004). Two groups have recently described the generation of mice in which Ret has been selectively deleted in dopaminergic neurons by crossing mice with floxed Ret alleles with mice expressing Cre driven by the dopamine-transport (DAT) promoter (Jain et al., 2006; Kramer et al., 2007). Surprisingly, given the robust pharmacological effects of at least three GFLs on these neurons, no effect of deletion of Ret during development or "middle" adulthood is observed by anatomical, neurochemical, or behavioral analysis. One group (Kramer et al., 2007) analyzed "old" mice and did observe anatomical and neurochemical (but not behavioral) deficits.

The GFLs exert pharmacological neurotrophic effects on a myriad of neuronal types of PNS and CNS origin. As noted above, GDNF was discovered based on its ability to affect the survival of cultured mesencephalic dopaminergic neurons and neurturin on its ability to maintain survival of dissociated sympathetic neurons. Similar survival-promoting effects occur in spinal and cranial motor neurons, Ret-expressing subpopulations of sensory neurons of both neural crest and placode origin, enteric neurons, and parasympathetic neurons. Indeed, at least a subpopulation of all major neuronal types in the periphery, or that project to the periphery, is *pharmacologically* responsive to GDNF and neurturin. The actions of artemin are largely restricted to those populations expressing GFRα3 (sympathetic and a subpopulation of sensory neurons). All neuronal populations (sympathetic, parasympathetic, enteric) residing exclusively in the periphery are non-responsive to persephin.

Based upon the actions of the factors, GFLs have been examined in animal models of neurodegenerative diseases or of nervous system injury. Here we shall consider four pathological settings:

Parkinson's disease: As noted above, the original search that lead to the discovery of GDNF (Lin, 1993) was predicated upon identifying a factor that exerted neurotrophic effects on mesencephalic neurons and, thus, would be a potential therapy for Parkinson's disease (PD). GDNF, and to a lesser extent the other GFLs, have been studied extensively in models of this disease. PD is currently the only condition in which GFLs have progressed to clinical trials. The cardinal motor systems of PD result from the degeneration of dopaminergic neurons in the substantia nigra pars compactor (SNpc). Shortly after the discovery of GDNF, several papers appeared demonstrating a protective effect of GDNF in dopaminergic neurons exposed to mechanical or chemical insults (Hoffer et al., 1994; Beck et al., 1995; Hudson et al., 1995; Kearns and Gash, 1995; Tomac et al., 1995). In the ensuing years, scores of papers have described the efficacy of GDNF administered in a variety of ways in animal models of PD. Based on the efficacy of GDNF in rodent and primate (Gash et al., 1996) models, a randomized, double-blind trial of GDNF was carried out in the late 1990s in patients with advanced PD (Nutt et al., 2003). GDNF was administered at various doses by monthly intracerebroventricular (ICV) injection via an indwelling catheter. The trial was not designed to determine whether GDNF was "neuroprotective," but rather whether it would improve symptoms as measured by the Unified Parkinson's Disease Rating Scale (UPDRS). GDNF did not improve parkinsonism, but did produce several side effects including weight loss, nausea, hyponatremia, and Lhermitte's sign (a shock-like sensation moving down the spine when flexing the neck). As noted previously, several potential drawbacks to this ICV infusion method of administration include the limited distribution of the factor into the parenchyma of the brain, especially the target area of nigral neurons, the putamen; and the increased likelihood of side effects associated with distribution of the factor throughout the entire CSF space, which likely accounted for the several adverse effects observed. Despite the caveats associated with this method of administration, the failure of this study dampened enthusiasm for the potential utilization of GDNF and other factors administered directly into the CNS as a therapy for PD and possibly other neurodegenerative diseases.

An important paper that sparked renewed enthusiasm for CNS-administered trophic factor as a potential treatment for PD was reported in 2000 (Kordower et al., 2000) in which investigators administered by lentiviral vector GDNF both into aged primates and into primates made parkinsonian by MPTP. Both models of virally expressed GDNF in the striatum and nigra produced neuroprotective/neurorestorative effects as assessed anatomically by histological and imaging methods. Significantly, reversal of parkinsonian symptoms occurred. These robust effects were not associated with adverse effects such as those seen in the clinical trial after ICV administration. This paper emphasized the therapeutic potential of GDNF administered into the *target* of nigral neurons, as opposed to into the CSF, and also suggested that the gene therapy approach was a potentially viable way to express in a controlled manner the factor in target areas. Further encouragement was provided by a subsequent report of a small Phase I (i.e., non-blinded, non-placebo controlled) clinical trial (Gill et al., 2003) in which five PD patients were infused with GDNF protein directly into the putamen for 1 year without significant adverse events. A substantial improvement in both the motor and activities-of-daily-living subscores of the UPDRS was reported, as well as reductions in medication-induced dyskinesias. Similar results were reported in a second Phase I study (Slevin et al., 2005) by another group.

Based on these results, an industry-sponsored double-blind, placebo-controlled, multi-center Phase II trial was performed with doses of GDNF protein and intra-putamen infusion protocols similar, but not identical, to those used in the two Phase I studies (Lang et al., 2006). Despite ^{18}F-DOPA imaging results indicating that GDNF exerted a biological effect, treated or placebo-control groups had no significant improvement in PD symptoms. In addition, the development of GDNF antibodies was observed in some GDNF-treated subjects. Based on the negative clinical results of this trial and potential safety issues, the industry sponsor discontinued clinical trials and required patients from the Phase I trials, who had continued on GDNF treatment, to discontinue treatment. The decision to discontinue the development of GDNF as a treatment for PD has been very controversial in the scientific, clinical, and PD-patient communities. The points of contention related to both the design (dose, infusion method, patient selection, power of trial) and the interpretation of the trial (see Sherer et al., 2006 for discussion of these issues).

As an alternative to the intra-putamenal infusion of GDNF protein, others have pursued the potential of neurturin gene therapy in the treatment of PD. As noted above, neurturin can also act via the GFRα1 receptor expressed on mesenecpohalic dopaminergic neurons. In both rodent and monkey, neurturin delivered intrastriatally via an adeno-associated viral vector-2 (AAV-2) gene therapy vector produces efficacy in PD models comparable to GDNF, without observable adverse effects (Kordower et al., 2006; Gasmi et al., 2007; Herzog et al., 2007). An open-label Phase-I trial of AAV-neurturin delivered by multiple bilateral injections into the putamen (Marks et al., 2006) has followed 12 subjects for at least 1 year. No significant adverse events were observed and improvements in motor function similar to those in the Phase-I GDNF trials (Gill et al., 2003) (Slevin et al., 2005) were described. A double-blind, placebo-controlled Phase II trial of neurturin gene therapy in PD is currently underway.

Huntington's Disease: Huntington's Disease (HD) is an autosomal dominant genetic disease caused by a triplet repeat mutation in the Huntington gene located on chromosome 4. The devastating motor, psychiatric, and cognitive symptoms are attributed to the death of striatal projection and cortical neurons. The molecular mechanism of the neuronal death is not precisely known. Currently, no treatment for this disease is available and, as a consequence, the majority of people at risk for this disease choose not be tested for the mutant gene. If a neurotrophic factor were available that directly or indirectly protected these neurons, then this may be an ideal arrangement for trophic-factor therapy in that treatment could be initiated years prior to onset of symptoms when little or no neuronal loss has occurred, thus delaying or preventing the onset of disease symptoms. Animal models of HD have focused on the loss of striatal neurons and the associated motor deficits; these models include toxin administration, particularly intrastriatal administration of excitatory amino acids (kainite or quinolinic acid), or systemic administration of the mitochondrial toxin 3-nitorpropionic acid. In addition, transgenic animal models in which the mutant human gene is expressed reproduce fundamental features of the human disease (for recent review, see Ramaswamy et al., 2007b).

The initial report of an effect of GFLs in an HD model (Perez-Navarro et al., 1996) used intrastriatal placement of grafts of fibroblasts expressing GDNF to partially protect striatal neurons from the effects of quinolinic acid. The effect was observed on the calbindin+ projection neurons, with no protective effect on the parvalbumin+ interneurons. Similar experiments by the same group (Perez-Navarro et al., 2000; Marco et al., 2002) reported more robust protective effects of neurturin-secreting fibroblasts on the calbindin+ neurons. These actions of the GFLs are particularly germane to HD, since the projection neurons are much more numerous in the striatum and are the population selectively lost in the disease. More recently, the efficacy of either GDNF or neurturin, administered intrastriatally via AAV-2 gene therapy vectors, has been demonstrated in both the systemic 3-nitropropionic acid and the transgenic mouse models by Kordower and colleagues (McBride et al., 2003, 2006; Ramaswamy et al., 2007a,b). These studies demonstrate both preservation of striatal neurons and diminution of functional deficits in both models.

A limitation of these published results is their focus on the anatomical lesions in the striatum and associated motor function deficits. These experiments address only those components of this devastating disorder. A critical issue that needs to be addressed in assessing the rationale for moving these observations into clinical trials is whether the cortical destruction, which may underlie the psychiatric or cognitive symptoms of the disease, is affected by these or similar interventions with GFLs.

Sensory Neuropathies and Neuropathic Pain: Ret is expressed in approximately 60% of adult sensory neurons (Molliveret al., 1997; Bennett et al., 1998; Kashiba et al., 2003); these neurons can be small, medium, or large. These Ret-expressing cells also express one or more of the GFRαs 1–3 (Molliver et al.,

1997; Bennett et al., 1998; Orozco et al., 2001; Kashiba et al., 2003), indicating responsiveness to GDNF, neurturin, and/or artemin. Many sensory neurons, particularly those in the large- or medium-diameter populations, may also express one of the Trk receptors responsive to the neurotrophins. Special attention has been paid to small diameter nociceptive neurons because of their interesting developmental pattern and because of their obvious pathophysiological significance in pain syndromes. During early development, virtually all of these small cells express TrkA and are NGF-dependent for survival (Johnson et al., 1980; Molliver et al., 1997). However, beginning in mid-gestation and continuing postnatally, about half of these cease to express TrkA and begin to express Ret. The Ret+ cells overlap completely with the IB4-lectin-binding population. In the adult, TrkA-positive neurons project primarily to laminas I and IIo, while the Ret-positive cells to lamina IIi. The ability of GFLs to act trophically on these neurons was shown by the ability of intrathecally administered GDNF to prevent biochemical and functional changes subsequent to sciatic nerve section in adult rats. More important to human pathology, intrathecal infusion of GDNF prevents and reverses hyperalgesia in neuropathic pain models because GDNF prevents alterations in the expression of several genes associated with this injury (Boucher et al., 2000; Wang et al., 2003). Neurturin also prevents the development of neuropathic pains in at least one of these models (J. Golden, personal communication). Treatment of uninjured animals with GDNF or neurturin was not associated with changes in normal sensory thresholds. The ability of artemin to prevent the development of neuropathic pain is somewhat controversial with reports of efficacy with systemic artemin in a peripheral nerve injury model (Gardell et al., 2003) and in a *herpes simplex* virus-induced model (Asano et al., 2006). However, others have failed to find efficacy in the nerve-injury models of artemin administered either peripherally or intrathecally (Bolon et al., 2004). The use of GFLs in neuropathic pain is worthy of pursuit given the limited efficacy and side effects of currently available treatments (Finnerup et al., 2007).

The GFLs have also been examined in an animal model of diabetic sensory neuropathy. Sensory neuropathies, including painful neuropathies, are a common consequence of diabetes. In the mouse model of streptozoticin-induced diabetes, severely reduced cutaneous innervation is observed similar to that seen in humans. Intrathecal GDNF or neurturin is able to significantly ameliorate the condition, increasing both axon branching and cutaneous innervation (Christianson et al., 2003).

These studies suggest that GDNF and neurturin, at least if administered intrathecally, may have efficacy in some neuropathies and neuropathic pain syndromes.

Addiction: As noted above, GDNF was initially discovered based on the ability to promote the survival of cultured mesencephalic dopaminergic neurons. Although the vast majority of interest in the literature has focused on the potential importance of this activity on the etiology or treatment of PD, it is important to remember that mesencephalic dopaminergic neurons are destined not just to become SNpc dopaminergic neurons, but also ventral tegmental area (VTA) dopaminergic neurons. Both populations of midbrain dopaminergic neurons express Ret and GFRα1 in the adult brain (Nosrat et al., 1997; Golden, et al., 1998). The VTA dopaminergic neurons project to the limbic forebrain (e.g., the nucleus accumbens, the NAc). These neurons have a major role in reward behavior and in addiction (for review, see Hyman et al., 2006). Chronic exposure to a variety of addicting drugs causes a series of sustained biochemical alterations in the VTA and the Nucleus Accumbens. In a seminal paper by Nestler and colleague (Messer et al., 2000), evidence was provided both for a physiological role of GDNF as modifier of adaptation to drugs-of–abuse and for the ability of exogenous GDNF to suppress addictive behavior in rodents. They demonstrate that loss of GDNF either via decrease of GDNF levels in $GDNF^{+/-}$ mice or via infusion of anti-GDNF into the Nucleus Accumbens suppresses reward-seeking behavior in response to cocaine. Administration of morphine causes a lowering of the level of Ret phosphorylation. When used pharmacologically, GDNF prevents permanent biochemical changes association with chronic exposure to addictive substances and decreased reward-seeking behaviors caused by addicting drugs. The ability of GDNF to suppress reward behaviors has been confirmed after central administration of GDNF, and by others using alternative treatment methods (Green-Sadan et al., 2003, 2005). The increased susceptibility to addicition of GDNF+/- animals has been confirmed (Airavaara et al., 2007; Yan et al., 2007).

Despite the enormous personal and societal burdens caused by drug addiction, and the confirmation of the striking results reported by Nestler and colleagues in 2000 (Messer et al., 2000), this aspect of the physiology and pharmacology of GFLs has received surprisingly little attention and concentrated effort. Obviously, the difficulties in supplying the GFLs acting via GFRα1 (GFNF, neurturin) represent a hurdle to the use of these agents. We hope improved methods to supply the factors to the CNS will make this a more attractive approach in the future.

OTHER MEMBERS OF THE TGFB-SUPERFAMILY

The GFLs (especially GDNF) represent by far the most thoroughly studied molecules as neurotrophic factors of the TGFβ-superfamily. However, several other members of that superfamily have been examined in a variety of contexts both for their physiological function and for their pharmacological potential in the nervous system. Unlike the GFLs, these other family members were initially identified, and have been largely studied, outside the nervous system. Many of the actions they have in the nervous system are mediated, not by neurotrophic actions, but via other effects (e.g., proliferative effects, effects of inflammatory processes). Such molecules include the TGFβs (for recent reviews, see Duenker, 2005; Gomes et al., 2005; Vivien and Ali, 2006) and certain members of the bone morphogenetic protein (BMP) subfamily. These neurotrophic factors exert effects on many neuronal types in culture (Hattori et al., 1999; Nonner et al., 2001; Gratacos et al., 2002; Yabe et al., 2002) and an in vivo model of dopaminergic injury (Zuch et al., 2004) and stroke (Harvey et al., 2005). Other members of the TGFß-superfamily include growth/differentiation factor 5 (GDF-5), which has been studied primarily for activity on dopaminergic neurons (Krieglstein et al., 1995); and growth/differentiation factor-15 (also named macrophage inhibitory cytokine-1; (Strelau et al., 2000). As noted above (indeed, inherent in the names), these other factors exert many other actions that seem to make them more problematic as potential therapeutic agents to exploit neurotrophic activities. However, further work with them is needed in models of injury or disease to more fully assess their potential.

THE CYTOKINE GROWTH FACTORS

A second class of nervous system growth factors is the *cytokine* family, consisting of leukemia inhibitory factory (LIF; also known as cholinergic differentiation factor) (Patterson, 1994; Taupin et al., 1998), ciliary neurotrophic factor (CNTF) (Barbin et al., 1984; Lin et al., 1989; Sendtner et al., 1994), and cardiotrophin-1 (Herrmann et al., 1993; Arce et al., 1998). The cytokines differ from the neurotrophins not only in molecular structure, but also in fundamental features of receptor physiology. CNTF is a protein of 200 amino acids (molecular weight 22 700 Da) that interacts with a tripartite receptor complex including a specific CNTF receptor alpha, an LIF receptor β, and a gp130 receptor (Ip et al., 1992b; Ip and Yancopoulos, 1992). Binding of CNTF activates Janus tyrosine kinases (JAK), which phosphorylate members of the STAT family of transcription factors. The phosphorylated STAT proteins then translocate to the nucleus where they activate CNTF-responsive genes (Segal and Greenberg, 1996). Unlike the classic neurotrophins, CNTF lacks a signal peptide for secretion from the cell. Since CNTF is produced by both Schwann cells and astrocytes, CNTF may act as a factor that is released from cells after injury in the CNS or PNS; however, its precise role remains to be established.

In vitro, CNTF can act as a survival factor for motor neurons and can prevent motor neuronal degeneration after neonatal axotomy and as a consequence of spontaneous degeneration in adult mice with mutant motor neuronopathy (Arakawa, Sendtner et al., 1990) (Sendtner et al., 1990, 1992b; Mitsumoto et al., 1994a,b; Lindsay, 1996). Mouse mutant motor neuronopathy resembles in some respects the spontaneous degeneration of motor neurons that occurs in ALS. Of note, the neuroprotection provided by combined treatment with both BDNF and CNTF in mouse mutant motor neuronopathy is greater than the protection resulting from delivery of either growth factor alone, suggesting that neurotrophins can act synergistically to promote neuronal protection even if targeting a single neuronal population (Mitsumoto et al., 1994a,b). Expression of CNTF is naturally upregulated in peripheral nerves after axotomy and may support motor and sensory neuronal survival and axonal regeneration after injury (Apfel Arezzo et al., 1993). CNTF also protects basal forebrain cholinergic neurons (Hagg et al., 1992) and substantia nigra dopaminergic neurons (Hagg and Varon, 1993) from lesion-induced degeneration without restoring the transmitter phenotype of the injured neurons. CNTF rescues striatal interneurons after excitotoxic lesions (Emerich et al., 1996). Further, in many instances neurons responding to CNTF do not express any components of the tripartite receptor complex, suggesting that CNTF may act by inducing the release of neuroprotective agents from neighboring neurons or glia. Additional neurons that exhibit CNTF responses include ciliary ganglion neurons, hippocampal neurons, neurons of the spinal nucleus of the bulbocavernosus (Forger et al., 1997), retinal photoreceptors (Ezzeddine et al., 1997), corticospinal neurons (Dale et al., 1995), and progenitor cells that can differentiate into oligodendrocytes or type II astrocytes (O2A progenitor cells). CNTF ameliorates neuronal loss and improves functional outcomes in rats after cerebral ischemia (Kumon et al., 1996). Despite this extensive characterization of neuronal responses to CNTF, its role in normal development remains unclear, since CNTF-null mutant mice exhibit few changes in the nervous system until postnatal ages, when motor neuronal loss becomes evident (see Richardson, 1994; Sendtner et al., 1994 for review). Of note, and possibly of concern with regard to the treatment of neurological disease, CNTF induces several components of reactive gliosis in the brain (Levison et al., 1996).

CNTF was the one of the first neurotrophic factors to undergo large-scale clinical trials for the treatment of a human neurological disease, ALS (Group, 1996). This disease was targeted for neurotrophic factor treatment because CNTF prevents progressive motor neuronal degeneration in mouse mutant neuronopathy (Sendtner et al., 1992b), an effect on motor neurons that was subsequently supported by the finding that targeted mutation of the CNTF gene induced motor neuron degeneration during development (Masu et al., 1993). CNTF was administered subcutaneously in this trial, thereby depending on uptake and retrograde transport of CNTF from the neuromuscular junction to rescue spinal motor neurons. CNTF failed to demonstrate efficacy in humans (Group, 1996). Reasons for this failure may include the following: (1) the extremely short half life of CNTF after peripheral injection (minutes), (2) the possibility that subcutaneous injections of CNTF fail to achieve concentrations in the spinal cord sufficient to rescue motor neurons, and (3) the absence of preclinical efficacy data from a species larger than the mouse; had negative outcomes resulted from experiments in larger mammals, human trials might have been designed differently or postponed until superior efficacy data and delivery methods had been developed. It is also possible, of course, that CNTF would fail to exhibit efficacy even in the presence of adequate central delivery of the growth factor because the underlying mechanism of neuronal toxicity in ALS cannot be affected by neurotrophic factors. Clearly, the adequate delivery of growth factors to their intended therapeutic targets remains an important factor that limits satisfactory clinical testing for efficacy. An attempt to deliver CNTF intrathecally in patients with ALS has been conducted by placing encapsulated cells genetically modified to produce CNTF into the thecal space of the spinal cord; a Phase I study with this approach reported safety, but the sample size was too small to address efficacy (Aebischer, Schluep et al., 1996). Results of intrathecal CNTF delivery from another 12-patient study indicated a possible immune response (Zurn et al., 2000) to the implanted cells or the bovine serum in which the cells were raised; efficacy data has not been reported. A problem that may be difficult to surmount in the therapeutic use of CNTF is the occurrence of cachexia (weight loss) after systemic administration (Henderson et al., 1994). Using this potentially adverse property to its advantage, the company owning CNTF performed a clinical trial of the growth factor to treat human obesity (Ettinger et al., 2003). While patients lost weight after receiving CNTF, the effect was not persistent and the drug was not pursued further for this indication.

More recently, CNTF has been examined as a neuroprotective factor for photoreceptor and retinal ganglion cell degeneration (Cayouette and Gravel, 1997; Leaver et al., 2006). CNTF prevents retinal ganglion cell loss after axotomy and other mechanisms of injury (Leaver et al., 2006), and separately, reduces photoreceptor loss in models of degeneration including retinitis pigmentosa mouse models (Cayouette and Gravel, 1997; Cayouette et al., 1998; Liang et al., 2001). However, lack of CNTF efficacy in ocular models and possible damage or adverse remodeling of cells and their patterns of protein production induced by the growth factor (Bok, 2005; Buch et al., 2006; Beltran et al., 2007) have been reported. Nonetheless, CNTF has begun clinical testing in retinitis pigmentosa using implants of encapsulated, genetically modified cells into the eye (Sieving et al., 2006).

The cytokine growth factor LIF also influences several classes of neurons. In vitro, LIF promotes survival of motor neurons (Patterson, 1994), and cortical neurons and glia (Taupin et al., 1998). In the injured adult peripheral nerve, LIF supports motor and sensory axonal regeneration (Arakawa et al., 1990; Cheema et al., 1994a,b), a function that may be mediated in part by upregulated expression of LIF in Schwann cells (Banner and Patterson, 1994). LIF upregulates NT-3 expression in the injured spinal cord and augments growth of corticospinal axons (Blesch et al., 1999). LIF is the first neurotrophic factor shown to upregulate expression of another trophic factor in the CNS after in vivo injury, a finding that correlates with enhanced axonal growth (Blesch et al., 1999). Thus, neurotrophins influence CNS function at yet another level by modulating production of growth factors by other cells. Similar findings are reported in in vitro preparations (Thoenen, 1995) and in the PNS, wherein NGF delivery augments BDNF expression in motor neurons (Apfel, Wright et al., 1996).

THE INSULIN-LIKE GROWTH FACTORS

Insulin and the insulin-like growth factors (IGF-I and IGF-II) constitute another family of molecules (reviewed in Roberts and LeRoith, 1988; Folli et al., 1996) with cell survival and growth-promoting qualities (Bothwell, 1982; Mill et al., 1985; Ishii, 1993; LeRoith et al., 1993). Most nervous-system studies of this family have focused on IGF-I and IGF-II, which are broadly expressed in adulthood and development. Patterns of IGF-I and IGF-II expression in the nervous system are regionally specific and temporally regulated (Ishii, 1993; LeRoith et al., 1993; Folli et al., 1996). In particular, IGF-I mRNA is expressed in neurons and glia, and IGF-II mRNA is expressed in cells of the choroid plexus, the leptomeninges, and to a lesser extent on glia (LeRoith et al., 1993). Although insulin and insulin-like growth factors modulate glucose utilization by many cell types, these molecules also act as specific, localized signals in the nervous system that modulate cell survival (Recio-Pinto et al., 1986; DiCicco and Black, 1988; McMorris and Dubois-Dalcq, 1988; Knusel et al., 1990; Beck et al., 1993), growth, differentiation (McMorris et al., 1986; McMorris and Dubois-Dalcq, 1988; Bondy et al., 1990), synapse function (Ishii, 1989) and neurotransmitter release ((Dahmer and Perlman, 1988); see (Ishii, 1993; LeRoith et al., 1993) for reviews).

The IGF receptor is a tetramer consisting of two alpha and two beta subunits (LeRoith, Roberts et al., 1993) linked by disulfide units. The α subunits lie outside the cell and bind IGF ligands at cysteine-rich domains. The β subunits are receptors containing a short extracellular domain, a transmembrane domain, and a cytoplasmic domain; the latter contains an AT-binding site and a tyrosine kinase domain. IGF family ligands bind to the α subunit, which in turn binds to β subunits. The binding of ligand-occupied α to β induces a conformational change in β that activates a receptor tyrosine kinase and induces autophosphorylation, leading to phosphorylation of additional substrates and cascades that mediate biological responses. Distinct from the IGF-I receptor, the IGF-II receptor consists of a single protein with a large extracellular and a short intracellular domain, and signal transduction occurs through activation of glycine. A family of insulin-like growth factor binding proteins also exist, presumably modulating the binding of IGF-I and IGF-II to the IGF receptor (Yee et al., 1987). Intracellular mechanisms influenced by IGF-1 include both the ERK pathway (Subramaniam et al., 2005) and the AKT pathways (Zheng and Quirion, 2006), which modulate cell function and survival.

mRNA for IGF-I is detectable and is differentially regulated during development in several regions of the nervous system (LeRoith et al., 1993). These regions include trigeminal and sympathetic neurons, sensory and cerebellar relay systems, non-pyramidal cells of the cortex and hippocampus, Purkinje cells of the cerebellum, and muscle (Ishii, 1993; LeRoith et al., 1993). In the postnatal and mature brain, IGF-I mRNA is present in frontal cortex, the hippocampus, amygdala, suprachiasmatic nucleus, and cerebellar and sensory relay systems. Expression of IGF-I receptors peaks during development and is found in cerebellum, neocortex, olfactory bulb, retina, and choroid plexus (LeRoith et al., 1993). In vitro, IGF-I promotes the survival or proliferation of cortical neurons (Aizenman and de Vellis, 1987), basal forebrain neurons (Knusel, Michel et al.,

1990), ventral mesencephalon neurons (Knusel et al., 1990; Beck et al., 1993), primary motor neurons (Yin et al., 1994), sympathetic neurons (Zackenfels et al., 1995), and oligodendrocytes (Aizenman et al., 1986), and induces neurite outgrowth from motor (Caroni and Grandes, 1990; Caroni et al., 1994), sensory (Bothwell, 1982), sympathetic (Zackenfels et al., 1995), and cortical neurons (Aizenman and de Vellis, 1987). It may also influence oligodendrocyte differentiation and myelination (LeRoith et al., 1993). Overexpression of IGF-I in transgenic mice during development results in a disturbance in the formation of normal barrel fields in primary sensory cortex: an increase is observed in the number of neurons, neuronal area and volume per barrel, but total neuronal density decreases (Gutierrez-Ospina et al., 1996). Myelin synthesis increases in IGF-I-overexpressing mice (Carson et al., 1993). These data support a role for IGF-I in modulating neuronal number and nervous system patterning.

The effects of IGF-I in in vivo models of nervous system disease have been characterized in several systems (Leinninger and Feldman, 2005). IGF-I immunoreactivity increases in Schwann cells after nerve crush (Hansson et al., 1986). Injections of IGF-I or IGF-II into rat muscle induces sprouting of motor neurons (Caroni and Grandes, 1990), and IGF-I infusions into the lesioned peripheral nerve or dorsal root ganglion improve sensitivity to sensory stimuli post-injury, an effect that is blocked by injection of anti-IGF-I (Kanje et al., 1989). Systemic administration of IGF-I improves the extent of functional recovery in the hindlimbs of mice after sciatic nerve lesions (Contreras et al., 1995). In addition, IGF-I reduces the number and extent of demyelination sites in mice with experimental autoimmune encephalomyelitis, a model of multiple sclerosis, and improves their functional outcome (Liu et al., 1995b; Yao et al., 1995). Altered IGF-1-related signaling also occurs in Alzheimer's disease, and IGF-1 may modulate processing of amyloid peptides (Liang et al., 2001; Niikura et al., 2001).

On the basis of IGF-I effects on motor neurons, clinical trials of IGF-I in patients with ALS were conducted in two separate, prospective, randomized, controlled studies. One study demonstrated significant reduction of disease progression (Laiet al., 1997) whereas the other study did not (Borasio et al., 1998). These findings, reported a decade ago, were the subject of considerable discussion in the clinical community and resulted in the re-initiation of another large controlled clinical study that is currently underway. IGF has been administered peripherally in these trials as subcutaneous injections, raising familiar questions regarding access of IGF to degenerating motor neurons in the spinal cord in quantities sufficient to influence motor neuron survival and function. Better targeted delivery of the growth factor to motor neurons might be required to ultimately determine the value of IGF-I in treating ALS. Pursuing this possibility, an AAV vector expressing IGF-1 was injected into the muscles of superoxide dismutase mice; administered by this route, the AAV-IGF-1 vector is both expressed at the neuromuscular junction and undergoes some retrograde transport into the spinal cord leading to IGF-1 expression precisely in the motor neurons that are undergoing degeneration (Kaspar et al., 2003). This led to marked prolongation of survival in ALS mice. Attempts are underway currently to enhance the efficiency of retrograde transport as a potential prelude to clinical trials of AAV-IGF-1 gene delivery in ALS.

The distribution of IGF-II mRNA is less extensive than that of IGF-I in the CNS and is associated primarily with the choroid plexus, leptomeninges, and glia. In vitro, IGF-II promotes survival and neurite extension from sensory and sympathetic neurons (Recio-Pinto et al., 1986; Zackenfels et al., 1995). IGF-I and IGF-II also stimulate proliferation of rat inner ear epithelial cells (Zheng et al., 1997). These limited findings have not served as the basis for clinical trials.

HEMATOPOIETIC CYTOKINES

The hemotopoietic cytokines, erythropoietin (EPO) and granulocyte-colony stimulating factor (G-CSF) have been long-studied and clinically exploited for their ability to induce the proliferation of populations of hematopoietic linage cells. However, data have emerged over the last decade demonstrating the expression of the receptors for these factors within the central and peripheral nervous systems and for the ability of the factors to mitigate the effects of certain insults to the nervous system (for a recent extensive review, see Tonges et al., 2007). Very recently, similar neurotrophic activity has been reported for a third such factor, granulocyte-macrophage colony stimulating factor (GM-CSF) (Schabitz and Schneider, 2007). Given the extensive literature on these factors in other settings, and their potential clinical availability, it is an attractive notion to expand their potential utility to the nervous system.

EPO and G-CSF act via distinctly different receptors, but are ultimately able to activate similar signal transduction pathways directly leading to activation of JAK/STAT pathways and pathways associated with traditional neurotrophic factors (e.g., MAPK and PI-3 kinase pathways; (Tonges et al., 2007)),. Although the receptors for these factors are found on certain neuronal populations, the mechanisms by which they may ameliorate neuronal injury in vivo are complex and probably multifaceted. In addition to direct anti-apoptotic/trophic roles analogous to classical neurotrophic factors, they also exert anti-inflammatory effects by dampening the release of inflammatory cytokines implicated in stroke, autoimmune disease of the nervous system (e.g., EAE), and in models of chronic neurodegenerative disease (Agnello et al., 2002; Villa et al., 2003; Zhang et al., 2005). Both EPO and G-CSF exert angiogenic activity, which may be particularly important in the positive effects in ischemia and stroke models (Sadamoto et al., 1998) (Nittaet al., 1999). Lastly, both factors augment stem cell proliferation and neuronal differentiation (Wang et al., 2004; Schneider et al., 2005), which may provide another potential mechanism for long-term positive effects of these factors.

This constellation of potential activities has attracted the most attention as a potential therapy after stroke. Several groups report positive results in rodent models with these factors (reviewed in Lu and Xiao, 2007). Clinical trials carried out thus far have been limited, with some suggestion of, but certainly not definitive support for, positive effects (Bath and Sprigg, 2007). Ongoing blinded trials will be required to assess safety and to determine any efficacy. For information on these ongoing trials, see (Bath and Sprigg, 2007) and www.strokecenter.org

As noted above, positive results have also been reported in the animal model of multiple sclerosis, experimental autoimmune encephalomyelitis (EAE) and in rodent models of PD. Several possible mechanisms, in addition to classic neurotrophism, may have a role in the efficacy reported. Given that these proteins are already widely used clinically, they represent attractive candidates as neuroprotective agents in neurodegenerative disease. The many issues inherent in treatment of the chronic diseases, as opposed to stroke or adjuncts to chemotherapy, obviously have to be addressed.

FIBROBLAST GROWTH FACTORS

The fibroblast growth factors encompass, in mammals, a 22-member family. In humans, these include FGF-1–14 and FGF-16–23 (FGF-19 is the human ortholog of mouse FGF-15). Although the FGFs vary somewhat in size, they have an approximately 120 amino acid core with a 30–60% amino acid identity (Itoh and Ornitz, 2004). The family can be separated phylogenetically into seven subfamilies. Most, but not all, members of the family have mitogenic activity on myriad cells types. Interestingly, one subfamily (FGF 11–14) appears to be intracellular proteins in neurons and other cells types. The actions of extracellular FGFs are mediated by a family of transmembrane tyrosine kinase receptors (FGFR1–FGFR4) that exist in multiple isoforms generated by alternative mRNA splicing. In addition, the actions of FGFs are modified by complex interactions with extracellular glycosaminoglycan heparin and cell surface haparan sulfate glycosaminoglycans (reviewed by Ornitz, 2000).

Several members of the FGF family have important functions in the very early development of the nervous system in processes such as the induction and early patterning of the neural plate, and as mediators of neuroepithelial organization. They also act as regulators of later organization and development of the CNS by actions such as regulation of fate determination, differential growth, and patterning in along the neuroaxis (reviewed by Dono, 2003).

From the neurotrophic factor perspective, by far the most interest has been placed in the biology, pharmacology, and potential therapeutic activity of the first described subfamily of FGFs comprised of FGF1 (originally known as acidic FGF) and particularly FGF-2 (originally known as basic FGF). As the name of the family implies, these initial members of the family were originally isolated based upon their ability to induce fibroblast proliferation (Gospodarowicz, 1974). Historically, activity of these factors was observed in extracts of many tissues and fluids based on actions in a myriad of cell types of mesodermal and ectodermal origin and based upon their angiogenic actions. They were accordingly assigned a myriad of names prior to the elucidation of amino acid and cDNA sequence (see Gospodarowicz, 1990 for review). FGF-1 and FGF-2 lack classical signal sequences and, thus, do not undergo efficient ER-Golgi-mediated secretion, unlike the classical neurotrophic factors families (neurotrophins and GFLs). Rather, the FGFs are transported less efficiently into the extracellular space. They may have more of a role after release subsequent to injury to modulate CNS plasticity.

The importance of FGF-2 in CNS development is exemplified by developmental defects in FGF-2-null animals in several structures including the cortex, hippocampus, substantia nigra, and in the spinal cord (Gospodarowicz, 1990; Dono et al., 1998; Ortega et al., 1998; Timmer et al., 2007). In adult nervous system, FGFs and FGF receptors continue to be expressed, and FGF1 is retrogradely transported in a receptor-mediated manner in many CNS neuronal populations (Wanaka et al., 1990; Ferguson and Johnson, 1991; Stock et al., 1992; Eckenstein, 1994), suggesting that FGFs, particularly FGF-1, would be capable of producing neurotrophic effects directly on neurons. After an initial report (Morrison et al., 1986) of apparent neurotrophic activity to enhance survival and outgrowth in primary cortical cultures and in axotomized retinal ganglion cells in adult animals (Sievers et al., 1987), the subsequent two decades have provided many examples of such activity (for review, see Dono, 2003; Reuss and von Bohlen und Halbach, 2003), including on neuronal types of particular interest from the point of view of human disease and nervous system injury.

Based upon the many effects of FGFs that may be potentially ameliorative, the factors, especially FGF-2, have been examined in a variety neuronal injury, stroke, or neurodegeneratrive disease models (reviewed in Tuszynski, 1999; Dono, 2003; Reuss and von Bohlen und Halbach, 2003). For example, neuronal survival is enhanced after lesion of retinal ganglion cells (Sievers et al., 1987), medial septal cholinergic neurons after fornix/fimbria transection, and cortical neurons after entorhinal lesions (Ramirez et al., 1999). Effects are also seen in both toxin models (e.g., dopamine neuron injury in MPTP-treated mice; Otto and Unsicker, 1990) and genetic models (e.g, HD; (Jin et al., 2005)) of human neurodegerative disease.

Studies in the 1990s show efficacy of intraventricularly infused FGF-1 and FGF-2 in animal models of CNS ischemia (Sasaki et al., 1992; Nakata et al., 1993). This was followed by demonstration of efficacy of FGF-2 when administered intravenously subsequent to a permanent middle cerebral artery occlusion (Fisher et al., 1995), thus providing a rationale for the use the factor in a clinically relevant paradigm. Based on these encouraging results, two double-blind, placebo-controlled clinical trials were initiated. Unfortunately both were terminated early because of interim analyses indicating little likelihood of favorable risk:benefit emerging from the completed study (Bogousslavsky et al., 2002).

GROWTH FACTOR ANALOGS

Small molecules have been generated that are capable of binding to neurotophic factor receptors to act either as agonists or as antagonists to trophic actions (see Longo and Massa, 2005). These molecules are relatively short peptides of 9–20 amino acids and possess the advantage that they can cross the BBB after peripheral administration. For example, Longo and colleagues have identified small-molecule ligands that bind to the p75 neurotrophin receptor and are capable of preventing pro-neurotrophin-induced cell degeneration (Massa et al., 2006). Other peptides bind to trkA (Xie et al., 2000) or trkB (Fletcher and Hughes, 2006) and mediate neurotrophin-related signal transduction. Some ligands to G protein-coupled receptors are capable of trans-activating trophic signaling, including adenosine (Jeanneteau and Chao, 2006). Small analogs of nitric oxide (NO) exhibit neuroprotective properties by acting through neurotrophic signaling mechanisms (Thatcher et al., 2006). Reports regarding the effectiveness of these compounds in in vivo models of disease are limited. The advantages of small-molecule ligands to growth factor receptors are clear: central administration can be avoided and dosing can be adjusted simply. The hypothetical disadvantage of a peripheral, non-targeted growth factor approach to treating neurological disease is that the effects of the ligands would be distributed throughout the nervous system, causing the same adverse effects as non-targeted central delivery of growth factors.

Another strategy for delivering neurotrophic molecules across the BBB is to link a trophic factor to a molecule that is actively transported into the CNS (Friden et al., 1993). With this type of approach, NGF has been linked to transferrin-receptor antibodies and crosses the BBB. In a quinolinic acid lesion model of HD in rodents, striatal neurons are protected after peripheral administration of an NGF-transferrin-receptor antibody conjugate in rodent (Charles et al., 1996). Similarly, NGF-transferrin-receptor antibody conjugates protect basal forebrain cholinergic neurons from degeneration in aged rats (Bäckman et al., 1996) and after excitotoxic lesions of the cortex (Kordower et al., 1994). However, development of these compounds has not continued; once again, they are subject to the hypothetical problem that they would elicit adverse effects of non-targeted delivery.

Finally, pan-neurotrophins and neurotrophic factor "fusion" proteins are synthetic peptides that contain active sites of more than one neurotrophic factor within a single molecule (Ibanez et al., 1993). Pan-neurotrophin-1, a fusion of NGF, BDNF, and NT-3, exhibits transport properties and neurotrophic properties in vitro that are consistent with the known activities of its three trophic-factor components independently (Ilag et al., 1995). Pan-neurotrophic factors are of potential use in neurological disorders characterized by the degeneration of multiple types of neurons, including stroke, trauma, and Alzheimer's disease. Once again, their targeted delivery to the nervous system would be a challenge.

SUMMARY OF GROWTH FACTOR EFFECTS AND POTENTIAL USES

From the preceding summary of growth-factor effects in the nervous system, it is evident that different neurotrophic proteins influence nervous system function at many structural and functional levels, and throughout life. Growth factors are retrogradely or anterogradely transported in axons, and can act in either an autocrine or an paracrine manner. Under physiological conditions and in a therapeutic context, growth factors provide a means of preventing cell loss, promoting axonal growth, influencing synaptic transmission, modulating injury responses, and generating pluripotent stem cells for neural repair.

Perhaps the most straightforward clinical target for growth-factor therapy is the prevention of neuronal loss. Applications to progressive neurodegenerative disorders in which cell loss occurs include Alzheimer's disease, PD, HD, ALS, and multiple sclerosis. A cell rescue strategy is also logical for disorders such as stroke and neural trauma. Based on the results of preclinical studies, candidate growth factors for each of these disorders can be identified. However, delivery of growth factors to the CNS has remained a substantial obstacle in assessing their potential benefits for neurological disease: it is essential that growth factors are delivered in effective doses across the BBB, but their distribution must be limited to targeted neurons to avoid adverse effects. At present, gene delivery or intraparenchmyal infusions of growth factors are the only means of achieving these goals. Ongoing clinical trials in PD and Alzheimer's disease are testing these possibilities.

Growth factors may also find application for the treatment of other types of nervous system disease. Peripheral neuropathy remains a compelling target for growth-factor therapy, yet the delivery of adequate doses continues to be a challenge. Further research will be required to enhance targeting of growth factors in these conditions.

CONSIDERATIONS AND FUTURE DIRECTIONS

The abundant data summarized above generally provide a strong basis for continuing to pursue neurotrophic factors for neuroprotective and regenerative therapy of neurological disease. However, the results of these studies have a vital role in directing future avenues in trophic-factor research. Issues that have been brought into focus by experiments performed to date include the following: (a) the importance of understanding the pharmacokinetics and bioavailability of trophic factors after various routes of delivery, (b) the restriction of neurotrophic factor actions to targeted cell populations, and (c) the development of effective delivery systems for introducing trophic agents to humans.

Ensuring that exogenously delivered trophic factors remain biologically active is a primary concern in designing successful trophic-factor therapy. Most neurotrophic factors are natural proteins and various elements can affect their bioavailability after systemic administration in pharmacological doses. Proteolytic cleavage by non-specific peptidases may deactivate the trophic factor molecule, a possibility that may partially account for the negative outcome of CNTF in ALS trials (Group, 1996). In addition, various factors may bind to exogenously delivered trophic agents rendering them unavailable to neurons. For instance, serum-born alpha-2 macroglobulin may bind circulating NGF and diminish its bioavailability (Hefti, 1997). The binding of trophic factors to endogenous receptors on non-targeted cell populations should also be taken into account when evaluating pharmacokinetics, since such interactions would further reduce bioavailability for interacting with targeted cells.

Perhaps one of the most crucial lessons gleaned from clinical trials of neurotrophic factors for nervous system disease thus far is the importance of restricting the biological actions of exogenously delivered trophic factors to the cell population targeted for therapy, as indicated in the preceding sections. The widespread presence of trophic factor receptors on various neuronal and non-neuronal cells throughout the body makes the goal of restricting trophic factor actions a challenging one. Other trophic-factor-mediated actions, such as axonal sprouting, alterations in neurotransmitter systems, and modulation of synaptic function (Lewin and Barde, 1996), heighten concerns regarding non-specific trophic-factor delivery. In addition, NGF affects the immune system by stimulating mast cell migration and histamine release, and can influence cells of the reproductive system (Rohrer and Sommer, 1983). The functional impact of long-term, systemic trophic-factor administration on these responsive systems remains to be fully evaluated.

Thus, the development of techniques for delivering trophic factors in a restricted manner to targeted cell populations will have a key role in determining the success of neurotrophic factor therapy as a clinical tool. One means of achieving targeted and regionally restricted growth-factor delivery in the CNS is intraparenchymal growth-factor infusion (Hu et al., 1997; Gill et al., 2003; Lang et al., 2006). NGF infusion to the rat brain can elevate NGF levels in tissues near the infusion site sufficiently to rescue cholinergic neurons (Hu et al., 1997) but does not result in the widespread diffusion of NGF throughout the CNS. Alternatively, gene therapy is another means of specifically targeting trophic factor to localized neuronal populations, as noted above and in subsequent chapters (Tuszynski et al., 2005).

In conclusion, neurotrophic factors provide an unprecedented opportunity to treat currently untractable neurological diseases. Most of the promise of the growth factors is based on the remarkable potency of these proteins in animal models; however, adequate clinical testing requires methods that accurately target and restrict their administration to regions of degenerating neurons in the brain. Once these goals are achieved, neurotrophic factors could radically alter the landscape of neurological therapy.

ACKNOWLEDGMENTS

Supported by the NIH (NS09881, NS42291, AG10435, and AG13730), Veterans Affairs, and the Adelson Program in Neural Rehabilitation and Repair.

REFERENCES

(1999). A controlled trial of recombinant methionyl human BDNF in ALS: the BDNF Study Group (Phase III). *Neurology* 52(7): 1427–1433.

Acheson A, Conover JC et al. (1995). A BDNF autocrine loop in adult sensory neurons prevents cell death. *Nature* 374: 450–453.

Aebischer P, Schluep M et al. (1996). Intrathecal delivery of CNTF using encapsulated genetically modified xenogenic cells in amyotrophic lateral sclerosis patients. *Nat Med* 2: 1041.

Agnello D Bigini P et al. (2002). Erythropoietin exerts an anti-inflammatory effect on the CNS in a model of experimental autoimmune encephalomyelitis. *Brain Res* 952(1): 128–134.

Airaksinen MS and Saarma M (2002). The GDNF family: signalling, biological functions and therapeutic value. *Nat Rev Neurosci* 3(5): 383–394.

Airavaara M, Tuomainen H et al. (2007). Effects of repeated morphine on locomotion, place preference and dopamine in heterozygous glial cell line-derived neurotrophic factor knockout mice. *Genes Brain Behav* 6(3): 287–298.

Aizenman Y and de Vellis J (1987). Brain neurons develop in a serum and glial free environment: effects of transferrin, insulin, insulin-like growth factor-I and thyroid hormone on neuronal survival, growth and differentiation. *Brain Res* 406: 32–42.

Aizenman YJ, Weischsel ME et al. (1986). Changes in insulin and transferrin requirements of pure brain neuronal cultures during embryonic development. *Proc Natl Acad Sci U S A* 83: 2263–2266.

Altar CA Cai N et al. (1997). Anterograde transport of brain-derived neurotrophic factor and its role in the brain. *Nature* 389(6653): 856–860.

Angeletti RH, Hermodson MA et al. (1973). Amino acid sequences of mouse 2.5S nerve growth factor. II. Isolation and characterization of the thermolytic and peptic peptides and the complete covalent structure. *Biochemistry* 12: 100–115.

Ankeny DP McTigue DM et al. (2001). Pegylated brain-derived neurotrophic factor shows improved distribution into the spinal cord and stimulates locomotor activity and morphological changes after injury. *Exp Neurol* 170(1): 85–100.

Apfel S (1997). Treatment of peripheral nervous system disorders. *Clinical Applications of Neurotrophic Factors*. Lippincott-Raven, Philadelphia. pp. 63–81.

Apfel SC (2002). Nerve growth factor for the treatment of diabetic neuropathy: what went wrong, what went right, and what does the future hold? *Int Rev Neurobiol* 50: 393–413.

Apfel SC, Lipton RB et al. (1991). Nerve growth factor prevents toxic neuropathy in mice. *Ann Neurol* 29: 87–90.

Apfel SC, Arezzo JC et al. (1993). Effects of administration of ciliary neurotrophic factor on normal motor and sensory peripheral nerves in vivo. *Brain Res* 604: 1–6.

Apfel SC, Arezzo JC et al. (1994). Nerve growth factor administration protects against experimental diabetic sensory neuropathy. *Brain Res* 634: 7–12.

Apfel SY, Wright DE et al. (1996). Nerve growth factor upregulates expression of brain-derived neurotrophic factor in the peripheral nervous system. *Mol Cell Neurosci* 7: 134–142.

Arakawa Y, Sendtner M et al. (1990). Survival effect of ciliary neurotrophic factor on chick embryonic motoneurons in culture: comparison with other neurotrophic factors and cytokines. *J. Neurosci*. 10: 3507–3515.

Arce V, Pollock RA et al. (1998). Synergistic effects of Schwann- and muscle-derived factors on motoneuron survival involve GDNF and cardiotropin-1 (CT-1). J Neurosci 18: 1440–1448.

Asano K, Asahina S et al. (2006). Attenuating effect of artemin on herpes-related pain responses in mice infected with herpes simplex. In Vivo 20(4): 533–537.

Bäckman C, Rose GM et al. (1996). Systemic administration of a nerve growth factor conjugate reverses age-related cognitive dysfunction and prevents cholinergic neuron atrophy. *J Neurosci* 16(17): 5437–5442.

Baloh RH, Tansey MG et al. (1998). Artemin, a novel member of the GDNF ligand family, supports peripheral and central neurons and signals through the GFRalpha3-RET receptor complex. *Neuron* 21(6): 1291–1302.

Baloh RH, Enomoto H et al. (2000). The GDNF family ligands and receptors - implications for neural development. *Curr Opin Neurobiol* 10(1): 103–110.

Banner LR and Patterson PH (1994). Major changes in the expression of the mRNAs for cholinergic differentiation factor/leukemia inhibitory factor and its receptor after injury to adult peripheral nerves and ganglia. *Proc Natl Acad Sci U S A* 91: 7109–7913.

Barbacid M (1994). The trk family of neurotrophin receptors. *J Neurobiol* 25: 1386–1403.

Barbin G, Manthorpe M et al. (1984). Purification of the chick eye ciliary neurotrophic factor. *J Neurochem* 43: 1468–1478.

Barde YA, Edgar et al. (1982). Purification of a new neurotrophic factor from mammalian brain. *Eur Mol Biol Organ* 1: 549–553.

Barnett J, Baecker P et al. (1990). Human beta nerve growth factor obtained from a baculovirus expression system has potent in vitro and in vivo neurotrophic activity. *Exp Neurol* 110: 11–24.

Bath KG and Lee FS (2006). Variant BDNF (Val66Met) impact on brain structure and function. *Cogn Affect Behav Neurosci* 6(1): 79–85.

Bath PM and Sprigg N (2007). Colony stimulating factors (including erythropoietin, granulocyte colony stimulating factor and analogues) for stroke. *Cochrane Database Syst Rev* (2): CD005207.

Beck KD, Knusel B et al. (1993). The nature of the trophic action of brain-derived neurotrophic factor, des(1-3)-insulin-like growth factor-1, and basic fibroblast growth factor on mesencephalic dopaminergic neurons developing in culture. *Neuroscience* 52: 855–866.

Beck KD, Valverde J et al. (1995). Mesencephalic dopaminergic neurons protected by GDNF from axotomy-induced degeneration in the adult brain. *Nature* 373(6512): 339–341.

Beltran WA, Wen R et al. (2007). Intravitreal injection of ciliary neurotrophic factor (CNTF) causes peripheral remodeling and does not prevent photoreceptor loss in canine RPGR mutant retina. *Exp Eye Res* 84(4): 753–771.

Bennett DL, Michael GJ et al. (1998). A distinct subgroup of small DRG cells express GDNF receptor components and GDNF is protective for these neurons after nerve injury. *J Neurosci* 18(8): 3059–3072.

Berkmeier LR, Winslow JW et al. (1991). Neurotrophin-5: A novel neurotrophic factor that activates trk and trkB. *Neuron* 7: 857–866.

Berninger B, Garcia DE et al. (1993). BDNF and NT-3 induce intracellular calcium elevation in hippocampal neurones. *Neuroreport* 4: 1303–1306.

Berton O, McClung CA, DiLeone RJ et al. (2006). Essential role of BDNF in the mesolimbic dopamine pathway in social defeat stress. *Science* 311: 864–868.

Bespalov MM and Saarma M (2007). GDNF family receptor complexes are emerging drug targets. *Trends Pharmacol Sci* 28(2): 68–74.

Blesch A, Uy HS, Grill RJ et al. (1999). Leukemia inhibitory factor augments neurotrophin expression and corticospinal axon growth after adult CNS injury. *J Neurosci* 19: 3556–3566.

Blochl A and Thoenen H (1996). Localization of cellular storage compartments and sites of constitutive and activity-dependent release of nerve growth factor (NGF) in primary cultures of hippocampal neurons. *Mol Cell Neurosci* 7: 173–190.

Blondet B, Murawsky M et al. (1997). Brain-derived neurotrophic factor fails to arrest neuromuscular disorders in the paralyse mouse mutant, a model of motoneuron disease. *J Neurol Sci* 153: 20–24.

Bogousslavsky J, Victor SJ et al. (2002). Fiblast (trafermin) in acute stroke: results of the European-Australian phase II/III safety and efficacy trial. *Cerebrovasc Dis* 14(3–4): 239–251.

Bok D (2005). Ciliary neurotrophic factor therapy for inherited retinal diseases: pros and cons. *Retina* 25(8 Suppl): S27–S28.

Bolon B, Jing S et al. (2004). The candidate neuroprotective agent artemin induces autonomic neural dysplasia without preventing peripheral nerve dysfunction. *Toxicol Pathol* 32(3): 275–294.

Bondy CA, Werner H et al. (1990). Cellular pattern of insulin-like growth factor-I (IGF-I) and type I IGF receptor gene expression in early organogenesis: comparison with IGF-II gene expression. *Mol Endocrinol* 4: 1386–1398.

Bonhoeffer T, (1996). Neurotrophins and activity-dependent development of the neocortex. *Curr Opin Neurobiol* 6: 119–126.

Borasio GD, Robberecht W et al. (1998). A placebo-controlled trial of insulin-like growth factor-I in amyotrophic lateral sclerosis. European ALS/IGF-I Study Group. *Neurology* 51(2): 583–586.

Bothwell M (1982). Insulin and somatomedin MSA promote nerve growth factor-independent neurite formation by cultured chick dorsal root ganglionic sensory neurons. *J Neurosci Res* 8: 225–231.

Boucher TJ, Okuse K et al. (2000). Potent analgesic effects of GDNF in neuropathic pain states. *Science* 290(5489): 124–127.

Buch PK, MacLaren RE et al. (2006). In contrast to AAV-mediated Cntf expression, AAV-mediated Gdnf expression enhances gene replacement therapy in rodent models of retinal degeneration. *Mol Ther* 14(5): 700–709.

Bueller JA, Aftab M et al. (2006). BDNF Val66Met allele is associated with reduced hippocampal volume in healthy subjects. *Biol Psychiatry* 59(9): 812–815.

Buzsaki G, Ponomareff GL et al. (1988). Suppression and induction of epileptic activity by neuronal grafts. *Proc Nat Acad Sci* 85: 9327–9330.

Cabelli RJ, Hohn A et al. (1995). Inhibition of ocular dominance column formation by infusion of NT-4/5 or BDNF. *Science* 267: 1662–1666.

Cabelli RJ, Shelton DL et al. (1997). Blockade of endogenous ligands of trkB inhibits formation of ocular dominance columns. *Neuron* 19: 63–76.

Caroni P and Grandes P (1990). Nerve sprouting in innervated adult skeletal muscle induced by exposure to elevated levels of insulin-like growth factors. *J Cell Biol* 110: 1307–1317.

Caroni P, Schneider C et al. (1994). Role of insulin-like growth factors in nerve sprouting: suppression of terminal sprouting in paralyzed muscles by IGF-binding protein 4. *J Cell Biol* 125: 893–902.

Carson MJ, Behringer RR et al. (1993). Insulin-like growth factor increases brain growth and central nervous system myelination in transgenic mice. *Neuron* 10: 729–740.

Carter BD, Kaltschmidt C et al. (1996). Selective activation of NF-kB by nerve growth factor through the neurotrophin receptor p75. *Science* 272: 542.

Castren E, Pitkanen M et al. (1993). The induction of LTP increases BDNF and NGF mRNA but decreases NT-3 mRNA in the dentate gyrus. *Neuroreport* 4: 895–898.

Cayouette M and Gravel C (1997). Adenovirus-mediated gene transfer of ciliary neurotrophic factor can prevent photoreceptor degeneration in the retinal degeneration (rd) mouse. *Hum Gene Ther* 8(4): 423–430.

Cayouette M, Behn D et al. (1998). Intraocular gene transfer of ciliary neurotrophic factor prevents death and increases responsiveness of rod photoreceptors in the retinal degeneration slow mouse. *J Neurosci* 18(22): 9282–9293.

Chao MV (1994). The p75 neurotrophin receptor. *J Neurobiol* 25: 1373–1385.

Charles V, Mufson EJ et al. (1996). Atrophy of cholinergic basal forebrain neurons following excitotoxic cortical lesions is reversed by intravenous administration of an NGF conjugate. *Brain Res* 728: 193–203.

Cheema SS, Richards L et al. (1994a). Leukemia inhibitory factor prevents the death of axotomised sensory neurons in the dorsal root ganglia of the neonatal rat. *J Neurosci Res* 37: 213–218.

Cheema SS, Richards LJ et al. (1994b). Leukaemia inhibitory factor rescues motoneurones from axotomy-induced cell death. *Neuroreport* 5: 989–992.

Chen KS, Nishimura MC et al. (1997). Disruption of a single allele of the nerve growth factor gene results in atrophy of basal forebrain cholinergic neurons and memory deficits. *J Neurosci* 17: 7288–7296.

Chiu AY, Chen EW et al. (1994). Distinct neurotrophic responses of axotomized motor neurons to BDNF and CNTF in adult rats. *Neuroreport.* 5: 693–696.

Christianson JA, Riekhof JT, Wright DE, (2003). Restorative effects of neurotrophin treatment on diabetes-induced cutaneous axon loss in mice. *Exp Neurol* 179: 188–199.

Clatterbuck RE, Price DL et al. (1994). Further characterization of the effects of brain-derived neurotrophic factor and ciliary neurotrophic factor on axotomized neonatal and adult mammalian motor neurons. *J Comp Neurol* 342: 45–56.

Cohen RI, Marmur R et al. (1996). Nerve growth factor and neurotrophin-3 regulate the proliferation and survival of developing rat brain oligodendrocytes. *J Neurosci* 16: 6433–6442.

Cohen S and Levi-Montalcini R (1957). Purification and properties of a nerve growth-promoting factor isolated from mouse sarcoma. *Cancer Res* 17: 15–20.

Cohen S, Levi-Montalcini R et al. (1954). A Nerve growth-stimulating factor isolated from sarcom as 37 and 180. *Proc Natl Acad Sci U S A* 40(10): 1014–1018.

Conner JM, Lauterborn JC et al. (1997). Distribution of brain-derived neurotrophic factor (BDNF) protein and mRNA in the normal adult rat CNS: evidence for anterograde axonal transport. *J Neurosci* 17(7): 2295–2313.

Conner JM, Darracq MA, Roberts J, and Tuszynski MH (2001). Non-tropic actions of neurotrophins: subcortical NGF gene delivery reverses age-related degeneration of primate cortical cholinergic innervation. *Proc Nat Acad Sci* 98: 1941–1946.

Conner KM, Franks KM, Olson AK, et al. (2007). NGF is essential for hippocampal plasticity and learning. *J Neurosci* In press.

Conover JC, Erickson JT et al. (1995). Neuronal deficits, not involving motor neurons, in mice lacking BDNF and/or NT4. *Nature* 375: 235–238.

Contreras PC, Steffler C et al. (1995). Systemic administration of rhIGF-1 enhanced regeneration after sciatic nerve crush in mice. *J Pharmacol Exp Ther* 274: 1443–1449.

Cooper JD, Salehi A et al. (2001). Failed retrograde transport of NGF in a mouse model of Down's syndrome: reversal of cholinergic neurodegenerative phenotypes following NGF infusion. *Proc Natl Acad Sci U S A* 98(18): 10439–10444.

Cordon-Cardo C, Tapley P et al. (1991). The trk tyrosine protein kinase mediates the mitogenic properties of nerve growth factor and neurotrophin-3. *Cell* 66: 173–183.

Crowley C, Spencer SD et al. (1994). Mice lacking nerve growth factor display perinatal loss of sensory and sympathetic neurons yet develop basal forebrain cholinergic neurons. *Cell* 76: 1001–1011.

Crutcher KA (1987). Sympathetic sprouting in the central nervous system: a model for studies of axonal growth in the mature mammalian brain. *Brain Res Rev* 12: 203–233.

Dahmer MK and Perlman RL (1988). Bovine chromaffin cells have insulin-like growth factor-I (IGF-I) receptors: IGF-I enhances catecholamine secretion. *J Neurochem* 51: 321–323.

Dale SM, Kuang RZ et al. (1995). Corticospinal motor neurons in the adult rat: degeneration after intracortical axotomy and protection by ciliary neurotrophic factor. *Exp Neurol* 135: 67–73.

DiCicco E and Black IB (1988). Insulin growth factors regulate the mitotic cycle in cultured rat sympathetic neuroblasts. *Proc Natl Acad Sci U S A* 85: 4066–4070.

DiStefano PS, Freidman B et al. (1992). The neurotrophins BDNF, NT-3 and NGF display distinct patterns of retrograde axonal transport in peripheral and central neurons. *Neuron* 8: 983–993.

Dobrowsky RT, Werner MH et al. (1994). Activation of the sphingomyelin cycle through the low-affinity neurotrophin receptor. *Science* 265: 1596–1599.

Domenici L, Cellerino A et al. (1995). Antibodies to nerve growth factor (NGF) prolong the sensitive period for monocular deprivation in the rat. *Neuroreport* 5: 2041–2044.

Dono R (2003). Fibroblast growth factors as regulators of central nervous system development and function. *Am J Physiol Regul Integr Comp Physiol* 284(4): R867–R881.

Dono R, Texido G et al. (1998). Impaired cerebral cortex development and blood pressure regulation in FGF-2-deficient mice. *Embo J* 17(15): 4213–4225.

Duenker N (2005). Transforming growth factor-beta (TGF-beta) and programmed cell death in the vertebrate retina. *Int Rev Cytol* 245: 17–43.

Eckenstein FP (1994). Fibroblast growth factors in the nervous system. *J Neurobiol* 25(11): 1467–1480.
Egan MF, Kojima M et al. (2003). The BDNF val66met polymorphism affects activity-dependent secretion of BDNF and human memory and hippocampal function. *Cell* 112(2): 257–269.
Emerich DW, Winn S et al. (1994). Implants of polymer-encapsulated human NGF-secreting cells in the nonhuman primate: rescue and sprouting of degenerating cholinergic basal forebrain neurons. *J Comp Neurol* 349: 148–164.
Emerich DF, Lindner MD et al. (1996). Implants of encapsulated human CNTF-producing fibroblasts prevent behavioral deficits and striatal degeneration in a rodent model of Huntington's disease. *J Neurosci* 16(16): 5168–5181.
Enomoto H, Araki T et al. (1998). GFR alpha1-deficient mice have deficits in the enteric nervous system and kidneys. *Neuron* 21(2): 317–324.
Ernfors P, Lee K-F et al. (1994a). Mice lacking brain-derived neurotrophic factor develop with sensory deficits. *Nature* 368: 147–150.
Ernfors P, Lee et al. K-F (1994b). Lack of neurotrophin-3 leads to deficiencies in the peripheral nervous system and loss of limb proprioception. *Cell* 77: 503–512.
Ernfors P, Van de Water T et al. (1995). Complementary roles of BDNF and NT-3 in vestibular and auditory development. *Neuron* 14: 1153–1164.
Ettinger MP, Littlejohn TW et al. (2003). Recombinant variant of ciliary neurotrophic factor for weight loss in obese adults: a randomized, dose-ranging study. *JAMA* 289(14): 1826–1832.
Ezzeddine ZD, Yang X et al. (1997). Postmitotic cells fated to become photoreceptors can be respecified by CNTF treatment of the retina. *Development* 124: 1055–1067.
Fagan AM, Zhang H et al. (1996). TrkA, but not TrkC, receptors are essential for survival of sympathetic neurons in vivo. *J Neurosci* 16: 6208–6218.
Fahnestock M, Yu G et al. (2004). ProNGF: a neurotrophic or an apoptotic molecule? *Prog Brain Res* 146: 101–110.
Farinas I, Jones KR et al. (1994). Severe sensory and sympathetic deficits in mice lacking neurotrophin-3. *Nature* 369: 658–661.
Ferguson IA and Johnson EM Jr (1991). Fibroblast growth factor receptor-bearing neurons in the CNS: identification by receptor-mediated retrograde transport. *J Comp Neurol* 313(4): 693–706.
Finnerup NB, Sindrup SH et al. (2007). Chronic neuropathic pain: mechanisms, drug targets and measurement. *Fundam Clin Pharmacol* 21(2): 129–136.
Fischer W, Wictorin K et al. (1987). Amelioration of cholinergic neuron atrophy and spatial memory impairment in aged rats by nerve growth factor. *Nature* 329: 65–68.
Fisher M, Meadows ME et al. (1995). Delayed treatment with intravenous basic fibroblast growth factor reduces infarct size following permanent focal cerebral ischemia in rats. *J Cereb Blood Flow Metab* 15(6): 953–959.
Fletcher JM and Hughes RA (2006). Novel monocyclic and bicyclic loop mimetics of brain-derived neurotrophic factor. *J Pept Sci* 12(8): 515–524.
Folli F, Ghidella S et al. (1996). The early intracellular signaling pathway for the insulin/insulin-like growth factor receptor family in the mammalian central nervous system. *Mol Neurobiol* 13: 155–183.
Forger NG, Howell ML et al. (1997). Sexual dimorphism in the spinal cord is absent in mice lacking the ciliary neurotrophic factor receptor. *J Neurosci* 17: 9605–9612.
Friden PM, Walus LR et al. (1993). Blood-brain barrier penetration and in vivo activity of an NGF conjugate. *Science* 259: 373–377.
Friedman B, Kleinfeld D et al. (1995). BDNF and NT-4/5 exert neurotrophic influences on injured adult spinal motor neurons. *J Neurosci* 15: 1044–1056.
Fryer RH, Kaplan DR et al. (1996). Developmental and mature expression of full-length and truncated TrkB receptors in the rat forebrain. *J Comp Neurol* 374(1): 21–40.
Funakoshi H, Frisen J et al. (1993). Differential expression of mRNAs from neurotrophins and their receptors after axotomy of the sciatic nerve. *J Cell Biol* 123: 455–465.

Gao WQ, Dybdal N et al. (1995). Neurotrophin-3 reverses experimental cisplatin-induced peripheral sensory neuropathy. *Ann Neurol* 38: 30–37.

Gardell LR, Wang R et al. (2003). Multiple actions of systemic artemin in experimental neuropathy. *Nat Med* 9(11): 1383–1389.

Garofalo L, Ribeiro-da-Silva A et al. (1992). Nerve growth factor-induced synaptogenesis and hypertrophy of cortical cholinergic terminals. *Proc Natl Acad Sci U S A* 89: 2639–2643.

Gash DM, Zhang Z et al. (1996). Functional recovery in parkinsonian monkeys treated with GDNF. *Nature* 380(6571): 252–255.

Gasmi M, Brandon EP et al. (2007). AAV2-mediated delivery of human neurturin to the rat nigrostriatal system: Long-term efficacy and tolerability of CERE-120 for Parkinson's disease. *Neurobiol Dis.*

Giehl KM and Tetzlaff W (1996). BDNF and NT-3, but not NGF, prevent axotomy-induced death of rat corticospinal neurons in vivo. *Eur J Neurosci* 8: 1167–1175.

Gill SS, Patel NK et al. (2003). Direct brain infusion of glial cell line-derived neurotrophic factor in Parkinson disease. *Nat Med* 9(5): 589–595.

Golden JP, Baloh RH et al. (1998). Expression of neurturin, GDNF, and their receptors in the adult mouse CNS. *J Comp Neurol* 398(1): 139–150.

Gomes FC, Sousa Vde O et al. (2005). Emerging roles for TGF-beta1 in nervous system development. *Int J Dev Neurosci* 23(5): 413–424.

Gomes RA, Hampton C et al. (2006). The dynamic distribution of TrkB receptors before, during, and after synapse formation between cortical neurons. *J Neurosci* 26(44): 11487–11500.

Gorin PD and Johnson EM (1980). Effects of long term nerve growth factor deprivation on the nervous system of the adult rat: an autoimmune approach. *Brain Res* 198: 27–42.

Gospodarowicz D (1974). Localisation of a fibroblast growth factor and its effect alone and with hydrocortisone on 3T3 cell growth. *Nature* 249(453): 123–127.

Gospodarowicz D (1990). Fibroblast growth factor. Chemical structure and biologic function. *Clin Orthop Relat Res* (257): 231–148.

Gotz R, Koster R et al. (1994). Neurotrophin-6 is a new member of the nerve growth factor family. *Nature* 372: 266–269.

Gratacos E, Gavalda N et al. (2002). Bone morphogenetic protein-6 is a neurotrophic factor for calbindin-positive striatal neurons. *J Neurosci Res* 70(5): 638–644.

Green-Sadan T, Kinor N et al. (2003). Transplantation of glial cell line-derived neurotrophic factor-expressing cells into the striatum and nucleus accumbens attenuates acquisition of cocaine self-administration in rats. *Eur J Neurosci* 18(7): 2093–2098.

Green-Sadan T, Kuttner Y et al. (2005). Glial cell line-derived neurotrophic factor-conjugated nanoparticles suppress acquisition of cocaine self-administration in rats. *Exp Neurol* 194(1): 97–105.

Grill R, Murai K et al. (1997). Cellular delivery of neurotrophin-3 promotes corticospinal axonal growth and partial functional recovery after spinal cord injury. *J Neurosci* 17: 5560–5572.

Group A-C TS (1996). A double-blind placebo-controlled trial of subcutaneous recombinant human ciliary neurotrophic factor (rhCNTF) in amyotrophic lateral sclerosis. *Neurology* 46: 1244–1249.

Gutierrez Ospina G, Calikoglu AS et al. (1996). In vivo effects of insulin-like growth factor-I on the development of sensory pathways: analysis of the primary somatic sensory cortex (S1) of transgenic mice. *Endocrinology* 137: 5484–5492.

Hagg T, Quon D et al. (1992). Ciliary neurotrophic factor prevents neuronal degeneration and promotes low affinity NGF receptor expression in the adult rat CNS. *Neuron* 8: 145–158.

Hagg T and Varon S (1993). Ciliary neurotrophic factor prevents neuronal degeneration of adult rat substantia nigra dopaminergic neurons in vivo. *Proc Natl Acad Sci U S A* 90: 6315–6319.

Hall D, Dhilla A et al. (2003). Sequence variants of the brain-derived neurotrophic factor (BDNF) gene are strongly associated with obsessive-compulsive disorder. *Am J Hum Genet* 73(2): 370–376.

Hansson H-A, Dahlin LB et al. (1986). Evidence indicating trophic importance of IGF-I in regenerating peripheral nerves. *Acta Physiol Scand* 126: 609–614.

Harrington AW, Leiner B et al. (2004). Secreted proNGF is a pathophysiological death-inducing ligand after adult CNS injury. *Proc Natl Acad Sci U S A* 101(16): 6226–6230.

Harvey BK, Hoffer BJ et al. (2005). Stroke and TGF-beta proteins: glial cell line-derived neurotrophic factor and bone morphogenetic protein. *Pharmacol Ther* 105(2): 113–125.

Hattori A, Katayama M et al. (1999). Bone morphogenetic protein-2 promotes survival and differentiation of striatal GABAergic neurons in the absence of glial cell proliferation. *J Neurochem* 72(6): 2264–2271.

He X-L, Garcia C (2004). Structure and nerve growth factor complexed with the shared neurotrophin receptor p75. *Science* 304: 870–875.

Hefti F (1986a). Nerve growth factor promotes survival of septal cholinergic neurons after fimbrial transections. *J Neurosci* 6(8): 2155–2162.

Hefti F (1986b). Nerve growth factor promotes survival of septal cholinergic neurons after injury. In A Fisher, Hanin I and Lachman C, eds. *Alzheimer's and Parkinson's Disease. Strategies for Research and Development.* Plenum Press, New York. pp. 615–625.

Hefti F (1997). Pharmacology of Neurotrophic Factors. *Annu Rev Pharmacol Toxicol* 37: 239–267.

Heldt SA, Stanek L et al. (2007). Hippocampus-specific deletion of BDNF in adult mice impairs spatial memory and extinction of aversive memories. *Mol Psychiatry* 12(7): 656–670.

Helgren ME, Cliffer KD et al. (1997). Neurotrophin-3 administration attenuates deficits of pyridoxine-induced large-fiber sensory neuropathy. *J Neurosci* 17: 372–382.

Hempstead BL, Martin-Zanca D et al. (1991). High affinity NGF binding requires coexpression of the trk proto-oncogene and the low-affinity NGF receptor. *Nature* 350: 678–682.

Henderson JT, Seniuk NA et al. (1994). Localization of CNTF immunoreactivity to neurons and astroglia in the CNS. *Brain Res Mol Brain Res* 22: 151–165.

Herrmann JL, Menter DG et al. (1993). Mediation of NGF-stimulated extracellular matrix invasion by the human melanoma low-affinity p75 neurotrophin receptor: melanoma p75 functions independently of trkA. *Mol Biol Cell* 4: 1205–1216.

Herzog CD, Dass B et al. (2007). Striatal delivery of CERE-120, an AAV2 vector encoding human neurturin, enhances activity of the dopaminergic nigrostriatal system in aged monkeys. *Mov Disord.*

Hofer MM and YA Barde (1988). Brain-derived neurotrophic factor prevents neuronal death in vivo. *Nature* 331: 261–262.

Hoffer BJ, Hoffman A et al. (1994). Glial cell line-derived neurotrophic factor reverses toxin-induced injury to midbrain dopaminergic neurons in vivo. *Neurosci Lett* 182(1): 107–111.

Hohn A, Leibrock J et al. (1990). Identification and characterization of a novel member of the nerve growth factor/brain derived neurotrophic factor family. *Nature* 344: 339–341.

Holtzman DM, Li Y et al. (1993). Nerve growth factor reverses neuronal atrophy in a Down syndrome model of age-related neurodegeneration. *Neurology* 43: 2668–2673.

Hu L, Cote SL et al. (1997). Differential modulation of the cholinergic phenotype of the nucleus basalis magnocellularis neurons by applying NGF at the cell body or cortical terminal fields. *Exp Neurol* 143: 162–171.

Hudson J, Granholm AC et al. (1995). Glial cell line-derived neurotrophic factor augments midbrain dopaminergic circuits in vivo. *Brain Res Bull* 36(5): 425–432.

Hyman C, Hofer M et al. (1994). Brain derived neurotrophic factor is a neurotrophic factor for dopaminergic neurons of the substantia nigra. *Nature* 350: 230–232.

Hyman SE, Malenka RC et al. (2006). Neural mechanisms of addiction: the role of reward-related learning and memory. *Annu Rev Neurosci* 29: 565–598.

Ibanez CF (2002). Jekyll-Hyde neurotrophins: the story of proNGF. *Trends Neurosci* 25(6): 284–286.

Ibanez CF, Ilag LL et al. (1993). An extended surface of binding to Trk tyrosine kinase receptors in NGF and BDNF allows the engineering of a multifunctional pan-neurotrophin. *Embo J* 12(6): 2281–2293.

Ikeda K, Klinkosz B et al. (1995). Effects of brain-derived neurotrophic factor on motor dysfunction in wobbler mouse motor neuron disease. *Ann Neurol* 37: 505–511.

Ilag LL, Curtis R et al. (1995). Pan-neurotrophin 1: a genetically engineered neurotrophic factor displaying multiple specificities in peripheral neurons in vitro and in vivo. *Proc Natl Acad Sci U S A* 92(2): 607–611.

Ip NY and Yancopoulos GD (1992). Ciliary neurotrophic factor and its receptor complex. *Prog Growth Factor Res* 4: 139–155.

Ip NY, Ibanez CF et al. (1992a). Mammalian neurotrophin-4: Structure, chromosonal localization, tissue distribution and receptor specificity. *Proc Natl Acad Sci U S A* 89: 3060–3064.

Ip NY, Nye SH et al. (1992b). CNTF and LIF act on neuronal cells via shared signalling pathways that involve IL-6 signal transducing receptor component gp130. *Cell* 69: 1121–1132.

Ip NY, Li Y et al. (1993). Cultured hippocampal neurons show response to BDNF, NT-3 and NT-4, but not NGF. *J Neurosci* 13: 3394–3405.

Ishii DN (1989). Relationship of insulin-like growth factor II gene expression in muscle to synaptogenesis. *Proc Natl Acad Sci U S A* 86: 2898–2902.

Ishii DN (1993). Neurobiology of insulin and insulin-like growth factors. In SE Loughlin and JH Fallon eds. *Neurotrophic Factors*. Academic Press, San Diego. pp. 415–442.

Itami C, Kimura F et al. (2007). Brain-derived neurotrophic factor regulates the maturation of layer 4 fast-spiking cells after the second postnatal week in the developing barrel cortex. *J Neurosci* 27(9): 2241–2252.

Itoh N and DM Ornitz (2004). Evolution of the Fgf and Fgfr gene families. *Trends Genet* 20(11): 563–569.

Jain S, Naughton CK et al. (2004). Mice expressing a dominant-negative Ret mutation phenocopy human Hirschsprung disease and delineate a direct role of Ret in spermatogenesis. *Development* 131(21): 5503–5513.

Jain S, Golden JP et al. (2006). RET is dispensable for maintenance of midbrain dopaminergic neurons in adult mice. *J Neurosci* 26(43): 11230–11238.

Jeanneteau F and Chao MV (2006). Promoting neurotrophic effects by GPCR ligands. *Novartis Found Symp* 276: 181–189; [discussion 189–192, 233–237, 275–281].

Jin X, Hu H et al. (2003). Brain-derived neurotrophic factor mediates activity-dependent dendritic growth in nonpyramidal neocortical interneurons in developing organotypic cultures. *J Neurosci* 23(13): 5662–5673.

Jin K, LaFevre-Bernt M et al. (2005). FGF-2 promotes neurogenesis and neuroprotection and prolongs survival in a transgenic mouse model of Huntington's disease. *Proc Natl Acad Sci U S A* 102(50): 18189–18194.

Johnson DG, Gorden P et al. (1971). A sensitive radioimmunoassay for p75 nerve growth factor antigen in serum and tissues. *J Neurochem* 18: 2355–2362.

Johnson EM Jr, Gorin PD et al. (1980). Dorsal root ganglion neurons are destroyed by exposure in utero to maternal antibody to nerve growth factor. *Science* 210(4472): 916–918.

Jones KR, Farinas I et al. (1994). Targeted disruption of the BDNF gene perturbs brain and sensory neuron development but not motor neuron development. *Cell* 76: 989–999.

Jonhagen ME, Nordberg A, Amberla K et al. (1998). Intracerebroventricular infusion of nerve growth factor in three patients with Alzheimer's disease. *Dement Geriatr Cogn Disord* 9: 246–257.

Kang HJ and Schuman EM (1995). Long lasting neurotrophin-induced enhancement of synaptic transmission in the adult hippocampus. *Science* 267: 1658–1662.

Kanje M, Skottner A et al. (1989). Insulin-like growth factor I (IGF-I) stimulates regeneration of the rat sciatic nerve. *Brain Res* 486: 396–398.

Kaplan DR, Hempstead BL et al. (1991). The trk proto-oncogene product: a signal transducing receptor for nerve growth factor. *Science* 252: 554–560.

Kaplan DR and Miller FD (2000). Neurotrophin signal transduction in the nervous system. *Curr Opin Neurobiol* 10: 381–391.

Kashiba H, Uchida Y et al. (2003). Distribution and colocalization of NGF and GDNF family ligand receptor mRNAs in dorsal root and nodose ganglion neurons of adult rats. *Brain Res Mol Brain Res* 110(1): 52–62.

Kaspar BK, Llado J et al. (2003). Retrograde viral delivery of IGF-1 prolongs survival in a mouse ALS model. *Science* 301(5634): 839–842.

Katz A and Meiri N (2006). Brain-derived neurotrophic factor is critically involved in thermal-experience-dependent developmental plasticity. *J Neurosci* 26(15): 3899–3907.

Kearns CM and Gash DM (1995). GDNF protects nigral dopamine neurons against 6-hydroxydopamine in vivo. *Brain Res* 672(1–2): 104–111.

Kin HG, Wang T et al. (1994). Neurotrophin 3 potentiates neuronal activity and inhibits gamma-aminobutyratergic synaptic transmission in cortical neurons. *Proc Natl Acad Sci U S A* 91: 12341–12345.

Klein R, Smeyne RJ et al. (1993). Targeted disruption of the trkB neurotrophin receptor gene results in nervous system lesions and neonatal death. *Cell* 75: 113–122.

Klein R, Silos-Santiago I et al. (1994). Disruption of the neurotrophin-3 receptor gene trkC eliminates Ia muscle afferents and results in abnormal movements. *Nature* 368: 249–251.

Knipper M, Leung LS et al. (1994). Short-term modulation of glutamatergic synapses in adult rat hippocampus by NGF. *Neuroreport* 5: 2433–2436.

Knusel B, Michel PP et al. (1990). Selective and nonselective stimulation of central cholinergic and dopaminergic development in vitro by nerve growth factor, basic fibroblast growth factor, epidermal growth factor, insulin and the insulin-like growth factors I and II. *J Neurosci* 10: 558–570.

Kobayashi NR, Fan D-P et al. (1997). BDNF and NT-4/5 prevent atrophy of rat subrospinal neurons after cervical axotomy, stimulate GAP-43 and Alpha-1-tubulin mRNA expression, and promote axonal regeneration. *J Neurosci* 17: 9583–9595.

Kordower JH (2003). In vivo gene delivery of glial cell line – derived neurotrophic factor for Parkinson's disease. *Ann Neurol* 53 Suppl 3: S120–S132; [discussion S132–S134].

Kordower JH, Charles V et al. (1994). Intravenous administration of a transferrin receptor antibody-nerve growth factor conjugate prevents the degeneration of cholinergic striatal neurons in a model of Huntington's disease. *Proc Nat Acad Sci U S A* 91: 9077–9080.

Kordower JH, Chen EY et al. (1996). Intrastriatal implants of polymer encapsulated cells genetically modified to secrete human nerve growth factor: trophic effects upon cholinergic and noncholinergic striatal neurons. *Neuroscience*. 72: 63–77.

Kordower JH, Emborg ME et al. (2000). Neurodegeneration prevented by lentiviral vector delivery of GDNF in primate models of Parkinson's disease. *Science* 290(5492): 767–773.

Kordower JH, Herzog CD et al. (2006). Delivery of neurturin by AAV2 (CERE-120)-mediated gene transfer provides structural and functional neuroprotection and neurorestoration in MPTP-treated monkeys. *Ann Neurol* 60(6): 706–715.

Korsching S, Auburger G et al. (1985). Levels of nerve growth factor and its mRNA in the central nervous system of the rat correlate with cholinergic innervation. *EMBO J* 4: 1389–1393.

Korte M, Carroll P et al. (1995). Hippocampal long-term potentiation is impaired in mice lacking brain-derived neurotrophic factor. *Proc Natl Acad Sci U S A* 192: 8856–8860.

Kotzbauer PT, Lampe PA et al. (1996). Neurturin, a relative of glial-cell-line-derived neurotrophic factor. *Nature* 384(6608): 467–470.

Kramer ER, Aron L et al. (2007). Absence of Ret signaling in mice causes progressive and late degeneration of the nigrostriatal system. *PLoS Biol* 5(3): e39.

Krieglstein K, Suter-Crazzolara C et al. (1995). Trophic and protective effects of growth/differentiation factor 5, a member of the transforming growth factor-beta superfamily, on mid-brain dopaminergic neurons. *J Neurosci Res* 42(5): 724–732.

Kromer LF (1987). Nerve growth factor treatment after brain injury prevents neuronal death. *Science* 235: 214–216.

Kucera J, Ernfors P et al. (1995a). Reduction in the number of spinal motor neurons in neurotrophin-3 deficient mice. *Neuroscience* 69: 321–330.

Kucera J, Fan G et al. (1995b). Dependence of developing group Ia afferents on neurotrophin-3. *J Comp Neurol* 363: 307–320.

Kumon Y, Sakaki S et al. (1996). Ciliary neurotrophic factor attenuates spatial cognition impairment, cortical infarction and thalamic degeneration in spontaneously hypertensive rats with focal cerebral ischemia. *Neurosci Lett* 206: 141–144.

Lai EC, Felice KJ et al. (1997). Effect of recombinant human insulin-like growth factor-I on progression of ALS. A placebo-controlled study. The North America ALS/IGF-I Study Group. *Neurology* 49(6): 1621–1630.

Lamballe F, Klein R et al. (1991). Trk C, a new member of the trk family of tyrosine protein kinases, is a receptor for neurotrophin-3. *Cell* 66: 967–979.

Lamballe F, Smeyne RJ et al. (1994). Developmental expression of trkC, the neurotrophin-3 receptor, in the mammalian nervous system. *J Neurosci* 14: 14–28.

Lang AE, Gill S et al. (2006). Randomized controlled trial of intraputamenal glial cell line-derived neurotrophic factor infusion in Parkinson disease. *Ann Neurol* 59(3): 459–466.

Leaver SG, Cui Q et al. (2006). AAV-mediated expression of CNTF promotes long-term survival and regeneration of adult rat retinal ganglion cells. *Gene Ther* 13(18): 1328–1341.

Leßmann V, Gottman K et al. (1994). BDNF and NT-4/5 enhance glutamatergic synaptic transmission in cultured hippocampal neurones. *Neuroreport* 6: 21–25.

Lee R, Kermani P et al. (2001). Regulation of cell survival by secreted proneurotrophins. *Science* 294(5548): 1945–1948.

Leinninger GM and EL Feldman (2005). Insulin-like growth factors in the treatment of neurological disease. *Endocr Dev* 9: 135–159.

Leirock J, Lottspeich F et al. (1989). Molecular cloning and expression of brain derived neurotrophic factor. *Nature* 341: 149–152.

LeRoith D, Roberts CTJ et al. (1993). Insulin-like growth factors in the brain. In S Loughlin and J Fallon eds. *Neurotrophic Factors*. Academic Press, San Diego. pp.: 391–414.

Levi-Montalcini R (1987). The nerve growth factor 35 years later. *Science* 237(4819): 1154–1162.

Levi-Montalcini R and Angeletti PU (1963). Essential role of the nerve growth factor on the survival and maintenance of dissociated sensory and sympathetic embryonic nerve cells in vitro. *Dev Biol* 7: 653–659.

Levi-Montalcini R and Hamburger V (1951). Selective growth stimulating effects of mouse sarcoma on the sensory and sympathetic nervous system of the chick embryo. *J Exp Zool* 116: 321–362.

Levi-Montalcini R and Hamburger Y (1953). A diffusible agent of mouse sarcoma, producing hyperplasia of sympathetic ganglia and hyperneurotization of viscera in the chick embryo. *J Exp Zool* 123: 233–288.

Levi-Montalcini R, Meyer H et al. (1954). In vitro experiments on the effects of mouse sarcoma 180 and 37 on the spinal and sympathetic ganglia of the chick embryo. *Cancer Res* 14: 49–57.

Levison SW, Ducceschi MH et al. (1996). Acute exposure to CNTF in vivo induces multiple components of reactive gliosis. *Exp Neurol* 141: 256–268.

Lewin GR and Barde Y-A (1996). Physiology of the neurotrophins. *Annu Rev Neurosci* 19: 289–317.

Lewin GR and Mendell LM (1993). Nerve growth factor and nociception. *Trends Neurosci* 16: 353–359.

Lewin GR, Ritter AM et al. (1992). On the role of nerve growth factor in the development of myelinated nociceptors. *J Neurosci* 12: 1896–1905.

Liang FQ, Dejneka NS et al. (2001). AAV-mediated delivery of ciliary neurotrophic factor prolongs photoreceptor survival in the rhodopsin knockout mouse. *Mol Ther* 3(2): 241–248.

Lin LF, Mismer D et al. (1989). Purification, cloning and expression of ciliary neurotrophic factor (CNTF). *Science* 246: 1023–1025.

Lin LF, Doherty DH et al. (1993). GDNF: a glial cell line-derived neurotrophic factor for midbrain dopaminergic neurons. *Science* 260(5111): 1130–1132.

Lindsay RM (1996). Therapeutic potential of the neurotrophins and neurotrophin-CNTF combinations in peripheral neuropathies and motor neuron diseases. *Ciba Found Symp* 196: 39–48; discussion.

Lindsay RM and Harmar AJ (1989). Nerve growth factor regulates expression of neuropeptide genes in adult sensory neurons. *Nature* 337: 362–364.

Liu X, Ernfors P et al. (1995a). Sensory but not motor neuron deficits in mice lacking NT4 and BDNF. *Nature* 375: 238–241.
Liu X, Yao D-L et al. (1995b). Insulin-like growth factor I treatment reduces clinical deficits and lesion severity in acute demyelinating experimental autoimmune encephalomyelitis. *Mult Scler* 1: 2–9.
Lohof AM, Ip NY et al. (1993). Potentiation of developing neuromuscular synapses by the neurotrophins NT-3 and BDNF. *Nature* 363: 350–353.
Longo FM and Massa SM (2005). Neurotrophin receptor-based strategies for Alzheimer's disease. *Curr Alzheimer Res* 2(2): 167–169.
Lu B (2003). Pro-region of neurotrophins: role in synaptic modulation. *Neuron* 39: 735–738.
Lu CZ and Xiao BG (2007). Neuroprotection of G-CSF in cerebral ischemia. *Front Biosci* 12: 2869–2875.
Lu P, Yang H et al. (2004). Combinatorial therapy with neurotrophins and cAMP promotes axonal regeneration beyond sites of spinal cord injury. *J Neurosci* 24(28): 6402–6409.
Lu B, Pang PT et al. (2005). The yin and yang of neurotrophin action. *Nat Rev Neurosci* 6(8): 603–614.
Luikart BW, Nef S et al. (2005). TrkB has a cell-autonomous role in the establishment of hippocampal Schaffer collateral synapses. *J Neurosci* 25(15): 3774–3786.
Ma L, Harada T et al. (2002). Neurotrophin-3 is required for appropriate establishment of thalamocortical connections. *Neuron* 36(4): 623–634.
Mahadeo D, Kaplan L et al. (1994). High affinity nerve growth factor binding displays a faster rate of association than p140trk binding. Implications for multisubunit polypedtide receptors. *J Biol Chem* 269: 6884–6891.
Maisonpierre PC, Belluscio L et al. (1990). Neurotrophin-3: a neurotrophic factor related to NGF and BDNF. *Science* 247: 1446–1451.
Marco S, Perez-Navarro E et al. (2002). Striatopallidal neurons are selectively protected by neurturin in an excitotoxic model of Huntington's disease. *J Neurobiol* 50(4): 323–332.
Marks WJ, Metman LV et al. (2006). Trophic factor gene transfer in Parkinson's Disease: preliminary outcomes from the Phase I CERE-120 study. *ANA T1-T114*, American Neurological Association.
Massa SM, Xie Y et al. (2006). Small, nonpeptide p75NTR ligands induce survival signaling and inhibit proNGF-induced death. *J Neurosci* 26(20): 5288–5300.
Masu Y, Wolf E et al. (1993). Disruption of the CNTF gene results in motor neuron degeneration. *Nature* 365: 27–32.
McAllister AK, DC Lo et al. (1995). Neurotrophins regulate dendritic growth in developing visual cortex. *Neuron* 15: 791–803.
McAllister AK, Katz LC et al. (1996). Dendritic regulation of cortical dendritic growth requires activity. *Neuron* 17: 1057–1064.
McAllister AK, Katz LC, Lo DC (1999). Neurotrophins and synaptic plasticity. *Annu Rev Neurosci* 22: 295–318.
McBride JL, During MJ et al. (2003). Structural and functional neuroprotection in a rat model of Huntington's disease by viral gene transfer of GDNF. *Exp Neurol* 181(2): 213–223.
McBride JL, Ramaswamy S et al. (2006). Viral delivery of glial cell line-derived neurotrophic factor improves behavior and protects striatal neurons in a mouse model of Huntington's disease. *Proc Natl Acad Sci U S A* 103(24): 9345–9350.
McMorris FA and Dubois-Dalcq M (1988). Insulin-like growth factor I promotes cell proliferation and oligodendroglial commitment in rat glial progenitor cells developing in vitro. *J Neurosci Res* 21: 199–209.
McMorris FA, Smith TM et al. (1986). Insulin-like growth factor I/somatomedin C: a potent inducer of oligodendrocyte development. *Proc Natl Acad Sci U S A* 83: 822–826.
Meng X, Lindahl M et al. (2000). Regulation of cell fate decision of undifferentiated spermatogonia by GDNF. *Science* 287(5457): 1489–1493.
Messer CJ, Eisch AJ et al. (2000). Role for GDNF in biochemical and behavioral adaptations to drugs of abuse. *Neuron* 26(1): 247–257.

Meyer M, Matsuoka I et al. (1992). Enhanced synthesis of brain-derived neurotrophic factor in the lesioned peripheral nerve: different mechanisms are responsible for the regulation of BDNF and NGF mRNA. *J Cell Biol* 119: 45–54.

Milbrandt J, de Sauvage FJ et al. (1998). Persephin, a novel neurotrophic factor related to GDNF and neurturin. *Neuron* 20(2): 245–253.

Mill JF, Chao MV et al. (1985). Insulin, insulin-like growth factor II, and nerve growth factor effects on tubulin mRNA levels and neurite formation. *Proc Natl Acad Sci U S A* 82: 7126–7130.

Mitsumoto H, Ikeda K et al. (1994a). The effects of ciliary neurotrophic factor (CNTF) on motor neuron dysfunction in Wobbler mouse motor neuron disease. *Ann Neurol* 36: 142–148.

Mitsumoto H, Ikeda K et al. (1994b). Arrest of motor neuron disease in wobbler mice cotreated with CNTF and BDNF. *Science* 265: 1107–1110.

Molliver DC, Wright DE et al. (1997). IB4-binding DRG neurons switch from NGF to GDNF dependence in early postnatal life. *Neuron* 19(4): 849–861.

Moore MW, Klein RD et al. (1996). Renal and neuronal abnormalities in mice lacking GDNF. *Nature* 382(6586): 76–79.

Morrison RS, Sharma A et al. (1986). Basic fibroblast growth factor supports the survival of cerebral cortical neurons in primary culture. *Proc Natl Acad Sci U S A* 83(19): 7537–7541.

Nakata N, Kato H et al. (1993). Protective effects of basic fibroblast growth factor against hippocampal neuronal damage following cerebral ischemia in the gerbil. *Brain Res* 605(2): 354–356.

Nawa H, Carnahan J et al. (1995). BDNF protein measured by a novel enzyme immunoassay in normal brain and after seizure: partial disagreement with mRNA levels. *Eur J Neurosci* 7: 1527–1536.

Nebes RD and Brady CB (1989). Focused and divided attention in Alzheimer's disease. *Cortex* 25: 305–315.

Niikura T, Hashimoto Y et al. (2001). Insulin-like growth factor I (IGF-I) protects cells from apoptosis by Alzheimer's V642I mutant amyloid precursor protein through IGF-I receptor in an IGF-binding protein-sensitive manner. *J Neurosci* 21(6): 1902–1910.

Nilsson AS, Fainziliber M et al. (1998). Neurotrophin-7: a novel member of the neurotrophin family from the zebrafish. *FEBS Lett* 424: 285–290.

Nitta K, Uchida K et al. (1999). Recombinant human erythropoietin stimulates vascular endothelial growth factor release by glomerular endothelial cells. *Eur J Pharmacol* 373(1): 121–124.

Nonner D, Barrett EF et al. (2001). Bone morphogenetic proteins (BMP6 and BMP7) enhance the protective effect of neurotrophins on cultured septal cholinergic neurons during hypoglycemia. *J Neurochem* 77(2): 691–699.

Nosrat CA, Tomac A et al. (1997). Cellular and developmental patterns of expression of Ret and glial cell line-derived neurotrophic factor receptor alpha mRNAs. *Exp Brain Res* 115(3): 410–422.

Nutt JG, Burchiel KJ et al. (2003). Randomized, double-blind trial of glial cell line-derived neurotrophic factor (GDNF) in PD. *Neurology* 60(1): 69–73.

Ochs G, Penn RD et al. (2000). A phase I/II trial of recombinant methionyl human brain derived neurotrophic factor administered by intrathecal infusion to patients with amyotrophic lateral sclerosis. *Amyotroph Lateral Scler Other Motor Neuron Disord* 1(3): 201–206.

Ornitz DM (2000). FGFs, heparan sulfate and FGFRs: complex interactions essential for development. *Bioessays* 22(2): 108–112.

Orozco OE, Walus L et al. (2001). GFRalpha3 is expressed predominantly in nociceptive sensory neurons. *Eur J Neurosci* 13(11): 2177–2182.

Ortega S, Ittmann M et al. (1998). Neuronal defects and delayed wound healing in mice lacking fibroblast growth factor 2. *Proc Natl Acad Sci U S A* 95(10): 5672–5677.

Otten U (1984). Nerve growth factor and the peptidergic sensory neurons. *Trends Pharmacol Sci* 5: 307–310.

Otto D and Unsicker K (1990). Basic FGF reverses chemical and morphological deficits in the nigrostriatal system of MPTP-treated mice. *J Neurosci* 10(6): 1912–1921.

Otten U, Goedert M et al. (1979). Immunization of adult rats against 2.5 S NGF: effects on the peripheral sympathetic nervous system. *Brain Res* 176(1): 79–90.

Pang PT, Teng HK et al. (2004). Cleavage of proBDNF by tPA/plasmin is essential for long-term hippocampal plasticity. *Science* 306(5695): 487–491.

Patterson PH (1994). Leukemia inhibitory factor, a cytokine at the interface between neurobiology and immunology. *Proc Natl Acad Sci U S A* 91: 7833–7835.

Patterson SL, Grover LM et al. (1992). Neurotrophin expression in rat hippocampal slices: a stimulus paradigm inducing LTP in CA1 evokes increases in BDNF and NT-3 mRNAs. *Neuron* 9: 1081–1088.

Pearson RCA, Sofroniew MV et al. (1983). Persistence of cholinergic neurons in the basal nucleus in a brain with senile dementia of the Alzheimer's type demonstrated by immunohistochemical staining for choline acetyltransferase. *Brain Res* 289: 375–379.

Perez-Navarro E, Arenas E et al. (1996). Glial cell line-derived neurotrophic factor protects striatal calbindin-immunoreactive neurons from excitotoxic damage. *Neuroscience* 75(2): 345–352.

Perez-Navarro E, Akerud P et al. (2000). Neurturin protects striatal projection neurons but not interneurons in a rat model of Huntington's disease. *Neuroscience* 98(1): 89–96.

Petrides PE and Shooter EM (1986). Rapid isolation of the 7S-nerve growth factor complex and its subunits from murine submaxillary glands and saliva. *J Neurochem* 46: 721–725.

Pichel JG, Shen L et al. (1996). Defects in enteric innervation and kidney development in mice lacking GDNF. *Nature* 382(6586): 73–76.

Purves D (1988). *Body and Brain: A Trophic Theory of Neural Connections*. Harvard Univ. Press. Cambridge: MA.

Rabizadeh S, Oh J et al. (1993). Induction of apoptosis by the low-affinity NGF receptor. *Science* 261: 345–348.

Ramaswamy S, McBride JL et al. (2007a). Neurturin gene therapy improves motor function and prevents death of striatal neurons in a 3-nitropropionic acid rat model of Huntington's disease. *Neurobiol Dis* 26(2): 375–384.

Ramaswamy S, Shannon KM et al. (2007b). Huntington's disease: pathological mechanisms and therapeutic strategies. *Cell Transplant* 16(3): 301–312.

Ramer MS, Priestley JV et al. (2000). Functional regeneration of sensory axons into the adult spinal cord. *Nature* 403(6767): 312–316.

Ramirez JJ, Finklestein SP et al. (1999). Basic fibroblast growth factor enhances axonal sprouting after cortical injury in rats. *Neuroreport* 10(6): 1201–1214.

Rao P, Hsu KC et al. (1995). Upregulation of NF-kappa B-dependent gene expression mediated by the p75 tumor necrosis factor receptor. *J Interferon Cytokine Res* 15: 171–177.

Rattiner LM, Davis M, French CT, Ressler KJ (2004). Brain-derived neurotrophic factor and tyrosine kinase receptor B involvement in amygdala-dependent fear conditioning. *J Neurosci* 24: 4796–4806.

Recio-Pinto E, Rechler MM et al. (1986). Effects of insulin, insulin-like growth factor-II, and nerve growth factor on neurite formation and survival in cultured sympathetic and sensory neurons. *J Neurosci* 6: 1211–1219.

Reichardt LF (2006). Neurotrophin-regulated signalling pathways. *Philos Trans R Soc Lond B Biol Sci* 361(1473): 1545–1564.

Reuss B and von Bohlen O und Halbach (2003). Fibroblast growth factors and their receptors in the central nervous system. *Cell Tissue Res* 313(2): 139–157.

Richardson PM (1994). Ciliary neurotrophic factor: a review. *Pharmac Ther* 63: 187–198.

Ritter AM, Lewin GR et al. (1991). Requirement for nerve growth factor in the development of myelinated nociceptors in vivo. *Nature* 350: 500–502.

Roberts CT and LeRoith D (1988). Molecular aspects of the insulin-like growth factors, their binding proteins and receptors. *Balliere's Clin Endocrinol Metab* 2: 1069–1085.

Rohrer H and Sommer I (1983). Simultaneous expression of neuronal and glial properties by chick ciliary ganglion cells during development. *J Neurosci* 3: 1683–1699.

Rosenberg MB, Friedmann T et al. (1988). Grafting genetically modified cells to the damaged brain: restorative effects of NGF expression. *Science* 242: 1575–1578.

Rosenthal A, Goeddel DV et al. (1990). Primary structure and biological activity of a novel human neurotrophic factor. *Neuron* 4: 767–773.

Ruit KG, Osborne PA et al. (1990). Nerve growth factor regulates sympathetic ganglion cell morphology and survival in the adult mouse. *J Neurosci* 10: 2412–2419.

Sadamoto Y, Igase K et al. (1998). Erythropoietin prevents place navigation disability and cortical infarction in rats with permanent occlusion of the middle cerebral artery. *Biochem Biophys Res Commun* 253(1): 26–32.

Sahenk Z, Nagaraja HN et al. (2005). NT-3 promotes nerve regeneration and sensory improvement in CMT1A mouse models and in patients. *Neurology* 65(5): 681–689.

Sanchez MP, Silos-Santiago I et al. (1996). Renal agenesis and the absence of enteric neurons in mice lacking GDNF. *Nature* 382(6586): 70–73.

Sasaki K, Oomura Y et al. (1992). Acidic fibroblast growth factor prevents death of hippocampal CA1 pyramidal cells following ischemia. *Neurochem Int* 21(3): 397–402.

Schabitz WR and Schneider A (2007). New targets for established proteins: exploring G-CSF for the treatment of stroke. *Trends Pharmacol Sci* 28(4): 157–161.

Schecterson LC and Bothwell M (1992). Novel roles for neurotrophins are suggested by BDNF and NT-3 mRNA expression in developing neurons. *Neuron.* 9: 449–463.

Schneider A, Kuhn HG et al. (2005). A role for G-CSF (granulocyte-colony stimulating factor) in the central nervous system. *Cell Cycle* 4(12): 1753–1757.

Schnell L, Schneider R et al. (1994). Neurotrophin-3 enhances sprouting of corticospinal tract during development and after adult spinal cord lesion. *Nature* 367: 170–173.

Schuchardt A, D'Agati V et al. (1994). Defects in the kidney and enteric nervous system of mice lacking the tyrosine kinase receptor Ret *Nature* 367(6461): 380–383.

Schwab ME, Otten U et al. (1979). Nerve growth factor (NGF) in the rat CNS: Absence of specific retrograde axonal transport and tyrosine hydroxylase induction in locus coeruleus and substantia nigra. *Brain Res* 168: 473–483.

Schwartz PM, Borghesani PR et al. (1997). Abnormal cerebellar development and foliation in BDNF -/- mice reveals a role for neurotrophins in CNS patterning. *Neuron* 19: 269–281.

Scott J, Selby M et al. (1983). Isolation and nucleotide sequence of a cDNA encoding the precursor of mouse nerve growth factor. *Nature* 302: 538–540.

Segal RA and Greenberg ME (1996). Intracellular signalling pathways activated by multiple neurotrophic factors. *Ann Rev Neurosci* 45: 463–489.

Seidah NG, Benjannet S et al. (1996a). Cellular processing of the neurotrophin precursors of NT-3 and BDNF by the mammalian proprotein convertases. *FEBS Lett* 379: 247–250.

Seidah NG, Benjannet S et al. (1996b). Cellular processing of the nerve growth factor prescursor by the mammalian pro-protein convertases. *Biochem J* 314: 951–960.

Seiler M and Schwab ME (1984). Specific retrograde transport of nerve growth factor (NGF) from cortex to nucleus basalis in the rat. *Brain Res* 300: 33–39.

Sendtner M, Kreutzberg GW et al. (1990). Ciliary neurotrophic factor prevents the degeneration of motor neurons after axotomy. *Nature* 345: 440.

Sendtner M, Holtmann B et al. (1992a). Brain-derived neurotrophic factor prevents the death of motoneurons in newborn rats after nerve section. *Nature* 360: 757–759.

Sendtner M, Schmalbruch H et al. (1992b). Ciliary neurotrophic factor prevents degeneration of motor neurons in mouse mutant neuronopathy. *Nature* 358: 502–504.

Sendtner M, Carroll P et al. (1994). Ciliary neurotrophic factor. *J Neurobiol* 25: 1436–1453.

Sherer TB, Fiske BK et al. (2006). Crossroads in GDNF therapy for Parkinson's disease. *Mov Disord* 21(2): 136–141.

Sievers J, Hausmann B et al. (1987). Fibroblast growth factors promote the survival of adult rat retinal ganglion cells after transection of the optic nerve. *Neurosci Lett* 76(2): 157–162.

Sieving PA, Caruso RC et al. (2006). Ciliary neurotrophic factor (CNTF) for human retinal degeneration: phase I trial of CNTF delivered by encapsulated cell intraocular implants. *Proc Natl Acad Sci U S A* 103(10): 3896–3901.

Silos-Santiago I, Molliver DC et al. (1995). Non-TrkA-expressing small DRG neurons are lost in TrkA deficient mice. *J Neurosci* 15: 5929–5942.
Slevin JT, Gerhardt GA et al. (2005). Improvement of bilateral motor functions in patients with Parkinson disease through the unilateral intraputaminal infusion of glial cell line-derived neurotrophic factor. *J Neurosurg* 102(2): 216–222.
Smeyne RJ, Klein R et al. (1994). Severe sensory and sympathetic neuropathies in mice carrying a disrupted Trk/NGF receptor gene. *Nature* 368: 246–249.
Smith DE, Roberts J, Gage FH, and Tuszynski MH (1999). Age-associated neuronal atrophy occurs in the primate brain and is reversible by growth factor gene therapy. *Proc Nat Acad Sci U S A* 96: 10893–10898.
Sofroniew MV, Galletly NP et al. (1990). Survival of adult basal forebrain cholinergic neurons after loss of target neurons. *Science* 247: 338–342.
Sofroniew MV, Howe CL et al. (2001). Nerve growth factor signaling, neuroprotection, and neural repair. *Annu Rev Neurosci* 24: 1217–1281.
Stock A, Kuzis K et al. (1992). Localization of acidic fibroblast growth factor in specific subcortical neuronal populations. *J Neurosci* 12(12): 4688–4700.
Strelau J, Sullivan A et al. (2000). Growth/differentiation factor-15/macrophage inhibitory cytokine-1 is a novel trophic factor for midbrain dopaminergic neurons in vivo. *J Neurosci* 20(23): 8597–8603.
Subramaniam S, Shahani N et al. (2005). Insulin-like growth factor 1 inhibits extracellular signal-regulated kinase to promote neuronal survival via the phosphatidylinositol 3-kinase/protein kinase A/c-Raf pathway. *J Neurosci* 25(11): 2838–2852.
Takahashi M (2001). The GDNF/RET signaling pathway and human diseases. *Cytokine Growth Factor Rev* 12(4): 361–373.
Tansey MG, Baloh RH et al. (2000). GFRalpha-mediated localization of RET to lipid rafts is required for effective downstream signaling, differentiation, and neuronal survival. *Neuron* 25(3): 611–623.
Taupin JL, Pitard V et al. (1998). Leukemia inhibitory factor: part of a large ingathering family. *Int Rev Immunol* 16: 397–426.
Thatcher GR, Bennett BM et al. (2006). NO chimeras as therapeutic agents in Alzheimer's disease. *Curr Alzheimer Res* 3(3): 237–245.
Thoenen H (1995). Neurotrophins and neuronal plasticity. *Science* 270: 593–598.
Thoenen H and Barde YA (1980). Physiology of nerve growth factor. *Physiol Rev* 60: 1284–1335.
Timmer M, Cesnulevicius K et al. (2007). Fibroblast growth factor (FGF)-2 and FGF receptor 3 are required for the development of the substantia nigra, and FGF-2 plays a crucial role for the rescue of dopaminergic neurons after 6-hydroxydopamine lesion. *J Neurosci* 27(3): 459–471.
Tomac A, Lindqvist E et al. (1995). Protection and repair of the nigrostriatal dopaminergic system by GDNF in vivo. *Nature* 373(6512): 335–339.
Tonges L, Schlachetzki JC et al. (2007). Hematopoietic cytokines–on the verge of conquering neurology. *Curr Mol Med* 7(2): 157–170.
Tongiorgi E, Righi M et al. (1997). Activity-dependent dendritic targeting of BDNF and TrkB mRNAs in hippocampal neurons. *J Neurosci* 17(24): 9492–9505.
Tuszynski M (1999). Neurotrophic Factors. In J Kordower and M Tuszynski. *CNS Regeneration.* Elsevier Inc. Burlington, MA 01803, USA.
Tuszynski MH (2002). Growth-factor gene therapy for neurodegenerative disorders. *Lancet Neurol* 1(1): 51–57.
Tuszynski MH, U H-S et al. (1991). Recombinant human nerve growth factor infusions prevent cholinergic neuronal degeneration in the adult primate brain. *Ann Neurol* 30: 625–636.
Tuszynski MH, Peterson DA et al. (1994). Fibroblasts genetically modified to produce nerve growth factor induce robust neuritic ingrowth after grafting to the spinal cord. *Exp Neurol* 126: 1–14.
Tuszynski MH, Gabriel K et al. (1996). Nerve growth factor delivery by gene transfer induces differential outgrowth of sensory, motor and noradrenergic neurites after adult spinal cord injury. *Exp Neurol* 137: 157–173.

Tuszynski MH, Mafong E et al. (1996). BDNF and NT-4/5 prevent injury-induced motor neuron degeneration in the adult central nervous system. *Neuroscience* 71: 761–771.

Tuszynski MH, Thal L et al. (2005). A phase 1 clinical trial of nerve growth factor gene therapy for Alzheimer disease. *Nat Med* 11: 551–555.

Varon S, Nomura J et al. (1967a). The isolation of the mouse nerve growth factor protein in a high molecular weight form. *Biochemistry* 6: 2202–2209.

Varon S, J. Nomura et al. (1967b). Subunit structure of a high molecular weight form of the nerve growth factor from mouse submaxillary gland. *Proc Natl Acad Sci U S A* 57: 1782–1789.

Villa P, Bigini P et al. (2003). Erythropoietin selectively attenuates cytokine production and inflammation in cerebral ischemia by targeting neuronal apoptosis. *J Exp Med* 198(6): 971–975.

Vivien D and Ali C (2006). Transforming growth factor-beta signalling in brain disorders. *Cytokine Growth Factor Rev* 17(1–2): 121–128.

Wanaka A, Johnson EM Jr et al. (1990). Localization of FGF receptor mRNA in the adult rat central nervous system by in situ hybridization. *Neuron* 5(3): 267–281.

Wang R, Guo W et al. (2003). Glial cell line-derived neurotrophic factor normalizes neurochemical changes in injured dorsal root ganglion neurons and prevents the expression of experimental neuropathic pain. *Neuroscience* 121(3): 815–824.

Wang L, Zhang Z et al. (2004). Treatment of stroke with erythropoietin enhances neurogenesis and angiogenesis and improves neurological function in rats. *Stroke* 35(7): 1732–1737.

Wehrman T, He X et al. (2007). Structural and mechanistic insights into nerve growth factor interactions with the TrkA and p75 receptors. *Neuron* 53(1): 25–38.

Wellmer A, Misra VP et al. (2001). A double-blind placebo-controlled clinical trial of recombinant human brain-derived neurotrophic factor (rhBDNF) in diabetic polyneuropathy. *J Peripher Nerv Syst* 6(4): 204–210.

Widmer HR, Knusel B et al. (1993). BDNF protection of basal forebrain cholinergic neurons after axotomy: complete protection of p75 NGFR-positive cells. *Neuroreport* 4: 363–366.

Williams LR (1991). Hypophagia is induced by intracerebroventricular administration of nerve growth factor. *Exp Neurol* 113: 31–37.

Winkler J, Ramirez GA et al. (1997). Reversible Schwann cell hyperplasia and sprouting of sensory and sympathetic neurites after intraventricular administration of nerve growth factor. *Ann Neurol* 41: 82–93.

Xie Y, Tisi MA et al. (2000). Nerve growth factor (NGF) loop 4 dimeric mimetics activate ERK and AKT and promote NGF-like neurotrophic effects. *J Biol Chem* 275(38): 29868–29874.

Xu XM, Guenard V et al. (1995). A combination of BDNF and NT3 promotes supraspinal axonal regeneration in Schwann cell grafts in adult rat thoracic spinal cord. *Exp Neurol* 134: 261–272.

Yabe T, Samuels I et al. (2002). Bone morphogenetic proteins BMP-6 and BMP-7 have differential effects on survival and neurite outgrowth of cerebellar granule cell neurons. *J Neurosci Res* 68(2): 161–168.

Yan O, Gray K et al. (1991). Retrograde transport of BDNF and NT-3 by spinal motoneurons. *Soc Neurosci Abstr* 17: 1119.

Yan Q, Elliott J et al. (1992). Brain-derived neurotrophic factor rescues spinal motor neurons from axotomy-induced cell death. *Nature* 360: 753–755.

Yan Q, Matheson C et al. (1994). The biological responses of axotomized motoneurons to brain-derived neurotrophic factor. *J Neurosci* 14: 5281–5291.

Yan Y, Yamada K et al. (2007). Enduring vulnerability to reinstatement of methamphetamine-seeking behavior in glial cell line-derived neurotrophic factor mutant mice. *Faseb J* 1994–2004.

Yao DL, Liu X et al. (1995). Insulin-like growth factor I treatment reduces demyelination and upregulates gene expression of myelin-related proteins in experimental autoimmune encephalomyelitis. *Proc Natl Acad Sci USA* 92: 6190–6194.

Yee JK, DJ Jolly et al. (1987). Epitope insertion into the human hypoxanthine phosphoribosyltransferase protein and detection of the mutant protein by an anti-peptide antibody. *Gene* 53: 97–104.

Yin QW, Johnson J et al. (1994). Cell death of spinal motoneurons in the chick embryo following deafferentation: rescue effects of tissue extracts, soluble proteins, and neurotrophic agents. *J Neurosci* 14: 7629–7640.

Zackenfels K, Oppenheim RW et al. (1995). Evidence for an important tole of IGF-I and IGF-II for the early development of chick sympathetic neurons. *Neuron* 14: 731–741.

Zhang J, Li Y et al. (2005). Erythropoietin treatment improves neurological functional recovery in EAE mice. *Brain Res* 1034(1–2): 34–39.

Zheng WH and Quirion R (2006). Insulin-like growth factor-1 (IGF-1) induces the activation/phosphorylation of Akt kinase and cAMP response element-binding protein (CREB) by activating different signaling pathways in PC12 cells. *BMC Neurosci* 7: 51.

Zheng JL, Helbig C et al. (1997). Induction of cell proliferatio nby fibroblast and insulin-like growth factors in pure rat inner ear epithelial cell cultures. *J Neurosci* 17: 216–226.

Zhou P, Porcionatto M et al. (2007). Polarized signaling endosomes coordinate BDNF-induced chemotaxis of cerebellar precursors. *Neuron* 55(1): 53–68.

Zuch CL, David D et al. (2004). Beneficial effects of intraventricularly administered BMP-7 following a striatal 6-hydroxydopamine lesion. *Brain Res* 1010(1–2): 10–16.

Zurn AD, Henry H et al. (2000). Evaluation of an intrathecal immune response in amyotrophic lateral sclerosis patients implanted with encapsulated genetically engineered xenogeneic cells. *Cell Transplant* 9(4): 471–84.

5

STEM CELL THERAPY FOR BRAIN TUMORS

RAHUL JANDIAL*, ALLEN L. HO†, MICHAEL L. LEVY‡, AND EVAN Y. SNYDER§

*Division of Neurological Surgery – University of California, San Diego, California
†University of California, San Diego, California
‡Department of Pediatric Neurosurgery – Children's Hospital San Diego, California
§Center for Neurosciences and Aging – The Burnham Institute for Medical Research, La Jolla, California

ABSTRACT

Though currently available clinical treatments and therapies have clearly extended the prognostic outlook for many patients with brain tumors, many of these advances are short lived, particularly with respect to high-grade gliomas such as glioblastoma multiforme. The missing link to a efficacious treatment of these cancers is a more complete understanding of the basic molecular and cellular origin of brain tumors. However, new discoveries of stem cell and developmental neurobiology have now borne the cancer stem cell hypothesis, drawing off of intriguing similarities between neoplastic and malignant cells within the central nervous system. Investigation of this mechanism of brain tumor propagation is the current frontier of stem cell and cancer biology. Neurosurgery is also watching closely of this promising new area of focus. "Molecular neurosurgery," glioma treatments involving biologics, such as homing neural stem cells, that target the cancer at the level of individual migratory cell, is a quickly developing field. This coming progression of applied cancer stem cell research, coupled with current modalities, promises more comprehensive brain cancer interventions.

Brain tumors, and, in particular, high-grade aggressive gliomas remain a difficult and evasive foe to our current armamentarium of therapeutic weapons. Clearly, steroids, surgery, and adjuvant therapy (chemotherapy and radiation) have prolonged the longevity of patients. Yet the extension of life has been limited, and

Copyright © 2008 Elsevier Inc.
All rights of reproduction in any form reserved.

in the case of glioblastoma multiforme remains nearly 1 year after diagnosis and treatment with all available modalities (DeAngelis, 2001; Laws et al., 2003; Chang et al., 2005). Part of the challenge arises from our lack of basic molecular and cellular mechanisms of brain tumor origin. At the same time, our increasing knowledge of stem and developmental biology offers not only potential to understand the origin of brain tumors, but also the opportunity to exploit that knowledge to target and better treat brain tumors.

Classically, neurogenesis in the adult mammal and humans was thought not to exist. Modern methods and investigation have refuted this claim, and understanding of the germinal zones and their resident neural stem cells (NSCs) has come about. Remarkably, the behavior of NSCs and neoplastic cells of neuroectodermal origin (Aboody et al., 2000), at least in terms of migratory capacity, ability to insinuate themselves into normal tissue, self-renewal potential, and molecular signature, has strong overlap with malignant cells in the central nervous system (CNS). Furthermore, evidence is mounting that the proliferative and metastatic capacity of brain tumors is maintained by the presence within the tumors of a population of 'stem-like' cells (Hemmati et al., 2003; Singh et al., 2003, 2004; Galli et al., 2004; Yuan et al., 2004). It remains unclear whether the brain tumors and their 'cancer stem cells' emerge from the de-differentiation of mutated mature neural cells or emanate *de novo* from otherwise normal organogenic progenitors fated or predisposed to become neoplastic. Evidence supports both as potential etiologies, while not clearly accepting either scenario (Bachoo et al., 2002).

The biology and management of brain tumors as well as other malignancies may be linked to a better understanding of stem cell biology (Clarke, 2004; Berger et al., 2004). This could have dual benefits of understanding both how neoplastic stem cells emerge and how normal stem cells behave in response. The intricate and possibly overlapping potential of stem cells to create and populate the CNS in a physiological manner as well as be the source of malignant disease is the current frontier of stem and cancer biology. This investigation will be fostered by recent generation of more representative in vivo animal models that has helped to accelerate the development of new diagnostic technologies and novel therapeutic agents (Zhu and Parada, 2002; Romer et al., 2004).

Neurosurgery is being transformed by advances in imaging technologies, including high-resolution magnetic resonance imaging (MRI), MR spectroscopy, and positron emission tomography (PET) scans, as well as diffusion and perfusion imaging which permit better localization and characterization of lesions, and their relationship with normal brain architecture (Nelson and Cha, 2003). These advances have led to notable improvements in surgical resection. Yet the horizon for intervention to treat glioma cells that have the intrinsic capability to infiltrate local structures and to migrate over great distances rests on what is now being called "molecular neurosurgery". Treating patients with the biologics along with current modalities will allow tumor targeting at the level of individual migratory cancer cells. Fortunately, NSCs are excellent candidates based on their ability to track cancer cells, capacity to be genetically modified, and, if proven to be the cell of origin, could even be targeted in a pre-morbid state.

NEURAL STEM CELLS

The NSCs are the most primordial and least committed cells of the nervous system. Because immunocytochemically detectable markers that are sufficiently specific and sensitive have not yet been defined, a NSC is still best defined operationally. To be a neural "stem" cell, as opposed to a "progenitor" cell or "precursor" cell, a single clonal cell must have the following functional properties: (1) "multipotency", the ability to yield mature cells in all three fundamental neural lineages throughout the nervous system: neurons of all types; astrocytes of all types; and oligodendrocytes – in regional and developmental stage-appropriate manners; (2) the ability to populate a developing region and/or repopulate an ablated or degenerated region of the CNS with appropriate cell types; (3) "self-renewal", i.e., the ability to produce daughter cells with identical properties.

It has been demonstrated that a single neural cell with stem-like qualities could be isolated from and re-implanted back into the brain, where its progeny could integrate seamlessly, differentiate into integral members of the CNS, respond to prevalent developmental cues to yield appropriate multiple neural cell types (both neuronal and glial), and import foreign genes into the CNS (Campbell et al., 1995; Sidman et al., 1995). Neural cells with stem cell properties have been isolated from the embryonic, neonatal, and adult nervous system, and propagated in vitro by a variety of equally effective and safe means – both epigenetic and genetic. Populations of NSCs are derived from embryonic stem cells via induction of differentiation in vitro into progeny from the three embryonic germ layers. The control of this differentiation into NSCs has been described, yet remains to be completely understood (Reubinoff et al., 2001; Zhang et al., 2003). Different in vitro protocols are employed to obtain embryoid bodies and embryonic stem cell-derived NSCs, providing a non-somatic source for NSCs. While there is a great deal of debate as to the best source for neural progenitors, it is the behavior of the NSC derived directly from the neuroectoderm (Snyder et al., 1992; Palmer et al., 2001) which has established the "gold standard" for what can, and should be achievable by cells with normal stem-like attributes.

Engrafted exogenous NSCs manipulated ex vivo to express a variety of transgenes can integrate locally at the site of implantation, and, if applied to the proper germinal zone, be exploited to disseminate therapeutic genes, or to yield some desired neural cell types globally throughout the CNS (Snyder et al., 1992, 1997b; Rosario et al., 1997; Park et al., 1999; Zlomanczuk et al., 2002). The tremendous migratory capability of NSCs in conjunction with their innate tropism for intracranial pathologies make them ideal therapeutic agents in a variety of neurological diseases (Flax et al., 1998; Snyder and Loring, 2005; Yip et al., 2006).

EXOGENOUS AND ENDOGENOUS NSCS RESPOND TO GLIOMAS

Aboody and colleagues first described the behavior of exogenous NSCs. When NSCs were placed into a rodent host harboring an inoculated glioma, they evinced the ability to migrate toward and distribute within the glioma. These properties were demonstrated not only when the NSCs were implanted near the tumor, but also when implanted in the contralateral hemisphere of the tumor. Further, the migratory capacity existed when NSCs were implanted in the cerebral ventricles. The challenge of treating malignant gliomas lies in targeting the cancerous cells that migrate outside of the tumor bulk, and inevitably lead to progression and recurrence even when the great majority of the tumor bulk is resected or treated. Surprisingly, NSCs could be seen tracking along side individual cancerous cells suggesting this could be a powerful way to address the most vexing element of brain tumor treatment (Aboody et al., 2000). These findings were confirmed by other investigators, and reduction of tumor size was demonstrated by many, when NSCs were genetically modified to deliver chemotherapeutics, interleukins, and thymidine kinase (Benedetti et al., 2000; Ehtesham et al., 2002a,b; Barresi et al., 2003; Li et al., 2005; Uhl et al., 2005).

The existence of endogenous NSCs is well accepted, and more information about the germinal zones in which they reside continues to expand (Curtis et al., 2007). The endogenous NSCs, particularly from the subventricular zone, have been shown to migrate toward gliomas in mice, and evince the same gliotropism as exogenous NSCs (Glass et al., 2005). Clearly this does lead to some of the cellular heterogeneity in the tumor microenvironment (Fomchenko and Holland, 2005). Interestingly, the response of NSCs to gliomas is less as animals age, and may be one factor that leads to the higher incidence of malignant gliomas in elderly patients.

Whether it is an issue of pure numerical advantage of the rapidly dividing tumor cells overwhelming the small numbers of endogenous NSCs or other unidentified factors remains to be determined. Or perhaps some 'rogue' endogenous NSCs themselves, under some circumstances, are the culprits giving rise to the tumor from the onset. These are extremely important clinically relevant questions that might influence what type of new chemotherapeutic interventions should be devised – drugs that boost the response of endogenous NSCs to tumors, or perhaps those that eliminate some of those NSCs (Yip et al., 2006).

MECHANISMS FOR NSC HOMING TO GLIOMAS

Several studies (Snyder et al., 1995, 1997a; Lacorazza et al., 1996; Yandava et al., 1999; Riess et al., 2002) have demonstrated that NSCs home into regions of intracranial pathology, opening up an entire new field of NSC-based treatment options for intracerebral lesions. Numerous factors responsible for luring NSCs to intracranial pathology have been suggested previously (Yip et al., 2003, 2005). A variety of factors released and expressed by glioma cells, tumor stroma (comprised of adjacent reactive astrocytes, microglia, and oligendendrocytes), tumor-derived endothelium, and the damage surrounding normal brain tissue have been shown to induce NSC gliomatropism. While factors such as stem cell factor (SCF) and monocyte chemoattractant protein-1 (MCP1) have been acknowledged as agents of NSC gliomatropism (Erlandsson et al., 2004; Sun et al., 2004; Widera et al., 2004; Serfozo et al., 2006), other factors still have to be characterized, and their roles in NSC gliomatropism have yet to be defined (Werbowetski et al., 2004).

Chemokines are an intrinsic part of the normal development and function of many biological systems, especially in the CNS. Chemokine expression may be disturbed in diseased conditions (Gerard et al., 2001), and they are an important point of inquiry into NSC attraction to pathology. The first suggestion that products of inflammation might, in addition to being inimical, also function as a beacon for reparative cells was made by the observation of the haematopoeitic chemokine receptor CXCR4 being expressed in NSCs in tandem with the expression of its cognate ligand SDF-1a within regions of CNS injury and degeneration (particularly by reactive astrocytes and endothelium). In this way, products of inflammation would guide and direct the homing of NSCs to affected regions of the CNS (Imitola et al., 2004). This same binding pair is also crucial to normal CNS development (Lazarini et al., 2003). This implies that the homing behavior of NSCs to pathology within the developed brain may simply be a recapitulation of NSC migration during CNS organogenesis.

The role of CXCR4 and SDF-1a in tumor growth, migration, and angiogenesis reinforces the comparison between NSCs and neoplastic CNS cells. SDF-1a is expressed on tumor-derived endothelium, and CXCR4 is found on a variety of tumor growths, including GBMs, and medullablastomas. This indicates a probable paracrine growth loop, and an explanation for angiocentric growth of intracranial neoplasms on a molecular level (Rempel et al., 2000; Zhou et al., 2002). Rubin and colleagues (2003) have shown that small molecule antagonists of CXCR4 are successful in inhibiting the growth of GBM and medulloblastoma in experimental models (Rubin et al., 2003). Since expression of SDF-1a by the tumor-associated endothelium stimulates the migration of NSCs (Allport et al., 2004; Fears et al., 2004), it would follow that this same inhibition of the SDF-1a/CXCR4 interaction actually prevents the gliomatropic migration of NSCs (Ehtesham et al., 2004). In addition, this interaction also appears to be an integral part of the gliomatropism of circulating adult haematopoeietic progenitor cells (Tabatabai et al., 2005). Finally, because the receptor CXCR4 is also expressed by metastatic breast cancer and ligand SDF-1a is concurrently expressed by brain endothelium, a mechanism for the transmigration of circulating metastatic neoplastic cells through the blood brain barrier – another characteristic shared by stem and neoplastic cells (Aboody et al., 2000; Pluchino et al., 2005) – can be proposed through the CXCR4/SDF-1a interaction (Lee et al., 2004). Indeed, Liang and associates (2004) were able to prevent breast cancer metastasis by inhibiting CXCR4 (Liang et al., 2004).

Recently, vascular endothelial growth factor (VEGF) has also been targeted as a source of NSC gliomatropism (Schmidt et al., 2005). Since the VEGF, with other pro-angiogenic factors, is vital to maintaining the aggressive behavior of GBM (Kaur et al., 2004), this finding is especially significant. Also, epidermal growth factor (EGF) and receptor signaling over-expression, an important component of glioma malignancy, invasion, and migration (Chicoine and Silbergeld, 1997; Feldkamp et al., 1997; Dunn et al., 2000), has been linked to NSC migration (Boockvar et al., 2003). Mediation of NSC migration toward tumor cells could be instituted by EGF concentration gradients. But the possibility still remains that glioma cells will continue to "outrun" NSCs, branding ineffective any endogenous NSC response. Given that the extreme motility of malignant glioma cells is dependent upon specific gene sets (Lefranc et al., 2005; Tatenhorst et al., 2005), delineating the role of these genes and subsequent downstream signaling events is essential for the eventual development of therapeutic agents and suitably armed exogenous NSCs to compensate for and target these pathways to enhance their homing abilities.

EXPLOITING NSCS AS VEHICLES FOR DELIVERING TOXIC PAYLOADS

Previously, most mechanisms of delivering a toxic payload were either through direct injection or with the introduction of adeno-associated virus (AAV) engineered to express the anti-angiogenic protein angiostatin directly into the tumor (Ma et al., 2002). The effectiveness of viral-mediated gene delivery to aggressively virulent brain tumors was limited by the "halo" effect, i.e., only tumor cells within a limited radius of injected viral vector were eradicated; tumor cells situated beyond that radius could escape to set up new satellite tumors (Yip et al., 2006). Infiltrating migratory cancer cells remained elusive, and difficult to target. While studies continue to find increasing roles for NSCs in the treatment of neurological pathology, substantial work has been done by various groups exploiting the unique tropism of NSCs for gliomas in the context of genetic modification to deliver a variety of anti-glioma gene products. NSCs have the capacity to convey large amounts of genetic information (beyond the limits imposed by the relatively small genome of the viral vectors) that make NSC a much more powerful and adaptive anti-tumor agent.

Aboody and colleagues employed a model where nude mice were inoculated with glioma cells, and subsequently transplanted with human and murine NSCs at various locations (intratumoral, contralateral hemisphere, intraventricular, and tail vein) with clear demonstration of NSCs migrating to, and distributing within, the tumor (Aboody et al., 2000) (Figure 5.1). Subsequently, NSCs were transfected with a gene for cytosine deaminase (CD), a prodrug-converting enzyme that converts 5-FC to 5-FU, and transplanted at some distance from the tumor. This technique led to approximately 80% reduction in tumor burden. The CDA pro-drug system, in particular, engenders an extremely large "bystander effect", i.e., the killing of even a small number of tumor cells sends "ripples" of oncolytic factors emanating from that epicenter of cell death to kill an even broader region of tumor cells. This could be mediated by expression of connexin-43 in untransduced glioma cells (Uhl et al., 2005). Therefore, even if the CDA transgene were to be downregulated in some NSCs, the oncolytic action of the population of NSCs would remain effective. In addition, CDA would serve as a suicide gene within the NSC – hence a built-in safety mechanism (Li et al., 2005). Other pro-drug systems exist and have been tried successfully in NSCs (e.g., herpes simplex virus (HSV) thymidine kinase (TK)) (Herrlinger et al., 2000). Herpes simplex virus1-thymidine kinase (HSVtk) has also been shown to be effective at treating gliomas when expressed by NSCs carrying the HSVtk gene. Murine NSCs expressing HSCtk were transplanted intratumoral (rat glioma cell line C6) into mice and rats both during tumor inoculation and after tumor inoculation, and then treated with ganciclovir. Interestingly, there was a 70% increase in survival with NSC–HSVtk transplantation after tumor inoculation, and 100% survival with co-transplantation of tumor and NSC – HSVtk (Herrlinger et al., 2000; Li et al., 2005).

Cytokines have been used successfully to treat some human malignancies via tumor toxicity or growth arrest (Eklund and Kuzel, 2004; Smyth et al., 2004). Similarly, immunotherapy for brain tumors via the direct instillation of cytokines or via the use of genetically modified viral vectors has demonstrated some efficacy (Jean et al., 1998; Ehtesham et al., 2002c; Rhines et al., 2003). Ehtesham et al. transfected murine NSCs with the gene for pro-inflammatory cytokine interleukin-12 (IL-12), demonstrated stable expression, and found increased survival with intratumoral transplantation in tumor-bearing syngeneic mice (Ehtesham et al., 2002a). The authors also observed infiltration of the tumor by T lymphocytes in response to regional expression of IL-12; a finding corroborated by others (Yang et al., 2004). NSCs overcome the hurdle of achieving a high enough local concentration of therapeutic compounds by their specific homing ability.

Malignant gliomas are known to produce pro-angiogenic factors VEGF and basic fibroblast growth factor (bFGF), and the histopathological grading of malignancy includes assessment of vascular proliferation in the tumor sample. Accordingly, anti-angiogenic compounds are candidates for interfering with tumor growth as one mechanism to target gliomas, and genetically modified NSCs could express the toxic payload of choice. The delivery of such molecules preferentially to the vascular endothelium within the tumors is further supported by the observation that NSCs have a predilection for, and transmigrate

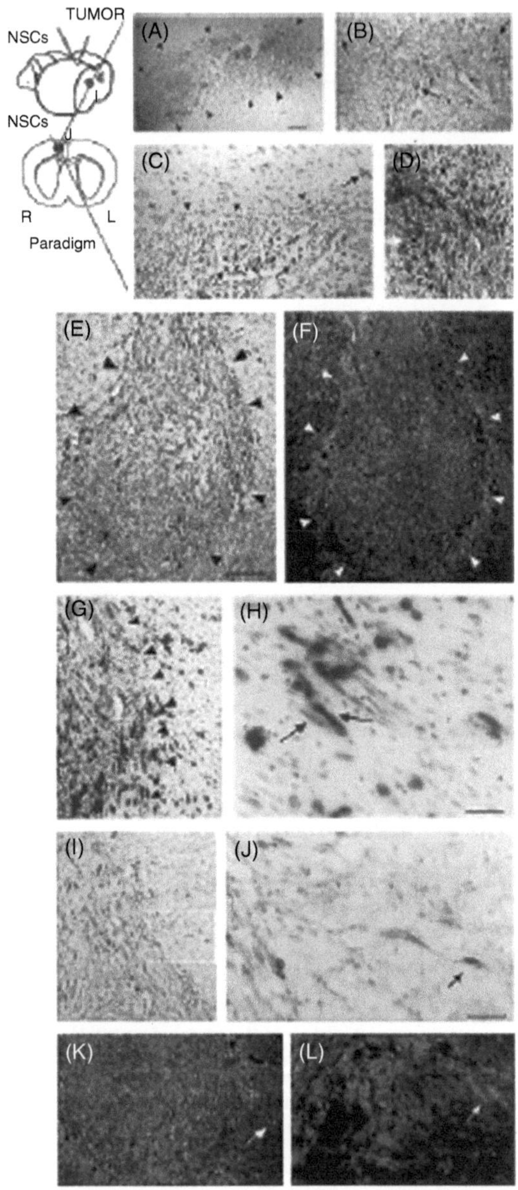

FIGURE 5.1 NCSs migrate extensively throughout a brain tumor mass in vivo and "trial" advancing tumor cells. Paradigm 1 is illustrated schematically. Section of brain under low (*A*) and high (*B*) power from an adult rat killed 48 h after NSC injection into an established D74 glioma, processed with X-Gal to detect blue-staining β-gal-producing NSCs and counterstained with neutral red to show dark red tumor cells; arrowheads demarcate approximate edges of the tumor mass where it interfaces with normal tisue. Donor X-Gal$^+$ blue NSCs (arrows) can be seen extensively distributed throughout the mass, interspersed among the red tumor cells. (C) Tumor, 10 days after NSC injection, illustrating that, although NSCs (arrows) have infiltrated the mass, they largely stop at the junction between tumor and normal tissue (arrowheads) except where a tumor cell

FIGURE 5.1 (*Continued*) (dark red, elongated) is entering normal tissue; then NSCs appear to "follow" the invading tumor cell into surrounding tissue (upper right arrow). This phenomenon becomes more dramatic when examining NSC behavior in a more virulent and aggressively invasive tumor, the CNS-1 glioblastoma in the adult nude mouse, pictured in *D*. This section illustrates extensive migration and distribution of blue NSCs (arrows) throughout the infiltrating glioblastoma up to and along the infiltrating tumor edge (red arrowheads) and into surrounding tissue in juxtaposition to many dark red$^+$ tumor cells invading normal tissue. The "tracking" of individual glioblastoma cells is examined in greater detail in *E–L*, where CNS-1 cells have been labeled ex vivo by transduction with GFP cDNA. (*E* and *F*) Sister sections showing a low power view of transgene-expressing NSCs distributed throughout the main tumor mass to the tumor edge (outlined by arrowheads). Sections were either costained with X-Gal (NSCs, blue) and neutral red (tumor cells, dark red and elongated) (*E*) or processed for double immunofluorescence using an anti-β-gal antibody (NSCs, red) and an FITC-conjugated anti-GFP antibody (glioblastoma cells, green) (*F*). Low (*G*) and high (*H*) power views of tumor edge (arrowheads) with blue NSCs (blue arrow) in immediate proximity to and intermixed with an invading tumor "island" (dard red spindle-shaped cells) (red arrow). (*I* and *J*) Low and high power views, respectively (boxed area in *I* is magnified in *J*), of a blue NSC in direct juxtaposition to a single migrating neutral red$^+$, spindle-shaped tumor cell (arrow), the NSC "riding" the glioma cell in "piggy-back" fashion. (*K* and *L*) low and high power views, respectively, under fluorescence microscopy, of single migrating GFP$^+$ tumor cells (green) in juxtaposition to β-gal$^+$ NSCs(red). Region indicated by white arrow in *K* and magnified in *L* illustrates NSCs apposed to tumor cells migrating away from the main tumor bed. (Scale bars: *A*, 40 μm, 30 μm in *B*; *C*, 30 μm, 25 μm in *D*; *E*, 90 μm, 100 μm in *F*; *H*, 15 μm, 60 μm in *G*; *J*, 30 μm, 60 μm in *I*, 70 μm in *K*, 35 μm in *L*.) (Reproduced from Aboody et al., 2000)

through, tumor endothelium via association of their CXCR4 receptors with endothelial-expressed SDF-1a, as well as via a4-integrin (Allport et al., 2004).

Most studies investigating the potential of NSCs in tumor therapy have employed murine cells. The ability to home and target gliomas has also been demonstrated with the use of human NSCs. Kim et al. employed PEX, a naturally occurring fragment of human metalloproteinase-2 that acts as an inhibitor of glioma and endothelial cell proliferation, migration, and angiogenesis, to target human glioma cells (Gli26). Using the HB1.F3 cell line, immortalized human NSCs were transduced by a vector with PEX, and subsequently transplanted intratumoral into the mice. The mice were evaluated for tumor size with MRI and histological examination revealing a profound reduction in tumor volume and angiogenesis (Kim et al., 2005).

With the demonstration of NSCs unique capability to migrate toward and distribute within brain tumors, the scenario of their potential use as cellular delivery vehicles arose. Further, they are capable of crossing the blood–brain barrier and finding even small micro-deposits of tumor cells, making them even more suitable to the unique challenges to effect getting therapeutics into the CNS (Table 5.1). As more is learned about brain repair and regenerative medicine, stem cells remain applicable in other clinical scenarios with the treatment of malignant brain disease.

TABLE 5.1 Strategies in neural stem cell (NSC)-mediated brain tumour therapeutics

	Immuno-modulatory	Growth regulatory and tumouricidal	Viral therapy	Pro-drug converting enzyme/suicide gene therapy	Biophysical agents
	Initiation of enhanced anti-tumour immune response via local delivery and expression of high concentrations of cytokines	Induction of tumour growth arrest via interaction with NSCs or binding to expressed growth regulatory factors	Introduction of virus into vicinity of tumour cells causing cytolysis	Enzymatic conversion of pro-drug into toxins. Cytotoxicity is amplified by the 'by-stander' effect	Delivery of agents to vicinity of tumour, which requires subsequent secondary activation
Published studies[b]	ILA, 12 IL2[a]	TRAIL IFN-β	HSV Adenovirus AAV	Cytosine deaminase HSV-TK	
Future?	GM-CSF	PF4 (platelet factor 4) TNF(73)	Reovirus type 3 VSV	Deoxycytidine kinase	Nanoshells photodynamic therapy

IL, interleukin; GM-CSF, Granulocyte-macrophage colony stimulating factor; AAV, adeno-associated virus; HSV-TK, Herpes simplex virus thymidine kinase; VSV, Vesicular stomatitis virus; TNF, tumour necrosis factor; IFN, interferon.

[a] Transduced MSCs.

[b] *Source*: Aboody et al., 2000; Benedetti et al., 2000; Ehtesham et al., 2002abc; Barresi et al., 2003; Chiocca et al., 2003, 2004; Gomez-Manzano et al., 2004; Kew and Levine, 2003; Ma et al., 2002; Kurihara et al., 2000; Fueyo et al., 2003; Stojdl et al., 2003; Herrlinger et al., 2000; Manome et al., 1996; Lynch et al., 1999; Arnhold et al., 2003; Jean et al., 1998; Rhines et al., 2003; Eklund and Kuzel, 2004; Smyth et al., 2004; Yang et al., 2004; Walczak et al., 1999; Kim et al., 2004; Lewin et al., 2004; Zhang et al., 2004; Anderson et al., 2005; Daldrup-Link et al., 2005; Jaiswal and Simon et al., 2004; Gao et al., 2004; Stroh et al., 2005; Loo et al., 2005; Hirsch et al., 2003.

HORIZON

More must be learned about developmental neurobiology, stem cell biology, molecular imaging, and gene regulation for the successful clinical translation of NSCs. In the setting of malignant brain tumors, the NSCs would not need to differentiate and integrate into the neural circuitry, a much more challenging proposition. Instead, if they were proven to be safe, and traceable, they would simply need to do what has been already demonstrated as their unique ability – migrate to, distribute within, and track individual tumor cells. Safety must be rigorously established. The ability to image and track cellular transplants with luciferase, superparamagnetic iron oxide particles, and quantum dots is evolving rapidly (Zhang et al., 2004; Daldrup-Link et al., 2005; Stroh et al., 2005). Well-characterized NSCs that can be genetically modified and banked would be another necessary component. Some have shown mesenchymal cells to be effective (Nakamizo et al., 2005), but we favor the view that stem cells derived from the lineage-of-origin of the cancer are best suited for "hunting it down", and eradicating it. In the future as we learn more about sources and methods for stem cell biology, the potential for autologous stem cell transplants may occur.

Through NSC-mediated treatment of malignant gliomas, practical protocols for preparing and delivering stem cells will be devised, safety will be proven, and the behavior of stem cells in a human brain will be observed, all of which will help the broader field of stem cell biology and regenerative medicine.

REFERENCES

Aboody KS, Brown A, Rainov NG, et al. (2000). Neural stem cells display extensive tropism for pathology in adult brain: evidence from intracranial gliomas. *Proc Natl Acad Sci U S A* 97(23): 12846–12851.

Allport JR, Shinde Patil VR, and Weissleder R (2004). Murine neuronal progenitor cells are preferentially recruited to tumor vasculature via alpha4-integrin and SDF-1alpha-dependent mechanisms. *Cancer Biol Ther* 3(9): 838–844.

Anderson SA, Glod J, Arbab AS, et al. (2005). Noninvasive MR imaging of magnetically labeled stem cells to directly identify neovasculature in a glioma model. *Blood* 105(1): 420–425.

Arnhold S, Hilgers M, Lenartz D, et al. (2003). Neural precursor cells as carriers for a gene therapeutical approach in tumor therapy. *Cell Transplant* 12(8): 827–837.

Bachoo RM, Maher EA, Ligon KL, et al. (2002). Epidermal growth factor receptor and Ink4a/Arf: governing terminal differentiation and transformation stem cell to astrocyte axis. *Cancer Cell* 1(3): 269–277.

Barresi V, Belluardo N, Sipione S, et al. (2003). Transplantation of prodrug-converting neural tumor therapy. *Cancer Gene Ther* 10(5): 396–402.

Benedetti S, Pirola B, Pollo B, et al. (2000). Gene therapy of experimental brain tumors using neural progenitor. *Nat Med* 6(4): 447–450.

Berger F, Gay E, Pelletier L, Tropel P, and Wion D. (2004). Development of gliomas: potential role of asymmetrical cell division of neural stem cells. *Lancet Oncol* 5(8): 511–514.

Boockvar JA, Kapitonov D, Kapoor G, et al. (2003). Constitutive EGFR signaling confers a motile phenotype to neural stem cells. *Mol Cell Neurosci* 24(4): 1116–1130.

Campbell K, et al. (1995). Regional incorporation and site-specific differentiation of striatal precursors transplanted to the embryonic forebrain ventricle. *Neuron* 15: 1259–1273.

Chang SM, Parney IF, Huang W, et al. (2005). Patterns of care for adults with newly diagnosed malignant glioma. *JAMA* 293(5): 557–564.

Chicoine MR and Silbergeld DL (1997). Mitogens as motogens. *J Neurooncol* 35(3): 249–257.

Chiocca EA, Aghi M, and Fulci G (2003). Viral therapy for glioblastoma. *Cancer J* 9(3): 167–179.

Chiocca EA, Broaddus WC, Gillies GT, et al. (2004). Neurosurgical delivery of chemotherapeutics, targeted toxins, genetic viral therapies in neuro-oncology. *J Neurooncol* 69(1–3): 101–117.

Clarke MF (2004). Neurobiology: at the root of brain cancer. *Nature* 432(7015): 281–282.

Curtis MA, Kam M, Nannmark U, et al. (2007). Human neuroblasts migrate to the olfactory bulb via a lateral ventricular extension. *Science* March 2, 315(5816): 1243–1249. Epub 2007 February 15.

Daldrup-Link HE, Rudelius M, Piontek G, et al. (2005). Migration of iron oxide-labeled human hematopoietic progenitor cells in a mouse model: in vivo monitoring with 1.5-T MR imaging equipment. *Radiology* 234(1): 197–205.
DeAngelis LM (2001). Brain tumors. *New Engl J Med* 344(2): 114–123.
Dunn IF, Heese O, and Black PM (2000). Growth factors in glioma angiogenesis: FGFs, PDGF, EGF, and TGFs. *J Neurooncol* 50(1–2): 121–137.
Ehtesham M, Kabos P, Kabosova A, Neuman T, et al. (2002a). The use of interleukin 12-secreting neural stem cells for the treatment of intracranial glioma. *Cancer Res* 62(20): 5657–5663.
Ehtesham M, Kabos P, Gutierrez MA, et al. (2002b). Induction of glioblastoma apoptosis using neural stem cell-mediated delivery of tumor necrosis factor-related apoptosis-inducing ligand. *Cancer Res* 62(24): 7170–7174.
Ehtesham M, Samoto K, Kabos P, et al. (2002c). Treatment of intracranial glioma with in situ interferon-gamma and necrosis factor-alpha gene transfer. *Cancer Gene Ther* 9(11): 925–934.
Ehtesham M, Yuan X, Kabos P, et al. (2004). Glioma tropic neural stem cells consist of astrocytic precursors and their migratory capacity is mediated by CXCR4. *Neoplasia* 6(3): 287–293.
Eklund JW and Kuzel TM (2004). A review of recent findings involving interleukin-2-based cancer therapy. *Curr Opin Oncol* 16(6): 542–546.
Erlandsson A, Larsson J, and Forsberg-Nilsson K (2004). Stem cell factor is a chemoattractant and a survival factor for CNS cells. *Exp Cell Res* 301(2): 201–210.
Fears CY, Sontheimer HW, Bullard DC, and Gladson CL (2004). Could labeled neuronal progenitor cells be used to target glioma tumor endothelium? *Cancer Biol Ther* 3(9): 845–846.
Feldkamp MM, Lau N, and Guha A (1997). Signal transduction pathways and their relevance in human astrocytomas. *J Neurooncol* 35(3): 223–248.
Flax JD, Aurora S, Yang C, et al. (1998). Engraftable human neural stem cells respond to developmental cues, replace neurons, and express foreign genes. *Nat Biotechnol* 16(11): 1033–1039.
Fomchenko EI and Holland EC (2005). Stem cells and brain cancer. *Exp Cell Res* 306(2): 323–329.
Fueyo J, Alemany R, Gomez-Manzano C, et al. (2003). Preclinical characterization of the antiglioma activity of a tropism-enhanced adenovirus targeted to the retinoblastoma pathway. *J Natl Cancer Inst* 95(9): 652–660.
Galli R, Binda E, Orfanelli U, et al. (2004). Isolation and characterization of tumorigenic, stem-like neural precursors from human glioblastoma. *Cancer Res* 64(19): 7011–7021.
Gao X, Cui Y, Levenson RM, Chung LW, et al. (2004). In vivo cancer targeting and imaging with semiconductor quantum dots. *Nat Biotechnol* 22(8): 969–976.
Gerard C and Rollins BJ (2001). Chemokines and disease. *Nat Immunol* 2(2): 108–115.
Glass R, Synowitz M, Kronenberg G, et al. (2005). Glioblastomainduced attraction of endogenous neural precursor cells is associated with improved survival. *J Neurosci* 25(10): 2637–2646.
Gomez-Manzano C, Yung WK, Alemany R, and Fueyo J (2004). Genetically modified adenoviruses against gliomas: from bench to bedside. *Neurology* 63(3): 418–426.
Hemmati HD, Nakano I, Lazareff JA, et al. (2003). Cancerous stem cells can arise from pediatric brain tumors. *Proc Natl Acad Sci U S A* 100(25): 15178–15183.
Herrlinger U, Woiciechowski C, Sena-Esteves M, et al. (2000). Neural precursor cells for delivery of replicationconditional HSV-1 vectors to intracerebral gliomas. *Mol Ther* 1(4): 347–357.
Hirsch LR, Stafford RJ, Bankson JA, et al. (2003). Nanoshell-mediated near-infrared thermal therapy of tumors undermagnetic resonance guidance. *Proc Natl Acad Sci U S A* 100(23): 13549–13554.
Imitola J, Raddassi K, Park KI, et al. (2004). Directed migration of neural stem cells to sites of CNS injury by the stromal cell-derived factor 1alpha/CXC chemokine receptor 4 pathway. *Proc Natl Acad Sci U S A* 101(52): 18117–18122.
Jaiswal JK and Simon SM (2004). Potentials and pitfalls of fluorescent quantum dots for biological imaging. *Trends Cell Biol* 14(9): 497–504.
Jean WC, Spellman SR, Wallenfriedman MA, et al. (1998). Interleukin-12-based immunotherapy against rat 9L glioma. *Neurosurgery* 42(4): 850–856. [discussion 856–857].
Kaur B, Tan C, Brat DJ, et al. (2004). Genetic and hypoxic regulation of angiogenesis in gliomas. *J Neurooncol* 70(2): 229–243.

Kew Y and Levin VA (2003). Advances in gene therapy and immunotherapy for brain tumors. *Curr Opin Neurol* 16(6): 665–670.

Kim I, Kim H, Im S, Snyder EY, et al. (2004). Induction of intracranial glioblastoma apoptosis by transplantation of TRAIL (tumor necrosis factor-related apoptosis-inducing ligand) expressing human neural stem cells (NSCs). Annual meeting of society of neuroscience. San Diego.

Kim SK, Cargioli TG, Machluf M, et al. (2005). PEX-producing human neural stem cells inhibit tumor growth in a mouse glioma model. *Clin Cancer Res* August 15, 11(16): 5965–5970.

Kurihara H, Zama A, Tamura M, et al. (2000). Glioma/glioblastoma-specific adenoviral gene expression using the nestin gene regulator. *Gene Ther* 7(8): 686–693.

Lacorazza HD, Flax JD, Snyder EY, and Jendoubi M (1996). Expression of human beta-hexosaminidase alpha-subunit gene (the gene defect of Tay-Sachs disease) in mouse brains upon engraftment of transduced progenitor cells. *Nat Med* 2(4): 424–429.

Laws ER, Parney IF, Huang W, et al. (2003). Survival following surgery and prognostic factors for recently diagnosed malignant glioma: data from the Glioma Outcomes Project. *J Neurosurg* 99(3): 467–473.

Lazarini F, Tham TN, Casanova P, et al. (2003). Role of the alpha-chemokine stromal cell-derived factor (SDF-1) in the developing and mature central nervous system. *Glia* 42(2): 139–148.

Lee BC, Lee TH, Avraham S, and Avraham HK (2004). Involvement of the chemokine receptor CXCR4 and its ligand stromal cell-derived factor 1alpha in breast cancer cell migration through human brain microvascular endothelial cells. *Mol Cancer Res* 2(6): 327–338.

Lefranc F, Brotchi J, and Kiss R (2005). Possible future issues in the treatment of glioblastomas: special emphasis on cell migration and the resistance of migrating glioblastoma cells to apoptosis. *J Clin Oncol* 23(10): 2411–2422.

Lewin M, Carlesso N, Tung CH, et al. (2000). Tat peptide-derivatized magnetic nanoparticles allow in recovery of progenitor cells. *Nat Biotechnol* 18(4): 410–414.

Li S, Tokuyama T, Yamamoto J, et al. (2005). Bystander effect-mediated gene therapy of gliomas using genetically engineered neural stem cells. *Cancer Gene Ther* 12(7): 600–607.

Liang Z, Wu T, Lou H, et al. (2004). Inhibition of breast cancer metastasis by selective synthetic polypeptide against CXCR4. *Cancer Res* 64(12): 4302–4308.

Loo C, Lowery A, Halas N, West J, et al. (2005). Immunotargeted nanoshells for integrated cancer imaging and therapy. *Nano Lett* 5(4): 709–711.

Lynch WP, Sharpe AH, and Snyder EY (1999). Neural stem cells as engraftable packaging lines can mediate gene delivery to microglia: evidence from studying retroviral env-related neurodegeneration. *J Virol* 73(8): 6841–6851.

Ma HI, Lin SZ, Chiang YH, et al. (2002). Intratumoral gene therapy of malignant brain tumor in a rat model with angiostatin delivered by adeno-associated viral (AAV) vector. *Gene Ther* 9(1): 2–11.

Manome Y, Wen PY, Dong Y, et al. (1996). Viral vector transduction of the human deoxycytidine kinase cDNA sensitizes glioma cells to the cytotoxic effects of cytosine arabinoside in vitro and in vivo. *Nat Med* 2(5): 567–573.

Nakamizo A, Marini F, Amano T, et al. (2005). Human bone marrow-derived mesenchymal stem cells in the treatment of gliomas. *Cancer Res* 65(8): 3307–33018.

Nelson SJ and Cha S (2003). Imaging glioblastoma multiforme. *Cancer J* 9(2): 134–145.

Palmer TD, Schwartz PH, Taupin P, et al. (2001) Progenitor cells from human brain after death. *Nature* 411(6833): 42–43.

Park KI, Liu S, Flax JD, et al. (1999). Transplantation of neural progenitor and stem cells: developmental insights may suggest new therapies for spinal cord and other CNS dysfunction. *J Neurotrauma* 16(8): 675–687.

Pluchino S, Zanotti L, Rossi B, et al. (2005). Neurosphere-derived multipotent precursors promote neuroprotection by an immunomodulatory mechanism. *Nature* 436(7048): 266–271.

Rempel SA, Dudas S, Ge S, and Gutierrez JA (2000). Identification and localization of the cytokine SDF1 and its receptor, CXC chemokine receptor 4, to regions of necrosis and angiogenesis in human glioblastoma. *Clin Cancer Res* 6(1): 102–111.

Reubinoff BE, Pera MF, Vajta G, and Trounson AO (2001). Effective cryopreservation of human embryonic stem cells by the open pulled straw vitrification method. *Hum Reprod* October, 16(10): 2187–2194.

Rhines LD, Sampath P, DiMeco F, et al. (2003). Local immunotherapy with interleukin-2 delivered from polymer microspheres combined with interstitial chemotherapy: a treatment for experimental malignant glioma. *Neurosurgery* 52(4): 872–879. [discussion 879–880].

Riess P, Zhang C, Saatman KE, et al. (2002). Transplanted neural stem cells survive, differentiate, and improve neurological motor function after experimental traumatic brain injury. *Neurosurgery* 51(4): 1043–1052. [discussion 1052–1054].

Romer JT, Kimura H, Magdaleno S, et al. (2004). Suppression of the Shh pathway using a small molecule inhibitor medulloblastoma in Ptc1(+/_)p53(_/_) mice. *Cancer Cell* 6(3): 229–240.

Rosario CM, Yandava BD, Kosaras B, et al. (1997). Differentiation of engrafted multipotent neural progenitors towards replacement of missing granule neurons in meander tail cerebellum may help determine the locus of mutant gene action. *Development* 124(21): 4213–4224.

Rubin JB, Kung AL, Klein RS, et al. (2003). A small-molecule antagonist of CXCR4 inhibits intracranial growth of primary brain tumors. *Proc Natl Acad Sci U S A* 100(23): 13513–13518.

Schmidt NO, Przylecki W, Yang W, et al. (2005). Brain tumor tropism of transplanted human neural stem cells is induced by vascular endothelial growth factor. *Neoplasia* 7(6): 623–629.

Serfozo P, Schlarman MS, Pierret C, et al. (2006). Selective migration of neuralized embryonic stem cells to stem cell factor and media conditioned by glioma cell lines. *Cancer Cell Int* 6:1.

Sidman RL, et al. (1995). Cell proliferation and migration in the primitive ependymal zone: an autoradiographic study of histogenesis in the nervous system. *Exp Neurol* 1: 322–333.

Singh SK, Clarke ID, Terasaki M, et al. (2003). Identification of a cancer stem cell in human brain. *Cancer Res* 63(18): 5821–5828.

Singh SK, Hawkins C, Clarke ID, et al. (2004). Identification of human brain tumour initiating cells. *Nature* 432(7015): 396–401.

Smyth MJ, Cretney E, Kershaw MH, and Hayakawa Y (2004). Cytokines in cancer immunity and immunotherapy. *Immunol Rev* 202: 275–293.

Snyder EY and Loring JF (2005). A role for stem cell biology in the physiological and pathological aspects of aging. *J Am Geriatr Soc* September, 53(9 Suppl): S287–S291.

Snyder EY, Deitcher DL, Walsh C, et al. (1992). Multipotent neural cell lines can engraft and participate in development of mouse cerebellum. *Cell* 68(1): 33–51.

Snyder EY, Taylor RM, and Wolfe JH (1995). Neural progenitor cell engraftment corrects lysosomal storage throughout the MPS VII mouse brain. *Nature* 374(6520): 367–370.

Snyder EY, Park KI, Flax JD, et al. (1997a). Potential of neural "stem-like" cells for gene therapy and repair of the degenerating central nervous system. *Adv Neurol* 72: 121–132.

Snyder EY, Yoon C, Flax JD, and Macklis JD (1997b). Multipotent neural precursors can differentiate toward replacement of neurons undergoing targeted apoptotic degeneration in adult mouse neocortex. *Proc Natl Acad Sci U S A* 94(21): 11663–11668.

Stojdl DF, Lichty BD, tenOever BR, Paterson JM, et al. (2003). VSV strains with defects in their ability to shutdown innate immunity are potent systemic anti-cancer agents. *Cancer Cell* 4: 263–275.

Stroh M, Zimmer JP, Duda DG, et al. (2005). Quantum dots spectrally distinguish multiple species within the tumor milieu in vivo. *Nat Med* 11(6): 678–682.

Sun L, Lee J, and Fine HA (2004). Neuronally expressed stem cell factor induces neural stem cell migration to areas of brain injury. *J Clin Invest* 113(9): 1364–1374.

Tabatabai G, Bahr O, Mohle R, et al. (2005). Lessons from the bone marrow: how malignant glioma cells attract adult haematopoietic progenitor cells. *Brain* 128(Pt 9): 2200–2211.

Tatenhorst L, Puttmann S, Senner V, and Paulus W (2005). Genes associated with fast glioma cell migration in vitro and in vivo. *Brain Pathol* 15(1): 46–54.

Uhl M, Weiler M, Wick W, et al. (2005). Migratory neural stem cells for improved thymidine kinase-based gene therapy of malignant gliomas. *Biochem Biophys Res Commun* 328(1): 125–129.

Werbowetski T, Bjerkvig R, and Del Maestro RF (2004). Evidence for a secreted chemorepellent that directs glioma cell invasion. *J Neurobiol* 60(1): 71–88.

Widera D, Holtkamp W, Entschladen F, et al. (2004). MCP-1 induces migration of adult neural stem cells. *Eur J Cell Biol* 83(8): 381–387.

Yandava BD, Billinghurst LL, and Snyder EY (1999). "Global" cell replacement is feasible via neural stem cell transplantation: evidence from the dysmyelinated shiverer mouse brain. *Proc Natl Acad Sci U S A* 96(12): 7029–7034.

Yang SY, Liu H, and Zhang JN (2004). Gene therapy of rat malignant gliomas using neural stem cells expressing IL-12. *DNA Cell Biol* 23(6): 381–389.

Yip S, Aboody KS, Burns M, et al. (2003). Neural stem cell biology may be well suited for improving brain tumor therapies. *Cancer J* 9(3): 189–204.

Yip S, Sidman RL, and Snyder E (2005). Stem cells for targeting CNS malignancy. *Principles of Molecular Neurosurgery*. Karger Publishers, Basel. p. 624–644.

Yip S, Sabetrasekh R, Sidman RL, and Snyder EY (2006). Neural stem cells as novel cancer therapeutic vehicles. *Eur J Cancer* June, 42(9): 1298–1308. Epub 2006 May 12. Review.

Yuan X, Curtin J, and Xiong Y (2004). Isolation of cancer stem cells from adult glioblastoma multiforme. *Oncogene* 23(58): 9392–9400.

Zhang ZG, et al. (2003). Magnetic resonance imaging and neurosphere therapy of stroke in rat. *Ann. Neurol* 53: 259–263.

Zhang Z, Jiang Q, Jiang F, et al. (2004). In vivo magnetic resonance imaging tracks adult neural progenitor cell targeting of brain tumor. *Neuroimage* 23(1): 281–287.

Zhou Y, Larsen PH, Hao C, and Yong VW (2002). CXCR4 is a major chemokine receptor on glioma cells and mediates their survival. *J Biol Chem* 277(51): 49481–49487.

Zhu Y and Parada LF (2002). The molecular and genetic basis of neurological tumours. *Nat Rev Cancer* 2(8): 616–626.

Zlomanczuk P, Mrugala M, de la Iglesia HO, et al. (2002). Transplanted clonal neural stem-like cells respond to remote photic stimulation following incorporation within the suprachiasmatic nucleus. *Exp Neurol* 174(2): 162–168.

6

STEM CELL THERAPIES FOR PARKINSON'S DISEASE

ANDREAS ANDROUTSELLIS-THEOTOKIS,
MARIA A. RUEGER, AND
RON D.G. MCKAY

Laboratory of Molecular Biology, National Institute of Neurological Disorders and Stroke, National Institutes of Health, Bethesda, MD 20892

ABSTRACT

Stem cell biology offers the possibility of new approaches in regenerative medicine. These include in vitro generation of neurons that die in degenerative diseases from stem cells and their subsequent transplantation to inflicted patients, as well as the pharmacological stimulation of endogenous stem cells which may promote the survival of compromised neurons, or even replace them. Because the hallmark of Parkinson's disease is the progressive loss of dopaminergic neurons in the nigrostriatal pathway, this disease is an ideal candidate for stem cell therapies. Research over the last decade shows that dopaminergic neurons can be generated from both fetal and embryonic stem cells with high efficiencies. Current approaches in the treatment of Parkinson's disease involve pharmacological treatments that increase the function of the remaining dopaminergic neurons but do not slow the progression of the degeneration and transplantation of fetal dopaminergic neurons. Here we review stem cell-based efforts to replace and/or protect dopaminergic neurons in vitro and in vivo.

Copyright © 2008 Elsevier Inc.
All rights of reproduction in any form reserved.

TRANSPLANTING DOPAMINE NEURONS IN PARKINSON'S DISEASE

For more than 30 years we have known that the distinct biochemistry of catecholamines is at the center of neurodegenerative and psychiatric disease (Ehringer and Hornykiewicz, 1960; Molinoff et al., 1971). The development of the Falck–Hillarp reaction (Falck and Hillarp, 1959) allowed a fluorescence visualization of monoamines in the brain and this technique also demonstrated decreased catecholamine metabolite levels in the cerebrospinal fluid of patients with Parkinson's disease (PD) (Brody et al., 1970). In PD, a progressive loss of dopaminergic (DA) neurons occurs in the substantia nigra (SN) and other brain stem nuclei. This neuronal loss is associated with the formation of intracellular inclusions, Lewy bodies, and leads to dopamine depletion in the projections from the reticular formation (pars compacta) to the striatum, with the putamen being most affected. Later in the disease, other transmitter systems involving serotonergic cells in the median raphe, noradrenergic cells in the locus ceruleus, and cholinergic cells in the nucleus basalis of Meynert also become affected by the neurodegenerative process. Therefore, patients with PD have not only the typical motor impairment with resting tremor, bradykinesia, and rigidity but also balance problems, autonomic nervous dysfunction, and cognitive and psychiatric features. Although the disease affects other neurons, the central role of dopamine neurons in PD is affirmed by the remarkable effects of levodopa in patients (Cotzias et al., 1967).

A breakthrough in PD research came with pharmacological models of parkinsonism developed by selectively targeting and killing the DA neurons of the nigrostriatal pathway by infusion with 6-hydroxy-dopamine (6-OHDA) (Sachs and Jonsson, 1975; Ungerstedt, 1968) or 1-methyl-4-phenyl-1,2,3,6-tetrahydropyridine (MPTP) (Burns et al., 1983, 1984; Langston et al., 1984). Around the same time, MPTP contamination in illicit drugs used by human addicts was identified as the cause for the parkinsonian symptoms they developed (Langston et al., 1983). The specific destruction of these cells was soon followed by their replacement by grafting the fetal precursors of dopamine neurons in rodent and primate models (Bjorklund and Stenevi, 1979; Perlow et al., 1979; Redmond et al., 1986; Ungerstedt and Arbuthnott, 1970). These studies showed that grafted dopamine neurons survive for long periods and modulate the behavior of the host animals.

The cells used for these transplants were acutely obtained from the region of the fetal nervous system that normally generates dopamine neurons. Human fetal mesencephalic tissue is a source rich in postmitotic DA neurons and has been used in multiple clinical trials, although the useful developmental window limits tissue availability (Brundin et al., 1988). Several open-label trials in which fetal mesencephalic tissue was transplanted to PD patients have shown that grafted neurons can survive and reinnervate the striatum (Kordower et al., 1995), normalize striatal dopamine release, and restore the impaired movement-related cortical activation underlying akinesia (Piccini et al., 2000). However, larger sham surgery-controlled trials showed only modest symptomatic improvement, despite the robust survival and fiber outgrowth of transplanted neurons assessed by ^{18}F-fluorodopa positron emission tomography ([^{18}F]DOPA-PET) as well as by postmortem examination (Olanow et al., 2003). Furthermore, severe dyskinesia has been observed following transplantation of human fetal mesencephalic tissue and has been implicated with tissue processing prior to implantation (Freed et al., 2001; Hagell et al., 2002; Olanow et al., 2003). Transplantation of human fetal mesencephalic tissue will not become a routine treatment in PD unless: (1) the outcome can be reliably assessed and is clearly superior to other treatments available, (2) appropriate ex vivo-derived cells can be sourced and (3) endogenous regenerative processes are recruited to allow and promote the integration of the graft into the otherwise hostile host environment. We review these subjects here and stress the importance of understanding the biology of the donor cells and the host environment which includes multiple cell types and signals.

IMAGING AND CURRENT TREATMENT OPTIONS

IMAGING

Any new therapy requires authoritative assessment of dopamine function in vivo. Imaging is being used in the diagnostic work-up for PD and plays an important role in assessing the efficiency of individual novel therapies (Piccini et al., 1999; Freed et al., 2001; Hilker et al., 2003, 2004, Whone et al., 2003) reviewed in Rueger et al., 2007. An especially appealing feature of imaging is the possibility to precisely characterize disease-specific changes in order to identify individual patients that will benefit from a specific therapy. This, however, requires an imaging paradigm fully validated to reflect the biological parameters of interest. We briefly introduce the commonly used imaging techniques and discuss their value in this context.

Historically, Magnetic Resonance Imaging (MRI) did not show gross abnormalities in patients with PD, whereas Positron-Emission-Tomography (PET) revealed specific changes in nigrostriatal function. With the advent of high-field MRI and improved ultrasound technology, changes in the pars compacta of the SN in PD patients (Hutchinson et al., 2003; Walter et al., 2003) and within the putamen, pons, midbrain, and cortex in patients with atypical parkinsonian syndromes may be detected (Schulz et al., 1999; Seppi et al., 2003). New PET and Single Photon Emission Computed Tomography (SPECT) probes are now able to assess a variety of pre- and postsynaptic processes that reflect the impaired function of the dopaminergic system in the striatum (Rueger et al., 2007). Together with MRI and [^{18}F]fluoro-2-deoxy-D-glucose ([^{18}F]FDG) PET, these tracers can be used to differentiate various stages of PD from multiple system atrophy (MSA) and normal controls. In the early stages, [^{18}F]-6-fluoro-dopa ([^{18}F]DOPA) is reduced unilaterally or bilaterally in the posterior putamen with relatively increased (upregulated) expression of D2 receptors and preserved glucose metabolism (Morrish et al., 1995; Ghaemi et al., 2002). In contrast, in MSA, pre- and postsynaptic alterations in [^{18}F]DOPA and [^{11}C]raclopride are accompanied by a reduction in glucose metabolism, indicating pre- and postsynaptic neurodegeneration. In advanced PD, ventral and anterior putamen and dorsal caudate dopaminergic function are altered (Nurmi et al., 2001).

Current imaging technologies characterize basal ganglia function with regard to general dopamine and glucose metabolism and are sensitive enough to distinguish early from late stages of disease. However, the assessment of more subtle features calls for the refinement of existing or the development of new imaging technologies. The prediction of clinical milestones such as loss of levodopa efficacy or involvement of telencephalic functions would be very valuable in selecting patients for specific treatments and defining response to therapy

NEUROTRANSMITTER PHARMACOLOGY

The normal function of the basal ganglia depends on the balance between the two main output pathways of the striatum, the direct and indirect projection systems (Gerfen, 2000). In brief, the striatum comprises two distinct populations of neurons: Direct-pathway neurons bear D1 receptors and provide inhibitory inputs to the internal segment of the globus pallidus (GPi) and substantia nigra pars reticulata (SNr). Inhibitory indirect-pathway neurons possess D2 receptors and project to the external segment of the globus pallidus (GPe), which in turn provides inhibitory projections to the subthalamic nucleus (STN). The STN provides excitatory inputs to all the output nuclei of the basal ganglia (GPe, GPi, and SNr). The nigrostriatal dopaminergic projection originating in the substantia nigra pars compacta (SNc) exerts a dual action on efferent striatal projection neurons. It inhibits D2 receptors in the indirect pathway and activates D1 receptors in the direct pathway. Additionally, dopamine D2 receptors are present on the terminals of dopamine neurons and therefore also function as autoreceptors. In the dopamine-deficient state of PD, the indirect pathway is therefore hyperactive, which is being used as an approach for some of the treatment options discussed below. Since dopamine receptors are functionally regulated, a compensatory increase of dopamine D2 receptor density – but no change in dopamine D1 receptor expression – is found in the striatum of patients with untreated PD. Following anti-parkinsonian drug treatment there is a downregulation of the elevated dopamine D2 receptors to normal levels (Hurley and Jenner, 2006).

In the past, pharmacological treatment for PD has mostly focused on symptomatic relief using drugs that regulate either dopaminergic or non-dopaminergic functions in the brain. Therapeutic efficacy has usually been evaluated using clinical scores such as the Unified Parkinson Disease Rating

Scale (UPDRS). Dopaminergic drugs include levodopa, dopamine receptor agonists, catechol-O-methyltransferase (COMT) inhibitors, and monoamine oxidase (MAO) B inhibitors. Levodopa is the drug most commonly used in PD and remains the "gold standard" that other drugs need to be measured against. The beneficial effects of levodopa decrease as the DA cells are lost.

Enzymes that degrade dopamine are also potential targets even though they may also lose efficacy as the source of dopamine declines. The efficacy of levodopa can be enhanced by co-administration of a COMT inhibitor, which increases the plasma half-life of levodopa and augments the anti-parkinsonian effects (Nutt et al., 1994). Inhibition of MAO-B, the major enzyme metabolizing dopamine in the human brain, has a beneficial clinical effect in patients with levodopa-associated fluctuating symptoms (Rascol et al., 2005). Serotonergic neurons and astrocytes contain predominantly MAO-B, whereas catecholaminergic neurons contain mainly MAO-A. In PD, selective MAO-B inhibitors are preferred because the human basal ganglia have higher MAO-B than MAO-A activity, and because dopamine is equally well metabolized by both isoenzymes in humans (Youdim et al., 2006). Some evidence from a recent clinical trial suggests a disease-modifying, neuroprotective effect of the MAO-B inhibitor rasagiline (Parkinson-Study-Group, 2004), but due to a relatively short follow-up time in this trial this result has not yet been unanimously accepted, making large trials with longer follow-up necessary.

Anticholinergic drugs have been used to treat PD for over 100 years. Their mechanism of action is largely unclear; their function is associated with a general imbalance between the dopaminergic and the cholinergic system. Muscarinic receptors are a subtype of cholinergic receptors, and today, selective muscarinic antagonists are occasionally used as a symptomatic treatment especially in tremor-dominant early PD. However, they are not drugs of the first choice due to limited efficacy and the propensity to cause neuropsychiatric side effects (Katzenschlager et al., 2003). Several other non-dopaminergic drugs including adenosine A2a receptors antagonists, alpha2 noradrenergic receptor antagonists, GABAergic drugs, and glutamate antagonists are also being investigated as additional treatments for PD, and have been reviewed elsewhere (Linazasoro, 2004). The exact function of most of those drugs is not entirely clear; most of their action seems to be mediated through inhibition of the indirect dopaminergic pathway, which is pathologically hyperactive in PD, as described above.

The dopamine receptors are obvious targets and dopamine receptor agonists are effective as a monotherapy for early PD as well as an adjunct to levodopa (Barone et al., 1999). Most currently used dopaminergic agonists are selective for D2 and D3 dopamine receptors (Schapira et al., 2006). Recent clinical trials compared different dopamine agonists to levodopa monotherapy (Parkinson-Study-Group, 2000; Rascol et al., 2000; Oertel et al., 2006). First results of those studies suggest a delayed onset of motor complications and reduced dyskinesias with dopamine agonist monotherapy compared to levodopa. Within the last years, D2 dopamine receptor agonists have been shown to modify disease progression in patients and animal models of PD (assessed by PET and SPECT imaging), and to protect cultured DA neurons from cytotoxic stresses (Clarke, 2004). The dopamine agonists pramipexole and ropinirole were studied in two separate trials; the former in CALM-PD with SPECT over 4 years, and the latter in REAL-PET, with PET over 2 years. Patients with early PD were randomly assigned to receive either the dopamine agonist or levodopa and followed up with serial scans. Patients could be supplemented with levodopa if symptoms needed treatment. Both studies showed a significant reduction in the loss of the respective imaging signal (about 35%) over 2 years in those patients taking the dopamine agonist, and this was maintained in the pramipexole group at 4 years. These results are potentially of great importance; however, they do not confirm a protective role for dopamine agonists, although such an action would be supported by the protective effect seen in laboratory studies (Schapira, 2002). Unfortunately, a placebo group was not included in either study and so it is not known whether the agonist was protective or levodopa toxic. The ELLDOPA trial randomized early PD patients to a placebo arm or one of the three levodopa treatment groups and performed an SPECT analysis before and at 40 weeks (Fahn et al., 2004). The levodopa arms showed a greater rate of transporter signal loss than did the placebo group over the period of the trial. The levodopa-treated patients had a better UPDRS score than placebo patients after washout, although this was only for 2 weeks, and a prolonged symptomatic effect could not be excluded. These trials set the standard for any new treatment aimed directly at the dopaminergic system, but our understanding of the precise effects of these treatments on the DA neurons must be expanded. New tools of ex vivo developmental biology such as the generation of DA neurons from progenitors and stem cells may allow a more detailed analysis of the effects of dopamine receptor agonists on nigral and striatal neurons.

DEEP BRAIN STIMULATION

Surgical treatment for PD is based on our knowledge about the functional basal ganglia circuitry introduced above (Obeso et al., 2000; Breit et al., 2004). The principle aim of the deep brain stimulation (DBS) procedures is to restore the equilibrium of neuronal activity in the basal ganglia network and to release thalamocortical projections from over-inhibition exerted by hyperactive basal ganglia output nuclei in PD (Benazzouz and Hallett, 2000). The STN is the most commonly used target for DBS in the treatment of PD, due to the marked improvement of all cardinal symptoms of the disease. Therapeutic effects of deep brain stimulation can be followed by PET imaging, revealing important information about the advantages and limitations of this method. Determining the regional cerebral metabolic rate of glucose (rCMRGlc) by means of [^{18}F]FDG PET, an activating effect of DBS upon its target structures and a central role of the STN in motor as well as associative, limbic, and cerebellar basal ganglia circuits, could be confirmed (Hilker et al., 2003). STN-DBS presents an increasingly used treatment option associated with significant improvement in motor function in patients with PD whose condition cannot be further improved with medical therapy (Deep-Brain-Stimulation-for-PD-Study-Group, 2001). However, the success of the therapy largely depends on the selection of the appropriate candidate patients and on the precise implantation of the stimulation electrode. Furthermore, initial speculations about a potential neuroprotective effect of DBS have been recently disputed in a prospective study in which disease progression as determined by means of serial [^{18}F]DOPA PET showed a continuous decline of dopaminergic function in patients with advanced PD under clinically effective bilateral STN stimulation (Hilker et al., 2005). Although there is still an empirical quality to this approach, the success of DBS points out the potential value of targeting non-DA cells. The stimulation of the STN reported here may be associated with decreased physiological output suggesting potential gene and cell therapy approaches that may reach the same therapeutic goal.

NEUROTROPHIC FACTORS AND GENE THERAPY

During central nervous system development, waves of neuronal death help refine the neuronal circuitry (Oppenheim, 1991). These apoptotic waves may claim up to 50% of all neurons that are thought to compete with each other for trophic support from their targets. Trophic support is provided in the form of target – secreted neurotrophic factors, and they are currently being exploited to promote the survival of DA neurons in the context of PD. The prototypical neurotrophic factor, Nerve Growth Factor (NGF) was isolated in the fifties (Cohen et al., 1954), but it took decades to identify additional members of the family and their receptors (Chao, 2003). Currently, several families of growth factors have been shown to possess neurotrophic properties and are being studied as potential therapeutics for PD. A particularly promising factor in PD is the Glial-Derived Neurotrophic Factor (GDNF), first isolated as a trophic factor for midbrain DA neuron cultures from a glial cell line (Lin et al., 1993). Despite its scarcity in the adult brain, GDNF levels are upregulated following insults such as ischemia, and may contribute to an innate neuroprotection response (Abe and Hayashi, 1997). In animal models, GDNF is strongly neuroprotective (Tomac et al., 1995; Gash et al., 1996). It should be pointed out that regeneration of nigrostriatal neurons reaching their target in the striatum and preservation of normal motor functions is promoted only when GDNF is injected into the striatum; intracerebroventricular (ICV) and intranigral administration have either no or deleterious effects, respectively (Kirik et al., 2000a). In a randomized controlled trial in human PD, monthly ICV bolus injections failed to provide clinical benefit but resulted in important adverse effects (Nutt et al., 2003). A single-center, open-label trial of continuous intraputamenal (Ipu) infusion in five patients reported excellent drug tolerability. After 6 months of treatment, mean off-motor scores on the Unified Parkinson's Disease Rating Scale (UPDRS) score improved by 32% with a statistically significant increase in putamenal [^{18}F]DOPA uptake as determined by PET in the vicinity of the catheter tip in the posterior putamen (Gill et al., 2003). However, a recent randomized controlled clinical trial did not detect a significant clinical improvement from bilateral lpu infusion of GDNF compared with placebo in patients with moderate to severe PD (Lang et al., 2006). Interestingly, the PET component of this study demonstrated a 23% increase of posterior putamen [^{18}F]DOPA influx constant in the GDNF-treated group, suggesting a consequence of nerve terminal sprouting and/or upregulation of tracer transport and decarboxylation by the remaining functioning terminals. Possibly, more significant clinical benefit could be obtained with improved distribution and/or higher doses of GDNF delivered for longer periods of time. Other methods of GDNF delivery, such as cell-based or viral vector systems, may also hold promise for the future. GDNF delivered by a lentiviral system, for example, prevented degeneration and induced regeneration in MPTP-lesioned monkeys (Kordower et al., 2000).

Gene therapy using viral delivery systems is extensively used in the context of experimental PD. The most common viral vectors to target post-mitotic cells in the CNS are Adeno-Associated viruses (AAV) and lentiruses (Costantini et al., 2000). These delivery systems exhibit low toxicity and their vector DNA integrates into the genome of the host cells, thereby providing sustained expression of the transgene. These properties render them strong candidates for human applications. Gene therapy for PD was first developed in rat models using transduction of a single gene encoding tyrosine hydroxylase (TH) (During et al., 1994; Kaplitt et al., 1994). In the past years, different approaches to gene therapy for PD have been developed. One paradigm is the simultaneous transduction of multiple genes essential in the production and turnover of dopamine in nigrostriatal nerve terminals, which include (i) TH converting tyrosine to levodopa in the presence of tetrahydrobiopterin (BH4); (ii) GTP cyclohydroxylase I (GCHI) as the rate-limiting enzyme in the biosynthesis of BH4; (iii) aromatic amino decarboxylase (AADC) converting Levodopa to dopamine; (iv) vesicular monoamine transporter type 2 (VMAT-2) transporting dopamine into synaptic vesicles. The following gene combinations have been successfully used in animal models of PD: TH and GCHI (Mandel et al., 1998; Kirik et al., 2002), TH, GCHI, and AADC (Shen et al., 2000; Azzouz et al., 2002; Muramatsu et al., 2002), and TH, GCHI, AADC, and VMAT-2 (Sun et al., 2004). A different approach is to use gene therapy to protect dopamine neurons, i.e., the transduction of genes encoding growth- and anti-apoptotic factors for the prevention of further degeneration of nigrostriatal neurons. It has been shown that intrastriatal transduction of GDNF can protect nigrostriatal neurons from neurotoxic stress in rat and primate models of PD (Choi-Lundberg et al., 1997; Kirik et al., 2000b; Kordower et al., 2000). Other "neuroprotective" genes tested in rodent models include JNK-interacting protein-1 (JIP-1; Xia et al., 2001), apoptosis protease activating factor-1 (APAF-1) dominant negative inhibitor (Mochizuki et al., 2001), neuronal apoptosis inhibitor protein (NAIP (Crocker et al., 2001)), Hsp70 (Dong et al., 2005), and Parkin (Lo Bianco et al., 2004). Such approaches have been reviewed elsewhere (Burton et al., 2003).

Another promising gene therapy paradigm is the conversion of excitatory to inhibitory output neurons arising from the STN by transduction of excitatory glutamatergic neurons of the STN with glutamic acid decarboxylase (GAD), the enzyme that catalyzes the synthesis of the neurotransmitter GABA. In a rat model of PD, this treatment resulted in strong neuroprotection of nigral dopamine neurons and rescue of the parkinsonian behavioral phenotype (Luo et al., 2002). This paradigm was applied in the first clinical gene therapy trial in humans suffering from PD, and PET imaging was performed to assess the therapeutic effects (During et al., 2001). Results of this study have not yet been published.

Whichever method is used to protect or replace compromised DA neurons, the field will benefit from a precise knowledge of the biology of the many growth factors present in the brain that will affect their functional integration and survival. Genetic data from the human population is currently going through an explosive growth as the technology that identifies polymorphisms becomes easier to apply. Mutations that compromise dopamine neuron survival have been identified in several genes but Fibroblast Growth Factor 20 (FGF-20) is currently the only soluble growth factor that genetic data implicate in PD (van der Walt et al., 2004). Recent work with DA neuron cultures demonstrated that FGF-20 stimulates dopamine release and promotes the survival of calbindin – negative DA neurons (Murase and McKay, 2006), the subset of DA neurons most at risk in PD (Yamada et al., 1990). This study demonstrates that defining the function of a gene that alters disease risk in dopamine neurons requires precise control of neuronal differentiation. We now discuss technologies required to assign direct and indirect roles for other growth factors on dopamine neurons.

EX VIVO AND IN VIVO STEM CELLS IN PD

MAKING DOPAMINE NEURONS IN THE LABORATORY

Considerable efforts have been made to derive large numbers of DA neurons in the laboratory. As neurons cannot proliferate, this work has focused expanding the numbers of precursor cells and generating large numbers of neurons by differentiating the precursors (Cattaneo and McKay, 1990). The first success with transplantation into hemiparkinsonian rats used DA neurons derived from midbrain precursor cells obtained from rat embryos (Studer et al., 1998). These precursor cells generate dopamine neurons for only short periods in culture perhaps because of an inherent limit on the expansion of somatic precursors. In contrast, mouse embryonic stem (ES) cells can proliferate extensively in an undifferentiated state and can generate a highly enriched population of neuroepithelial precursor cells (Okabe et al., 1996). Mouse ES cells efficiently differentiate into DA neurons in the laboratory (Lee et al., 2000). Many groups have pursued the use of mouse and human ES cells as source of dopamine neurons. We highlight three studies here to show the potential importance of this approach. Electrophysiology, behavior and PET imaging clearly demonstrate that ex vivo-derived dopamine neurons function for long periods in a rat model of PD (Kim et al., 2002). The derivation of functional dopamine neurons from new ES cells after nuclear transfer is one of the few clear examples of the extraordinary power of combining genetic manipulation and ex vivo development (Wakayama et al., 2001). Dopamine neurons that function in animal models derived from human ES cells raise the possibility that functional human neurons may soon be routinely obtained ex vivo (Roy et al., 2006). However, so far it remains to be shown that human ES cell-derived neurons fulfill the requirements of clinically successful grafts in vivo. Several aspects of this technology must be controlled to reach this end. Here we focus on two, the identity of midbrain dopamine neurons and the regenerative potential of the adult brain. As our understanding of these questions develops, it is clear that they are central to our understanding of PD and will likely influence any new therapy or assessment tool that is developed.

To generate DA neurons from precursor cells or ES cells ex vivo, one is faced with a crucial question: How do we define a DA neuron in the culture dish? Expression of tyrosine hydroxylase and dopamine release may not be adequate as DA cells can be found in other brain regions that lack other features that are required in subtantia nigra dopamine neurons. In addition, within the midbrain there are different classes of dopamine neuron with distinct biological functions and susceptibility to the stresses of PD (Gerfen et al., 1987; Yamada et al., 1990). Grafts of midbrain DA neurons from the lateral SNpc and the medial VTA exhibit distinct axon projection properties in the lesioned brain suggesting that these cells have distinct targeting and synaptic properties (Zuddas et al., 1991; Hudson et al., 1994) Here we ask if the initial success of ex vivo ES cell differentiation can be extended by a more precise understanding of the normal development of midbrain dopamine neurons.

THE IDENTITY OF MIDBRAIN DA NEURONS

The transcription factors Lmx1b and En1/2 are expressed from an early stage of development in precursors and DA neurons. These genes are required for the appropriate development of dopamine neurons and for their survival in the post-natal midbrain (Smidt et al., 2000; Simon et al., 2001; Alberi et al., 2004). The transcription factors Nurr1 and Pitx3 are expressed after the progenitor stage and promote the maturation and survival of DA neurons (Zetterstrom et al., 1997; Saucedo-Cardenas et al., 1998; Wallen A. et al., 1999; Wallen A. A. et al., 2001; Hwang et al., 2003; Nunes et al., 2003; Smits et al., 2003; van den Munckhof et al., 2003; Smidt et al., 2004; Maxwell et al., 2005). The transcription factors Lmx1a and Msx1 are thought to coordinate dopamine neuron neuronal differentiation by suppressing Nkx6.1 and inducing Ngn2 (Chizhikov and Millen, 2004; Andersson et al., 2006). For a review of the intricate relations between these and other factors affecting DA neuron specification, differentiation and survival, see Ang, 2006.

Progress in understanding the development of midbrain DA neurons has already influenced efforts to generate DA neurons from ES cells ex vivo. Overexpression of Lmx1a is thought to promote dopamine neuron differentiation (Andersson et al., 2006). Nurr1 overexpression also promotes the efficient production of functional DA neurons from ES cells (Lee et al., 2000; Kim et al., 2002). In both of these studies, sonic hedgehog (SHH) was used to promote the numbers of DA precursors generated in vitro. SHH is a secreted factor that is thought to commit precursors to the dopaminergic fate (Hynes et al., 1997; Ye et al., 1998; Lee et al., 2000). Foxa2 and Foxa1 have been proposed as markers of midbrain DA neuron progenitors (Puelles et al., 2003). The expression of SHH and Foxa2 are closely linked. Enhanced differentiation of dopamine neurons has also been observed when ES cells were co-cultured with stromal cells (Kawasaki et al., 2000). In vitro and in vivo models also allow the functional analysis of the specific class of dopamine neurons at most risk in PD (Kim et al., 2002; Murase and McKay, 2006). Establishing a strong link between the developmental mechanisms specifying distinct subtypes of dopamine neuron and disease sensitivity may generate a new series of tools to identify therapies that promote the survival of dopamine neurons.

ENDOGENOUS STEM CELLS

The ultimate goal of regenerative medicine is the restoration of function in the diseased adult brain. We now know that the adult brain maintains stem cells/progenitor cells and is able to generate new neurons (Altman and Das, 1965; Reynolds and Weiss, 1992; Lois and Alvarez-Buylla, 1993; Luskin, 1993; Palmer et al., 1995; van Praag et al., 2002). These endogenous cells have been proposed as the basis for regenerative therapies in models of multiple sclerosis models, Alzheimer's disease, stroke, and epilepsy (Hayashi et al., 1994; Bengzon et al., 1997; Magavi et al., 2000; Arvidsson et al., 2001; Nait-Oumesmar et al., 2007).

Currently, the evidence for adult DA neuron replacement from endogenous sources is sparse and controversial. An initial study (Zhao et al., 2003) suggested basal and lesion-induced neurogenesis in the adult SN on the basis of BrdU incorporation and dye uptake, but another laboratory failed to reproduce that data (Mohapel et al., 2005). A more recent study (Shan et al., 2006) suggested adult neurogenesis in the SN based on BrdU incorporation assays and the use of an LacZ reporter driven by the second intron enhancer of nestin, but the conclusions remain controversial due to the limits of these techniques. Two more studies (Van Kampen and Robertson, 2005; Van Kampen and Eckman, 2006) used a D3 dopamine receptor agonist with and without 6OHDA lesion and assessed BrdU incorporation in TH+ cells, but others (Borta and Hoglinger, 2007) have suggested that these may be false-positive cells due to nuclear membrane breakdown, an early feature of apoptotic cell death (Vila and Przedborski, 2003). Aberrant re-entry into the cell cycle during apoptosis may explain these observations, consistent with most reports that fail to confirm neurogenesis in the adult SN (El-Khodor *et al.*, 2003) (reviewed in (Borta and Hoglinger, 2007)).

However, the consensus is that the adult SN does harbor stem/progenitor cells. Progenitor cells have been isolated from the adult rodent SN (Lie et al., 2002). These were shown to have gliogenic potential in vivo and both gliogenic and neurogenic potential in vitro, suggesting that the environment in the adult SN may not be conducive to neurogenesis. The stem cell niche of the adult brain may be significantly activated in ischemia models by pharmacological manipulations (Nakatomi et al., 2002; Androutsellis-Theotokis et al., 2006). The regenerative effects of these indirect approaches have not yet been fully tested in PD models. It is possible that current lesion methods (6OHDA, MPTP) which induce much faster degeneration of DA neurons than in PD may not be an appropriate trigger for adult neurogenesis in the SN. A case in point is the first report to show adult neurogenesis in the cerebral cortex, where a very subtle lesion that specifically targeted a subset of pyramidal neurons that project from the neocortex to the thalamus induced neuronal replacement (Magavi et al., 2000). Another possibility is that progenitor cells are required in the SN to generate glia following insults, and therefore provide trophic support (e.g., GDNF, BDNF) and neuroprotection to existing, compromised neurons (Mao et al., 2001; Chen LW, et al., 2002, 2004; Chen Y, et al., 2005; Steiner et al., 2006).

Although in terms of PD there is a focus on the loss of DA neurons in the SN, other areas of the brain and non-motor functions are also affected in patients leading to symptoms that include olfactory dysfunction (Berendse et al., 2001), impaired spatial memory (Pillon et al., 1997), and depression (Oertel et al., 2001). Although the cause of these symptoms in not known, they seem to be unrelated to the nigrostiatal DA neuron deficit of PD because they do not respond to treatments that increase dopamine function. It is conceivable that these symptoms are due to impairments of the adult stem cell niche and subsequent neurogenesis (Borta and Hoglinger, 2007). Indeed, inhibition of neurogenesis can result in impaired olfaction and spatial memory, and depression (Nilsson et al., 1999; Santarelli et al., 2003; Enwere et al., 2004). These findings raise the possibility that PD reduces the neurogenic potential of the brain, and that, in turn, leads to secondary (non-dopamine-responsive) symptoms. Care is needed when using animal PD models because the acute insult of the toxins used to kill DA neurons may induce a short-lived stimulation of the stem cell niche that may not be found in PD patients (Hayakawa et al., 2007).

It is still not clear what the role and therapeutic potential of adult stem/progenitor cells in terms of PD is. But adult stem cells should not be viewed as dormant entities that may awake following brain insults to simply replenish lost neurons. Current research demonstrates that stem cells and DA neurons functionally communicate through dopamine receptors on the surface of progenitor cells. Neurotransmitters and neurotransmitter receptor agonists and antagonists can modulate the proliferative properties of precursor cells in the largest stem cell niche of the adult brain, the SVZ. The loss of DA neurons induces the loss of stem cells in the SVZ (Hoglinger et al., 2004). The implication is that PD is not just a disease of the nigrostriatal pathway but of the whole brain.

These findings suggest a new line of thought: If loss of DA neurons induces the loss of progenitor cells, and since these may be important in the survival of DA neurons, a vicious cycle of neuronal AND precursor cell death may be set up. To stop this loop, we should target both neuronal survival and the precursor cells. In the immediate future, it will be important to activate the endogenous stem cell niche and address whether this process protects neurons. Blocking neuronal degeneration and stimulating precursor cells may allow new regenerative therapies in PD.

REFERENCES

Abe K and Hayashi T (1997). Expression of the glial cell line-derived neurotrophic factor gene in rat brain after transient MCA occlusion. *Brain Res* 776: 230–234.

Alberi L, Sgado P, and Simon HH (2004). Engrailed genes are cell-autonomously required to prevent apoptosis in mesencephalic dopaminergic neurons. *Development* 131: 3229–3236.

Altman J and Das GD (1965). Autoradiographic and histological evidence of postnatal hippocampal neurogenesis in rats. *J Comp Neurol* 124: 319–335.

Andersson E, Tryggvason U, Deng Q et al. (2006). Identification of intrinsic determinants of midbrain dopamine neurons. *Cell* 124: 393–405.

Androutsellis-Theotokis A, Leker RR, Soldner F et al. (2006). Notch signalling regulates stem cell numbers in vitro and in vivo. *Nature* 442: 823–826.

Ang SL (2006). Transcriptional control of midbrain dopaminergic neuron development. *Development* 133: 3499–3506.

Arvidsson A, Kokaia Z, and Lindvall O (2001). *N*-methyl-D-aspartate receptor-mediated increase of neurogenesis in adult rat dentate gyrus following stroke. *Eur J Neurosci* 14: 10–18.

Azzouz M, Martin-Rendon E, Barber RD et al. (2002). Multicistronic lentiviral vector-mediated striatal gene transfer of aromatic L-amino acid decarboxylase, tyrosine hydroxylase, and GTP cyclohydrolase I induces sustained transgene expression, dopamine production, and functional improvement in a rat model of Parkinson's disease. *J Neurosci* 22: 10302–10312.

Barone P, Bravi D, Bermejo-Pareja F et al. (1999). Pergolide monotherapy in the treatment of early PD: a randomized, controlled study. Pergolide Monotherapy Study Group. *Neurology* 53: 573–579.

Benazzouz A and Hallett M (2000). Mechanism of action of deep brain stimulation. *Neurology* 55: S13–S16.

Bengzon J, Kokaia Z, Elmer E et al. (1997). Apoptosis and proliferation of dentate gyrus neurons after single and intermittent limbic seizures. *Proc Natl Acad Sci U S A* 94: 10432–10437.

Berendse HW, Booij J, Francot CM et al. (2001). Subclinical dopaminergic dysfunction in asymptomatic Parkinson's disease patients' relatives with a decreased sense of smell. *Ann Neurol* 50: 34–41.

Bjorklund A and Stenevi U (1979). Reconstruction of the nigrostriatal dopamine pathway by intracerebral nigral transplants. *Brain Res* 177: 555–560.

Borta A and Hoglinger GU (2007). Dopamine and adult neurogenesis. *J Neurochem* 100: 587–595.

Breit S, Schulz JB, and Benabid AL (2004). Deep brain stimulation. *Cell Tissue Res* 318: 275–288.

Brody JA, Chase TN, and Gordon EK (1970). Depressed monoamine catabolite levels in cerebrospinal fluid of patients with parkinsonism dementia of Guam. *N Engl J Med* 282: 947–950.

Brundin P, Barbin G, Strecker RE et al. (1988). Survival and function of dissociated rat dopamine neurones grafted at different developmental stages or after being cultured in vitro. *Brain Res* 467: 233–243.

Burns RS, Chiueh CC, Markey SP et al. (1983). A primate model of parkinsonism: selective destruction of dopaminergic neurons in the pars compacta of the substantia nigra by *N*-methyl-4-phenyl-1,2,3,6-tetrahydropyridine. *Proc Natl Acad Sci U S A* 80: 4546–4550.

Burns RS, Markey SP, Phillips JM, and Chiueh CC (1984). The neurotoxicity of 1-methyl-4-phenyl-1,2,3,6-tetrahydropyridine in the monkey and man. *Can J Neurol Sci* 11: 166–168.

Burton EA, Glorioso JC, and Fink DJ (2003). Gene therapy progress and prospects: Parkinson's disease. *Gene Ther* 10: 1721–1727.

Cattaneo E and McKay R (1990). Proliferation and differentiation of neuronal stem cells regulated by nerve growth factor. *Nature* 347: 762–765.

Chao MV (2003). Neurotrophins and their receptors: a convergence point for many signalling pathways. *Nat Rev Neurosci* 4: 299–309.

Chen LW, Wei LC, Qiu Y et al. (2002). Significant up-regulation of nestin protein in the neostriatum of MPTP-treated mice. Are the striatal astrocytes regionally activated after systemic MPTP administration? *Brain Res* 925: 9–17.

Chen LW, Hu HJ, Liu HL et al. (2004). Identification of brain-derived neurotrophic factor in nestin-expressing astroglial cells in the neostriatum of 1-methyl-4-phenyl-1,2,3,6-tetrahydropyridine-treated mice. *J Neurosci* 126: 941–953.

Chen Y, Ai Y, Slevin JR et al. (2005). Progenitor proliferation in the adult hippocampus and substantia nigra induced by glial cell line-derived neurotrophic factor. *Exp Neurol* 196: 87–95.

Chizhikov VV and Millen KJ (2004). Control of roof plate formation by Lmx1a in the developing spinal cord. *Development* 131: 2693–2705.

Choi-Lundberg DL, Lin Q, Chang YN et al. (1997). Dopaminergic neurons protected from degeneration by GDNF gene therapy. *Science* 275: 838–841.

Clarke CE (2004). Neuroprotection and pharmacotherapy for motor symptoms in Parkinson's disease. *Lancet Neurol* 3: 466–474.

Cohen S, Levi-Montalcini R, and Hamburger V. (1954). A Nerve Growth-Stimulating Factor Isolated from Sarcom as 37 and 180. *Proc Natl Acad Sci U S A* 40: 1014–1018.

Costantini LC, Bakowska JC, Breakefield XO, and Isacson O (2000). Gene therapy in the CNS. *Gene Ther* 7: 93–109.

Cotzias GC, Van Woert MH, and Schiffer LM (1967). Aromatic amino acids and modification of parkinsonism. *N Engl J Med* 276: 374–379.

Crocker SJ, Wigle N, Liston P et al. (2001). NAIP protects the nigrostriatal dopamine pathway in an intrastriatal 6-OHDA rat model of Parkinson's disease. *Eur J Neurosci* 14: 391–400.

Deep-Brain-Stimulation-for-PD-Study-Group (2001). Deep-brain stimulation of the subthalamic nucleus or the pars interna of the globus pallidus in Parkinson's disease. *N Engl J Med* 345: 956–963.

Dong Z, Wolfer DP, Lipp HP, and Bueler H (2005). Hsp70 gene transfer by adeno-associated virus inhibits MPTP-induced nigrostriatal degeneration in the mouse model of Parkinson disease. *Mol Ther* 11: 80–88.

During MJ, Naegele JR, O'Malley KL, and Geller AI (1994). Long-term behavioral recovery in parkinsonian rats by an HSV vector expressing tyrosine hydroxylase. *Science* 266: 1399–1403.

During MJ, Kaplitt MG, Stern MB, and Eidelberg D (2001). Subthalamic GAD gene transfer in Parkinson disease patients who are candidates for deep brain stimulation. *Hum Gene Ther* 12: 1589–1591.

Ehringer H and Hornykiewicz O (1960). Distribution of noradrenaline and dopamine (3-hydroxytyramine) in the human brain and their behavior in diseases of the extrapyramidal system. *Klin Wochenschr* 38: 1236–1239.

El-Khodor BF, Oo TF, Kholodilov N, and Burke RE (2003). Ectopic expression of cell cycle markers in models of induced programmed cell death in dopamine neurons of the rat substantia nigra pars compacta. *Exp Neurol* 179: 17–27.

Enwere E, Shingo T, Gregg C et al. (2004). Aging results in reduced epidermal growth factor receptor signaling, diminished olfactory neurogenesis, and deficits in fine olfactory discrimination. *J Neurosci* 24: 8354–8365.

Fahn S, Oakes D, Shoulson I et al. (2004). Levodopa and the progression of Parkinson's disease. *N Engl J Med* 351: 2498–2508.

Falck B and Hillarp NA (1959). On the cellular localization of catechol amines in the brain. *Acta Anat (Basel)* 38: 277–279.

Freed CR, Greene PE, Breeze RE et al. (2001). Transplantation of embryonic dopamine neurons for severe Parkinson's disease. *N Engl J Med* 344: 710–719.

Gash DM, Zhang Z, Ovadia A et al. (1996). Functional recovery in parkinsonian monkeys treated with GDNF. *Nature* 380: 252–255.

Gerfen CR (2000). Molecular effects of dopamine on striatal-projection pathways. *Trends Neurosci* 23: S64–S70.

Gerfen CR, Herkenham M, and Thibault J (1987). The neostriatal mosaic: II. Patch- and matrix-directed mesostriatal dopaminergic and non-dopaminergic systems. *J Neurosci* 7: 3915–3934.

Ghaemi M, Hilker R, Rudolf J et al. (2002). Differentiating multiple system atrophy from Parkinson's disease: contribution of striatal and midbrain MRI volumetry and multi-tracer PET imaging. *J Neurol Neurosurg Psychiatry* 73: 517–523.

Gill SS, Patel NK, Hotton GR et al. (2003). Direct brain infusion of glial cell line-derived neurotrophic factor in Parkinson disease. *Nat Med* 9: 589–595.

Hagell P, Piccini P, Bjorklund A et al. (2002). Dyskinesias following neural transplantation in Parkinson's disease. *Nat Neurosci* 5: 627–628.

Hayakawa H, Hayashita-Kinoh H, Nihira T et al. (2007). The isolation of neural stem cells from the olfactory bulb of Parkinson's disease model. *Neurosci Res* 57: 393–398.

Hayashi Y, Kashiwagi K, Ohta J et al. (1994). Alzheimer amyloid protein precursor enhances proliferation of neural stem cells from fetal rat brain. *Biochem Biophys Res Commun* 205: 936–943.

Hilker R, Voges J, Ghaemi M et al. (2003). Deep brain stimulation of the subthalamic nucleus does not increase the striatal dopamine concentration in parkinsonian humans. *Mov Disord* 18: 41–48.

Hilker R, Voges J, Weisenbach S et al. (2004). Subthalamic nucleus stimulation restores glucose metabolism in associative and limbic cortices and in cerebellum: evidence from a FDG-PET study in advanced Parkinson's disease. *J Cereb Blood Flow Metab* 24: 7–16.

Hilker R, Portman AT, Voges J et al. (2005). Disease progression continues in patients with advanced Parkinson's disease and effective subthalamic nucleus stimulation. *J Neurol Neurosurg Psychiatry* 76: 1217–1221.

Hoglinger GU, Rizk P, Muriel MP et al. (2004). Dopamine depletion impairs precursor cell proliferation in Parkinson disease. *Nat Neurosci* 7: 726–735.

Hudson JL, Bickford P, Johansson M et al. (1994). Target and neurotransmitter specificity of fetal central nervous system transplants: importance for functional reinnervation. *J Neurosci* 14: 283–290.

Hurley MJ and Jenner P (2006). What has been learnt from study of dopamine receptors in Parkinson's disease? *Pharmacol Ther* 111: 715–728.

Hutchinson M, Raff U, and Lebedev S (2003). MRI correlates of pathology in parkinsonism: segmented inversion recovery ratio imaging (SIRRIM). *Neuroimage* 20: 1899–1902.

Hwang DY, Ardayfio P, Kang UJ et al. (2003). Selective loss of dopaminergic neurons in the substantia nigra of Pitx3-deficient aphakia mice. *Brain Res Mol Brain Res* 114: 123–131.

Hynes M, Stone DM, Dowd M et al. (1997). Control of cell pattern in the neural tube by the zinc finger transcription factor and oncogene Gli-1. *Neuron* 19: 15–26.

Kaplitt MG, Leone P, Samulski RJ et al. (1994). Long-term gene expression and phenotypic correction using adeno-associated virus vectors in the mammalian brain. *Nat Genet* 8: 148–154.

Katzenschlager R, Sampaio C, Costa J, and Lees A (2003). Anticholinergics for symptomatic management of Parkinson's disease. *Cochrane Database Syst Rev* CD003735.

Kawasaki H, Mizuseki K, Nishikawa S et al. (2000). Induction of midbrain dopaminergic neurons from ES cells by stromal cell-derived inducing activity. *Neuron* 28: 31–40.

Kim JH, Auerbach JM, Rodriguez-Gomez JA et al. (2002). Dopamine neurons derived from embryonic stem cells function in an animal model of Parkinson's disease. *Nature* 418: 50–56.

Kirik D, Rosenblad C, and Bjorklund A (2000a). Preservation of a functional nigrostriatal dopamine pathway by GDNF in the intrastriatal 6-OHDA lesion model depends on the site of administration of the trophic factor. *Eur J Neurosci* 12: 3871–3882.

Kirik D, Rosenblad C, Bjorklund A, and Mandel RJ (2000b). Long-term rAAV-mediated gene transfer of GDNF in the rat Parkinson's model: intrastriatal but not intranigral transduction promotes functional regeneration in the lesioned nigrostriatal system. *J Neurosci* 20: 4686–4700.

Kirik D, Georgievska B, Burger C et al. (2002). Reversal of motor impairments in parkinsonian rats by continuous intrastriatal delivery of L-dopa using rAAV-mediated gene transfer. *Proc Natl Acad Sci U S A* 99: 4708–4713.

Kordower JH, Freeman TB, Snow BJ et al. (1995). Neuropathological evidence of graft survival and striatal reinnervation after the transplantation of fetal mesencephalic tissue in a patient with Parkinson's disease. *N Engl J Med* 332: 1118–1124.

Kordower JH, Emborg ME, Bloch J et al. (2000). Neurodegeneration prevented by lentiviral vector delivery of GDNF in primate models of Parkinson's disease. *Science* 290: 767–773.

Lang AE, Gill S, Patel NK et al. (2006). Randomized controlled trial of intraputamenal glial cell line-derived neurotrophic factor infusion in Parkinson disease. *Ann Neurol* 59: 459–466.

Langston JW, Ballard P, Tetrud JW, and Irwin I (1983). Chronic Parkinsonism in humans due to a product of meperidine-analog synthesis. *Science* 219: 979–980.

Langston JW, Forno LS, Rebert CS, and Irwin I (1984). Selective nigral toxicity after systemic administration of 1-methyl-4-phenyl-1,2,5,6-tetrahydropyrine (MPTP) in the squirrel monkey. *Brain Res* 292: 390–394.

Lee SH, Lumelsky N, Studer L et al. (2000). Efficient generation of midbrain and hindbrain neurons from mouse embryonic stem cells. *Nat Biotechnol* 18: 675–679.

Lie DC, Dziewczapolski G, Willhoite AR et al. (2002). The adult substantia nigra contains progenitor cells with neurogenic potential. *J Neurosci* 22: 6639–6649.

Lin LF, Doherty DH, Lile JD et al. (1993). GDNF: a glial cell line-derived neurotrophic factor for midbrain dopaminergic neurons. *Science* 260: 1130–1132.

Linazasoro G (2004). Recent failures of new potential symptomatic treatments for Parkinson's disease: causes and solutions. *Mov Disord* 19: 743–754.

Lo Bianco C, Schneider BL, Bauer M et al. (2004). Lentiviral vector delivery of parkin prevents dopaminergic degeneration in an alpha-synuclein rat model of Parkinson's disease. *Proc Natl Acad Sci U S A* 101: 17510–17515.

Lois C and Alvarez-Buylla A (1993). Proliferating subventricular zone cells in the adult mammalian forebrain can differentiate into neurons and glia. *Proc Natl Acad Sci U S A* 90: 2074–2077.

Luo J, Kaplitt MG, Fitzsimons HL et al. (2002). Subthalamic GAD gene therapy in a Parkinson's disease rat model. *Science* 298: 425–429.

Luskin MB (1993). Restricted proliferation and migration of postnatally generated neurons derived from the forebrain subventricular zone. *Neuron* 11: 173–189.

Magavi SS, Leavitt BR, and Macklis JD (2000). Induction of neurogenesis in the neocortex of adult mice. *Nature* 405: 951–955.

Mandel RJ, Rendahl KG, Spratt SK et al. (1998). Characterization of intrastriatal recombinant adeno-associated virus-mediated gene transfer of human tyrosine hydroxylase and human GTP-cyclohydrolase I in a rat model of Parkinson's disease. *J Neurosci* 18: 4271–4284.

Mao L, Lau YS, Petroske E, and Wang JQ (2001). Profound astrogenesis in the striatum of adult mice following nigrostriatal dopaminergic lesion by repeated MPTP administration. *Brain Res Dev Brain Res* 131: 57–65.

Maxwell SL, Ho HY, Kuehner E et al. (2005). Pitx3 regulates tyrosine hydroxylase expression in the substantia nigra and identifies a subgroup of mesencephalic dopaminergic progenitor neurons during mouse development. *Dev Biol* 282: 467–479.

Mochizuki H, Hayakawa H, Migita M et al. (2001). An AAV-derived Apaf-1 dominant negative inhibitor prevents MPTP toxicity as antiapoptotic gene therapy for Parkinson's disease. *Proc Natl Acad Sci U S A* 98: 10918–10923.

Mohapel P, Frielingsdorf H, Haggblad J et al. (2005). Platelet-derived growth factor (PDGF-BB) and brain-derived neurotrophic factor (BDNF) induce striatal neurogenesis in adult rats with 6-hydroxydopamine lesions. *J Neurosci* 132: 767–776.

Molinoff PB, Brimijoin S, and Axelrod J (1971). Trans-synaptic induction of dopamine (3,4-dihydroxyphenethylamine)-hydroxylase in adrenergic tissues of the rat. *Biochem J* 123: 32P–33P.

Morrish PK, Sawle GV, and Brooks DJ (1995). Clinical and [18F] dopa PET findings in early Parkinson's disease. *J Neurol Neurosurg Psychiatry* 59: 597–600.

Muramatsu S, Fujimoto K, Ikeguchi K et al. (2002). Behavioral recovery in a primate model of Parkinson's disease by triple transduction of striatal cells with adeno-associated viral vectors expressing dopamine-synthesizing enzymes. *Hum Gene Ther* 13: 345–354.

Murase S and McKay RD (2006). A specific survival response in dopamine neurons at most risk in Parkinson's disease. *J Neurosci* 26: 9750–9760.

Nait-Oumesmar B, Picard-Riera N, Kerninon C et al. (2007). Activation of the subventricular zone in multiple sclerosis: evidence for early glial progenitors. *Proc Natl Acad Sci U S A* 104: 4694–4699.

Nakatomi H, Kuriu T, Okabe S et al. (2002). Regeneration of hippocampal pyramidal neurons after ischemic brain injury by recruitment of endogenous neural progenitors. *Cell* 110: 429–441.

Nilsson M, Perfilieva E, Johansson U et al. (1999). Enriched environment increases neurogenesis in the adult rat dentate gyrus and improves spatial memory. *J Neurobiol* 39: 569–578.

Nunes I, Tovmasian LT, Silva RM et al. (2003). Pitx3 is required for development of substantia nigra dopaminergic neurons. *Proc Natl Acad Sci U S A* 100: 4245–4250.

Nurmi E, Ruottinen HM, Bergman J et al. (2001). Rate of progression in Parkinson's disease: a 6-[18F]fluoro-L-dopa PET study. *Mov Disord* 16: 608–615.

Nutt JG, Woodward WR, Beckner RM et al. (1994). Effect of peripheral catechol-O-methyltransferase inhibition on the pharmacokinetics and pharmacodynamics of levodopa in parkinsonian patients. *Neurology* 44: 913–919.

Nutt JG, Burchiel KJ, Comella CL et al. (2003). Randomized, double-blind trial of glial cell line-derived neurotrophic factor (GDNF) in PD. *Neurology* 60: 69–73.

Obeso JA, Rodriguez-Oroz MC, Rodriguez M et al. (2000). Pathophysiologic basis of surgery for Parkinson's disease. *Neurology* 55: S7–S12.

Oertel WH, Hoglinger GU, Caraceni T et al. (2001). Depression in Parkinson's disease. An update. *Adv Neurol* 86: 373–383.

Oertel WH, Wolters E, Sampaio C et al. (2006). Pergolide versus levodopa monotherapy in early Parkinson's disease patients: The PELMOPET study. *Mov Disord* 21: 343–353.

Okabe S, Forsberg-Nilsson K, Spiro AC et al. (1996). Development of neuronal precursor cells and functional postmitotic neurons from embryonic stem cells in vitro. *Mech Dev* 59: 89–102.

Olanow CW, Goetz CG, Kordower JH et al. (2003). A double-blind controlled trial of bilateral fetal nigral transplantation in Parkinson's disease. *Ann Neurol* 54: 403–414.

Oppenheim RW (1991). Cell death during development of the nervous system. *Annu Rev Neurosci* 14: 453–501.

Palmer TD, Ray J, and Gage FH (1995). FGF-2-responsive neuronal progenitors reside in proliferative and quiescent regions of the adult rodent brain. *Mol Cell Neurosci* 6: 474–486.

Parkinson-Study-Group (2000). A randomized controlled trial comparing pramipexole with levodopa in early Parkinson's disease: design and methods of the CALM-PD Study. Parkinson Study Group. *Clin Neuropharmacol* 23: 34–44.

Parkinson-Study-Group (2004). A controlled, randomized, delayed-start study of rasagiline in early Parkinson disease. *Arch Neurol* 61: 561–566.

Perlow MJ, Freed WJ, Hoffer BJ et al. (1979). Brain grafts reduce motor abnormalities produced by destruction of nigrostriatal dopamine system. *Science* 204: 643–647.

Piccini P, Brooks DJ, Bjorklund A et al. (1999). Dopamine release from nigral transplants visualized in vivo in a Parkinson's patient. *Nat Neurosci* 2: 1137–1140.

Piccini P, Lindvall O, Bjorklund A et al. (2000). Delayed recovery of movement-related cortical function in Parkinson's disease after striatal dopaminergic grafts. *Ann Neurol* 48: 689–695.

Pillon B, Ertle S, Deweer B et al. (1997). Memory for spatial location in d'e novo' parkinsonian patients. *Neuropsychologia* 35: 221–228.

Puelles E, Acampora D, Lacroix E et al. (2003). Otx dose-dependent integrated control of antero-posterior and dorso-ventral patterning of midbrain. *Nat Neurosci* 6: 453–460.

Rascol O, Brooks DJ, Korczyn AD et al. (2000). A five-year study of the incidence of dyskinesia in patients with early Parkinson's disease who were treated with ropinirole or levodopa. 056 Study Group. *N Engl J Med* 342: 1484–1491.

Rascol O, Brooks DJ, Melamed E et al. (2005). Rasagiline as an adjunct to levodopa in patients with Parkinson's disease and motor fluctuations (LARGO, Lasting effect in Adjunct therapy with Rasagiline Given Once daily, study): a randomised, double-blind, parallel-group trial. *Lancet* 365: 947–954.

Redmond DE, Sladek JR Jr, Roth RH et al. (1986). Fetal neuronal grafts in monkeys given methylphenyltetrahydropyridine. *Lancet* 1: 1125–1127.

Reynolds BA and Weiss S (1992). Generation of neurons and astrocytes from isolated cells of the adult mammalian central nervous system. *Science* 255: 1707–1710.

Roy NS, Cleren C, Singh SK et al. (2006). Functional engraftment of human ES cell-derived dopaminergic neurons enriched by coculture with telomerase-immortalized midbrain astrocytes. *Nat Med* 12: 1259–1268.

Rueger MA, Kracht LW, Hilker R et al. (2007). Role of in vivo imaging of the central nervous system for developing novel drugs. *Q J Nucl Med Mol Imaging* 51: 164–181.

Sachs C and Jonsson G (1975). Mechanisms of action of 6-hydroxydopamine. *Biochem Pharmacol* 24: 1–8.

Santarelli L, Saxe M, Gross C et al. (2003). Requirement of hippocampal neurogenesis for the behavioral effects of antidepressants. *Science* 301: 805–809.

Saucedo-Cardenas O, Quintana-Hau JD, Le WD et al. (1998). Nurr1 is essential for the induction of the dopaminergic phenotype and the survival of ventral mesencephalic late dopaminergic precursor neurons. *Proc Natl Acad Sci U S A* 95: 4013–4018.

Schapira AH (2002). Dopamine agonists and neuroprotection in Parkinson's disease. *Eur J Neurol* 9 Suppl 3: 7–14.

Schapira AH, Bezard E, Brotchie J et al. (2006). Novel pharmacological targets for the treatment of Parkinson's disease. *Nat Rev Drug Discov* 5: 845–854.

Schulz JB, Skalej M, Wedekind D et al. (1999). Magnetic resonance imaging-based volumetry differentiates idiopathic Parkinson's syndrome from multiple system atrophy and progressive supranuclear palsy. *Ann Neurol* 45: 65–74.

Seppi K, Schocke MF, Esterhammer R et al. (2003). Diffusion-weighted imaging discriminates progressive supranuclear palsy from PD, but not from the parkinson variant of multiple system atrophy. *Neurology* 60: 922–927.

Shan X, Chi L, Bishop M et al. (2006). Enhanced de novo neurogenesis and dopaminergic neurogenesis in the substantia nigra of 1-methyl-4-phenyl-1,2,3,6-tetrahydropyridine-induced Parkinson's disease-like mice. *Stem Cells* 24: 1280–1287.

Shen Y, Muramatsu SI, Ikeguchi K et al. (2000). Triple transduction with adeno-associated virus vectors expressing tyrosine hydroxylase, aromatic-L-amino-acid decarboxylase, and GTP cyclohydrolase I for gene therapy of Parkinson's disease. *Hum Gene Ther* 11: 1509–1519.

Simon HH, Saueressig H, Wurst W et al. (2001). Fate of midbrain dopaminergic neurons controlled by the engrailed genes. *J Neurosci* 21: 3126–3134.

Smidt MP, Asbreuk CH, Cox JJ et al. (2000). A second independent pathway for development of mesencephalic dopaminergic neurons requires Lmx1b. *Nat Neurosci* 3: 337–341.

Smidt MP, Smits SM, Bouwmeester H et al. (2004). Early developmental failure of substantia nigra dopamine neurons in mice lacking the homeodomain gene Pitx3. *Development* 131: 1145–1155.

Smits SM, Ponnio T, Conneely OM et al. (2003). Involvement of Nurr1 in specifying the neurotransmitter identity of ventral midbrain dopaminergic neurons. *Eur J Neurosci* 18: 1731–1738.

Steiner B, Winter C, Hosman K et al. (2006). Enriched environment induces cellular plasticity in the adult substantia nigra and improves motor behavior function in the 6-OHDA rat model of Parkinson's disease. *Exp Neurol* 199: 291–300.

Studer L, Tabar V, and McKay RD (1998). Transplantation of expanded mesencephalic precursors leads to recovery in parkinsonian rats. *Nat Neurosci* 1: 290–295.

Sun M, Kong L, Wang X et al. (2004). Coexpression of tyrosine hydroxylase, GTP cyclohydrolase I, aromatic amino acid decarboxylase, and vesicular monoamine transporter 2 from a helper virus-free herpes simplex virus type 1 vector supports high-level, long-term biochemical and behavioral correction of a rat model of Parkinson's disease. *Hum Gene Ther* 15: 1177–1196.

Tomac A, Lindqvist E, Lin LF et al. (1995). Protection and repair of the nigrostriatal dopaminergic system by GDNF in vivo. *Nature* 373: 335–339.

Ungerstedt U (1968). 6-Hydroxy-dopamine induced degeneration of central monoamine neurons. *Eur J Pharmacol* 5: 107–110.

Ungerstedt U and Arbuthnott GW (1970). Quantitative recording of rotational behavior in rats after 6-hydroxy-dopamine lesions of the nigrostriatal dopamine system. *Brain Res* 24: 485–493.

van den Munckhof P, Luk KC, Ste-Marie L et al. (2003). Pitx3 is required for motor activity and for survival of a subset of midbrain dopaminergic neurons. *Development* 130: 2535–2542.

van der Walt, JM, Noureddine MA, Kittappa R et al. (2004). Fibroblast growth factor 20 polymorphisms and haplotypes strongly influence risk of Parkinson disease. *Am J Hum Genet* 74: 1121–1127.

Van Kampen JM and Eckman CB (2006). Dopamine D3 receptor agonist delivery to a model of Parkinson's disease restores the nigrostriatal pathway and improves locomotor behavior. *J Neurosci* 26: 7272–7280.

Van Kampen JM and Robertson HA (2005). A possible role for dopamine D3 receptor stimulation in the induction of neurogenesis in the adult rat substantia nigra. *J Neurosci* 136: 381–386.

van Praag H, Schinder AF, Christie BR et al. (2002). Functional neurogenesis in the adult hippocampus. *Nature* 415: 1030–1034.

Vila M and Przedborski S (2003). Targeting programmed cell death in neurodegenerative diseases. *Nat Rev Neurosci* 4: 365–375.

Wakayama T, Tabar V, Rodriguez I, Perry AC et al. (2001). Differentiation of embryonic stem cell lines generated from adult somatic cells by nuclear transfer. *Science* 292: 740–743.

Wallen A, Zetterstrom RH, Solomin L et al. (1999). Fate of mesencephalic AHD2-expressing dopamine progenitor cells in NURR1 mutant mice. *Exp Cell Res* 253: 737–746.

Wallen AA, Castro DS, Zetterstrom RH et al. (2001). Orphan nuclear receptor Nurr1 is essential for Ret expression in midbrain dopamine neurons and in the brain stem. *Mol Cell Neurosci* 18: 649–663.

Walter U, Niehaus L, Probst T et al. (2003). Brain parenchyma sonography discriminates Parkinson's disease and atypical parkinsonian syndromes. *Neurology* 60: 74–77.

Whone AL, Watts RL, Stoessl AJ et al. (2003). Slower progression of Parkinson's disease with ropinirole versus levodopa: The REAL-PET study. *Ann Neurol* 54: 93–101.

Xia XG, Harding T, Weller M et al. (2001). Gene transfer of the JNK interacting protein-1 protects dopaminergic neurons in the MPTP model of Parkinson's disease. *Proc Natl Acad Sci U S A* 98: 10433–10438.

Yamada T, McGeer PL, Baimbridge KG, and McGeer EG (1990). Relative sparing in Parkinson's disease of substantia nigra dopamine neurons containing calbindin-D28K. *Brain Res* 526: 303–307.

Ye W, Shimamura K, Rubenstein JL et al. (1998). FGF and Shh signals control dopaminergic and serotonergic cell fate in the anterior neural plate. *Cell* 93: 755–766.

Youdim MB, Edmondson D, and Tipton KF (2006). The therapeutic potential of monoamine oxidase inhibitors. *Nat Rev Neurosci* 7: 295–309.
Zetterstrom RH, Solomin L, Jansson L et al. (1997). Dopamine neuron agenesis in Nurr1-deficient mice. *Science* 276: 248–250.
Zhao M, Momma S, Delfani K et al. (2003). Evidence for neurogenesis in the adult mammalian substantia nigra. *Proc Natl Acad Sci U S A* 100: 7925–7930.
Zuddas A, Corsini GU, Barker JL et al. (1991). Specific reinnervation of lesioned mouse striatum by grafted mesencephalic dopaminergic neurons. *Eur J Neurosci* 3: 72–85.

7

ADULT NEURAL PROGENITOR CELLS IN CNS FUNCTION AND DISEASE

SEBASTIAN JESSBERGER, STEFAN AIGNER, JAMES B. AIMONE, AND FRED. H. GAGE

The Salk Institute for Biological Studies, La Jolla, CA 92037

ABSTRACT

Neural stem cells (NSCs) give rise to the central nervous system (CNS) during development and contribute to the cellular plasticity of the adult mammalian brain. The isolation and in vitro culture of NSCs provides the unique opportunity to address fundamental questions about cell proliferation, fate choice, differentiation, and synaptic maturation under experimentally defined conditions. The theoretical foundation of NSC biology rests basically on the three defining properties that a single NSC must fulfill: self-renewal, multipotency, and clonality. A popular cell culture tool, the neurosphere assay, has been used to study these key features of NSCs since 1992. Similar assays are now being used for other somatic stem cells including cancer stem cells. Recent progress in the stem cell field has been made by revisiting the neurosphere concept and demonstrating its actual potential and limits. This article discusses the immediate consequences of these findings and emphasizes the need for an improved and commonly accepted protocol for the characterization of NSCs in vitro.

Copyright © 2008 Elsevier Inc.
All rights of reproduction in any form reserved.

INTRODUCTION

Dividing neural stem/progenitor cells (NPCs) give rise to all three neural lineages during embryonic and early postnatal development. It had been a long-held belief that the birth of neuronal cells is restricted to this period and its rate tapers off in the early postnatal phase. Despite early evidence suggesting the generation of new neurons even in the adult brain (Altman and Das, 1965), the concept of adult neurogenesis – meaning the development of a mature neuron from a dividing progenitor – remained controversial for decades (Hinds, 1968; Kaplan and Hinds, 1977; Rakic, 1985, 2002). However, today there is ample and conclusive evidence from many mammalian species that NPCs persist throughout adulthood and are capable of giving rise to new neurons in two restricted areas of the adult brain: the hippocampal dentate gyrus (DG) and the subventricular zone (SVZ) of the lateral ventricle (Figure 7.1) (Ming and Song, 2005). Whereas neurogenesis appears to occur in the human hippocampus (Eriksson et al., 1998), the existence of SVZ neurogenesis in humans remains controversial (Bedard and Parent, 2004; Sanai et al., 2004; Curtis et al., 2007).

The finding that cells with certain features of NPCs persist in the adult brain and that new neurons are continuously added throughout life sparked tremendous scientific interest in adult neurogenesis and provided fresh hope for novel therapeutic avenues treating neurological disease. Adult NPCs might be used to replenish damaged brain structures in neurodegenerative or traumatic brain injuries, either by targeted activation of endogenous NPCs or by transplantation of in vitro propagated NPCs. In addition, adult neurogenesis might be a critical part of disease processes themselves. In this chapter, which is mostly focused on adult hippocampal neurogenesis, we will discuss recent advances in the

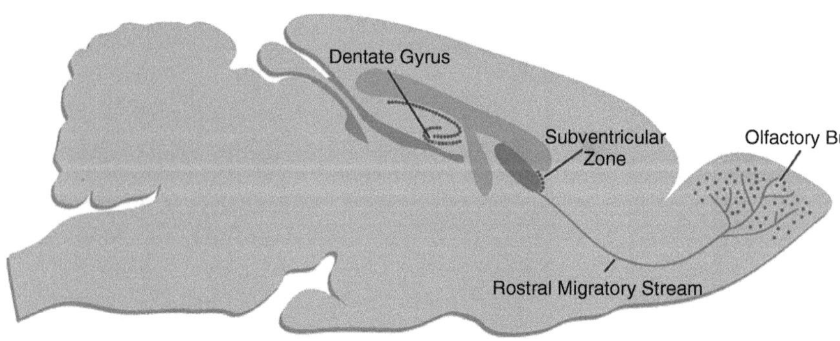

FIGURE 7.1 Neurogenesis occurs in two distinct regions of the adult brain. A schematic drawing of a sagittal section through a rat brain shows these two neurogenic regions of the adult brain. NPCs in the dentate gyrus generate new granule cells throughout life. In addition, new neuronal cells are born in the subventricular zone and migrate through the rostral migratory stream into the olfactory bulb, where they differentiate into olfactory granule cells and periglomerular neurons. (see Plate 1)

understanding of NPC biology and the potential function that adult neurogenesis might have in physiology and disease.

BIOLOGY OF NEURAL STEM CELLS

The identification and in vitro characterization of proliferative cells isolated from adult brain tissue (Reynolds and Weiss, 1992) was a crucial step in subsequently identifying a subpopulation of them as stem cells. The most commonly used definition of a stem cell is a cell that is able to generate daughter cells different from it and to renew itself. Most tissue stem cells are multipotent, which means they can give rise to all cell types of a given tissue. Even though self-renewal and multipotency have been shown convincingly in cells that have been isolated from the adult brain and propagated in vitro, there is only little, if any, evidence of those stem cell features for dividing cells within the adult brain (Gage, 2000).

The standard technique to culture NPCs is to maintain them as *neurospheres*, free-floating cell aggregates that form from isolated cells (Reynolds and Weiss, 1992; Lie et al., 2004). Even though NPCs do not grow clonally in neurosphere cultures (Singec et al., 2006; Jessberger et al., 2007a), their clonal self-renewal capacity can be tested by the formation of secondary spheres, provided the cultures were started with single cells growing in individual culture vessels. Alternatively, NPCs can be grown as *monolayers* with cells growing adherent to the culture dish surface (Figure 7.2) (Palmer et al., 1995; Ray et al., 1995; Ray and Gage, 2006). Notably, cells with NPC properties can be also isolated from fresh and post-mortem human tissue (Palmer et al., 2001).

The NPC cultures are extensively used to study the potency of NPCs in vitro (e.g., Seaberg and van der Kooy, 2002; Babu et al., 2007) and to identify molecular mechanisms of proliferation and differentiation (Lie et al., 2004; Ming and Song, 2005). We are only beginning to understand the molecular mechanisms that determine whether an adult NPC continues to proliferate or differentiates toward the glial or neuronal lineages. However, it is becoming increasingly clear that the balance of mitogenic, neurogenic, and gliogenic factors within a given adult neurogenic niche ultimately determines the fate of adult NPCs. A number of signaling cascades initiated by the action of hormones, growth factors, and neurotransmitters has been suggested to regulate neurogenesis (Lie et al., 2004; Ming and Song, 2005). As most studies examine single pathways in isolation, we do not currently have a good understanding of the interplay of different signaling pathways.

Regulatory factors include vascular endothelial growth factor (VEGF), epidermal growth factor (EGF), transforming growth factor α (TGFα), and basic fibroblast growth factor (bFGF or FGF-2) (Ming and Song, 2005). For NPCs isolated from the hippocampal SGZ, FGF-2–dependent cell proliferation has been shown to require the glycosylated form of cystatin C (CCg) as a cofactor (Taupin et al., 2000). A similar role has been ascribed to membrane-bound heparan-sulfate

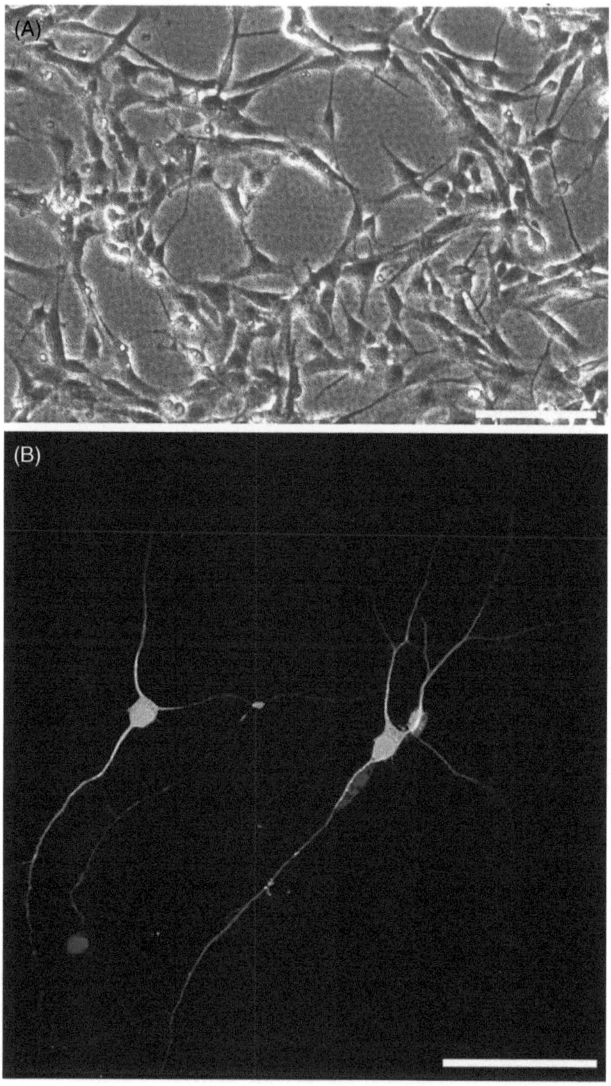

FIGURE 7.2 The NPCs can be propagated and differentiated in vitro. (A) Example of NPCs under proliferating conditions derived from the mouse SVZ and grown as a monolayer. (B) Neuronal differentiation of rat NPCs derived from the DG. Shown is an example of cells that differentiated into MAP2ab-expressing neuronal cells (red) following retroviral overexpression of the neurogenic bHLH transcription factor, NeuroD1 (transduced cells are green). Scale bars in A and B: 100 μm. (see Plate 2)

proteoglycans (Ray and Gage, 2006). Another example is sonic hedgehog (Shh), a morphogenetic signaling factor with complex roles in patterning during CNS development, that has been shown to also be involved in adult NPC maintenance both in vitro and in vivo (Lai et al., 2003).

Additional signaling pathways have been identified that are involved in fate specification of adult NPCs toward the glial or neuronal lineage. Fate choice appears to be mediated by the compound effect of competing gliogenic and neurogenic signaling molecules. Bone morphogenetic protein (BMP) instructs adult NPCs to adopt a glial fate. In neurogenic brain niches, however, the gliogenic influence of BMP is counteracted by antagonists. These include noggin, secreted by ependymal cells in the SVZ, and neurogenesin-1, secreted by astrocytes in the SGZ. However, it appears that the action of BMP antagonists alone is not sufficient to induce neuronal differentiation of adult NPCs. Instead, canonical Wnt signaling has been identified as a major player in positively regulating neurogenesis of adult NPCs of the hippocampus (Lie et al., 2005). Insights from studies in which exogenous brain-derived neurotrophic factor (BDNF) was delivered to neurogenic brain areas have implicated this factor in promoting adult NPC proliferation, neuronal differentiation, and survival (Abrous et al., 2005). It is not clear, however, if BDNF exerts these effects by directly stimulating NPCs or whether this neurotrophic factor alters the cellular microenvironment in which they reside. In addition to these established signaling pathways that are involved during embryonic and early postnatal neuronal development, a small non-coding RNA has also been implicated in specifying the neuronal fate of adult hippocampal NPCs (Kuwabara et al., 2004).

Surprisingly, little is known regarding how differentiation toward glial cells is regulated, and only a few signaling pathways have been identified that induce NPCs to differentiate into the oligodendrocytic or astrocytic lineage (Hsieh et al., 2004; Bonaguidi et al., 2005).

In vitro cultures of NPCs provided a great tool to get first insights into molecular mechanisms governing cellular potency and differentiation. However, there is also increasing evidence that NPCs behave very differently in vivo than in the "naked" in vitro situation. NPCs isolated from brain regions that are non-neurogenic in vivo (e.g., spinal cord) can properly differentiate into neuronal cells when differentiated under in vitro conditions. These discrepancies emphasize the potentially vast differences between in vivo and in vitro behavior of NPCs.

NEUROGENESIS IN THE ADULT BRAIN

Despite the above-mentioned lack of clear evidence of cells displaying stem cell characteristics in vivo, we will refer below to the dividing cells in the neurogenic areas that presumably give rise to new neurons and/or glial cells as NPCs. How can we visualize NPCs in vivo and what are the molecular characteristics of NPCs within their respective niches?

Proliferative activity within the dentate area occurs largely in the zone just below the granule cell layer (GCL) and the hilus. This area, called the subgranular zone (SGZ), is not strictly defined by anatomical boundaries but rather stretches about two cell layers deep into the hilus below the GCL (corresponding to approximately 20–30 μm in the rodent hippocampus) (Figure 7.3). Dividing cells can be detected by visualizing endogenous cell cycle markers such as Ki-67, phosphorylated histone H3, proliferative cell nuclear antigen (PCNA), or minichromosome maintenance deficient 2 (Mcm2). Exogenous labels such as the nucleotide analogues bromo-deoxyuridine (BrdU) and tritiated (^{3}H)-thymidine (which are incorporated into newly synthesized DNA) and marker-expressing retroviruses (which integrate into the genome) also allow tracking the progeny of dividing cells (Figure 7.4). Today the most commonly used label is BrdU, which is recognized by specific antibodies and thus permits relatively easy and reliable quantification and phenotyping of newborn cells when used in combination with cell type-specific markers such as NeuN (neurons), S100ß (astrocytes), or NG-2 (oligodendrocytic cells). However, BrdU detection requires immunohistochemical processing and visualizes only the nucleus of newborn cells. Those drawbacks can be overcome with the use of retroviral vectors expressing fluorescent proteins, thus enabling the morphological and physiological characterization of newborn cells in the adult brain (Figure 7.4) (van Praag et al., 2002; Carleton et al., 2003; Zhao et al., 2006; Toni et al., 2007).

Using thymidine analogues, an astounding picture emerges, showing that several types of dividing cells exist within the adult hippocampus (Figure 7.3) (Kempermann et al., 2004; Encinas et al., 2006). The so-called type-1 cells have a radial glia-like morphology and express the stem/progenitor markers nestin and Sox-2 as well as astrocytic markers such as GFAP and BLBP (Seri et al., 2001; Filippov et al., 2003; Kronenberg et al., 2003; Steiner et al., 2006). In contrast to the slowly dividing type-1 cells, type-2 cells exhibit a much higher proliferative activity. Type-2 cells are negative for GFAP but still express nestin and Sox-2, and a fraction of type-2 cells expresses early neuronal markers such as doublecortin (DCX) and the homeobox transcription factor Prox-1 (Kronenberg et al., 2003). Recent studies suggest that some of the DCX-expressing cells retain multipotentiality and self-renewal capacity in vitro (Walker et al., 2007). Type-3 cells express DCX but lack expression of nestin and Sox-2 and might represent committed neuroblasts (Jessberger et al., 2005).

Basically, the situation in the SVZ seems to be comparable to the DG (for details, see: Doetsch et al., 1999; Alvarez-Buylla and Garcia-Verdugo, 2002) even though contradictory concepts have been proposed describing the "neural stem cell" in the SVZ: several studies suggest that GFAP-positive astrocytic cells in the subependymal zone (e.g., Doetsch et al., 1999; Laywell et al., 2000) represent the stem cells of the adult SVZ, whereas others find that ependymal cells lining the ventricles possess stem cell properties (e.g., Johansson et al., 1999). More detailed lineage tracing strategies are underway that will hopefully add new data to finally resolve this controversy.

Notably, cells that express NPC markers such as nestin exist throughout the brain (Kronenberg et al., 2003). Cells with stem cells properties can be isolated and propagated in vitro from many areas that are non-neurogenic under

FIGURE 7.3 Expression of proteins associated with adult neurogenesis. (A) The somata of dividing NPCs in the adult hippocampus are localized in the SGZ and express a distinct set of marker proteins (Kempermann et al., 2004). Sox-2 (green in A, B) is expressed in the SGZ and labels cells that also co-label with the intermediate filament nestin (red in A, B) showing a radial glia-like morphology (arrowheads). (B) Similar markers are expressed in the adult SVZ, where a large number of cells express Sox-2 and nestin. (C) Dividing progenitors and immature newborn granule cells in the DG transiently express the microtubuli-associated protein DCX (red in B). The granule cell-specific transcription factor, Prox1 (green in C), is expressed early on (note the partial co-expression with DCX) but continues to label granule cells once they mature. GCL: granule cell layer, ML: molecular layer. V: ventricle. Scale bars in A–C are 50 μm. (see Plate 3)

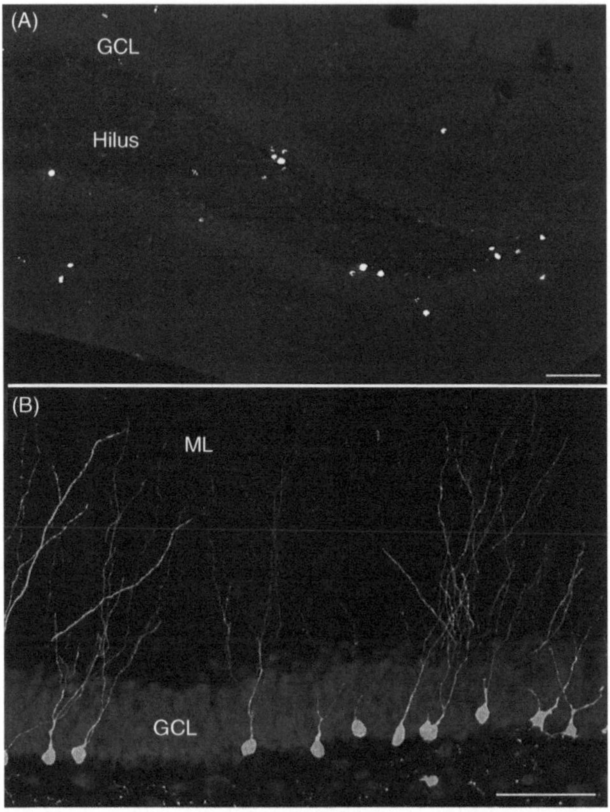

FIGURE 7.4 Labeling of newborn cells in the adult brain using thymidine-analogues or retroviral vectors. (A) BrdU is integrated into the DNA during s-phase and can be later visualized using specific antibodies. The example shown is from a mouse DG 6 weeks following a series of BrdU injections. Note that the majority of the cells are still localized in the inner third of the DG. (B) Intracerebral injections of fluorescence label-expressing retroviral vectors (GFP in green) that specifically integrate into dividing cells and label their progeny allow the whole cell visualization (e.g., Zhao et al., 2006), physiological characterization (e.g., Laplagne et al., 2006), and genetic manipulation (e.g., Tashiro et al., 2006) of newborn cells in the adult brain. Note the highly polarized morphology of newborn granule cells extending an apical dendrite into the ML. GCL: granule cell layer, ML: molecular layer. Scale bars in A and B: 100 μm. (see Plate 4)

physiological conditions in vivo (e.g., Horner et al., 2000; Lie et al., 2002). Thus, it seems that neurogenesis is either actively suppressed or not sufficiently supported in non-neurogenic areas, even though the cells (i.e., NPCs) are present. What distinguishes the NPC niche of the two neurogenic areas from the rest of the adult brain is unclear, but vascular architecture (Palmer et al., 2000) and the type of astrocytic cells and the signaling molecules they provide might be critically involved in the neurogenic permissiveness of the DG and the SVZ (Song et al., 2002; Lie et al., 2005).

MATURATION OF NEWBORN NEURONS IN THE ADULT HIPPOCAMPUS

Once dividing cells exit cell cycle, a complex maturation process starts that eventually leads to the integration of newborn neurons into the preexisting circuitry. Several laboratories have identified a set of molecular markers that are subsequently expressed over the first weeks following neuronal birth (Kempermann et al., 2004; Couillard-Despres et al., 2005; Piatti et al., 2006). Importantly, only a fraction of newborn neurons survives an initial critical period that appears to be regulated by an activity-dependent selection process (Tashiro et al., 2006). Once newborn granule cells survive this first phase, they become stably integrated into the DG circuitry (Kempermann et al., 2003). The integration of newborn neurons has been shown to occur on a functional (e.g., receiving input, participating in IEG responses to stimuli) and morphological level (e.g., forming synapses, extending axons to area CA3) (van Praag et al., 2002; Jessberger and Kempermann, 2003; Zhao et al., 2006; Toni et al., 2007). Interestingly, emerging evidence has suggested that newborn cells possess very distinct functional features in the first weeks after they are born compared to mature granule cells, such as a lower threshold for activation and increased levels of long-term potentiation (Wang et al., 2000; Schmidt-Hieber et al., 2004; Laplagne et al., 2006, Ge et al., 2007). We will discuss these special features, which might hold the key to understanding the biological relevance of adult neurogenesis, in more detail below.

MATURATION OF NEWBORN NEURONS IN THE ADULT OLFACTORY BULB

The anatomy of the neurogenic niche in the lateral ventricles is very distinct from that of the DG subgranular zone. Early after cell division of NPCs, neuronal cells migrate tangentially to the SVZ until they reach the rostral migratory stream (RMS) that leads newborn cells toward the olfactory bulb (OB) (Alvarez-Buylla and Garcia-Verdugo, 2002; Lie et al., 2004; Lledo et al., 2006). Two different types of neurons are generated in the adult OB. The majority of cells become OB granule cells, but a smaller fraction differentiates into OB periglomerular cells (Alvarez-Buylla and Garcia-Verdugo, 2002). Detailed electrophysiological analyses have identified several stages of neuronal maturation during which newborn neurons have to survive an activity-dependent selection process (Winner et al., 2002; Carleton et al., 2003; Lledo and Saghatelyan, 2005; Lledo et al., 2006). Finally, as with hippocampal neurogenesis, newborn OB neurons appear to form a highly plastic component of the adult OB (Magavi et al., 2005; Mizrahi, 2007).

REGULATION OF NEUROGENESIS

The number of neurons born in the adult hippocampus is not static but rather dynamically regulated by a variety of physiological as well as pathological factors. How these stimuli translate into changes in the number of newborn neurons

TABLE 7.1 Regulators of adult neurogenesis

Regulator	Proliferation	Survival	Neuronal Differentiation
Physiological Regulators of Adult Neurogenesis			
Genetic background	+/−	+/−	+/−
Enriched environment	?	+	+
Physical exercise	+	no change	+
Learning	no change	+	?
Aging	−	?	−
Dietary restriction	no change	+	?
Neurotransmitters	see Text		
Pathological Regulators of Adult Neurogenesis			
Stress	−	no change	?
Seizure activity	+	+(?)	+
Ischemia	+	+(?)	+
Irradiation	−	−	−
Neurodegenerative Diseases			
Alzheimer's Disease	+/−	?	?/−
Huntington's Disease	−	no change	no change
Parkinson's Disease	−	?	?
Drugs			
Opiates	−	?	?
Antidepressants	+	+	+
Ethanol	no change/−	−	?

is not clearly understood. Emerging evidence suggests that cellular depolarization of NPCs might initiate neuronal differentiation (Deisseroth et al., 2004; Tozuka et al., 2005). Elegant studies have also shown that nestin-expressing cells in the SGZ receive functional GABAergic input (Tozuka et al., 2005). Furthermore, NPCs have been shown to respond to neurotrophic factors that are dynamically regulated within the adult brain, such as VEGF, BDNF, and NT-3 (Cao et al., 2004; Schanzer et al., 2004; Duman and Monteggia, 2006). The list of known regulators of adult neurogenesis is constantly growing (for reviews, see Abrous et al., 2005; Ming and Song, 2005). These studies point to the functional involvement of adult neurogenesis in the disease processes that underlie depression, epilepsy, or age-related cognitive decline. Table 7.1 summarizes the most important regulators of SVZ and DG neurogenesis.

FUNCTIONAL SIGNIFICANCE OF HIPPOCAMPAL NEUROGENESIS

The hippocampus has long been thought to be involved in two, somewhat distinct brain functions: the generation of spatial maps and the formation of episodic memories (O'Keefe and Conway, 1978; Jung and McNaughton, 1993; Milner et al., 1998; Best et al., 2001; Squire et al., 2004). The DG appears to be critically involved in separating the input into the hippocampus, mainly from entorhinal cortex connections (Kesner et al., 2004; Rolls and Kesner, 2006). There are a growing number of theoretical and computational studies suggesting a role for adult neurogenesis in hippocampal function (Kempermann, 2002; Schinder and Gage, 2004; Becker, 2005; Aimone et al., 2006; Wiskott et al., 2006). Computational approaches might therefore prove to be powerful tools for predicting potential roles of newborn neurons in hippocampal function, which may then be tested experimentally.

But what is the current experimental evidence for a functional role of newborn neurons? Studies examining the expression of immediate early genes have shown that new neurons appear to respond to functional stimulation of the DG and might be specifically responsive to previous experiences or in the formation of spatial maps (Jessberger and Kempermann, 2003; Kee et al., 2007; Tashiro et al., 2007). Furthermore, there is an impressive list of studies that have correlated the amount of neurogenesis with performance in hippocampus-dependent learning tasks (Abrous et al., 2005). The most prominent enhancers/determinants of adult neurogenesis being accompanied by improved cognitive performance are enriched environment (Kempermann et al., 1997), voluntary physical exercise (van Praag et al., 1999), and genetic background (Kempermann and Gage, 2002). Likewise, stress (Gould and Tanapat, 1999) and aging (Zyzak et al., 1995), lead to decreased neurogenesis and impaired cognition. A strong positive correlation between the levels of hippocampal neurogenesis and performance in behavioral tasks is also apparent in transgenic and knockout mice, such as the NPY knockouts (Abrous et al., 2005; Howell et al., 2005).

A growing number of studies have attempted to provide causal evidence of a function of adult neurogenesis by specifically ablating newborn neurons in the adult hippocampus. The most commonly used approaches to reduce the levels of adult neurogenesis include administration of cytostatic drugs, irradiation, and the generation of transgenic mice that allow the inducible ablation of dividing cells in a tissue-specific manner. However, all of these strategies also affect non-neurogenic cells and brain regions, mainly due to the lack of cell specificity of ablation or subsequent inflammation. A broadly accepted approach that is devoid of any confounding adverse effects remains to be developed. New strategies will likely be developed in the near future that target key genes in the process of adult neurogenesis, such as WNT-signaling (Lie et al., 2005), or critical transcription factors, such as NeuroD (Miyata et al., 1999) or TLX (Shi et al., 2004). The hope is that these approaches result in more specific ablation of neurogenesis, allowing a more conclusive interpretation of behavioral effects.

Despite the technical shortcomings of existing techniques, ablation studies have shown impairment in a trace eyeblink-conditioning paradigm using the cytostatic drug methylazoxymethanol acetate (MAM) (Shors et al., 2001), in the Morris water maze using irradiation (Snyder et al., 2005), in an object-recognition task using MAM (Bruel-Jungerman et al., 2005), and in contextual fear conditioning using irradiation and genetic ablation in GFAP-TK mice (Saxe et al., 2006; Winocur et al., 2006). Interestingly, recent evidence suggests that the ablation of neurogenesis might not necessarily result in learning impairment, as mice with ablated neurogenesis actually performed better in a task testing some aspects of working memory (Saxe et al., 2007).

Besides an involvement of adult neurogenesis in processes underlying certain forms of learning and memory, experiments using irradiation have shown that intact hippocampal neurogenesis appears to be required for the effectiveness of the antidepressant fluoxetine in certain mouse strains (Santarelli et al., 2003; Holick et al., 2007). However, neurogenesis does not appear to be a critical component per se in emotional control, because the anxiolytic effects of environmental enrichment were completely unaffected by the ablation of new neurons (Meshi et al., 2006), leaving the precise role of new neurons in emotional processing unclear.

FUNCTIONAL SIGNIFICANCE OF OB NEUROGENESIS

Similar to newborn neurons in the DG, new olfactory neurons have been shown to respond to external stimuli (Carlen et al., 2002) and correlative evidence suggests a link between the numbers of newborn neurons and olfactory learning (Mandairon et al., 2006). There is also evidence that newborn OB neurons appear to be critically involved in odor discrimination (Alonso et al., 2006). However, causal evidence for a functional role of OB neurogenesis in the adult brain is sparse (Lledo et al., 2006).

ADULT NEUROGENESIS AND DISEASE

ALTERED NEUROGENESIS AS A POTENTIAL CAUSE OF DISEASE

As adult neurogenesis is apparently a component of brain physiology, it seems reasonable to speculate that alteration of adult neurogenesis might cause neurological disease. But what is the current evidence for this assumption? Again, most data rely on correlations. Numerous studies have shown that stress, which is believed to be a key component in the disease process of major depression, dramatically reduces the number of newborn neurons in the adult hippocampus by a corticosteroid-dependent mechanism (for a review, see Abrous et al., 2005). Human patients suffering from major depressive disorders have altered serum levels of corticosteroids and at the same time show a reduction in hippocampal volume (Sapolsky, 2000). Interestingly, some antidepressant treatment regimens do actually regulate the number of newborn neurons in the adult hippocampus (Duman and Monteggia, 2006; Warner-Schmidt and Duman, 2006). Furthermore, the effectiveness of certain antidepressants appears to be dependent on the levels of adult neurogenesis (Santarelli et al., 2003; but see also: David et al., 2007). Modern imaging tools, such as magnetic resonance imaging, that provide an indirect measure of adult neurogenesis and are also applicable in humans

(Pereira et al., 2007) might further spur our understanding of the connection between neurogenesis and major depression.

Adult neurogenesis is also greatly altered following seizure activity (Bengzon et al., 1997; Parent et al., 1997; Scott et al., 1998; Jessberger et al., 2005). Notably, not only is the number of newborn neurons increased, but the morphology and localization of newborn granule cells are also altered, with cells ectopically localized in the hilus (Parent et al., 1997; Scharfman et al., 2000) and extending aberrant basal dendrites (Shapiro and Ribak, 2006). This abnormal integration pattern has led to the hypothesis that seizure-induced neurogenesis might be involved in epileptogenesis or seizure-associated cognitive decline (Parent, 2002). This notion is supported by several studies showing that inhibition of seizure-induced neurogenesis is beneficial for the epileptic brain (Jung et al., 2004; Jessberger et al., 2007b). However, a recent study showed that seizure-generated granule cells might be less excitable than newborn cells in running animals under certain circumstances, thus suggesting a more positive role for newborn neurons following seizures (Jakubs et al., 2006). More studies are needed to elucidate the role of seizure-induced neurogenesis in the epileptic disease process.

The number of newborn neurons is not only regulated by physiologic or pathologic events (see above) but also dramatically declines with age (Kuhn et al., 1996; Cameron and McKay, 1999; Rao et al., 2006). Levels of neurogenesis correlate with cognitive performance in old age (Montaron et al., 2006), and enriched environment or physical activity are also potent enhancers of neurogenesis and learning and memory in aged animals (Kempermann et al., 2002; van Praag et al., 2005). Thus, adult neurogenesis might emerge as a promising target in the treatment of age-associated cognitive decline (also see below).

NPCS AS A POTENTIAL TREATMENT SOURCE FOR NEUROLOGICAL DISEASE

In addition to a potential role in some diseases, the capability of the adult brain to produce new neurons from dividing NPCs might be used to treat neurological disease. Besides representing a therapeutic target in the above-mentioned cases of depression or aging, there is now growing evidence that the generation of new neurons from NPCs might be utilized in regions that are, under normal conditions, non-neurogenic (such as the cortex and striatum). For instance, occlusion of the middle cerebral artery in rodents leads to neuronal damage in the striatum (and potentially cortical areas). Interestingly, neuroblasts from the adjacent SVZ appear to be attracted by the lesion site and migrate toward the damaged area, where a small percentage differentiates into striatal neurons (Arvidsson et al., 2002; Parent et al., 2002; Chopp et al., 2007). The contribution of ischemia-induced striatal neurogenesis to functional recovery after stroke remains unclear, but the finding that new neurons can, in principle, be formed in ischemia-damaged areas opens up novel approaches to restore brain function following ischemia. There is also some evidence showing de novo generation of cortical neurons following targeted ablation of individual cells (Magavi et al., 2000). The functional relevance of this induction of neurogenesis in cortical areas as a consequence of disease remains unclear.

Obviously, the replacement of damaged neuronal structures with newborn neurons that are generated by endogenous NPCs might also be utilized as a therapeutic strategy in a variety of neurodegenerative diseases such as Alzheimer's, Parkinson's, or Huntington's disease (for review, see Lie et al., 2004). Here, amplification or induction of neurogenesis to replace neurons lost in neurodegenerative disease may provide therapeutic benefits. Whether the high expectations vested in neural stem biology can be fulfilled is far from clear, but, without a doubt, the identification of endogenous neurogenetic potential in the adult has added a completely new perspective to the existing toolbox of treatment options.

A different line of therapeutic intervention that is currently being ambitiously pursued is the transplantation of stem/progenitor cells, stem cell-derived progenitors, or stem cell-derived neurons to cure neurological disease (for a review, see Bjorklund and Lindvall, 2000; Carson et al., 2006). Despite promising initial results (e.g., Lindvall et al., 1994), there is no indication so far that cell transplantation will become the therapeutic standard of care, as was originally hoped. The reason for that setback does not necessarily lie in the transplantation strategy per se but rather in the difficulty of standardizing in vitro stem cell cultures and transplantation conditions (Dunnett et al., 2001; Carson et al., 2006).

CONCLUSIONS

The finding that NPCs persist in the adult brain and new neurons are continuously generated in two distinct brain areas has forced us to re-think and adjust traditional views of adult brain function. The exciting possibility that the adult brain might be able to repair itself has also opened a completely new level of treatment options for neurological disease. Several key basic questions remain unanswered:

- What is the exact function of neurogenesis in the adult brain?
- Why does neurogenesis occur only in two areas under normal conditions?
- What are the mechanisms that regulate differentiation, maturation, and integration of newborn neurons in the adult brain?

The answers to these questions will not only increase our understanding of the basic biology underlying adult neurogenesis but will also increase the chances for successful cell replacement interventions using endogenous NPCs or cell transplantation into the diseased brain.

ACKNOWLEDGMENTS

The authors would like to thank Mary Lynn Gage for editing this manuscript. We apologize to all the authors whose work we could not cite due to space limitations.

REFERENCES

Abrous DN, Koehl M, and Le Moal M (2005). Adult neurogenesis: from precursors to network and physiology. *Physiol Rev* 85: 523–569.

Aimone JB, Wiles J, and Gage FH (2006). Potential role for adult neurogenesis in the encoding of time in new memories. *Nat Neurosci* 9: 723–727.

Alonso M, Viollet C, Gabellec MM, et al. (2006). Olfactory discrimination learning increases the survival of adult-born neurons in the olfactory bulb. *J Neurosci* 26: 10508–10513.

Altman J and Das GD (1965). Autoradiographic and histologic evidence of postnatal neurogenesis in rats. *J Comp Neurol* 124: 319–335.

Alvarez-Buylla A and Garcia-Verdugo JM (2002). Neurogenesis in adult subventricular zone. *J Neurosci* 22: 629–634.

Arvidsson A, Collin T, Kirik D, et al. (2002). Neuronal replacement from endogenous precursors in the adult brain after stroke. *Nat Med* 5: 5.

Babu H, Cheung G, Kettenmann H, Palmer TD, Kempermann G (2007). Enriched monolayer precursor cell cultures from micro-dissected adult mouse dentate gyrus yield functional granule cell-like neurons. *PLoS ONE* 2: e388.

Becker S (2005). A computational principle for hippocampal learning and neurogenesis. *Hippocampus* 15: 722–738.

Bedard A and Parent A (2004). Evidence of newly generated neurons in the human olfactory bulb. *Brain Res Dev Brain Res* 151: 159–168.

Bengzon J, Kokaia Z, Elmér E, et al. (1997). Apoptosis and proliferation of dentate gyrus neurons after single and intermittent limbic seizures. *Proc Natl Acad Sci U S A* 94: 10432–10437.

Best PJ, White AM, and Minai A (2001). Spatial processing in the brain: the activity of hippocampal place cells. *Annu Rev Neurosci* 24: 459–486.

Bjorklund A and Lindvall O (2000). Cell replacement therapies for central nervous system disorders. *Nat Neurosci* 3: 537–544.

Bonaguidi MA, McGuire T, Hu M, et al. (2005). LIF and BMP signaling generate separate and discrete types of GFAP-expressing cells. *Development* 132: 5503–5514.

Bruel-Jungerman E, Laroche S, and Rampon C (2005). New neurons in the dentate gyrus are involved in the expression of enhanced long-term memory following environmental enrichment. *Eur J Neurosci* 21: 513–521.

Cameron HA and McKay RD (1999). Restoring production of hippocampal neurons in old age. *Nat Neurosci* 2: 894–897.

Cao L, Jiao X, Zuzga DS, et al. (2004). VEGF links hippocampal activity with neurogenesis, learning and memory. *Nat Genet* 36: 827–835.

Carlen M, Cassidy RM, Brismar H, Smith GA, et al. (2002). Functional integration of adult-born neurons. *Curr Biol* 12: 606–608.

Carleton A, Petreanu LT, Lansford R, et al. (2003). Becoming a new neuron in the adult olfactory bulb. *Nat Neurosci* 6: 507–518.

Carson CT, Aigner S, and Gage FH (2006). Stem cells: the good, bad and barely in control. *Nat Med* 12: 1237–1238.

Chopp M, Zhang ZG, and Jiang Q (2007). Neurogenesis, angiogenesis, and MRI indices of functional recovery from stroke. *Stroke* 38: 827–831.

Couillard-Despres S, Winner B, Schaubeck S, et al. (2005). Doublecortin expression levels in adult brain reflect neurogenesis. *Eur J Neurosci* 21: 1–14.

Curtis MA, Kam M, Nannmark U, et al. (2007). Human neuroblasts migrate to the olfactory bulb via a lateral ventricular extension. Science 315: 1243–1249.

David DJ, Klemenhagen KC, Holick KA, et al. (2007). Efficacy of the MCHR1 antagonist N-[3-(1-{[4-(3,4-difluorophenoxy)phenyl]methyl}(4-piperidyl))-4-methylphenyl]-2-methylpropanamide (SNAP 94847) in mouse models of anxiety and depression following acute and chronic administration is independent of hippocampal neurogenesis. *J Pharmacol Exp Ther* 321: 237–248.

Deisseroth K, Singla S, Toda H, et al. (2004). Excitation-neurogenesis coupling in adult neural stem/progenitor cells. *Neuron* 42: 535–552.

Doetsch F, Caille I, Lim DA, et al. (1999). Subventricular zone astrocytes are neural stem cells in the adult mammalian brain. *Cell* 97: 703–716.

Duman RS and Monteggia LM (2006). A neurotrophic model for stress-related mood disorders. *Biol Psychiatry* 59: 1116–1127.

Dunnett SB, Bjorklund A, and Lindvall O (2001). Cell therapy in Parkinson's disease – stop or go? *Nat Rev Neurosci* 2: 365–369.

Encinas JM, Vaahtokari A, and Enikolopov G (2006). Fluoxetine targets early progenitor cells in the adult brain. *Proc Natl Acad Sci U S A* 103: 8233–8238.

Eriksson PS, Perfilieva E, Björk-Eriksson T, et al. (1998). Neurogenesis in the adult human hippocampus. *Nat Med* 4: 1313–1317.

Filippov V, Kronenberg G, Pivneva T, et al. (2003). Subpopulation of nestin-expressing progenitor cells in the adult murine hippocampus shows electrophysiological and morphological characteristics of astrocytes. *Mol Cell Neurosci* 23: 373–382.

Gage F (2000). Mammalian neural stem cells. *Science* 287: 1433–1438.

Ge S, Yang CH, Hsu KS, et al. (2007). A critical period for enhanced synaptic plasticity in newly generated neurons of the adult brain. *Neuron* 54: 559–566.

Gould E and Tanapat P (1999). Stress and hippocampal neurogenesis. *Biol Psychiatry* 46: 1472–1479.

Hinds JW (1968). Autoradiographic study of histogenesis in the mouse olfactory bulb. II. Cell proliferation and migration. *J Comp Neurol* 134: 305–322.

Holick KA, Lee DC, Hen R, and Dulawa SC (2007). Behavioral effects of chronic fluoxetine in BALB/cJ Mice do not require adult hippocampal neurogenesis or the serotonin 1A receptor. *Neuropsychopharmacology*. 2007 Apr 11; [Epub ahead of print].

Horner PJ, Power AE, Kempermann G, et al. (2000). Proliferation and differentiation of progenitor cells throughout the intact adult rat spinal cord. *J Neurosci* 20: 2218–2228.

Howell OW, Doyle K, Goodman JH, et al. (2005). Neuropeptide Y stimulates neuronal precursor proliferation in the post-natal and adult dentate gyrus. *J Neurochem* 93: 560–570.

Hsieh J, Aimone JB, Kaspar BK, et al. (2004). IGF-I instructs multipotent adult neural progenitor cells to become oligodendrocytes. *J Cell Biol* 164: 111–122.

Jakubs K, Nanobashvili A, Bonde S, et al. (2006). Environment matters: synaptic properties of neurons born in the epileptic adult brain develop to reduce excitability. *Neuron* 52: 1047–1059.

Jessberger S and Kempermann G (2003). Adult-born hippocampal neurons mature into activity-dependent responsiveness. *Eur J Neurosci* 18: 2707–2712.

Jessberger S, Romer B, Babu H, and Kempermann G (2005). Seizures induce proliferation and dispersion of doublecortin-positive hippocampal progenitor cells. *Exp Neurol* 196: 342–351.

Jessberger S, Clemenson GD, Jr, and Gage FH (2007a). Spontaneous fusion and nonclonal growth of adult neural stem cells. *Stem Cells* 25: 871–874.

Jessberger S, Nakashima K, Clemenson GD, Jr, et al. (2007b). Epigenetic miodulation of seizure-induced neurogenesis and cognitive decline. *J Neurosci* 22: 5967–5975

Johansson CB, Momma S, Clarke DL, et al. (1999). Identification of a neural stem cell in the adult mammalian central nervous system. *Cell* 96: 25–34.

Jung MW and McNaughton BL (1993). Spatial selectivity of unit activity in the hippocampal granular layer. *Hippocampus* 3: 165–182.

Jung KH, Chu K, Kim M, et al. (2004). Continuous cytosine-b-D-arabinofuranoside infusion reduces ectopic granule cells in adult rat hippocampus with attenuation of spontaneous recurrent seizures following pilocarpine-induced status epilepticus. *Eur J Neurosci* 19: 3219–3226.

Kaplan MS and Hinds JW (1977). Neurogenesis in the adult rat: electron microscopic analysis of light radioautographs. *Science* 197: 1092–1094.

Kee N, Teixeira CM, Wang AH, and Frankland PW (2007). Preferential incorporation of adult-generated granule cells into spatial memory networks in the dentate gyrus. *Nat Neurosci* 10: 355–362.

Kempermann G (2002). Why new neurons? Possible functions for adult hippocampal neurogenesis. *J Neurosci* 22: 635–638.

Kempermann G and Gage FH (2002). Genetic determinants of adult hippocampal neurogenesis correlate with acquisition, but not probe trial performance, in the water maze task. *Eur J Neurosci* 16: 129–136.

Kempermann G, Kuhn HG, and Gage FH (1997). More hippocampal neurons in adult mice living in an enriched environment. *Nature* 386: 493–495.

Kempermann G, Gast D, and Gage FH (2002). Neuroplasticity in old age: sustained fivefold induction of hippocampal neurogenesis by long-term environmental enrichment. *Ann Neurol* 52: 135–143.

Kempermann G, Gast D, Kronenberg G, et al. (2003). Early determination and long-term persistence of adult-generated new neurons in the hippocampus of mice. *Development* 130: 391–399.

Kempermann G, Jessberger S, Steiner B, and Kronenberg G (2004). Milestones of neuronal development in the adult hippocampus. *Trends Neurosci* 27: 447–452.

Kesner RP, Lee I, and Gilbert P (2004). A behavioral assessment of hippocampal function based on a subregional analysis. *Rev Neurosci* 15: 333–351.

Kronenberg G, Reuter K, Steiner B, et al. (2003). Subpopulations of proliferating cells of the adult hippocampus respond differently to physiologic neurogenic stimuli. *J Comp Neurol* 467: 455–463.

Kuhn HG, Dickinson-Anson H, and Gage FH (1996). Neurogenesis in the dentate gyrus of the adult rat: age-related decrease of neuronal progenitor proliferation. *J Neurosci* 16: 2027–2033.

Kuwabara T, Hsieh J, Nakashima K, et al. (2004). A small modulatory dsRNA specifies the fate of adult neural stem cells. *Cell* 116: 779–793.

Lai K, Kaspar BK, Gage FH, and Schaffer DV (2003). Sonic hedgehog regulates adult neural progenitor proliferation in vitro and in vivo. *Nat Neurosci* 6: 21–27.

Laplagne DA, Esposito MS, Piatti VC, et al. (2006). Functional convergence of neurons generated in the developing and adult hippocampus. *PLoS Biol* 4: e409.

Laywell ED, Rakic P, Kukekov VG, et al. (2000). Identification of a multipotent astrocytic stem cell in the immature and adult mouse brain. *Proc Natl Acad Sci U S A* 97: 13883–13888.

Lie DC, Dziewczapolski G, Willhoite AR, et al. (2002). The adult substantia nigra contains progenitor cells with neurogenic potential. *J Neurosci* 22: 6639–6649.

Lie DC, Song H, Colamarino SA, et al. (2004). Neurogenesis in the adult brain: new strategies for central nervous system diseases. *Annu Rev Pharmacol Toxicol* 44: 399–421.

Lie DC, Colamarino SA, Song HJ, et al. (2005). Wnt signalling regulates adult hippocampal neurogenesis. *Nature* 437: 1370–1375.

Lindvall O, Sawle G, Widner H, et al. (1994). Evidence for long-term survival and function of dopaminergic grafts in progressive Parkinson's disease. *Ann Neurol* 35: 172–180.

Lledo PM and Saghatelyan A (2005). Integrating new neurons into the adult olfactory bulb: joining the network, life-death decisions, and the effects of sensory experience. *Trends Neurosci* 28: 248–254.

Lledo PM, Alonso M, and Grubb MS (2006). Adult neurogenesis and functional plasticity in neuronal circuits. *Nat Rev Neurosci* 7: 179–193.

Magavi S, Leavitt B, and Macklis J (2000). Induction of neurogenesis in the neocortex of adult mice. *Nature* 405: 951–955.

Magavi SS, Mitchell BD, Szentirmai O, et al. (2005). Adult-born and preexisting olfactory granule neurons undergo distinct experience-dependent modifications of their olfactory responses in vivo. *J Neurosci* 25: 10729–10739.

Mandairon N, Sacquet J, Garcia S, et al. (2006). Neurogenic correlates of an olfactory discrimination task in the adult olfactory bulb. *Eur J Neurosci* 24: 3578–3588.

Meshi D, Drew MR, Saxe M, Ansorge MS, David D, Santarelli L, Malapani C, Moore H, Hen R (2006). Hippocampal neurogenesis is not required for behavioral effects of environmental enrichment. *Nat Neurosci* 9: 729–731.

Milner B, Squire LR, and Kandel ER (1998). Cognitive neuroscience and the study of memory. *Neuron* 20: 445–468.

Ming GL and Song H (2005). Adult neurogenesis in the mammalian central nervous system. *Annu Rev Neurosci* 28: 223–250.

Miyata T, Maeda T, and Lee JE (1999). NeuroD is required for differentiation of the granule cells in the cerebellum and hippocampus. *Genes Dev* 13: 1647–1652.

Mizrahi A (2007). Dendritic development and plasticity of adult-born neurons in the mouse olfactory bulb. *Nat Neurosci* 10: 444–452.

Montaron MF, Drapeau E, Dupret D, et al. (2006). Lifelong corticosterone level determines age-related decline in neurogenesis and memory. *Neurobiol Aging* 27: 645–654.

O'Keefe J and Conway DH (1978). Hippocampal place units in the freely moving rat: why they fire where they fire. *Exp Brain Res* 31: 573–590.

Palmer TD, Ray J, and Gage FH (1995). FGF-2-responsive neuronal progenitors reside in proliferative and quiescent regions of the adult rodent brain. *Mol Cell Neurosci* 6: 474–486.

Palmer TD, Willhoite AR, and Gage FH (2000). Vascular niche for adult hippocampal neurogenesis. *J Comp Neurol* 425: 479–494.

Palmer TD, Schwartz PH, Taupin P, et al. (2001). Cell culture. Progenitor cells from human brain after death. *Nature* 411: 42–43.

Parent JM (2002). The role of seizure-induced neurogenesis in epileptogenesis and brain repair. *Epilepsy Res* 50: 179–189.

Parent JM, Yu TW, Leibowitz RT, et al. (1997). Dentate granule cell neurogenesis is increased by seizures and contributes to aberrant network reorganization in the adult rat hippocampus. *J Neurosci* 17: 3727–3738.

Parent JM, Vexler ZS, Gong C, Derugin N, and Ferriero DM (2002). Rat forebrain neurogenesis and striatal neuron replacement after focal stroke. *Ann Neurol* 52: 802–813.

Pereira AC, Huddleston DE, Brickman AM, et al. (2007). An in vivo correlate of exercise-induced neurogenesis in the adult dentate gyrus. *Proc Natl Acad Sci U S A* 104: 5638–5643.

Piatti VC, Esposito MS, and Schinder AF (2006). The timing of neuronal development in adult hippocampal neurogenesis. *Neuroscientist* 12: 463–468.

Rakic P (1985). Limits of Neurogenesis in Primates. *Science* 227: 1054–1056.

Rakic P (2002). Adult neurogenesis in mammals: an identity crisis. *J Neurosci* 22: 614–618.

Rao MS, Hattiangady B, and Shetty AK (2006). The window and mechanisms of major age-related decline in the production of new neurons within the dentate gyrus of the hippocampus. *Aging Cell* 5: 545–558.

Ray J and Gage FH (2006). Differential properties of adult rat and mouse brain-derived neural stem/progenitor cells. *Mol Cell Neurosci* 31: 560–573.

Ray J, Raymon HK, and Gage FH (1995). Generation and culturing of precursor cells and neuroblasts from embryonic and adult central nervous system. *Methods Enzymol* 254: 20–37.

Reynolds BA and Weiss S (1992). Generation of neurons and astrocytes from isolated cells of the adult mammalian central nervous system [see comments]. *Science* 255: 1707–1710.

Rolls ET and Kesner RP (2006). A computational theory of hippocampal function, and empirical tests of the theory. *Prog Neurobiol* 79: 1–48.

Sanai N, Tramontin AD, Quinones-Hinojosa A, et al. (2004). Unique astrocyte ribbon in adult human brain contains neural stem cells but lacks chain migration. *Nature* 427: 740–744.

Santarelli L, Saxe M, Gross C, et al. (2003). Requirement of hippocampal neurogenesis for the behavioral effects of antidepressants. *Science* 301: 805–809.

Sapolsky RM (2000). Glucocorticoids and hippocampal atrophy in neuropsychiatric disorders. *Arch Gen Psychiatry* 57: 925–935.

Saxe MD, Battaglia F, Wang JW, et al. (2006). Ablation of hippocampal neurogenesis impairs contextual fear conditioning and synaptic plasticity in the dentate gyrus. *Proc Natl Acad Sci U S A* 103: 17501–17506.

Saxe MD, Malleret G, Vronskaya S, et al. (2007). Paradoxical influence of hippocampal neurogenesis on working memory. *Proc Natl Acad Sci U S A* 104: 4642–4646.

Schanzer A, Wachs FP, Wilhelm D, et al. (2004). Direct stimulation of adult neural stem cells in vitro and neurogenesis in vivo by vascular endothelial growth factor. *Brain Pathol* 14: 237–248.

Scharfman HE, Goodman JH, and Sollas AL (2000). Granule-like neurons at the hilar/CA3 border after status epilepticus and their synchrony with area CA3 pyramidal cells: functional implications of seizure-induced neurogenesis. *J Neurosci* 20: 6144–6158.

Schinder AF and Gage FH (2004). A hypothesis about the role of adult neurogenesis in hippocampal function. Physiology (Bethesda) 19: 253–261.

Schmidt-Hieber C, Jonas P, Bischofberger J (2004). Enhanced synaptic plasticity in newly generated granule cells of the adult hippocampus. *Nature* 429: 184–187.

Scott BW, Wang S, Burnham WM, et al. (1998). Kindling-induced neurogenesis in the dentate gyrus of the rat. *Neurosci Lett* 248: 73–76.

Seaberg RM and van der Kooy D (2002). Adult rodent neurogenic regions: the ventricular subependyma contains neural stem cells, but the dentate gyrus contains restricted progenitors. *J Neurosci* 22: 1784–1793.

Seri B, Garcia-Verdugo JM, McEwen BS, and Alvarez-Buylla A (2001). Astrocytes give rise to new neurons in the adult mammalian hippocampus. *J Neurosci* 21: 7153–7160.

Shapiro LA and Ribak CE (2006). Newly born dentate granule neurons after pilocarpine-induced epilepsy have hilar basal dendrites with immature synapses. *Epilepsy Res* 69: 53–66.

Shi Y, Chichung Lie D, Taupin P, et al. (2004). Expression and function of orphan nuclear receptor TLX in adult neural stem cells. *Nature* 427: 78–83.

Shors TJ, Miesegaes G, Beylin A, Zhao M, Rydel T, Gould E (2001). Neurogenesis in the adult is involved in the formation of trace memories. *Nature* 410: 372–376.

Singec I, Knoth R, Meyer RP, Maciaczyk J, Volk B, Nikkhah G, Frotscher M, Snyder EY (2006). Defining the actual sensitivity and specificity of the neurosphere assay in stem cell biology. *Nat Methods* 3: 801–806.

Snyder JS, Hong NS, McDonald RJ, and Wojtowicz JM (2005). A role for adult neurogenesis in spatial long-term memory. *Neuroscience* 130: 843–852.

Song H, Stevens CF, and Gage FH (2002). Astroglia induce neurogenesis from adult neural stem cells. *Nature* 417: 39–44.

Squire LR, Stark CE, and Clark RE (2004). The medial temporal lobe. *Annu Rev Neurosci* 27: 279–306.

Steiner B, Klempin F, Wang L, et al. (2006). Type-2 cells as link between glial and neuronal lineage in adult hippocampal neurogenesis. *Glia* 54: 805–814.

Tashiro A, Makino H, and Gage FH (2007). Experience-specific functional modification of the dentate gyrus through adult neurogenesis: a critical period during an immature stage. *J Neurosci* 27: 3252–3259.

Tashiro A, Sandler VM, Toni N, et al. (2006). NMDA-receptor-mediated, cell-specific integration of new neurons in adult dentate gyrus. *Nature* 442: 929–933.

Taupin P, Ray J, Fischer WH, Suhr ST, et al. (2000). FGF-2-responsive neural stem cell proliferation requires CCg, a novel autocrine/paracrine cofactor. *Neuron* 28: 385–397.

Toni N, Teng EM, Bushong EA, and (2007). Synapse formation on neurons born in the adult hippocampus. *Nat Neurosci* 10(6): 727–34.

Tozuka Y, Fukuda S, Namba T, et al. (2005). GABAergic excitation promotes neuronal differentiation in adult hippocampal progenitor cells. *Neuron* 47: 803–815.

van Praag H, Christie BR, Sejnowski TJ, and Gage FH (1999). Running enhances neurogenesis, learning and long-term potentiation in mice. *Proc Natl Acad Sci U S A* 96: 13427–13431.

van Praag H, Schinder AF, Christie BR, et al. (2002). Functional neurogenesis in the adult hippocampus. *Nature* 415: 1030–1034.

van Praag H, Shubert T, Zhao C, and Gage FH (2005). Exercise enhances learning and hippocampal neurogenesis in aged mice. *J Neurosci* 25: 8680–8685.

Walker TL, Yasuda T, Adams DJ, and Bartlett PF (2007). The doublecortin-expressing population in the developing and adult brain contains multipotential precursors in addition to neuronal-lineage cells. *J Neurosci* 27: 3734–3742.

Wang S, Scott BW, and Wojtowicz JM (2000). Heterogenous properties of dentate granule neurons in the adult rat. *J Neurobiol* 42: 248–257.

Warner-Schmidt JL and Duman RS (2006). Hippocampal neurogenesis: opposing effects of stress and antidepressant treatment. *Hippocampus* 16: 239–249.

Winner B, Cooper-Kuhn CM, Aigner R, et al. (2002). Long-term survival and cell death of newly generated neurons in the adult rat olfactory bulb. *Eur J Neurosci* 16: 1681–1689.

Winocur G, Wojtowicz JM, Sekeres M, et al. (2006). Inhibition of neurogenesis interferes with hippocampus-dependent memory function. *Hippocampus* 16: 296–304.

Wiskott L, Rasch MJ, and Kempermann G (2006). A functional hypothesis for adult hippocampal neurogenesis: avoidance of catastrophic interference in the dentate gyrus. *Hippocampus* 16: 329–343.

Zhao C, Teng EM, Summers RG, Jr., et al. (2006). Distinct morphological stages of dentate granule neuron maturation in the adult mouse hippocampus. *J Neurosci* 26: 3–11.

Zyzak DR, Otto T, Eichenbaum H, and Gallagher M (1995). Cognitive decline associated with normal aging in rats: a neuropsychological approach. *Learn Mem* 2: 1–16.

8

NEUROTROPHIC FACTORS IN ALZHEIMER'S DISEASE

ARMIN BLESCH* AND MARK H. TUSZYNSKI*,†

*Departments of Neurosciences-0626, University of California-San Diego-La Jolla, CA 92093-0626

†Veterans Affairs Medical Center, San Diego, CA 92161

ABSTRACT

Since the discovery of neurotrophic factors in the early 1950s, and the subsequent finding in the 1980s that they can prevent neuronal death in the adult brain, it has been clear that these molecules represent potential means of treating progressive neuronal loss in the context of neurodegenerative disorders. The first neurotrophic factor identified, Nerve Growth Factor (NGF), prevents the death and stimulates the function of basal forebrain cholinergic neurons. Degeneration of cholinergic neurons occurs early in Alzheimer's disease (AD), raising the possibility that NGF could be a means of preventing or reducing cholinergic dysfunction and loss in AD. However, means of effectively and safely delivering NGF to the central nervous system became a limiting factor in transitioning to clinical trials. This chapter will review the development of NGF therapy for AD, from early rodent studies through the demonstration that NGF gene delivery exhibits trophic influences on cholinergic neurons in the AD brain.

INTRODUCTION

Since the discovery that neurotrophic factors can stimulate neuronal function and enhance neuron survival, there has been speculation that this potent class of growth factors could prevent or slow neuronal degeneration in Alzheimer's disease (AD). The first neurotrophin identified, nerve growth factor (NGF), stimulates the survival and function of basal forebrain cholinergic neurons (BFCNs), suggesting that NGF could be a means for reducing the cholinergic component of cell degeneration in AD. Other neurotrophic factors might be able to enhance

Copyright © 2008 Elsevier Inc.
All rights of reproduction in any form reserved.

the function of cortical and hippocampal circuitry. This chapter will discuss the preclinical basis for neurotrophic factor therapy in AD and the transition from preclinical studies to human clinical trials in AD. The implementation of clinical testing of neurotrophic factor therapy in AD has been constrained by the dual need to achieve adequate concentrations of these proteins in specific brain regions containing degenerating neurons, and preventing growth factor spread to non-targeted regions to avoid adverse effects. Gene therapy is one of a limited number of potential methods for achieving these requirements.

RATIONALE FOR GROWTH FACTOR APPROACHES

Despite substantial progress in elucidating the pathophysiological causes underlying AD, current approved treatments merely aim to compensate for neuronal loss and accompanying loss of synaptic function. Ultimately, neuroprotective strategies aimed to slow or halt the degeneration of neurons and synapses could provide urgently needed novel therapies to treat this progressive neurodegenerative disorder affecting more than 4.5 million people in the United States alone.

The potent activities of neurotrophic factors in preventing neuronal death and stimulating neuronal function in various yet specific neuronal populations that are affected in human neurodegenerative disease provide a unique possibility to intervene early in the course of neurodegenerative disease to slow cell loss and fundamentally alter disease progression.

In particular, studies on NGF, identified more than 55 years ago (Levi-Montalcini and Hamburger, 1951; Levi-Montalcini, 1987) have provided ample evidence that degeneration of BFCNs severely affected in AD can be entirely prevented in rodent and primate models by delivery of NGF. In addition, NGF administration improves learning and memory in lesioned and aged rats, improves neuronal function, and prevents cell death in a mouse model of amyloid overexpression.

While these studies highlight the potent neuroprotection that can be achieved by neurotrophic factors, it has also been recognized over the last several years that clinical testing of growth factors requires adequate concentrations of these proteins that are administered specifically in brain regions containing degenerating neurons. In addition, the spread of growth factors to non-targeted regions has to be prevented to avoid adverse effects. A means to satisfy these requirements is essential for true testing of the hypothesis that growth factors could be of benefit in human neurological disorders. The lack of adequate delivery methods therefore limited the therapeutic potential of neurotrophic factors in human disorders and only recent research has identified means of meeting these requirements, resulting in a renewal of efforts to test growth factors in neurodegenerative disorders.

NERVE GROWTH FACTOR AND BASAL FOREBRAIN CHOLINERGIC NEURONS

The NGF was the first nervous system growth factor to be identified, serendipitously discovered more than 50 years ago in the course of studying chick sarcoma extracts, and was initially found to potently stimulate the outgrowth of embryonic dorsal root ganglion neurites (Levi-Montalcini and Hamburger, 1951; Levi-Montalcini, 1987). Subsequent studies demonstrated that sympathetic and sensory nociceptive neurons required NGF to survive embryonic development. However, another 35 years passed before human NGF was cloned (Ullrich et al., 1983), and neurotrophic factor effects in the adult brain were discovered. Only in the 1980s it was recognized that a population of neurons in the adult central nervous system (CNS), cholinergic neurons in the basal forebrain (nucleus basalis of Meynert, medial septum, vertical and horizontal limb of the diagonal band), also express receptors for NGF and more importantly respond to NGF (Hefti et al., 1984; Will and Hefti, 1985). In 1986, it was reported that NGF infusions into the ventricular system of the adult rat brain could prevent basal forebrain cholinergic neuronal death after lesions of their projections to the hippocampus (Kromer et al., 1981; Hefti, 1986; Williams et al., 1986) (Figure 8.1). These findings revolutionized the dogma that CNS injury inevitably results in neuronal loss that is untreatable, and led rapidly to the idea that NGF might be of therapeutic value for the treatment of the cholinergic deficit in AD (Hefti and Weiner, 1986).

An extensive expansion of research into the properties and potential therapeutic applications of NGF followed. In 1987, Fischer and colleagues reported that *spontaneous* atrophy of BFCNs in aged rats was also prevented by intraventricular infusion of NGF (Fischer et al., 1987), thereby improving learning

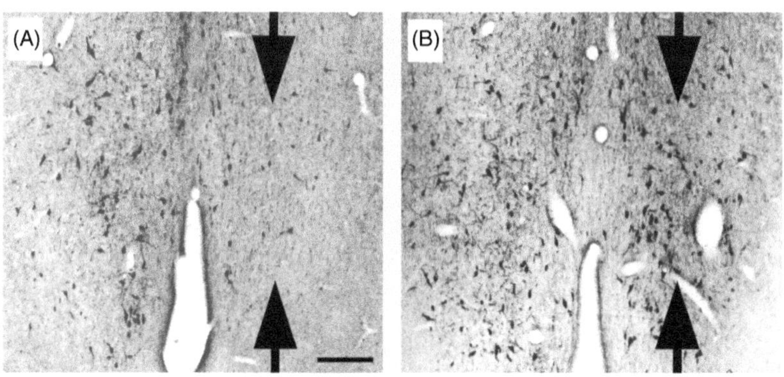

FIGURE 8.1 NGF prevents cholinergic neuronal death in rodents and primates (A) Fornix transections on the right side of the brain (arrows) cause cholinergic neurons to degenerate and die 1 month after the lesion. Cholinergic neurons on the left, intact side of the brain are unaffected by the lesion. Neurons are labeled with choline acetyltransferase (ChAT); images are from adult primate brain studies. (B) NGF infusions prevent the death of the majority of cholinergic neurons after fornix lesions in rodents and primates (shown) on the lesioned side of the brain. Scale bar, 98 μm.

and memory (Fischer et al., 1987; Markowska et al., 1994). Spatial memory was also improved by NGF infusion in adult, cholinergic-lesioned rats (Tuszynski and Gage, 1995). In mutant trisomy 16 mice bearing an extra copy of amyloid precursor protein (APP), NGF infusion also reversed basal forebrain cholinergic neuronal atrophy and improved learning and memory (Holtzman et al., 1993; Cooper et al., 2001). Showing NGF efficacy in yet another model of neuronal degeneration, basal forebrain cholinergic neuronal caused by excitotoxicity was also prevented by NGF infusion (Dekker et al., 1991, 1992; Liberini et al., 1993).

To examine the relevance of NGF-mediated neuroprotection to the larger human brain, NGF effects were further investigated in the adult non-human primate brain. In rhesus macaque and cynomolgous monkeys, NGF prevented degeneration of BFCNs after lesions (Figure 8.1) and reversed spontaneous, age-related cholinergic neuronal atrophy (Figure 8.2) (Tuszynski et al., 1990, 1991;

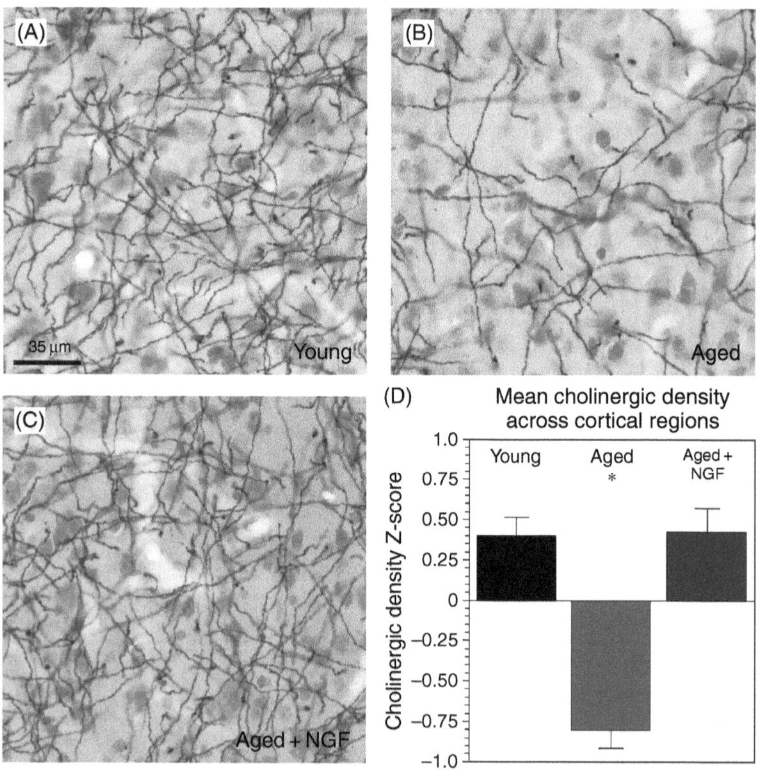

FIGURE 8.2 NGF restores cortical cholinergic terminal density in aged primates. (A) The normal architecture and density of cholinergic neurons in the temporal cortex of an adult rhesus monkey is shown by acetylcholine esterase staining. (B) Aging in rhesus monkeys is associated with reductions in cholinergic axon density. (C) NGF gene delivery restores AChE labeling in aged primates to levels of young adults. (D) Quantification of cholinergic axon density. From Conner et al., PNAS, 2001.

Koliatsos et al., 1991; Smith et al., 1999; Emerich et al., 1994; Kordower et al., 1994a; Conner et al., 2001).

Thus, cholinergic neuronal degeneration elicited by diverse mechanisms of neuronal damage including injury, excitotoxicity, aging, and amyloid overproduction could be prevented by NGF. NGF was equally effective in all species tested, including mice, rats, and non-human primates. These findings supported the possibility that NGF could also be useful in ameliorating the death and dysfunction of cholinergic neurons in AD (Appel, 1981; Hefti and Weiner, 1986; Holtzman and Mobley, 1994; Tuszynski and Gage, 1994), despite the fact the precise mechanism of broad neuronal loss in AD is not fully established.

NERVE GROWTH FACTOR AND ALZHEIMER'S DISEASE

Although loss of NGF expression is not an underlying cause of AD, several lines of evidence indicate that changes in NGF signaling and transport may contribute to the degeneration of BFCNs in AD. NGF is expressed throughout life in targets of cholinergic forebrain neurons, the neocortex and hippocampus (Korsching et al., 1985; Whittemore et al., 1986), where NGF is taken up by receptors on cholinergic axon terminals (Dawbarn et al., 1988; Kiss et al., 1988; Holtzman et al., 1992), and retrogradely transported to the cell soma (Howe and Mobley, 2004) to support neuronal function and activity (Sofroniew et al., 1990).

Levels of NGF decline within BFCNs in AD (Mufson et al., 1995; Scott et al., 1995), although production of NGF in the hippocampus or cortex is not reduced (Mufson et al., 1989a, 2003; Crutcher et al., 1993; Scott et al., 1995; Fahnestock et al., 1996; Hellweg et al., 1998; Savaskan et al., 2000). These findings suggest that retrograde NGF transport in signaling endosomes in the AD brain may be deficient, potentially from the cytoskeletal transport dysfunction that is a hallmark of AD pathology (Dai et al., 2002; Stokin et al., 2005). Diminished retrograde transport of NGF has also been shown in aged rodents (Cooper et al., 1994), a correlational animal model for the cholinergic atrophy observed in AD. Further support for this hypothesis comes from trisomy 16 mutant mice with increased APP production. This mice demonstrate a marked age-related atrophy of BFCNs, which can be reversed by NGF administration (Holtzman et al., 1993). The degeneration of cholinergic neurons is strongly associated with a highly diminished retrograde transport of NGF in these animals (Cooper et al., 2001), which can be directly linked to the presence of an extra copy of APP (Salehi et al., 2006).

Therapeutic strategies to enhance NGF signaling should therefore be aimed at cholinergic cell bodies (Emerich et al., 1994; Hu et al., 1997; Smith et al., 1999; Conner et al., 2001) rather than axon terminal regions to bypass transport defects while allowing for NGF signaling at the soma.

Whereas perturbations in levels of the high-affinity NGF receptor trkA and the low-affinity neutorophin receptor p75 occur in AD (Mufson et al., 1989a,b; Mahadeo et al., 1994; Boissiere et al., 1997; Savaskan et al., 2000; Chu et al., 2001), sufficient quantities of both of these receptors remain expressed at the cell soma and axon to retain NGF responsiveness in early AD (Counts and Mufson, 2005). It is particularly important that trkA receptors are expressed in considering NGF as a therapy for AD, because in the presence of the p75 receptor alone, NGF can lead to cell death rather than survival (Barrett and Bartlett, 1994; Chao, 1994). It is also important that NGF is therapeutically administered as the mature, secreted growth factor rather than as a pro-neurotrophin, because the latter form of NGF can mediate cell death (Lee et al., 2001).

While cholinergic neurons comprise only a small sub-population of all neurons undergoing degeneration in AD, several lines of evidence indicate that cholinergic decline, particularly in early and mid-stages of AD, may in fact contribute substantially to cognitive dysfunction. Virtually all cholinergic innervation to the hippocampus and neocortex arises from BFCNs (Mesulam and Geula, 1988; Heckers et al., 1994), and these projections exert a major role in modulating neural activity (Sofroniew et al., 1990; Howe and Mobley, 2001; Conner et al., 2003, 2005) and attention (Voytko et al., 1994; Wenk, 1997; Ridley et al., 2005) in the cortex and hippocampus. Dysfunction of cholinergic systems in humans would therefore be predicted to cause widespread deficits in executive processing, a type of cognitive deficit that is prominent in early and mid-stage AD (Perry and Hodges, 1999; Buckner, 2004), as cholinergic degeneration begins early and progresses rapidly (Perry et al., 1978, 1985; Candy et al., 1983; Bartus, 2000; Mufson et al., 2003a,b). Indeed, of all the neuronal systems that degenerate in AD, the loss of cholinergic neurons correlates best with reductions in synapse number and cognitive impairment (Perry et al., 1985; Masliah et al., 1991). The strategy of targeting cholinergic systems with cholinesterase inhibitors has shown statistically significant benefits on cognition in numerous trials (Davis et al., 1992; Grundman and Thal, 2000); the magnitude of this impact has been limited, however. The modest impact of cholinesterase inhibitors on cognition in AD may reflect the limitations in the dose of acetylcholine esterase inhibitors that can be achieved. Higher doses, which could be of greater benefit, cannot be tolerated due to peripheral adverse events, and the indiscriminate actions of cholinesterase inhibitors on all cholinergic synapses, rather than those active in performing on-line demand tasks. NGF, on the other hand, restores acetylcholine production and increases acetylcholine release from cells in a physiological manner (Dekker et al., 1991; Maysinger et al., 1992), augments downstream signal transduction systems including Erk, MAP kinase, and CREB, and prevents neuronal death in specifically targeted neuronal population if adequately delivered. Thus, the impact of NGF in AD could substantially exceed that of the cholinesterase inhibitors. NGF would not "cure" AD because the widespread neuronal decline that occurs in later stages of the disease would overwhelm a potential effect of cholinergic rescue in early and mid-stage disease. Nonetheless, cognitive rescue or substantial slowing in the rate of decline for even a few years would represent a major improvement over existing symptomatic therapies for AD.

A clinical program of NGF treatment in AD would therefore test two key hypotheses. First, the hypothesis would be tested that NGF will effectively block or ameliorate the mechanisms underlying cholinergic neuronal cell death in AD. Second, it remains to be determined whether targeting only the cholinergic component of neuronal decline in AD will result in meaningful cognitive improvements.

The mechanism causing cholinergic death in AD could be a lack of NGF availability or transport (as described above), but could also result from a number of other degenerative mechanisms unrelated to NGF levels. The broad array of animal models of cholinergic neuronal degeneration that can effectively be ameliorated by NGF administration support the hypothesis that NGF could be effective in AD; however, this remains to be established in AD itself. Based on cognitive improvements observed after NGF delivery in different animal models and effects of cholinesterase inhibitors, it is reasonable to hypothesize that NGF would be effective in AD, but this questions can only be fully addressed in clinical studies.

Despite the strong rationale for NGF treatment in AD, clinic translation of findings in animal model has been slow. For NGF and other nervous system growth factors to be effective, direct delivery into the CNS is required, as the blood–brain barrier prevents the transition of relatively large and polar molecules. Many early animal studies demonstrating efficacy in ameliorating cholinergic neuronal function infused NGF intracerebroventricularly, thereby exposing broad CNS regions to the growth factor. NGF rescues degenerating cholinergic neurons when administered by this route, but it also results in a number of adverse affects including pain from stimulation of dorsal root ganglion nociceptive neurons (Emmett et al., 1996; Winkler et al., 1997), weight loss (Williams, 1991), sympathetic axon sprouting in the cerebral vasculature (Isaacson et al., 1990), and migration and proliferation of Schwann cells into a thick layer surrounding the medulla and spinal cord (Emmett et al., 1996; Winkler et al., 1997). These adverse affects arise from NGF receptor expressing cells and reflect the continuing potency of NGF in the adult nervous system. Indeed, three AD patients in Sweden received intracerebroventricular NGF infusions, and developed pain and weight loss after several months (Eriksdotter Jonhagen et al., 1998). Thus, unacceptable, adverse effects from NGF-mediated stimulation of non-targeted structures in the peripheral and central nervous system make intracerebroventricular infusion of NGF an unsuitable method for the treatment of AD. Instead, to be adequately tested in human clinical trials in a well-tolerated and long-term manner, NGF must be administered in sufficient doses to achieve efficacy on targeted neurons, while preventing NGF spread to non-targeted regions to avoid adverse effects.

PRECLINICAL STUDIES OF EX VIVO NGF GENE THERAPY

One means of restricting NGF to a small region in the CNS, allowing accurate targeting and delivery of biologically active levels over prolonged time periods, is ex vivo and in vivo *gene delivery*. Autologous cells can be cultured, genetically modified to produce NGF, and subsequently grafted into the cholinergic basal forebrain region (**ex vivo** gene therapy), with the cells functioning as biological minipumps for local delivery of the growth factor. Alternatively, viral or non-viral vectors carrying the NGF gene can be injected in the basal forebrain (in vivo gene therapy), modifying cells already within the basal forebrain region to produce elevated levels of NGF (Figure 8.3). Gene therapy can specifically target the brain parenchyma, and avoids distribution of the growth factor throughout the ventricular system. Using intraparenchymal ex vivo gene therapy, lesion-induced death of BFCNs was prevented in rodents and primates (Rosenberg et al., 1988; Emerich et al., 1994; Kordower et al., 1994a; Tuszynski et al., 1996), without detection of the adverse effects that occur after intracerebroventricular NGF infusions. Ex vivo NGF gene delivery also ameliorated cognitive deficits and age-related cholinergic atrophy in rats (Chen and Gage, 1995; Martinez-Serrano et al., 1995, 1996). Studies in aged monkeys using autologous fibroblasts as vehicles of ex vivo NGF gene delivery further demonstrated complete reversal of age-related neuronal atrophy, restoration of cortical cholinergic inputs, and maintenance of NGF gene expression for at least one year (Figure 8.2) (Smith et al., 1999; Conner et al., 2001).

In subsequent dose-escalation safety and toxicology studies in primates (summarized on the NIH Recombinant Advisory Committee website, http://www4.od.nih.gov/oba/rac/minutes/1299rac.pdf), intraparenchymal ex vivo NGF

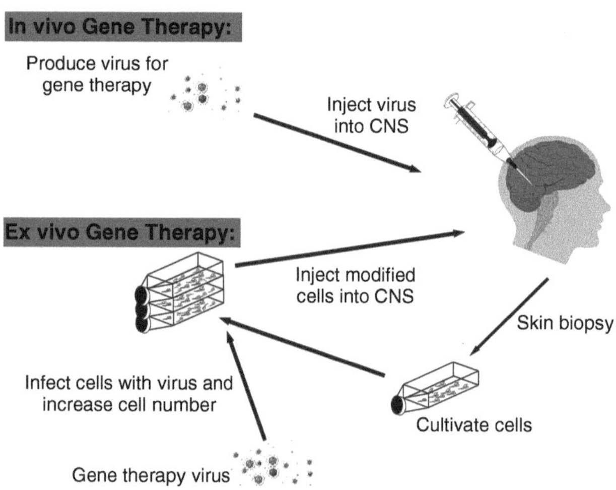

FIGURE 8.3 Gene transfer in the nervous system. In in vivo gene delivery, viral vectors capable of genetically modifying cells within the brain are injected into specific parenchymal targets, resulting in localized and sustained delivery of therapeutic genes such as growth factors. Ex vivo gene therapy transduces host cells in vitro to express a gene of interest; these cells are then injected into the brain to act as localized, cellular sources of growth factor secretion to neighboring cellular targets. Fibroblasts, Schwann cells, or stem cells can all hypothetically function as vehicles of ex vivo gene transfer into the nervous system.

gene therapy was found to be safe without causing adverse effects. In this study, autologous fibroblasts, cultivated from skin biopsies of adult rhesus monkeys, were genetically modified in vitro to produce and secrete human NGF, and expanded in number. These cells were then injected into the cholinergic basal forebrain over a range of doses up to 10-fold higher than the projected human dose that would be employed in a subsequent clinical trial. Over 1 year of observation, monkeys exhibited no weight loss, pain, or Schwann cell migration into the CNS. Gene expression persisted for at least 1 year in both primate and rodent brains: whereas gene expression over this time period declined by approximately 90% compared to 1-month post-injection, amounts of NGF protein in the brain remained fivefold greater than concentrations detected in intact brains. Further, persistent cellular effects of NGF on cholinergic cell size were detected after 1 year. Samples of cerebrospinal fluid collected regularly throughout the study revealed no detectable NGF within the ventricular system. These studies together with previous preclinical rodent and primate studies suggested that NGF gene delivery met the criteria for a clinical program: NGF could be accurately delivered to the basal forebrain region, exerted biological effects on cholinergic neurons, continued to express NGF for at least 1 year, and showed no leakage into the spinal fluid or toxicity at doses which exceeded those proposed for human trials.

The gene delivery vector used in these preclinical studies was derived from a murine Moloney leukemia (MLV) retrovirus (Rosenberg et al., 1988) expressing NGF constitutively from the 5′-long terminal repeat. The vector was incapable of replication and exhibited no evidence of host toxicity; however, in the event that an adverse event developed, it would not be possible to turn the vector "off". A regulatable vector system is highly desirable for the practice of gene therapy, but practical, non-immunogenic regulatable systems that have been sufficiently tested for human use are not currently available. Instead, adverse effects, if they developed, would need to be treated symptomatically. Because no adverse effects were detected as a result of the non-regulated MLV–NGF vector used in these studies among hundreds of rats and dozens of primates, it appeared that a clinically practical method existed for proceeding to a human clinical trial.

A PHASE 1 HUMAN TRIAL OF EX VIVO NGF GENE THERAPY FOR AD

In 2001, we initiated a Phase 1 human clinical trial of ex vivo NGF gene delivery in eight people with a diagnosis of early stage AD (Tuszynski et al., 2005). Early stage subjects were recruited both because they were judged capable of providing informed consent and because the hypothetical benefits of NGF in reducing neuronal loss and augmenting cell function would be most appreciable when the greatest number of cholinergic neurons are available to influence. Subject age ranged from 54 to 76 years, with an average of 67 years. Hypothetical risks of the study included surgical complications such as a hemorrhage from injection needle passage into the brain; mis-targeting of NGF-expressing cells leading to the spread of NGF into the ventricular system; tumor formation by grafted cells, and at least theoretically over-stimulation of cholinergic function. Any of the former could cause injury or death; however, none of these adverse effects were observed in preclinical studies. Subject enrollment occurred gradually, to allow a substantial period of time for observation before moving to the next dose group.

After obtaining informed consent, skin biopsies were performed to generate primary fibroblast cultures. These autologous cells were genetically modified (using the same MLV vectors employed in preclinical studies) to secrete human NGF within a range of 50–75 ng NGF/10^6cells/day. After in vitro expansion of cell number and verification of sterility, absence of mycoplasm and replication competent retrovirus, autologous NGF secreting fibroblasts were injected into the brain through one burr hole per side using stereotaxic neurosurgical targeting. Three dose cohorts were enrolled: Subjects 1 and 2 received a total of 1.25×10^6 autologous fibroblasts into the non-dominant, right nucleus basalis of Meynert; subjects 3–6 received a total of 2.5×10^6 cells, injected into both left and right basal nuclei; and subjects 7 and 8 received injection of 5×10^6 cells also injected both left and right basal nucleus. After gene delivery, cognitive function was assessed using standard clinical tools, together with serial MRI and PET scans.

At study initiation, the injection of genetically modified cells into the brain was performed on subjects in a sedated but wakeful state, a standard practice in other neurosurgical procedures. However, two subjects moved abruptly while the injection needle was in the brain, resulting in intraparenchymal hemorrhage. Neither hemorrhage required drainage, and one individual gradually improved and was discharged with a moderate worsening of his baseline language deficit. The second individual was initially stuporous from the hemorrhage and gradually improved; however, 5 weeks after treatment he sustained a pulmonary embolus leading to myocardial infarction and death. Subsequently, all subjects underwent general anesthesia or deep sedation to avoid unanticipated movement, and all remaining injections were completed safely. Subjects were discharged 1–2 days post-treatment.

The brain in the individual that died 5 weeks post-NGF-delivery was examined and confirmed the diagnosis of AD (Braak stage 4), with multiple plaques and neurofibrillary tangles distributed throughout the cortex, as well as cortical Lewy bodies. Robust NGF gene expression was evident by in situ hybridization within the NGF-secreting cell implants (Tuszynski et al., 2005). Notably, immunohistochemistry confirmed for the first time in the human that degenerating neurons retain sensitivity to neurotrophic factors: cholinergic axons robustly extended toward and into the site of NGF gene transfer, confirming a "trophic" response to the growth factor (Figure 8.4).

Over a mean follow-up period that now extends to 4 years, there have been no long-term adverse effects of NGF expression in the brain. Specifically, no weight loss, pain or other abnormality attributable to NGF or the MLV vector has been detected. In no case was there a clinical indication of NGF leakage into the ventricular system.

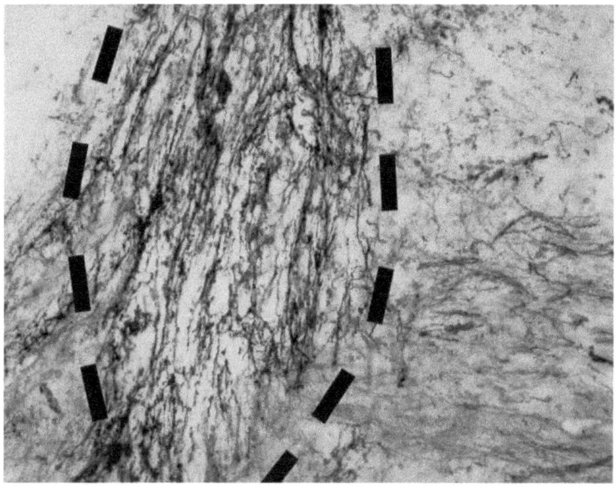

FIGURE 8.4 Trophic response of cholinergic neurons to NGF in the AD brain. Region within dashed lines contains graft of autologous fibroblasts genetically modified to secrete NGF and grafted into the human cholinergic basal forebrain of an AD patient. Five weeks after cell implantation, cholinergic axons immunolabeled for p75 sprout densely into the NGF-secreting graft, indicating that degenerating cholinergic neurons of the AD brain exhibit clear trophic responses to NGF.

This phase 1 study was neither designed nor powered to detect effects of NGF gene transfer on cognitive performance. Cognitive testing was performed in all subjects, but conclusions cannot be drawn from a small cohort of subjects lacking blinded assessments or comparison to a placebo control group. Bearing these caveats in mind, testing on the Mini-Mental Status Examination (MMSE) and Alzheimer's Disease Assessment Scale – Cognitive subcomponent (ADAS-Cog) scales suggested possible reductions in rate of decline over a 2-year period after NGF gene transfer (Tuszynski et al., 2005) (Figure 8.5). This appeared to be more pronounced at times more than 6 months after cell injection possibly reflecting the time needed to remodel cholinergic terminals in the cortex, the target region of the nucleus basalis of Meynert (NBM).

As another secondary outcome measure, serial PET scans were obtained in four bilaterally treated subjects to assess patterns of glucose utilization using 2-fluorodeoxyglucose PET. In AD, serial PET scans over time typically reveal a steady decline in glucose uptake reflecting decreased cortical activity (Potkin et al., 2001; Alexander et al., 2002). In contrast, the four bilaterally treated subjects in this trial demonstrated statistically significant increases in cortical glucose uptake by PET when measured 6–8 months after the first scan (Figure 8.6).

In summary, this small Phase 1 human trial of ex vivo NGF gene delivery demonstrated four points: (1) NGF can be delivered safely to the brain over an extended time period using gene delivery, but subjects must undergo

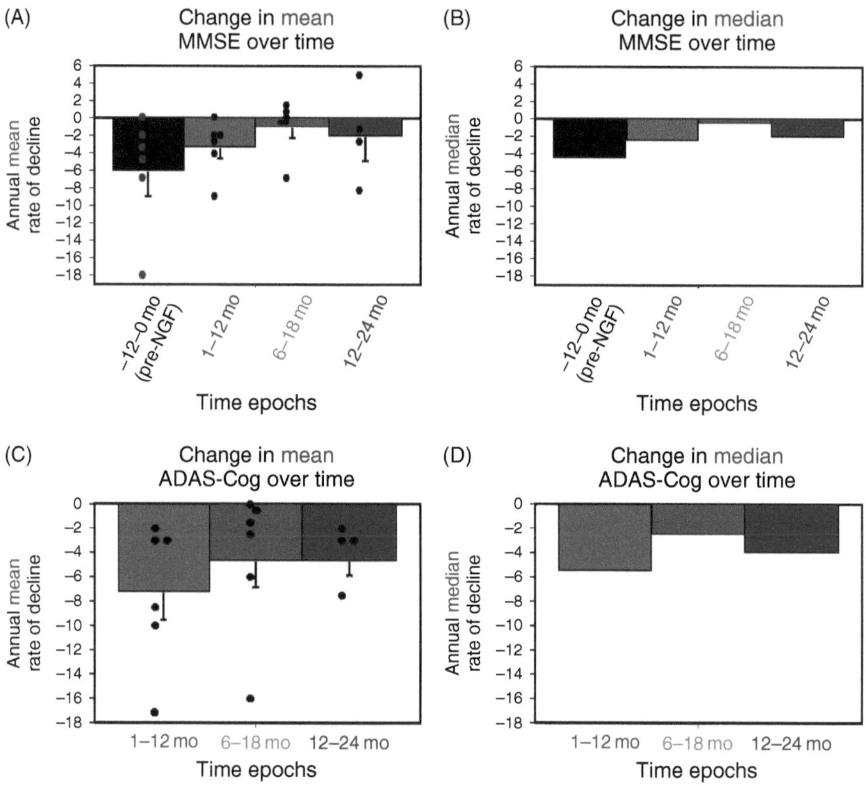

FIGURE 8.5 Cognitive outcomes (A) Mean annualized change in MMSE score in the year prior to treatment, and in time epochs of 1–12 months, 6–18 months, and 12–24 months post-treatment. Individual subject data points are shown in circles. Of note, during the 1-year period beginning 6 months after treatment, when sufficient time passed for NGF to enhance cholinergic projections, two of six patients show improved MMSE scores, one patient has no decline, and two patients decline by only one point. Overall decline was reduced 51% when compared to pre-operative rate for the mean 22 months post-treatment period. (B) Given wide range in MMSE scores, median data also are shown; findings parallel observations of mean scores. (C) Mean annualized changes in ADAS-Cog over time epochs of 1–12 months, 6–18 months, and 12–24 months post-treatment. Individual subject data points are shown in circles. As with MMSE, rate of decline slows after sufficient time has passed (6 months) for NGF to act upon cholinergic systems. Rate of decline is reduced by 36% at 6–18 months, compared to 1–12 months. (D) Given wide range in ADAS-Cog, median data also are shown. Rate of decline is slowed by 55% at 6–18 months compared to 1–12 months. From Tuszynski et al., *Nat Med*, 2005.

cell injections under general anesthesia or deep sedation. (2) Degenerating cholinergic neurons of the human brain exhibit clear trophic responses to growth factors. (3) Broad regions of the cortex demonstrate enhanced cortical glucose metabolism following NGF gene transfer. (4) Based on preliminary cognitive data and the preceding points, larger, controlled, and blinded clinical trials of NGF gene delivery in AD are warranted. Such studies are planned (see below).

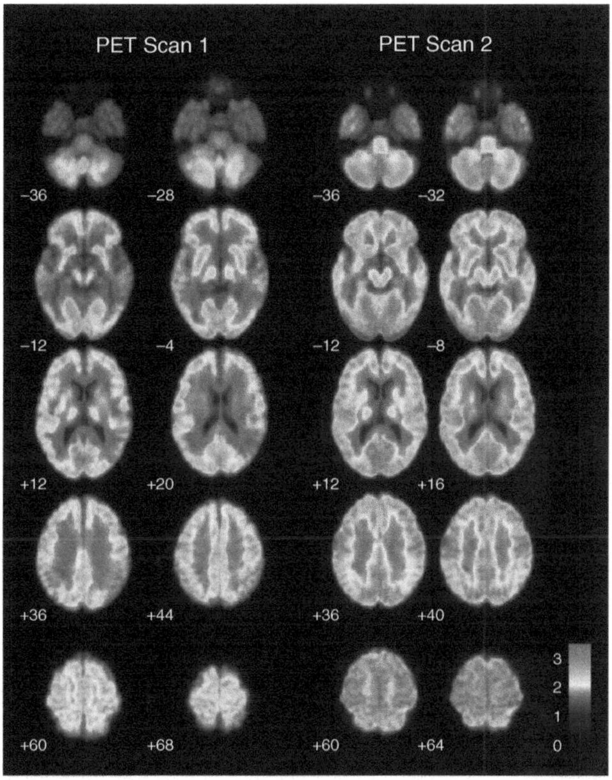

FIGURE 8.6 Effects of NGF Gene Transfer of PET Scans Glucose uptake by PET scan in four subjects treated with NGF, overlaid on standardized MRI templates. Representative axial sections, with 6–8 months between first and second scan, show interval increases in brain metabolism in diverse cortical regions, representing the broad cortical projection pattern of cholinergic systems. Images are average PET data from all four bilaterally treated patients that underwent serial PET scans. Flame scale indicates FDG use /100 g tissue/min; red color indicates more FDG use than blue. From Tuszynski et al., Nat Med, 2005.

IN VIVO ADENO-ASSOCIATED VIRUS-NGF GENE DELIVERY FOR AD

As noted above, there are two general methods for gene transfer into the CNS: ex vivo and in vivo gene delivery. The clinical program above utilized ex vivo NGF gene transfer because at the time studies for the Phase 1 trial were initiated, no safe vector system suitable for in vivo gene therapy was available and ex vivo gene delivery vectors such as MLV were superior to the commonly employed in vivo vectors of the time, such as adenovirus.

However, considerable progress has subsequently been made in generating safe, efficient, and potent in vivo gene delivery vectors, elevating in vivo gene delivery in the CNS as a superior alternative to ex vivo gene delivery. While the ex vivo NGF trial was useful for proof-of-principle, ex vivo techniques are generally cumbersome, labor-intensive, and expensive because they require extensive cell processing (cultivation from biopsies, cell expansion, retroviral transduction) and testing for sterility, mycoplasm, replication competent retrovirus, and NGF expression before gene transfer into each patient. In the case of in vivo gene delivery, vectors expressing a therapeutic gene of interest (such as NGF) are simply batch-manufactured, tested once, and the same vector batch can be injected into the brain of all subjects.

Among several viral "backbones" from which in vivo gene delivery vectors can be created, adeno-associated virus (AAV) has emerged as one of the most promising vectors for CNS gene therapy. Recombinant AAV does not express or contain any wild-type virus genes, is replication-incompetent, does not induce inflammation, and by all measures appears to be non-toxic. Further, they form stable extra-chromosomal elements (concatamers) that rarely integrate into the host genome, thereby substantially reducing the risk of insertional mutagenesis (i.e. are unlikely to cause a host cell to become cancerous). When injected into the rhesus monkey brain, these vectors appear to sustain therapeutic gene expression for at least 6 years, without evident loss of expression over time (Bankiewicz et al., 2006). AAV has now been used in seven human neurological gene therapy trials: a phase 1 trial of AAV-NGF gene transfer in AD, four clinical trials in Parkinson's disease (including a growth factor trial), and two trials in inborn errors of metabolism in the CNS (Batten's disease and Canavan's disease) (Mandel and Burger, 2004). AAV-mediated neurotrophin expression meets the requirements for specifically targeted and regionally restricted delivery of growth factors to the brain.

In preclinical studies, AAV–NGF and lentiviral–NGF gene delivery both prevent cholinergic neuronal death and enhance cognitive function as reported using ex vivo NGF gene transfer and NGF protein infusions (Klein et al., 2000; Mandel et al., 1999; Blesch et al., 2005; Blömer et al., 1998; Wu et al., 2004, 2005). Extensive preclinical safety studies also indicated that AAV–NGF gene delivery was safe in the CNS over extended time periods and at relative doses exceeding those proposed for human clinical trials in AD (see http://www.gemcris.od.nih.gov/Contents/GC_CLIN_TRIAL_RPT_VIEW.asp?WIN_TYPE=R&CTID=622). Based on the intriguing results of the ex vivo Phase 1 NGF clinical trial and these supportive preclinical safety and efficacy data, a recent phase 1 clinical trial of AAV–NGF in vivo gene delivery was initiated at Rush University Medical Center in early to mid-stage AD patients sponsored by Ceregene, Inc. Results from this phase 1 trial have recently been presented (Arvanitakis et al., 2007) and indicate that AAV-mediated NGF delivery is safe and well tolerated. Hypothetically, the effects of AAV–NGF gene transfer could be superior to those of ex vivo NGF gene transfer, by virtue of the fact that in vivo gene expression with AAV vectors using the chick-ß-actin hybrid cytomegalovirus (CAG) promoter is sustained for years, in contrast to gradually declining levels of expression with *ex vivo* vectors over time.

As the phase 1 AAV–NGF trial reaches a close, a multi-center phase 2 sham surgery controlled, double-blinded trial in AD is planned to start in 2007/2008. The phase 2 trial will explore the potential effect size of NGF gene delivery on cognitive decline in AD, advancing a program that, if successful, will provide a potential opportunity to fundamentally alter the progressive and tragic course of this common human disorder.

OTHER MEANS OF NGF DELIVERY TO THE CNS

As noted above, the reasonable testing of growth factors in CNS disorders requires a delivery method that achieves effective levels of growth factors deep within brain tissue, while *restricting* the growth factor to the actual target in order to avoid adverse effects of widespread delivery. Other methods for delivering neurotrophins to the CNS include binding active components of neurotrophic factors to carrier molecules that cross the blood–brain barrier such as the transferrin receptor (Friden et al., 1993; Kordower et al., 1994b), the use of small peptide analogs of the growth factors that can be peripherally administered and cross the blood–brain barrier (Longo et al., 1997; Xie et al., 2000), or direct intraparenchymal infusions of growth factors (Gill et al., 2003; Slevin et al., 2005). The benefit of peripheral neurotrophin administration includes elimination of the risk and expense of cranial injection, and the ability to adjust or terminate dosing. However, peripheral administration suffers from the serious shortcoming that neurotrophins, if reaching the brain in an adequate concentration, will be *non-targeted* and thereby risk adverse events from stimulation of non-targeted systems; this approach is not likely to succeed. Direct growth factor protein intraparenchymal infusions can accurately target growth factors to specific nuclear regions, deliver sufficient doses to elicit biological efficacy, limit spread to avoid adverse effects, and can be discontinued in the event of adverse effects (Tuszynski, 2000; Gill et al., 2003; Slevin et al., 2005). However, intraparenchymal infusions suffer from difficulty in evenly distributing growth factors in a target region from a single catheter tip, and from vulnerability to malfunction and infection (Gill et al., 2003). Efforts to develop this technology in Parkinson's disease face an uncertain future. On balance, gene delivery of growth factors currently appears to be the most practical means of testing their therapeutic potential in AD.

OTHER GROWTH FACTORS FOR AD

In addition to NGF, neurotrophic factors such as brain-derived neurotrophic factor (BDNF) or neurotrophin 4/5 might be of interest in the treatment of AD to address neuronal degeneration of neuronal populations other than BFCNs such as hippocampus and cortex. Ongoing studies are evaluating whether in vivo BDNF gene transfer can prevent neuronal cell death, loss of synapses and cognitive function in rodent and primate models of AD and neuronal degeneration. A number of other promising approaches for the treatment of AD are currently under development such as passive or active vaccination protocols targeting ß-amyloid accumulation in the brain or secretase inhibitors to prevent amyloid protein aggregation. Combination of these treatments with growth factor gene therapy could have significant impact on the progression of AD by protecting neurons from degeneration and enhancing their function while diminishing amyloid toxicity.

CONCLUSIONS

Growth factors potently influence neuronal survival and function. NGF exhibits broad activity against a multitude of toxic mechanisms in cholinergic basal forebrain neurons by influencing downstream common mechanisms of cell function, atrophy, and death. Thus, growth factors have the potential to provide novel neuroprotective treatments to slow or stop neuronal degeneration either alone or as combination therapies with other anti-AD therapies, including anti-amyloid approaches. Gene therapy appears to meet the need for accurately targeted, regionally restricted, safe and long-term neurotrophin delivery to the brain. Ongoing clinical programs are testing the hypotheses that NGF will protect cholinergic neurons in the pathogenic environment of the AD brain, and that targeting of the cholinergic system will be sufficient to meaningfully benefit quality of life in patients. Initial clinical findings from small, uncontrolled trials are intriguing, suggesting that larger scale, controlled, and blinded studies of NGF gene delivery in AD are warranted.

ACKNOWLEDGEMENTS

Supported by the NIH (AG10435), the Veterans Administration, the Shiley Family Foundation, and the Adelson Program in Neural Repair and Rehabilitation.

REFERENCES

Alexander GE, Chen K, Pietrini P et al. (2002). Longitudinal PET evaluation of cerebral metabolic decline in dementia: a potential outcome measure in Alzheimer's disease treatment studies. *Am J Psychiatry* 159(5): 738–745.

Appel SH (1981). A unifying hypothesis for the cause of amyotrophic lateral sclerosis, parkinsonism, and Alzheimer disease. *Ann Neurol* 10: 499–505.

Arvanitakis Z, Tuszynski MH, Bakay RD et al. (2007). Interim data from a phase 1 clinical trial of AAV-NGF (CERE-110) gene delivery in Alzheimer's disease. *American Acad Neurol Abstract* p05.071.

Bankiewicz KS, Forsayeth J, Eberling JL et al. (2006). Long-term clinical improvement in MPTP-lesioned primates after gene therapy with AAV-hAADC. *Mol Ther* 14(4): 564–570.

Barrett GL and Bartlett PF (1994). The p75 nerve growth factor receptor mediates survival or death depending on the stage of sensory neuron development. *Proc Natl Acad Sci U S A* 91: 6501–6505.

Bartus RT (2000). On neurodegenerative diseases, models, and treatment strategies: lessons learned and lessons forgotten a generation following the cholinergic hypothesis. *Exp Neurol* 163(2): 495–529.

Blesch A, Pfeiffer A, Conner JM et al. (2005). Regulated lentiviral NGF genetransfer controls rescue of medial septal cholinergic neurons. *Molec Ther* 11: 916–925.

Blömer U, Kafri T, Randolph-Moore L et al. (1998). Bcl-xL protects adult septal cholinergic neurons from axotomized cell death. *Proc Natl Acad Sci U S A* 95(5): 2603–2608.

Boissiere F, Hunot S, Faucheux B et al. (1997). Trk neurotrophin receptors in cholinergic neurons of patients with Alzheimer's disease. *Dement Geriatr Cogn Disord* 8(1): 1–8.

Buckner RL (2004). Memory and executive function in aging and AD: multiple factors that cause decline and reserve factors that compensate. *Neuron* 44(1): 195–208.

Candy JM, Perry RH, Perry EK et al. (1983). Pathological changes in the nucleus basalis of Meynert in Alzheimer's and Parkinson's diseases. *J Neurosci* 54: 277–289.

Chao MV (1994). The p75 neurotrophin receptor. *J Neurobiol* 25: 1373–1385.

Chen KS and Gage FH (1995). Somatic gene transfer of NGF to the aged brain: Behavioral and morphological amelioration. *J Neurosci* 15: 2819–2825.

Chu Y, Cochran EJ, Bennett et al. (2001). Down-regulation of trkA mRNA within nucleus basalis neurons in individuals with mild cognitive impairment and Alzheimer's disease. *J Comp Neurol* 437(3): 296–307.

Conner JM, Darracq MA, Roberts J et al. (2001). Non-tropic actions of neurotrophins: Subcortical NGF gene delivery reverses age-related degeneration of primate cortical cholinergic innervation. *Proc Nat Acad Sci USA* 98: 1941–1946.

Conner JM, Culberson A, Packowski C et al. (2003). Lesions of the basal forebrain cholinergic system impair task acquisition and abolish cortical plasticity associated with motor skill learning. *Neuron* 38: 819–829.

Conner JM, Chiba AA, and Tuszynski MH (2005). The basal forebrain cholinergic system is essential for cortical plasticity and functional recovery following brain injury. *Neuron* 46: 173–179.

Cooper JD, Lindholm D, and Sofroniew MV (1994). Reduced transport of [125I]nerve growth factor by cholinergic neurons and down-regulated TrkA expression in the medial septum of aged rats. *J Neurosci* 62(3): 625–629.

Cooper JD, Salehi A, Delcroix JD et al. (2001). Failed retrograde transport of NGF in a mouse model of Down's syndrome: Reversal of cholinergic neurodegenerative phenotypes following NGF infusion. *Proc Natl Acad Sci U S A* 98: 10439–10444.

Counts SE and Mufson EJ (2005). The role of nerve growth factor receptors in cholinergic basal forebrain degeneration in prodromal Alzheimer disease. *J Neuropathol Exp Neurol* 64(4): 263–272.

Crutcher KA, Scott SA, Liang S et al. (1993). Detection of NGF-like activity in human brain tissue: increased levels in Alzheimer's disease. *J Neurosci* 13: 2540–2550.

Dai J, Buijs RM, Kamphorst W et al. (2002). Impaired axonal transport of cortical neurons in Alzheimer's disease is associated with neuropathological changes. *Brain Res* 948(1–2): 138–144.

Davis KL, Thal LJ, Gamzu ER et al. (1992). A double-blind, placebo-controlled multicenter study of tacrine for Alzheimer's disease. The Tacrine Collaborative Study Group. *N Engl J Med* 327: 1253–1259.

Dawbarn D, Allen SJ, and Semenenko FM (1988). Coexistence of choline acetyltransferase and nerve growth factor receptors in the rat basal forebrain. *Neurosci Lett* 94: 138–144.

Dekker AJ, Langdon DJ, Gage FH, and Thal LJ (1991). NGF increases cortical acetylcholine release in rats with lesions of the nucleus basalis. *Neuroreport* 2: 577–580.

Dekker AJ, Gage FH, and Thal LJ (1992). Delayed treatment with nerve growth factor improves acquisition of a spatial task in rats with lesions of the nucleus basalis magnocellularis: evaluation of the involvement of different neurotransmitter systems. *J Neurosci* 48: 111–119.

Emerich DW, Winn S, Harper J et al. (1994). Implants of polymer-encapsulated human NGF-secreting cells in the nonhuman primate: rescue and sprouting of degenerating cholinergic basal forebrain neurons. *J Comp Neurol* 349: 148–164.

Emmett CJ, Stewart GR, Johnson RM et al. (1996). Distribution of radioiodinated recombinant human nerve growth factor in primate brain following intracerebroventricular infusion. *Exp Neurol* 140: 151–160.

Eriksdotter Jonhagen M, Nordberg A, Amberla K et al. (1998). Intracerebroventricular infusion of nerve growth factor in three patients with Alzheimer's disease. *Dement Geriatr Cogn Disord* 9(5): 246–257.

Fahnestock M, Scott SA, Jette N et al. (1996). Nerve growth factor mRNA and protein levels measured in the same tissue from normal and Alzheimer's disease parietal cortex. *Brain Res Mol Brain Res* 42(1): 175–178.

Fischer W, Wictorin K, Bjorklund A et al. (1987). Amelioration of cholinergic neuron atrophy and spatial memory impairment in aged rats by nerve growth factor. *Nature* 329: 65–68.

Friden PM, Walus LR, Watson P et al. (1993). Blood-brain barrier penetration and in vivo activity of an NGF conjugate. *Science* 259: 373–377.

Gill SS, Patel NK, Hotton GR et al. (2003). Direct brain infusion of glial cell line-derived neurotrophic factor in Parkinson disease. *Nat Med* 9(5): 589–595.

Grundman M and LJ Thal (2000). Treatment of Alzheimer's disease: rationale and strategies. *Neurol Clin* 18(4): 807–828.

Heckers S, Ohtake T, Wiley RG et al. (1994). Complete and selective cholinergic denervation of rat neocortex and hippocampus but not amygdala by an immunotoxin against the p75 NGF receptor. *J Neurosci* 14(3 Pt 1): 1271–1289.

Hefti F (1986). Nerve growth factor (NGF) promotes survival of septal cholinergic neurons after fimbrial transection. *J Neurosci* 6: 2155–2162.

Hefti F and WJ Weiner (1986). Nerve growth factor and Alzheimer's disease. *Ann Neurol* 20: 275–281.

Hefti F, Dravid A, and J Hartikka (1984). Chronic intraventricular injections of nerve growth factor elevate hippocampal choline acetyltransferase activity in adult rats with partial septo-hippocampal lesions. *Brain Res* 293(2): 305–311.

Hellweg R, von Richthofen S, Anders D (1998). The time course of nerve growth factor content in different neuropsychiatric diseases – a unifying hypothesis. *J Neural Transm* 105(8–9): 871–903.

Holtzman DM and Mobley CM (1994). Neurotrophic factors and neurologic disease. *West J Med* 161: 246–254.

Holtzman DM, Li Y, Parada LF et al. (1992). p140trk mRNA marks NGF-responsive forebrain neurons: evidence that trk gene expression is induced by NGF. *Neuron* 9: 465–478.

Holtzman DM, Li Y, Chen K et al. (1993). Nerve growth factor reverses neuronal atrophy in a Down syndrome model of age-related neurodegeneration. *Neurology* 43: 2668–2673.

Howe CL and Mobley WC (2001). Nerve growth factor effects on cholinergic modulation of hippocampal and cortical plasticity, In I Mocchetti, ed. *Neurobiology of the Neurotrophins*. FP Graham Company: Mountain Home, Tennessee.

Howe CL and Mobley WC (2004). Signaling endosome hypothesis: A cellular mechanism for long distance communication. *J Neurobiol* 58(2): 207–216.

Hu L, Cote SL, and Cuello AC (1997). Differential modulation of the cholinergic phenotype of the nucleus basalis magnocellularis neurons by applying NGF at the cell body or cortical terminal fields. *Exp Neurol* 143: 162–171.

Isaacson LG, Saffran BN, and Crutcher KA (1990). Intracerebral NGF infusion induces hyperinnervation of cerebral blood vessels. *Neurobiol Aging* 11: 51–55.

Kiss J, McGovern J, and Patel AJ (1988). Immunohistochemical localization of cells containing nerve growth factor receptors in the different regions of the adult rat forebrain. *Neurosci Lett* 27: 731–748.

Klein RL, Hirko AC, Meyers CA et al. (2000). NGF gene transfer to intrinsic basal forebrain neurons increases cholinergic cell size and protects from age-related, spatial memory deficits in middle-aged rats. *Brain Res* 875(1–2): 144–151.

Koliatsos VE, Clatterbuck RE, Nauta HJ et al. (1991). Human nerve growth factor prevents degeneration of basal forebrain cholinergic neurons in primates. *Ann Neurol* 30(6): 831–840.

Kordower J, Winn H, Liu SR et al. (1994a). The aged monkey basal forebrain: rescue and sprouting of axotomized basal forebrain neurons after grafts of encapsulated cells secreting human nerve growth factor. *Proc Nat Acad Sci U S A* 91: 10898–10902.

Kordower JH, Charles V, Bayer R et al. (1994b). Intravenous administration of a transferrin receptor antibody-nerve growth factor conjugate prevents the degeneration of cholinergic striatal neurons in a model of Huntington's disease. *Proc Nat Acad Sci U S A*, 91: 9077–9080.

Korsching S, Auburger G, Heumann R et al. (1985). Levels of nerve growth factor and its mRNA in the central nervous system of the rat correlate with cholinergic innervation. *EMBO J* 4: 1389–1393.

Kromer LF, Bjorklund A, and Stenevi U (1981). Regeneration of the septohippocampal pathways in adult rats is promoted by utilizing embryonic hippocampal implants was bridges. *Brain Res* 210: 173–200.

Lee R, Kermani P, Teng KK, and Hempstead BL (2001). Regulation of cell survival by secreted proneurotrophins. *Science* 294(5548): 1945–1948.

Levi-Montalcini R (1987). The nerve growth factor 35 years later. *Science* 237: 1154–1162.

Levi-Montalcini R and V Hamburger (1951). Selective growth stimulating effects of mouse sarcoma on the sensory and sympathetic nervous system of the chick embryo. *J Exp Zool* 116: 321–362.

Liberini P, Pioro EP, Maysinger D et al. (1993). Long-term protective effects of human recombinant nerve growth factor and monosialoganglioside GM1 treatment on primate nucleus basalis cholinergic neurons after neocortical infarction. *J Neurosci* 53: 625–637.

Longo FM, Manthorpe M, Xie YM, and Varon S (1997). Synthetic NGF peptide derivatives prevent neuronal death via a p75 receptor-dependent mechanism. *J Neurosci Res* 48: 1–17.

Mahadeo D, Kaplan L, Chao MV, and Hempstead BL (1994). High affinity nerve growth factor binding displays a faster rate of association than p140trk binding. Implications for multisubunit polypedtide receptors. *J Biol Chem* 269: 6884–6891.

Mandel RJ and C Burger (2004). Clinical trials in neurological disorders using AAV vectors: promises and challenges. *Curr Opin Mol Ther* 6(5): 482–490.

Mandel RJ, Gage FH, Clevenger DG et al. (1999). Nerve growth factor expressed in the medial septum following in vivo gene delivery using a recombinant adeno-associated viral vector protects cholinergic neurons from fimbria-fornix lesion-induced degeneration. *Exp Neurol* 55(1): 59–64.

Markowska AL, Koliatsos VE, Breckler SJ et al. (1994). Human nerve growth factor improves spatial memory in aged but not in young rats. *J Neurosci* 14: 4815–4824.

Martinez-Serrano A, Fischer W, and Bjorklund A (1995). Reversal of age-dependent cognitive impairments and cholinergic neuron atrophy by NGF-secreting neural progenitors grafted to the basal forebrain. *Neuron* (2): 473–484.

Martinez-Serrano A, Fischer W, Soderstrom S et al. (1996). Long-term functional recovery from age-induced spatial memory impairments by nerve growth factor gene transfer to the rat basal forebrain. *Proc Natl Acad Sci U S A* 93(13): 6355–6360.

Masliah E, Terry RD, Alford M et al. (1991). Cortical and subcortical patterns of synaptophysinlike immunoreactivity in Alzheimer's disease. *Am J Pathol* 138: 235–246.

Maysinger D, Piccardo P, Goiny M, and Cuello AC (1992). Grafting of genetically modified cells: effects of acetylcholine release in vivo. *Neurochem Int* 21(4): 543–548.

Mesulam MM and C Geula (1988). Nucleus basalis (Ch4) and cortical cholinergic innervation in the human brain: observations based on the distribution of acetylcholinesterase and choline acetyltransferase. *J Comp Neurol* 275(2): 216–240.

Mufson EJ, Bothwell M, Hersh LB, and Kordower JH (1989a). Nerve growth factor receptor immunoreactive profiles in the normal, aged human basal forebrain: Colocalization with cholinergic neurons. *J Comp Neurol* 285: 196–217.

Mufson EJ, Bothwell M, and Kordower JH (1989b). Loss of nerve growth factor receptor-containing neurons in Alzheimer's disease: a quantitative analysis across subregions of the basal forebrain. *Exp.Neurol* 105: 221–232.

Mufson EJ, Conner JM, and Kordower JH (1995). Nerve growth factor in Alzheimer's disease: defective retrograde transport to nucleus basalis. *Neuroreport* 6: 1063–1066.

Mufson EJ, Ikonomovic MD, and Styren SD et al. (2003a). Preservation of brain nerve growth factor in mild cognitive impairment and Alzheimer disease. *Arch Neurol* 60(8): 1143–1148.

Mufson EJ, Ginsberg SD, Ikonomovic MD, and DeKosky ST (2003b). Human cholinergic basal forebrain: chemoanatomy and neurologic dysfunction. *J Chem Neuroanat* 26(4): 233–242.

Narisawa-Saito M, Wakabayashi K, Tsuji S et al. (1996). Regional specificity of alterations in NGF, BDNF and NT-3 levels in Alzheimer's disease. *NeuroReport* 7(18): 2925–2928.

Perry RJ and Hodges JR (1999). Attention and executive deficits in Alzheimer's disease. A critical review. *Brain* 22 (Pt 3): 383–404.

Perry EK, Tomlinson BE, Blessed G et al. (1978). Correlation of cholinergic abnormalities with senile plaques and mental test scores in senile dementia. *BMJ* 2: 1457–1459.

Perry EK, Curtis M, Dick DJ et al. (1985). Cholinergic correlates of cognitive impairment in Parkinson's disease: comparison with Alzheimer's disease. *J Neurol Neurosurg Psychiatry* 48: 413–421.

Potkin SG, Anand R, Alva G et al., (2001). Brain metabolic and clinical effects of rivastigmine in Alzheimer's disease. *Int J Neuropsychopharm* 4: 223–230.

Ridley RM, Baker HF, Leow-Dyke A, and Cummings RM (2005). Further analysis of the effects of immunotoxic lesions of the basal nucleus of Meynert reveals substantial impairment on visual discrimination learning in monkeys. *Brain Res Bull* 65(5): 433–442.

Rosenberg MB, Friedmann T, Robertson RC et al. (1988). Grafting genetically modified cells to the damaged brain: restorative effects of NGF expression. *Science* 242: 1575–1578.

Salehi A, Delcroix JD, Belichenko PV et al. (2006). Increased App expression in a mouse model of Down's syndrome disrupts NGF transport and causes cholinergic neuron degeneration. *Neuron* 51(1): 29–42.

Savaskan E, Muller-Spahn F, Olivieri G et al. (2000). Alterations in trk A, trk B and trk C receptor immunoreactivities in parietal cortex and cerebellum in Alzheimer's disease. *Eur Neurol* 44(3): 172–180.

Scott SA, Mufson EJ, Weingartner JA et al. (1995). Nerve growth factor in Alzheimer's disease: Increased levels throughout the brain coupled with declines in nucleus basalis. *J. Neurosci* 15: 6213–6221.

Slevin JT, Gerhardt GA, Smith CD et al. (2005). Improvement of bilateral motor functions in patients with Parkinson disease through the unilateral intraputaminal infusion of glial cell line-derived neurotrophic factor. *J Neurosurg* 102(2): 216–22.

Smith DE, Roberts J, Gage FH, and Tuszynski MH (1999). Age-associated neuronal atrophy occurs in the primate brain and is reversible by growth factor gene therapy. *Proc Nat Acad Sci U S A* 96: 10893–10898.

Sofroniew MV, Galletly NP, Isacson O et al. (1990). Survival of adult basal forebrain cholinergic neurons after loss of target neurons. *Science* 247: 338–342.

Stokin GB, Lillo C, Falzone TL et al. (2005). Axonopathy and transport deficits early in the pathogenesis of Alzheimer's disease. *Science* 307(5713): 1282–1288.

Tuszynski MH (2000). Intraparenchymal NGF infusions rescue degenerating cholinergic neurons. *Cell Transplant* 9(5): 629–636.

Tuszynski MH and FH Gage (1994). Neurotrophic factors and Alzheimer's disease, In R Terry, R Katzman, and KL Bick, eds. Raven Press, New York. p. x.

Tuszynski MH and Gage FH (1995). Bridging grafts and transient NGF infusions promote long-term CNS neuronal rescue and partial functional recovery. *Proc Nat Acad Sci U S A* 92: 4621–4625.

Tuszynski MH, U HS, Amaral DG, and Gage FH (1990). Nerve growth factor infusion in primate brain reduces lesion-induced cholinergic neuronal degeneration. *J Neurosci* 10: 3604–3614.

Tuszynski MH, U HS, and Gage FH (1991). Recombinant human nerve growth factor infusions prevent cholinergic neuronal degeneration in the adult primate brain. *Ann Neurol* 30: 625–636.

Tuszynski MH, Roberts J, Senut MC et al. U HS (1996). Gene therapy in the adult primate brain: intraparenchymal grafts of cells genetically modified to produce nerve growth factor prevent cholinergic neuronal degeneration. *Gene Therapy* 3: 305–314.

Tuszynski MH, Thal L, Pay M, Salmon DP et al., (2005). A phase 1 clinical trial of nerve growth factor gene therapy for Alzheimer disease. *Nat Med* 11: 551–555.

Ullrich A, Gray A, Berman C, and Dull TJ (1983). Human beta-nerve growth factor gene sequence highly homologous to that of mouse. *Nature* 303: 821–825.

Voytko ML, Olton DS, Richardson RT et al. (1994). Basal forebrain lesions in monkeys disrupt attention but not learning and memory. *J Neurosci* 14: 167–186.

Wenk GL (1997). The nucleus basalis magnocellularis cholinergic system: one hundred years of progress. *Neurobiol Learn Memory* 67(2): 85–95.

Whittemore SR, Ebendal T, Larkfors L et al. (1986). Development and regional expression of beta nerve growth factor messenger RNA and protein in the rat central nervous system. *Proc Natl Acad Sci U S A* 83(3): 817–821.

Will B and Hefti F (1985). Behavioural and neurochemical effects of chronic intraventricular injections of nerve growth factor in adult rats with fimbria lesions. *Behav Brain Res* 17(1): 17–24.

Williams LR (1991). Hypophagia is induced by intracerebroventricular administration of nerve growth factor. *Exp Neurol* 113: 31–37.

Williams LR, Varon S, Peterson GM et al. (1986). Continuous infusion of nerve growth factor prevents basal forebrain neuronal death after fimbria-fornix transection. *Proc Natl Acad Sci U S A* 83: 9231–9235.

Winkler J, Ramirez GA, Kuhn HG et al. (1997). Reversible Schwann cell hyperplasia and sprouting of sensory and sympathetic neurites after intraventricular administration of nerve growth factor. *Ann Neurol* 41: 82–93.

Wu K, Meyers CA, Guerra NK et al. (2004). The effects of rAAV2-mediated NGF gene delivery in adult and aged rats. *Mol Ther* 9(2): 262–269.

Wu K, Meyer EM, Bennett JA et al. (2005). AAV2/5-mediated NGF gene delivery protects septal cholinergic neurons following axotomy. *Brain Res* 1061(2): 107–113.

Xie Y, Tisi MA, Yeo TT, and Longo FM (2000). Nerve growth factor (NGF) loop 4 dimeric mimetics activate ERK and AKT and promote NGF-like neurotrophic effects. *J Biol Chem* 275(38): 29868–29874.

PART

III

Novel Therapies for Parkinson's and Huntington's Disease

9

Direct Delivery of GDNF into the Non-Human Primate and Human Parkinsonian Brain: Success and Road Blocks

R Grondin[*,§], Z Zhang[*,§], Y Ai[*,§], J Slevin[†,§], AB Young[‡,§], DM Gash[*,§], and GA Gerhardt[*,†,§]

[*]*Departments of Anatomy & Neurobiology,*
[†]*Neurology,*
[‡]*Neurosurgery and*
[§]*Morris K. Udall Parkinson's Disease Research Center of Excellence, University of Kentucky Medical Center, Lexington, KY 40536*

Abstract

Trophic factors are proteins with enormous therapeutic potential in the treatment of neurodegenerative diseases like Parkinson's disease (PD), including the potential to (1) slow the degeneration of nigral dopaminergic neurons, (2) enhance the function of residual dopamine neurons or (3) restore function to injured neurons. Novel methods for sustained delivery of glial cell line-derived neurotrophic factor (GDNF) into the nigrostriatal pathway have been studied in non-human primates, including the use of computer-controlled infusion pumps. Using this approach, we have demonstrated that chronic, intracerebral infusions of GDNF promotes

Copyright © 2008 Elsevier Inc.
All rights of reproduction in any form reserved.

restoration of the nigrostriatal dopaminergic system and significantly improves motor functions in rhesus monkeys with neural deficits modeling PD.

However, translational studies from the laboratory to the clinic for the treatment of PD have been problematic. On one hand, GDNF was not efficacious when delivered intraventricularly in PD patients, likely failing because of poor drug penetration through the ventricular wall into the parenchyma of the basal ganglia. On the other hand, two independent open label Phase-1 studies have reported marked functional improvements in advanced PD patients receiving chronic intraputamenal infusion of GDNF, whereas a separate, randomized, blinded multicenter trial of intraputmenal GDNF infusion did not achieve the primary study endpoint. In addition, safety concerns have arisen including the presence of neutralizing antibodies to GDNF in some patients. The major difficulty with risk assessment at this time is the failure to have a Phase-2 trial replicating the successful Phase-1 trials in dose and methods of trophic factor delivery. Thus, before any definitive conclusions could be made regarding the use of GDNF as a therapy for PD, a properly designed, adequately powered multicenter Phase-2 clinical trial should be conducted in PD patients.

Keywords: Parkinson's disease, dopamine neurons, GDNF, clinical trials, intracerebral delivery, substantia nigra, putamen, lateral ventricle, programmable pumps

INTRODUCTION

Parkinson's disease (PD) is a common and devastating neurological disorder characterized mainly by impairment of motor function, due largely to a progressive degeneration of the substantia nigra pars compacta dopamine neurons that innervate the striatum (Lang and Lozano, 1998). The loss of striatal dopamine and the consequent dysfunction of the nigrostriatal pathway eventually lead to the cardinal symptoms of PD: resting tremor, cogwheel rigidity, bradykinesia, and loss of postural reflex. Therapeutic strategies for PD include replacing striatal dopamine using the dopamine precursor levodopa or dopamine receptor agonists, or both. As the disease worsens, patients that have become less responsive to pharmacological treatments may choose to undergo surgical treatments, such as deep-brain stimulation. These treatments provide symptomatic relief, but do not slow or halt continued degeneration of nigral dopaminergic neurons. One experimental approach that could potentially slow or reverse the progression of neuronal degeneration in parkinonsonian patients involves trophic factor administration.

Neurotrophic factors are endogenous proteins required for neuronal differentiation, guidance, and survival during development, and often for the maintenance of the adult nervous system. Classically, a neurotrophic factor is produced and secreted by target cells and is then taken up by the innervating nerve terminals to exert both local effects and, via retrograde axonal transport, trophic effects in the nerve cell bodies (Olson, 1996). Trophic factors are proteins with enormous therapeutic potential in the treatment of neurodegenerative diseases, including the potential of modifying neuronal dysfunctions. Neurotrophic factors may not only slow the degeneration of nigral dopaminergic neurons due to their neuroprotective properties, but may also enhance the function of residual dopamine neurons or even repair and restore function to injured dopamine neurons. Although there is little evidence that deficiencies of trophic factors are associated with the etiology of PD (Hornykiewicz, 1993), considerable effort has been devoted to the search for neurotrophic factors with survival-promoting activities on midbrain dopaminergic neurons that could potentially be of therapeutic value in the treatment of PD. Several factors have been shown to produce significant beneficial effects on dopamine neurons in culture and in animal models (Collier and Sortwell, 1999). However, of all the factors investigated to date, only glial cell line-derived neurotrophic factor (GDNF) was shown to dramatically protect and enhance the function of dopamine neurons in animal models of PD (Gash et al., 1998; Grondin and Gash, 1998).

The GDNF was identified over a decade ago as the first member of a new family of cytokines in the transforming growth factor β (TGF-β) superfamily. GDNF was originally isolated and purified from the conditioned medium of cultured rat glial cells from the B49 cell line (Lin et al., 1993). The monomeric form of this heparin-binding protein consists of 134 amino acid residues. The biologically active form is a glycosylated homodimer of $\sim$30 kDa (Lin et al., 1994). The GDNF is widely expressed throughout the body in many neuronal (e.g., striatum, cerebellum, cortex) and non-neuronal tissues (e.g., kidney, gut) (Strömberg et al., 1993; Springer et al., 1994; Suter-Crazzolara and Unsicker, 1994; Choi-Lundberg and Bohn, 1995; Trupp et al., 1995). The cellular responses to GDNF are mediated via a multisubunit receptor system consisting of a glycosyl-phosphatidylinositol-anchored membrane protein, termed by convention GFRα-1, that can bind GDNF on the cell surface and facilitate its interaction with the transmembrane tyrosine kinase Ret receptor (Jing et al., 1996; Treanor et al., 1996), which in turn will undergo transphosphorylation of specific tyrosine residues to activate intracellular signaling pathways (Sariola and Saarma, 2003)

In this chapter, we will review non-clinical data obtained in non-human primates modeling PD as well as clinical data recently obtained in PD patients following prolonged delivery of exogenous GDNF into the brain.

EFFECTS OF GDNF IN NON-HUMAN PRIMATES

Data collected in cell culture (Lin et al., 1993, 1994; Hou et al., 1996) and in rodent models of PD (Hoffer et al., 1994; Bowenkamp et al., 1995; Kearns and Gash, 1995; Tomac et al., 1995; Lapchak et al., 1997) have shown that GDNF can be both neuroprotective and neurorestorative for the dopaminergic system, providing strong support for a role of GDNF in treating PD. Although crucial and informative, studies involving GDNF treatment in rodent models are limited in their relevance to the human in that rodents have a much smaller nervous system, which differs significantly in numerous neuroanatomical and neurochemical parameters from the human. In contrast, non-human primates possess a central nervous system and behavioral repertoire much closer to the human than the rodent.

THE MPTP-LESIONED RHESUS MONKEY MODEL OF PARKINSON'S DISEASE

To assess the restorative actions of GDNF under conditions where neuroprotection would have only a minor role, the late stages of human PD can be modeled using rhesus monkeys (*Macaca mulatta*) administered 1-methyl-4-phenyl-1,2,3,6-tetrahydropyridine (MPTP) (Bankiewicz et al., 1986; Smith et al., 1993; Emborg-Knott and Domino, 1998). In humans and non-human primates, MPTP induces behavioral features with numerous similarities to those found in idiopathic PD. For instance, MPTP-treated non-human primates display the cardinal symptomology of PD: bradykinesia, rigidity, balance and gait abnormalities (Bankiewicz et al., 1986; Smith et al., 1993). In addition, histological and neurochemical alterations in the brain induced by MPTP administration also resemble those found in PD. Indeed, MPTP infusion through the right carotid artery results in an approximate 75% loss of dopamine neurons expressing the phenotypic marker tyrosine hydroxylase (TH) in the right substantia nigra and a greater than 99% depletion of dopamine in the right putamen (Gash et al., 1996). These reductions are comparable to those seen in advanced human PD where cell counts typically show a 60–70% loss of nigral dopamine neurons and 99% dopamine depletion in the putamen (Kish et al., 1988).

Over the past 10 years, our group has carried out an extensive series of experiments to study the restorative effects of GDNF in non-human primates expressing hemiparkinsonian features as a result of infusions of 2.4 mg MPTP per animal into the right carotid artery (Ovadia et al., 1995). Briefly, behavioral parameters associated with motor function are typically assessed using a primate parkinsonian scale (Ovadia et al., 1995) patterned after the human Unified Parkinson's Disease Rating Scale (UPDRS), and are scored in the following categories: rigidity, bradykinesia, posture, balance, tremor, and a food acquisition test. Immunohistochemical staining for TH, the rate-limiting enzyme in dopamine synthesis, is used to identify neuritric processes and cell bodies of dopamine neurons, which are estimated using unbiased stereological cell counting (Gash et al., 1996, 2005). Multiple tissue punches are taken from the basal ganglia to measure levels of dopamine and its metabolites using high-performance liquid chromatography (Grondin et al., 2002). All the procedures performed on non-human primates in our laboratory are in strict accordance with the *NIH Guide for the Care and Use of Laboratory Animals* and are approved by institutional animal care and use committees. The surgeries are conducted under sterile field conditions in a sterile surgical suite accredited by the Association for Assessment and Accreditation of Laboratory Animal Care.

ACUTE DELIVERY OF GDNF

Single Intracerebral Injections of GDNF

Because GDNF does not cross the blood–brain barrier, a challenge to its use is the difficulty associated with its delivery to the central nervous system. In our initial studies (Gash et al., 1996), sterile magnetic resonance imaging (MRI)-guided stereotaxic procedures were used to surgically deliver a single injection of GDNF directly into the right hemisphere of MPTP-lesioned hemiparkinsonian rhesus monkeys by one of the three routes: intranigral (150 μg), intracaudate (450 μg), and intracerebroventricular (450 μg). Treatment was not started until 2 months post MPTP administration. GDNF recipients showed significant functional improvements from all three routes of administration by 2 weeks post-treatment, which continued for the remainder of the 4-week test period. In these experiments, improvements were found in three of the cardinal features of PD: bradykinesia, rigidity, and postural instability. On the lesioned side of GDNF-treated animals, dopamine levels in the substantia nigra and globus pallidus were twice as high and nigral dopamine neurons were 20% larger on average, with an increase in fiber density.

GDNF Washout and Reinstatement Studies

Non-clinical studies to date have provided little information about the long-term effects from GDNF treatment. An important clinical issue is whether patients must take trophic factors like GDNF for the rest of their lives or if short-term treatment suffices to achieve long-term benefits. Thus, we then examined two important issues of which only limited non-clinical data are available: the duration of beneficial effects from GDNF following drug washout, and the response to reinstatement of trophic factor therapy.

Using an automated video tracking program, movement speed was analyzed in parkinsonian monkeys in response to intraventricular GDNF followed by drug washout and subsequent reinstatement of trophic factor treatment a year later (Grondin et al., 2003a). In this study, four hemiparkinsonian female rhesus monkeys were administered an intraventricular injection of 300 μg GDNF and monitored for 16 weeks from the injection. Monitoring began again at 52 weeks after the injection. At week 53, these monkeys received another intraventricular injection of 300 μg GDNF in 80 μl vehicle (10 mM citrate, 150 mM NaCl, pH 4.96) and were followed through week 56. The other four parkinsonian monkeys were used as controls. They were monitored over the same time course and at week 52 received an 80 μl intraventricular injection of vehicle. In all cases, injections were made into the right lateral ventricle using MRI-guided stereotaxic procedures.

The GDNF recipients were significantly faster for 16 weeks following drug withdrawal, slowing gradually from an average movement speed 65% faster than controls to average speeds about 16% faster (Figure 9.1). When GDNF treatment was reinstituted 1 year after the first injection, trophic factor recipients showed a significant increase in movement speed that paralleled previous GDNF responses (Figure 9.1). The time course of objectively quantified changes in movement speed paralleled the response pattern seen in subjective rating scale measurements (Figure 9.2). Movement speed is subjectively factored into the rating scale as an assessment of bradykinesia in the upper and lower limbs, but the scale also includes assessment of the other features of parkinsonism: rigidity, tremor, posture, and balance.

These results are consistent with data showing that increases in motor speed achieved in aged rhesus monkeys following GDNF infusion were maintained during a 2-month GDNF washout period (Grondin et al., 2003b). Increases in motor movement speed were seamlessly continued with reinstatement of GDNF

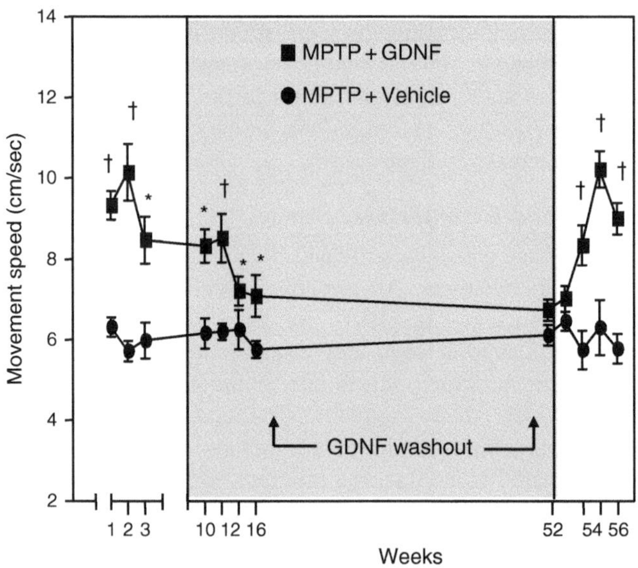

FIGURE 9.1 Movement speed. In the 3-week period following the initial intracerebral injection of GDNF, the trophic factor recipients displayed average movement speeds up to 65% faster than the control group. A significant level of improvement was maintained for up to 16 weeks post injection. When testing resumed 1 year after the initial injection (i.e., week 52), there was no significant difference in the movement speed between the two groups. The second intracerebral injection of GDNF was given at week 53. By the end of the first week after reinstitution of GDNF treatment (i.e., week 54), trophic factor recipients displayed a significant increase in movement speed that continued over the 3-week observation period. The pre-washout response resembled the post-washout response. There was not a significant change in vehicle-treated monkeys. $^*P \leq 0.05$; $^\dagger P \leq 0.01$; GDNF vs. controls, two-tailed unpaired *t*-tests comparing same time point between GDNF and control recipients.

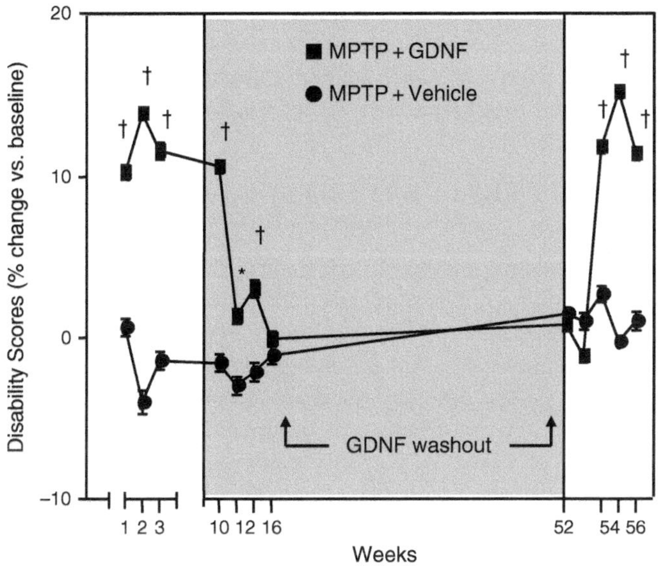

FIGURE 9.2 Rating scale. The time course of objectively quantified changes in movement speed was similar to the response pattern derived from PD rating scale scores. A maximum improvement in motor function of ~15% was seen in the GDNF recipients, both pre- and post-drug washout. Movement speed is subjectively factored into the rating scale as an assessment of bradykinesia in the limbs, but the scale also includes assessment of other parkinsonian features: rigidity, tremor, posture, and balance. $^*P \leq 0.05$; $^\dagger P \leq 0.01$; GDNF vs. controls, two-tailed unpaired *t*-tests comparing same time point between GDNF and control recipients.

in these 23–25-year old animals (Grondin et al., 2003b). These data provide evidence that GDNF responsiveness is retained in the nigrostriatal dopaminergic system for several months following cessation of exogenous trophic factor administration. The mechanisms underlying the long-term effects of GDNF lasting for months are unknown. GDNF has pronounced effects at the cellular level, promoting the growth of nigral dopamine neuron perikarya and the proliferation of TH+neurites (Gash et al., 1996; Kordower et al., 2000; Grondin et al., 2002). Both TH+fiber density and enlarged perikarya are present for at least 1 month following a GDNF injection (Gash et al., 1996). So both structural and functional effects from GDNF stimulation persist for some time after exposure to exogenous GDNF.

However, the motor response from GDNF treatment faded gradually over time to return to near baseline levels. Thus, while acute injection of GDNF directly into the brain is a feasible experimental procedure, sustained delivery of trophic factors may be necessary for optimal, long-term neuronal effects considering the progressive nature of PD. Consequently, novel methods for sustained delivery of GDNF into the nigrostriatal pathway have been studied in non-human primates, including implantation of encapsulated cells genetically engineered to produce and release trophic factors locally (Zurn et al., 2001), viral vector-mediated delivery (Kordower et al., 2000), and the use of computer-controlled infusion pumps. We have studied the latter method for almost a decade to lay the foundation for chronic delivery of trophic factors in the human brain.

CHRONIC DELIVERY OF GDNF USING COMPUTER-CONTROLLED PROGRAMMABLE PUMPS

In many published animal studies, it is difficult to distinguish between the results from protection (injury prevention) and restoration (recovery after an injury) because GDNF treatment is initiated in the hours to days following a lesion, while the injury sequelae are still unfolding. Another issue is the titer of biologically available GDNF necessary to produce beneficial effects. For instance, while significant beneficial effects can be quantified on host dopamine neurons and neuronal processes after viral vector GDNF transfection (Kordower et al., 2000), the levels of biologically available GDNF producing these effects are unclear. Therefore, to determine the titer of biologically available GDNF necessary to produce beneficial effects, a third series of experiments was undertaken in our laboratory to study the safety and efficacy of chronically infusing computer-controlled doses of GDNF into the primate brain using implantable, programmable pumps (Medtronic Inc., Minneapolis MN).

Programmable Pump and Catheter System

All our studies were conducted using SynchroMed™ Model 8616-10 pumps and a SynchroMed™ Model 8820 computer. The pump can dispense drugs in a variety of ways (e.g., continuous or timed infusion) according to instructions received by radio-frequency from the computer. The model 8616-10 pump is a round, titanium disk, about one-inch thick and three inches in diameter. The pump contains a collapsible 10-ml reservoir and a self-sealing silicone septum through which a needle is inserted to refill the pump reservoir. It also contains a bacterial retentive filter (0.22 μ) through which the drug passes as it leaves the reservoir. The implantable pump is connected to a catheter made of polyurethane, stereotactically implanted into the brain as per procedures described in detail elsewhere (Grondin et al., 2001). A removable stylet inserted in the catheter lumen provides additional stiffness and control during placement.

We have used different types of catheters (1 mm O.D.) for each of the three targets studied in our experiments: the lateral ventricle, the putamen, and the substantia nigra. The ventricular catheter has a hole in the tip with two adjacent side holes for drug delivery (model 8770AS). Three different catheters have been used for intraparenchymal delivery. To chronically deliver GDNF into the putamen, we have used a porous tip catheter (model 8770IP3) or a multiport catheter with a radiodense-closed tip (model 8770IP24A). The multiport catheter is composed of six laser holes that are placed radially over each 90 degrees of the catheter's circumference over a longitudinal distance of 3 mm, for a total of

24 laser holes (0.0015" or 37.5 μm in diameter). The most proximal set of radial holes are positioned 0.5 mm from the radiodense catheter tip. For placement into the substantia nigra, we have used catheters having a single opening (0.010" or 250 μm in diameter) at the distal tip (model 8770IP1A).

Typically, the catheter is surgically positioned into the brain after a minimum of 2 months following MPTP administration, when the parkinsonian features expressed by the animals have stabilized. The catheter is then seated in the groove of an L-shaped nylon device anchored against the skull using two nylon screws, and tunneled to the pump that is subcutaneously implanted in the lateral abdominal region of the animal (Grondin et al., 2001).

Intraventricular or Intraputamenal Delivery

Proteins diffuse slowly in brain tissue and can be rapidly degraded by proteases. As such, strategies for rapidly distributing the trophic factor through tissue to the targeted cell populations may be important. Convection enhanced delivery (CED) is one possibility, using bulk flow to significantly enhance tissue penetration and distribute macromolecules over larger volumes in the brain (Bobo et al., 1994). Thus, in addition to the basal infusion rate of 0.033 μl/min (i.e., 2 μl/h), which was necessary to keep the pumps operating properly, the pumps were programmed to deliver brief pulses at a CED rate of 10.5 μl over 30 s (equivalent to 21 μl/min), once every hour, for a total volume of 300 μl per day. Using the programmable pump and catheter system, we have demonstrated that chronic infusions of nominally 7.5 or 22.5 μg/day GDNF into the lateral ventricle or the putamen for 3 months promotes restoration of the nigrostriatal dopaminergic system and significantly improves motor functions in rhesus monkeys with neural deficits modeling the later stages of PD (Grondin et al., 2002). The functional improvements were associated with a pronounced up-regulation and regeneration of nigral dopamine neurons and their processes innervating the striatum. When compared to vehicle recipients, these functional improvements were associated with (1) >30% bilateral increase in nigral dopamine neuron cell size; (2) >20% bilateral increase in the number of nigral cells expressing the dopamine marker TH; (3) >70 and >50% bilateral increase, respectively, in dopamine metabolite levels in the striatum and the pallidum; (4) 233 and 155% increases in dopamine levels in the periventricular striatal region and in the globus pallidus, respectively, on the lesioned side; and (5) a fivefold increase in TH positive fiber density in the periventricular striatal region, on the lesioned side (Grondin et al., 2002).

These latter results emphasize the importance of residual dopamine fibers in the striatum for GDNF-induced recovery. In the vehicle recipients, the highest levels of dopamine fibers and dopamine were present in the periventricular region of the striatum, the area where significant dopamine fiber regeneration and increases in dopamine levels were found in the GDNF recipients. As such, the effect of GDNF on the lesioned right side of the striatum occurred where surviving elements of the nigrostriatal fibers were concentrated. The actions of GDNF in restoring striatal dopaminergic innervation may have been one of the principle components of the recovery seen in this series of experiments. The effects of GDNF on pallidal dopamine and dopamine metabolite levels are also of importance, considering that the globus pallidus receives dopaminergic input from the substantia nigra and is involved in regulating motor functions by sending outputs to the motor cortex via the thalamus (Smith and Kieval, 2000). It may be that changes in dopamine levels in extra-striatal areas of the basal ganglia and/or changes in the functions of dopamine uptake and release are as equally important as striatal tissue levels of dopamine for promoting behavioral improvements (Gash et al., 1996; Grondin et al., 2002). All of these effects from chronic administration of GDNF were greater than those previously seen from single injections of GDNF in MPTP-lesioned rhesus monkeys (Gash et al., 1996).

Intranigral Delivery

As degeneration of the nigrostriatal dopaminergic system is the principal pathophysiological feature of PD, the dopamine neurons in the substantia nigra of the midbrain and striatal projections to the caudate nucleus and the putamen constitute logical targets for GDNF treatment. Most attention to-date has focused on the putamen because it is the most dopamine-depleted nucleus in PD and because GDNF administered in this region in animal models is effective. However, prolonged trophic factor administration into only the substantia nigra has not been evaluated and the effects are not known. Therefore, we have recently evaluated intranigral delivery of GDNF by measuring and analyzing the effects of GDNF on motoric behavior and tissue levels of dopamine and its metabolites in MPTP-lesioned hemiparkinsonian monkeys (Gash et al., 2005). In this study, stereotaxic procedures guided by MRI were used to implant a single-port catheter into the right substantia nigra of 10 rhesus monkeys, 2 months following the induction of parkinsonian features via MPTP administration. Using a SynchroMed™ programmable pump, five animals received chronic infusions of GDNF (7.5 μg/day for 4 weeks and 22.5 μg/day for 8 weeks), at a CED rate similar to that described in "Intraventricular or Intraputamenal Delivery." Five control animals received vehicle in parallel.

This study showed that chronic intranigral delivery of 7.5–22.5 μg/day of exogenous GDNF, via programmable pumps, significantly improved motor functions in MPTP-lesioned rhesus monkeys (Gash et al., 2005). Improvement of the five GDNF recipients peaked on the fourth week of treatment at 2.5 points on the PD rating scale, equivalent to an average improvement of approximately 25% on the subjective rating scale. Functional improvements seen with intranigral delivery were comparable to those seen with putamenal delivery (Grondin et al., 2002). Although functional improvements were similar with both sites, one possibility is that GDNF is activating different signaling pathways, via axon terminal receptors in the putamen and somatodendritic receptors in the substantia nigra, similar to the site-specific actions reported for neurotrophins

(Whitmarsh and Davis, 2001). If so, then simultaneous GDNF administration to the substantia nigra and putamen could have additive effects on regeneration and functional improvements.

Tissue punches taken from 4-mm-thick coronal sections through the basal ganglia after 12 weeks of GDNF administration showed that motor improvements were associated with increases in dopamine and dopamine metabolite levels in the striatum and in the globus pallidus (Gash et al., 2005). The principal finding in this study was that the tissue distribution of intranigrally administered GDNF tightly correlates with improved motor functions and increased dopaminergic activity. This supports the hypothesis that increasing the volume of GDNF distribution in brain parenchyma is important for maximizing the efficacy of trophic factor therapy.

Also, both ventral tegmental area (VTA) and substantia nigra (SN) dopamine neurons in the midbrain showed evidence of dopaminergic upregulation in response to GDNF treatment, as evidenced by increased number of neurons expressing the dopamine marker TH (41% increase in the VTA, 44% in the SN) and increased cell size (15% increase in the VTA, 39% in the SN). Motor function improvements were directly correlated with the number of TH-positive dopamine neurons in the substantia nigra and TH-positive fiber density in the putamen, the latter indicating dopaminergic reinnervation. Most of the attention in GDNF studies to date has focused on substantia nigra dopamine neurons, with little information available on the trophic factor's effects on VTA dopamine neurons and pathways. Our results suggest that GDNF also promotes recovery of VTA dopamine neurons and their projections to the nucleus accumbens. The responses seen in our study are consistent with GDNF normalizing the VTA limbic dopaminergic pathway and do not suggest VTA over activation, as abnormal behaviors (e.g., hyperactive and/or stereotypic movements, self-mutilation) were not observed during the course of the study.

Taken together, data from our non-clinical studies provided support that chronic, intraparenchymal delivery of GDNF promotes restoration of the nigrostriatal dopaminergic system and significantly improves motor functions in MPTP-lesioned hemiparkinsonian rhesus monkeys modeling the advanced stages of PD, and laid the foundation for using this technique in PD patients.

DIRECT BRAIN DELIVERY OF GDNF IN HUMAN PARKINSONIAN SUBJECTS

VENTRICULAR DELIVERY

Based on the promising studies of the effects of GDNF in animal models of PD, an initial randomized, double-blind, placebo-controlled, dose-escalation clinical trial was conducted in 50 parkinsonian patients (n = 13 women, n = 37 men; Nutt et al., 2003). GDNF was administered monthly via ventricular delivery using an indwelling reservoir for up to 8 months. While the doses of GDNF (25–4000 μg/month) were in excess of those employed for non-human primate studies, little therapeutic efficacy was observed in these levodopa-responsive patients with moderately advanced PD. Instead, significant adverse effects were observed including nausea, anorexia, vomiting, and weight loss. In addition, one 65-year-old patient with a 23-year history of PD who came to autopsy while in the open extension phase of the study showed no significant effects of GDNF on dopamine neurons (Kordower et al., 1999). The problem may have been with the site and method of delivery; that is, monthly injections of the trophic factor into the lateral ventricle. Sufficient titers of GDNF may not have diffused through the ventricular wall and brain parenchyma to the targeted dopamine neurons in the substantia nigra and their afferent projections to the putamen. Accordingly, we have demonstrated the limited penetration of GDNF through the ventricular wall as compared to the adequate diffusion of GDNF observed with intraparenchymal infusion into the striatum of nonhuman primates (Ai et al., 2003).

CHRONIC DELIVERY OF GDNF USING COMPUTER-CONTROLLED PROGRAMMABLE PUMPS

Ventricular delivery theoretically allows access to large regions of the brain, but GDNF delivery to this site produced pronounced side effects, including nausea, anorexia, and vomiting, without significant therapeutic benefits in the clinical tests conducted to date as described in "Ventricular Delivery." This suggests that other delivery methods may be necessary to produce therapeutic effects while reducing side effects.

Bristol Phase-1 Trial

The excellent control achieved by pumps and infusion catheters in animal models of PD (Grondin et al., 2002; Gash et al., 2005) have made this approach the best for continued trials in humans. In a study conducted in England (Gill et al., 2003), five advanced PD patients with a previous history of good responses to levodopa underwent unilateral (n = 1) or bilateral (n = 4) insertion of drug infusion cannulae into the dorsal putamen. Human recombinant GDNF (14.4–43.2 μg per day) was chronically infused via indwelling 0.6-mm intraparenchymal catheter(s) and SynchroMed™ pumps implanted in the abdominal region. Patients were assessed pre- and post-operatively (3, 6, and 12 months) according to the core assessment program for intracerebral transplantations (CAPIT), in order to document changes in disease severity and medication requirements (Langston et al., 1992).

Chronic GDNF infusion resulted in improved motor function in all patients. After 1-year, there was an average 39% improvement in the OFF-medication motor scores (UPDRS III) and a 61% improvement in the activities of daily living (UPDRS II) scores. Although levodopa equivalents were maintained in three of the five patients during GDNF infusion, the overall dyskinesia scores were significantly reduced by over 60% while on medication. The patients also underwent [^{18}F]dopa positron emission tomography (PET) scans at baseline, 6, 12, and 18 months after GDNF infusion to assess putamen dopamine terminal function and correlate this with any symptomatic benefits. The GDNF infusion was associated with an average 28% increase in [^{18}F]dopa uptake around the catheter tip after 18 months, suggesting an increase in putamen dopamine storage and supporting changes in dopamine function post GDNF treatment. Chronic GDNF infusion was tolerated well in all patients and limited side effects were observed. There was no nausea, anorexia, vomiting, or weight loss reported, only that patients experienced tingling passing from the neck down the arms while flexing the neck (Lhermitte's sign), which was mild, intermittent, and non-distressing to the patients.

The same group also reported on the continuation of this Phase-1 study (Patel et al., 2005). Between 12 and 18 months, all patients continued to receive 14.4 μg GDNF per day. Between 18 and 24 months, the dose was increased to 28.8 μg GDNF per day (only Patient 4 reverted back to 14.4 μg at 20 months). After 2 years of GDNF infusion, there were still no serious clinical side effects and patients improved even further showing an average 57% improvement in their OFF-medication motor scores and 63% improvement in their activities of daily living scores (Patel et al., 2005).

GDNF Induces Neuronal Sprouting in the Human Brain

One of the five subjects in the Bristol Phase-1 trial recently died of a myocardial infarct and was autopsied (Love et al., 2005). This 62-year-old man received intraputamenal infusion of 14.4–43.2 μg per day GDNF for almost 2 years prior to his death. During that time, his UPDRS-III OFF-medication motor score had improved by 38% (24 months), and was associated with an 91% increase in [^{18}F]dopa uptake in the posterior putamen in the infused hemisphere (18 months). In comparison with the left putamen, TH+ fibers were increased fivefold in the posterior one-third of the right putamen. These findings suggest that chronic infusion of GDNF promotes sprouting of dopaminergic fibers in the human brain similar to what was reported in MPTP-lesioned non-human primates treated with GDNF (Gash et al., 1996; Grondin et al., 2002). It is not clear whether this increase in TH+ fibers is due to the repair of damaged and dysfunctional fibers or to the upregulation of healthy remaining fibers or both. Nonetheless, these data are the first to provide a substrate for the sustained clinical improvement in disability and enhanced [^{18}F]dopa uptake measured in GDNF-treated PD patients (Gill et al., 2003, see Slevin et al., 2005 in "University of Kentucky FDA-approved Phase-1 Trial" below).

University of Kentucky FDA-approved Phase-1 Trial

This exciting initial trial was followed up in the United States, at the University of Kentucky, with an FDA-approved open-label Phase-1 safety trial in 10 patients with advanced bilateral PD who underwent unilateral putamenal administration of GDNF for 6 months (Slevin et al., 2005). Each patient had an MRI-based stereotactic implantation of an intraputamenal multi-port catheter (1.65 mm O.D. diameter, 40 holes, 4 per 0.5 mm placed every 90°, 5 mm length; Model 10532, Medtronic, Inc.) inserted into the medial putamen contralateral to the most affected side. The SynchroMed infusion pump (Model 8626L-18, Medtronic, Inc.) was implanted subcutaneously in the ipsilateral lower quadrant abdominal wall and a catheter was tunneled to connect the pump to the indwelling intraputamenal catheter. The pump was programmed to infuse at a continuous basal infusion rate of 2 μl/h. Small bolus injections (21.3 μl delivered in 117 s) were delivered every 6 h to supplement the basal rate and to increase the brain area affected by the drug. Each patient was placed on a dose-escalation regimen of GDNF: 3, 10, and 30 g/day at successive 8-week intervals, followed by a 1-month washout period. Testing was conducted in an operationally defined OFF-medication condition: anti-parkinson medication withheld for 2.5 times the duration of each drug's predicted serum half-life. No changes were made in patients' anti-PD drug regimens during the course of the study.

Total UPDRS OFF- and ON-medication scores were significantly improved 33 and 34%, respectively, at 24 weeks of treatment compared to baseline scores. Also, both motor UPDRS OFF- and ON-medication scores were significantly improved by 30% at 24 weeks compared to baseline scores. All significant improvements of motor function continued through the 1-month washout period as predicted by non-clinical studies in non-human primates (see Figures 9.1 and 9.2; Grondin et al., 2003a,b). The only side effect observed was transient Lhermitte's symptoms in two subjects. In addition, improvements were bilateral as measured by balance and gait and increased speed of hand movements. Bilateral effects from unilateral GDNF administration have been seen repeatedly in non-human primates (Gash et al., 1996; Grondin et al., 2003b). The neural circuitry responsible has been partially worked out and involves pathways from the substantia nigra pars reticulata to the thalamus, thalamo-cortical projections, and bilateral glutaminergic cortico-striatal projections (Barbeito et al., 1989).

Taken together, these promising results obtained in all 10 patients in our open-label Phase-1 study, together with those described by Gill and colleagues (2003) in five additional patients led to further investigation of GDNF delivery as a therapy for PD.

FDA-approved Phase-2, Randomized, Placebo-Controlled Trial

Largely based on the Phase-1 trial conducted by Gill and colleagues (2003), an FDA-approved Phase-2, multicenter, randomized, double-blind, placebo-controlled study was conducted to evaluate the efficacy and safety of intraputamenal GDNF infusion in PD patients (Lang et al., 2006). The primary endpoint was the percentage change from baseline in the UPDRS OFF-medication motor score after 6 months of treatment. Baseline scores were recorded after 12 h without anti-parkinson medication. Also, patients underwent [^{18}F]dopa PET scans at baseline and 6 months after GDNF infusion.

Thirty-four patients were enrolled in the trial and divided into two treatment groups: GDNF- and vehicle-treated. Bilateral single-port intraparenchymal catheters (Model 8760, Medtronic, Inc.) were stereotactically implanted into the posterior dorsal putamen and attached to separate SynchroMed® infusion pumps (Medtronic, Inc.) placed in the patient's abdomen. The pumps were programmed to infuse 15 μg/putamen/day at a continuous rate of 6.25 μl/h. Due to failure to tolerate the scanning protocol and head motion artifacts, [^{18}F]dopa influx constant was only analyzed in 22 patients. GDNF-treated patients ($n = 9$) showed a median increase of 23.1% in the posterior putamen compared with a median reduction of –8.8% in the placebo group ($n = 13$) for a between group difference of ~32% after 6 months of infusion. However, despite increased [^{18}F]dopa uptake, mean percentage changes in UPDRS OFF-medication motor scores were only 10% in the GDNF group compared with 4.5% in the placebo group. This treatment difference was not significant. This is in sharp contrast with the two Phase-1 trials documenting pronounced long-lasting benefits from intraputamenal GDNF infusion in a total of 15 PD patients (Gill et al., 2003; Slevin et al., 2005).

The disparity between the Phase-2 trial and the two successful Phase-1 trials may be explained by differences in patient selection and treatment strategies, including drug dosage and infusion method. First, the patients enrolled in the open-label trials had generally milder disease than those enrolled in the Phase-2 trial, and as such may have had a stronger potential to respond to GDNF treatment due to a greater number of remaining nigrostriatal fibers. Second, baseline UPDRS OFF motor scores were determined under different conditions: baseline scores in the Phase-2 trial were recorded after anti-parkinson medication were withheld for 12 h, whereas baseline scores in the Kentucky Phase-1 trial were determined after anti-parkinson medications were withheld for 2.5 times the duration of each drug's predicted serum half-life (i.e., >36 h in some cases). Third, the patients in the Phase-2 trial received a lower dose of GDNF (15 μg/putamen/day) during the study, whereas the patients in the open studies received up to 30 μg/putamen/day (Slevin et al., 2005) and 43.2 μg/putamen/day (Gill et al., 2003). Fourth, the Phase-2 study emphasized point source delivery of GDNF using single-port catheters and a low infusion rate (0.1 μl/min), whereas the Kentucky Phase-1 study emphasized tissue distribution using a multi-port catheter and bolus infusion (21.3 μl/min, once every 6 h). Thus, the volume of GDNF distribution in the brain parenchyma needed to induce functional improvements may have been insufficient in the Phase-2 trial due to the alteration of these parameters. Last, the different clinical results could be due to a placebo effect since both the Bristol and the Kentucky Phase-1 trials were conducted in an open-label fashion. However, this seems unlikely considering that the double-blind study did not reveal a significant placebo effect (Lang et al., 2006), and because progressive clinical improvements were observed for up to 3 years in the Phase-1 trials, including sustained improvements during GDNF withdrawal for several weeks (Gill et al., 2003; Slevin et al., 2005, 2006).

POINT SOURCE CONCENTRATION OF GDNF MAY EXPLAIN FAILURE OF PHASE-2 CLINICAL TRIAL

As described above in "FDA-approved Phase-2, Randomized, Placebo-Controlled Trial", there were major technical differences between the three clinical trials in the catheters and infusion protocols used to evaluate intraputamenal infusion of GDNF for the treatment of PD (Sherer et al., 2006). In a recent series of experiments, we have tested the hypothesis that GDNF bioavailability in the brain was not optimal in the Phase-2 study and may have contributed to the failure to replicate results from the Phase-1 trials and animal studies. To determine if problems in drug bioavailability could have contributed to the discrepancies between studies, we have analyzed the distribution of intraputamenally infused GDNF in the rhesus monkey brain (Salvatore et al., 2006). I^{125}-GDNF was unilaterally infused into the putamen of three adult rhesus monkeys using the delivery system and infusion protocol followed in the Phase-2 clinical trial that failed to achieve its primary endpoint (Lang et al., 2006). Three age-matched animals received vehicle infusions following identical procedures.

The volume of distribution (Vd) of GDNF around the catheter was determined by immunocytochemistry. After 7 days of treatment, GDNF was found to be concentrated around the catheter tip with distribution ranging from as low as $87\,mm^3$ in one animal to $369\,mm^3$ in the monkey with the highest Vd. Considering that the normal volume of the human putamen in one hemisphere ranges from 4000 to $5000\,mm^3$ (White et al., 2002; Cousins et al., 2003; Tamagaki et al., 2005), the limited Vd of GDNF seen in non-human primates would only cover between 2 and 9% of the average human putamen. This study provides a snapshot at one time of treatment. However, available data suggest that GDNF Vd would not increase over time. Indeed, the range of infusion in the current study is nearly identical with the limited Vd found using the same catheter for infusing GDNF into the rhesus monkey substantia nigra (Gash et al., 2005). This is in sharp contrast to the much greater distribution achieved with multiport catheters and pulsatile delivery somewhat analogous to one Phase-1 trial (Ai et al., 2003; Slevin et al., 2005). At this time, the delivery protocol and catheter used in the Phase-2 GDNF intraputamenal infusion study should be considered as significant factor in the failure of this clinical trial to replicate the results of numerous animal studies and the two Phase-1 trials using different catheters and treatment protocols. In fact, the major difficulty with evaluating GDNF therapeutic potential at this time is the failure to have a Phase-2 trial replicating the successful Phase-1 trial in dose and methods of trophic factor delivery.

CONCERNS OVER TWO SAFETY ISSUES DERAIL ALL GDNF TRIALS

Since the completion of the Phase-2 clinical trial, two safety concerns have emerged, which eventually led to all testing of human recombinant GDNF being stopped in PD patients. First, antibodies to GDNF have been identified in the blood of some patients. The presence of neutralizing antibodies, which could potentially cross-react with endogenous GDNF, has been detected in three patients enrolled in the Phase-2 trial (Lang et al., 2006) and one patient enrolled in the Kentucky Phase-1 trial (Slevin et al., 2005, 2006). Although the long-term consequences of these observations are unknown, GDNF antibody-induced problems have not been detected in these patients at this time.

Second, 4 of 15 rhesus monkeys receiving high doses of 100 μg GDNF per putamen/day during a 6-month toxicology study presented focal cerebellar injury (Lang et al., 2006; Sherer et al., 2006). All, but one animal, had been withdrawn from GDNF for 3 months prior to terminating the study. The lesions were characterized by multisegmental Purkinje cell loss associated with variable atrophy/disruption of the overlying molecular layer and a variable loss of underlying granule cells. At this time, it is not clear whether abrupt GDNF withdrawal contributed to the cerebellar lesions seen in the toxicology study, since Purkinje cells have receptors for GDNF, or whether the lesions were the results of an experimental artifact (Sherer et al., 2006). Clearly, additional toxicology studies would be needed to address these questions. To better understand the conditions giving rise to the lesions, we have evaluated brain-banked cerebellar tissue from 40 rhesus monkeys in our non-clinical studies receiving chronic intracerebral infusions of GDNF ($n = 25$) or vehicle ($n = 15$; Ai et al., 2005). The studies involved infusions into either the putamen, lateral ventricle, or dorsal substantia nigra. GDNF doses ranged from 7.5 to 30 μg/day for periods up to 8 months. All animals were middle-aged to aged; many with MPTP-induced parkinsonian features. Coronal cerebellar sections from comparable levels in each monkey were histochemically stained with Giemsa or Hematoxylin & Eosin, or immunostained for glial fibrillary acidic protein or Calbindin D-28K. In the detailed histopathological analysis conducted to date, no detectable differences in the cerebellum have been found between vehicle and GDNF recipients (Ai et al., 2005). Specifically, focal cerebellar lesions involving the loss of Purkinje cells and granule cells along with molecular layer injury have not been identified in monkeys receiving GDNF in the clinically relevant dose range.

After the Phase-1 trial conducted at the University of Kentucky was halted by the sponsor due to the safety concerns mentioned above, all 10 patients were monitored for an additional year to evaluate the effects of drug withdrawal (Slevin et al., 2006). The delivery system was reprogrammed to deliver sterile saline at the basal infusion rate of 2 μl/h. Benefits from treatment were lost by 9–12 months after GDNF infusion was halted. Antibodies to GDNF have developed in seven patients with no evidence of clinical sequelae. In addition, there was no evidence of GDNF-induced cerebellar toxicity, as evaluated using MRI analysis and clinical testing (Slevin et al., 2006). Again, the potential safety risks of GDNF antibodies and cerebellar injuries are unknown and should be carefully monitored.

FUTURE DIRECTIONS

Studies using viral vector delivery of GDNF support the notion that protection of dopamine neurons from MPTP toxicity and improvements in the function of aged dopamine neurons can be achieved in non-human primates with combined striatal and substantia nigra expression of GDNF (Kordower et al., 2000). As such, simultaneous delivery of GDNF or other molecules with trophic properties into the putamen and the substantia nigra may have a synergistic effect on the nigrostriatal dopaminergic system, promoting greater restoration of function than single site administration. Application of GDNF directly to the striatum may affect populations of dopamine fibers that transport GDNF back to the substantia nigra and produce improved function of dopamine neurons, whereas application of GDNF directly to the substantia nigra may affect a second population of dopamine neurons, which project to the globus pallidus and other basal ganglia dopamine projections. This is also of importance considering that the globus pallidus is involved in regulating motor functions by sending outputs to the motor cortex via the thalamus.

A second major issue in repairing damaged dopamine neurons in patients with PD involves the question of when neurorestorative therapies should be started? It has long been postulated that early intervention in PD patients with less damage to dopamine neurons may yield the greatest outcome with neurotrophic factors such as GDNF due to the presence of a greater number of remaining nigrostriatal fibers. Additional studies are needed to test this hypothesis.

Also, little is known at the present time about the intracellular phosphorylation signaling mechanisms occurring in vivo following activation of Ret via GDNF or other related factors (Sariola and Saarma, 2003). New approaches for the treatment of PD may be developed by targeting GDNF receptors or the signaling pathway that is initiated when these receptors are activated. For example, understanding the mechanism of action of these trophic factors could lead to the development of low-molecular-weight drugs that could mimic the action of the endogenous proteins and be used as an alternative to recombinant neurotrophic factors.

Finally other members of the GDNF family, namely neurturin (Kotzbauer et al., 1996), persephin (Milbrandt et al., 1998), and artemin (Baloh et al., 1998), may also have therapeutic potential. Persephin was shown to be as potent as GDNF when delivered to the brain using neural stem cells in a rat model of PD (Akerud et al., 2002; Sariola and Saarma, 2003). We are currently testing exogenous delivery of neurturin using our catheter and pump system in MPTP-lesioned monkeys modeling PD. Also, non-clinical and clinical trials are underway to test the safety and efficacy of neurturin using viral-vector delivery as a therapy for PD (Dass et al., 2005). Alternative approaches such as gene therapy hold great promise for the local and prolonged delivery of trophic factors.

SUMMARY

Trophic factors are proteins with enormous therapeutic potential in the treatment of neurodegenerative diseases like PD, including the potential to not only slow the degeneration of nigral dopaminergic neurons due to their neuroprotective properties, but also to enhance the function of residual dopamine neurons or even repair and restore function to injured dopamine neurons. Due to the progressive nature of PD, sustained delivery of trophic factors such as GDNF appears necessary for optimal, long-term neuronal effects. Novel methods for sustained delivery of GDNF into the nigrostriatal pathway have been studied in non-human primates, including the use of computer-controlled infusion pumps. Using this approach, we have demonstrated that chronic infusions of nominally 7.5 or 22.5 μg/day GDNF into either the lateral ventricle, the putamen or the substantia nigra promote restoration of the nigrostriatal dopaminergic system and significantly improve motor functions in MPTP-lesioned rhesus monkeys with neural deficits modeling PD.

Based on the promising studies of the chronic effects of GDNF in non-human primate models of PD, a series of studies were conducted in PD patients. However, translational studies from the laboratory to the clinic for the treatment of advanced PD have been problematic. On one hand, GDNF was not efficacious when delivered intraventricularly in advanced PD patients, likely failing because of poor trophic factor penetration through the ventricular wall into the parenchyma of the basal ganglia. On the other hand, two independent open label Phase-1 studies have reported marked functional improvements in 15 advanced PD patients receiving intraputamen infusion of GDNF. However, a separate, randomized, blinded, multicenter trial of intraputmenal GDNF infusion in advanced PD patients failed to reach a predetermined endpoint of 25% difference in UPDRS OFF-medication motor scores between GDNF and placebo treatment. In addition, safety concerns have arisen including the presence of neutralizing antibodies to GDNF in some patients and the presence of focal cerebellar lesions in rhesus monkeys chronically receiving high dose of GDNF. The potential safety risks of GDNF antibodies and cerebellar injuries are unknown and should be carefully evaluated. The major difficulty with risk assessment at this time is the failure to have a Phase-2 trial replicating the successful Phase-1 trial in dose and methods of trophic factor delivery. In addition, the Phase-2 trial may have been statistically underpowered (Hutchinson et al., 2006). Thus, before any definitive conclusions could be made regarding the use of GDNF as a therapy for PD, a properly designed, adequately powered multicenter Phase-2 clinical trial should be conducted in PD patients to fully evaluate its potential.

ACKNOWLEDGMENTS

Supported by USPHS grants NS39787, AG13494, and contracts with Amgen Inc., Thousand Oaks, CA. We thank Medtronic Inc., Minneapolis, MN, for providing the pumps and associated hardware and software used for chronic infusion of GDNF.

REFERENCES

Akerud P, Holm PC, Castelo-Branco G, Sousa K et al. (2002). Persephin-overespressing neural stem cells regulate the function of nigral dopaminergic neurons and prevent their degeneration in a model of Parkinson's disease. *Mol Cell Neurosci* 21: 205–222.

Ai Y, Markesbery W, Zhang Z et al. (2003). Intraputamenal infusion of GDNF in aged rhesus monkeys: distribution and dopaminergic effects. *J. Comp. Neurol* 461: 250–261.

Ai Y, Carter N, Zhang AM et al. (2005). Cerebellar histopathological analysis in normal and parkinsonian rhesus monkeys following chronic intracerebral infusion of GDNF. Program No. 899.8. Abstract Viewer/Itinerary Planner. Washington, DC: Society for Neuroscience, 2005. Online.

Baloh RH, Tansey MG, Lampe PA et al. (1998). Artemin, a novel member of the GDNF ligand family, supports peripheral and central neurons and signals through the GFRα3-Ret receptor complexe. *Neuron* 21: 1291–1302.

Bankiewicz KS, Oldfield EH, Chiueh CC et al. (1986). Hemiparkinsonism in monkeys after unilateral internal carotid artery infusion of 1-methyl-4-phenyl-1,2,3,6-tetrahydropyridine. *Life Sci* 39: 7–16.

Barbeito L, Girault JA, Godeheu G et al. (1989). Activation of the bilateral corticostriatal glutamatergic projection by infusion of GABA into thalamic motor nuclei in the cat: an in vivo release study. *J Neurosci* 28: 365–374.

Bobo RH, Laske DW, Akbasak A et al. (1994). Convection-enhanced delivery of macromolecules in the brain. *Proc Natl Acad Sci U S A* 91: 2076–2080.

Bowenkamp KE, Hoffman AF, Gerhardt GA et al. (1995). Glial cell line-derived neurotrophic factor supports survival of injured midbrain dopaminergic neurons. *J Comp Neurol* 355: 479–489.

Choi-Lundberg DL and Bohn MC (1995). Ontogeny and distribution of glial cell line-derived neurotrophic factor (GDNF) mRNA in rat. *Dev Brain Res* 85: 80–88.

Collier TJ and Sortwell CE (1999). Therapeutic potential of nerve growth factors in Parkinson's disease. *Drugs Aging*, 14: 261–287.

Cousins DA, Burton EJ, Burn D et al. (2003). Atrophy of the putamen in dementia with Lewy bodies but not Alzheimer's disease: an MRI study. *Neurology* 61: 1191–1195.

Dass B, Herzog CD, Bakay RAE et al. (2005). Adeno-associated virus mediated gene delivery of neurturin (CERE-120) prevents MPTP-induced motor disability following injection into the striatum of non-human primates. Program No. 545.6. Abstract Viewer/Itinerary Planner. Washington, DC: Society for Neuroscience, 2005. Online.

Emborg-Knott ME and Domino EF (1998). MPTP-induced hemiparkinsonism in nonhuman primates 6–8 years after a single unilateral intracarotid dose. *Exp Neurol* 152: 214–220.

Gash DM, Zhang Z, Ovadia A et al. (1996). Functional recovery in parkinsonian monkeys treated with GDNF. *Nature* 380: 252–255.

Gash DM, Zhang Z, and Gerhardt GA (1998). Neuroprotective and neurorestorative properties of GDNF. *Ann. Neurol* 44: S121–S125.

Gash DM, Zhang Z, Ai Y et al. (2005). GDNF Distribution predicts dopaminergic restoration in parkinsonian monkeys. *Ann Neurol* 58: 224–233.

Gill SS, Patel NK, Hotton GR et al. (2003). Direct brain infusion of glial cell line-derived neurotrophic factor in Parkinson disease. *Nat Med* 9: 589–595.

Grondin R, and Gash DM (1998). Glial cell line-derived neurotrophic factor (GDNF) as a drug candidate for the treatment of Parkinson's disease. *J Neurol* 245: 35–42.

Grondin R, Zhang Z, Elsberry D et al. (2001). Chronic intracerebral delivery of trophic factors via a programmable pump as a treatment for parkinsonism. In MM Mouradian ed. *Parkinson's Disease-Methods & Protocols* Vol. 62, pp. 257–267. Humana Press.

Grondin R, Zhang Z, Cass WA et al. (2002). Chronic, controlled GDNF infusion promotes structural and functional recovery in advanced parkinsonian monkeys. *Brain* 125: 2191–2201.

Grondin R, Zhang Z, Ai Y et al. (2003a). GDNF improves motor function in aged and MPTP-lesioned monkeys: Effect of trophic factor withdrawal. Program No. 299.4. Abstract Viewer/Itinerary Planner. Washington, DC: Society for Neuroscience, 2003. Online.

Grondin R, Cass WA, Zhang Z et al. (2003b). GDNF Increases Stimulus-Evoked Dopamine Release and Motor Speed in Aged Rhesus Monkeys. *J Neurosci* 23: 1974–1980.

Hoffer BJ, Hoffman A, Bowenkamp K et al. (1994). Glial cell line-derived neurotrophic factor reverses toxin-induced injury to midbrain dopaminergic neurons in vivo. *Neurosci Lett* 182: 107–111.

Hornykiewicz O (1993). Parkinson's disease and the adaptive capacity of the nigrostriatal dopamine system: possible neurochemical mechanisms. *Adv Neurol* 60: 140–147.

Hou J-GG, Lin L-FH, and Mytilinenou C (1996). GDNFexerts neurotrophic effects on dopaminergic neurons in vitro and promotes their survival and regrowth after damage by 1-methyl-4-phenylpyridinium. *J Neurochem* 66: 74–82.

Hutchinson M, Gurney S, and Newson R (2007). GDNF in Parkinson disease: an object lesson in the tyranny of type II. *J Neurosci Methods* 163: 190–192.

Jing S, Wen D, Yu Y et al. (1996). GDNF-induced activation of the Ret protein tyrosine kinase is mediated by GDNFR-α, a novel receptor for GDNF. *Cell* 85: 1113–1124.

Kearns C and Gash DM (1995). GDNF protects nigral dopamine neurons against 6-hydroxydopamine in vivo. *Brain Res.* 672: 104–111.

Kish SJ, Shannak K, and Hornykiewicz O (1988). Uneven pattern of dopamine loss in the striatum of patients with idiopathic Parkinson's disease. Pathophysiology and clinical implications. *N Eng J Med* 318: 876–880.

Kordower JH, Palfi S, Chen EY et al. (1999). Clinicopathological findings following intraventricular glial-derived neurotrophic factor treatment in a patient with Parkinson's disease. *Ann Neurol* 46: 419–424.

Kordower JH, Emborg ME, Bloch J et al. (2000). Neurodegeneration prevented by lentiviral vector delivery of GDNF in primate models of Parkinson's disease. *Science* 290: 767–773.

Kotzbauer PT, Lampe PA, Heuckeroth RO et al. (1996). Neurturin, a relative of glial cell line-derived neurotrophic factor. *Nature* 384: 467–470.

Lang AE and Lozano AM (1998). Parkinson's disease. First of two parts. *N Engl J Med* 339: 1044–1053.

Lang AE, Gill S, Patel NK et al. (2006). Randomized controlled trial of intraputamenal glial cell line-derived neurotrophic factor infusion in Parkinson's disease. *Ann Neurol* 59: 459–466.

Langston JW, Widner H, Goetz CG et al. (1992). Core assessment program for intracerebral transplantations (CAPIT). *Mov Disord* 7: 2–13.

Lapchak PA, Miller PJ, Collins F, and Jiao S (1997). Glial cell line-derived neurotrophic factor attenuates behavioural deficits and regulates nigrostriatal dopaminergic and peptidergic markers in 6-hydroxydopamine-lesioned adult rats: Comparison of intraventricular and intranigral delivery. *J Neurosci* 78: 61–72.

Love S, Plaha P, Patel NK et al. (2005). Glial cell line-derived neurotrophic factor induces neuronal sprouting in human brain. *Nat Med* 11: 703–704.

Lin L-FH, Doherty DH, Lile JD et al. (1993). GDNF: A glial cell line-derived neurotrophic factor for midbrain dopaminergic neurons. *Science* 260: 1130–1132.

Lin L-FH, Zhang TJ, Collins F, and Armes LG (1994). Purification and initial characterization of rat B49 glial cell line-derived neurotrophic factor. *J Neurochem* 63: 758–768.

Milbrandt J, de Sauvage FJ, Fahrner TJ et al. (1998). Persephin, a novel neurotrophic factor related to GDNF and neurturin. *Neuron* 20: 245–253.

Nutt JG, Burchiel KJ, Comella CL et al. (2003). Randomized, double-blind trial of glial cell line-derived neurotrophic factor (GDNF) in PD. *Neurology* 60: 69–73.

Olson L (1996). Toward trophic treatment in parkinsonism: A primate step. *Nat Med* 2: 400–401.

Ovadia A, Zhang Z, and Gash DM (1995). Increased susceptibility to MPTP toxicity in middle-aged rhesus monkeys. *Neurobiol Aging* 16: 931–937.

Patel NK, Bunnage M, Plaha P et al. (2005). Intraputamenal infusion of glial cell line-derived neurotrophic factor in PD: A two-year outcome study. *Ann. Neurol* 57: 298–302.

Salvatore MF, Ai Y, Wong S, Fischer B et al. (2006). Point Source Concentration of GDNF may Explain Failure of Phase II Clinical Trial. *Exp Neurol* 202: 497–505.

Sariola H and Saarma M (2003). Novel functions and signaling pathways for GDNF. *J Cell Sci* 116: 3855–3862.

Sherer TB, Fiske BK, Svendsen CN et al. (2006). Crossroads in GDNF therapy for Parkinson's disease. *Mov Disord* 21: 136–141.

Slevin JT, Gerhardt GA, Smith CD et al. (2005). Unilateral Intraputamenal GDNF Improves Bilateral Motor Functions in Patients with Parkinson's disease. *J Neurosurg* 102: 216–222.

Slevin JT, Gash DM, Smith CD et al. (2006). Unilateral intraputamenal glial cell-line-derived neurotrophic factor in patients with Parkinson's disease: response to 1 year each of treatment and withdrawal. *Neurosurg. Focus* 20: E1

Smith Y and Kieval JZ (2000). Anatomy of the dopamine system in the basal ganglia. *Trends Neurosci* 23: S28–S33.

Smith RD, Zhang Z, Kurlan R et al. (1993). Developing a stable bilateral model of parkinsonism in rhesus monkeys. *J Neurosci* 52: 7–16.

Springer JE, Mu X, Bergmann LW, and Trojanoski JQ (1994). Expression of GDNF mRNA in rat and human nervous tissue. *Exp Neurol* 127: 167–170.

Strömberg I, Bjorklund L, Johansson M et al. (1993). Glial cell line-derived neurotrophic factor is expressed in the developing but not adult striatum and stimulates developing dopamine neurons in vivo. *Exp Neurol* 124: 401–412.

Suter-Crazzolara C and Unsicker K. (1994). GDNF is expressed in two forms in many tissues outside the CNS. *NeuroReport* 5: 2486–2488.

Tamagaki C, Sedvall GC, Jonsson EG et al. (2005). Altered white matter/gray matter proportions in the striatum of patients with schizophrenia: a volumetric MRI study. *Am J Psychiatry* 162: 2315–2321.

Tomac A., Lindqvist E, Lin L-FH et al. (1995). Protection and repair of the nigrostriatal dopaminergic system by GDNF in vivo. *Nature* 373: 335–339.

Treanor JJS, Goodman L, de Sauvage F et al. (1996). Charaterization of a multicomponent receptor for GDNF. *Nature* 382: 80–83.

Trupp M, Rydén M, Jörnavall C et al. (1995). Peripheral expression and biological activities of GDNF, a new neurotrophic factor for avian and mammalian peripheral neurons. J *Cell Biol* 130: 137–148.

White T, Andreasen NC, and Nopoulos P (2002). Brain volumes and surface morphology in monozygotic twins. *Cereb Cortex* 12: 486–493.

Whitmarsh AJ and Davis RJ (2001). Signal transduction by target-derived neurotrophins. *Nat Neurosci* 4: 963–964.

Zurn AD, Widmer HR, and Aebischer P (2001). Sustained delivery of GDNF: towards a treatment for Parkinson's disease. *Brain Res Rev* 36: 222–229.

10

STIMULATING AND ABLATIVE TREATMENT FOR PARKINSON'S DISEASE

JULIE G. PILITSIS AND ROY A.E. BAKAY

Department of Neurosurgery, Rush University Medical Center, Chicago, IL 60612

ABSTRACT

Since the advent of levodopa in the 1960s, medication has been the mainstay of treatment for Parkinson's Disease (PD). However, dyskinesias and motor fluctuations develop in up to 80% of patients with PD treated with levodopa for more than 5 years. Thus, chronic treatment with medication coupled with advancing disease results in severe disability for many patients. Surgical therapies offer symptom relief for a subset of these patients. Surgical interventions for PD have been practiced for decades, beginning with lesions geared at eliminating activity in specific basal ganglia structures. Renewed surgical interest in the 1980s was stimulated by a long history of medical failures and increased safety of ablative procedures. The development of deep brain stimulation in the 1990s offered even greater interest in the surgical treatment of movement disorders, by providing a reversible and bilateral surgical means of treatment. This chapter examines the therapeutic mechanisms associated with surgical treatment of PD, describes clinical indications for surgery, and discusses the efficacy and complications of procedures.

Keywords: Parkinson's Disease, deep brain stimulation, pallidotomy, subthalamic nucleus

Copyright © 2008 Elsevier Inc.
All rights of reproduction in any form reserved.

INTRODUCTION

The pathologic hallmark of Parkinson's Disease (PD) is the degeneration of dopaminergic cells within the substantia nigra and subsequent dopamine depletion of the striatum. Clinical manifestations of PD occur when nigrostriatal dopamine loss reaches approximately 60% (Brooks, 1998). Since the 1960s, the mainstay of PD treatment has been the medication levodopa, which works by being broken down into dopamine in the central nervous system. However, several limitations prevent its long-term clinical effectiveness (Lewitt and Nyholm, 2004). Dyskinesias and motor fluctuations develop in up to 80% of PD patients treated with levodopa for >5 years. Thus, chronic treatment with levodopa coupled with advancing disease results in severe disability for many patients. Treatment of levodopa-induced motor complications include manipulation of the dose, frequency and type of formulation, and/or addition of other medications such as dopamine agonists, catechol-O-methyl transferase inhibitors, monoamine oxidase-B (MAO-B) inhibitors, and N-methyl-D-aspartic acid (NMDA) receptor antagonists (Pahwa et al., 2006a). Surgical therapies offer symptom relief for a subset of patients.

Surgical interventions for PD have been practiced for decades, beginning with lesions geared at eliminating activity in specific basal ganglia structures. Even very early there was an appreciation that stimulation could affect the symptoms (Pool, 1954). In the 1990s, with improved equipment, it was final proven that chronic stimulation resulted in clinical benefits similar to those achieved with surgical lesioning, but that were also reversible (Benabid et al., 1991). Since the development of deep brain stimulation (DBS), there has been a general shift in thought paradigm from creating lesions to stimulation in the treatment of movement disorders including PD (Pollak et al., 2002). This change in thinking led to a focus in neurosurgery toward neuromodulation of symptoms.

There are several advantages of DBS over lesioning. First, DBS does not destroy the target; thus, as new treatment options emerge, patients are still candidates. That is, patients may still receive genetic therapy and/or cellular transplantation as those options become available. DBS also allows bilateral procedures without the high incidence of side effects associated with ablative procedures. The vast majority of side effects that do occur with DBS are from inadvertent stimulation of surrounding structures and can often be modified by adjustment of stimulation parameters and the use of different electrode contacts (Rizzone et al., 2001; Volkmann et al., 2002). Thus, adverse effects are for the most part reversible.

Lesioning may be advantageous in a certain subset of patients, especially in regions where personnel and technological infrastructure is limited. Programming DBS devices can be a difficult and time-consuming process and requires an experienced clinician for optimal results. Furthermore, hardware-associated complications occur in up to 8–14% of patients with DBS (Oh et al., 2002; Voges et al., 2006). This chapter will examine the therapeutic mechanisms associated with surgical treatment of PD, describe clinical indications for surgery, and discuss the efficacy and complications of procedures.

PATHOPHYSIOLOGY

In the 1980s, the work of Albin, Alexander, Delong, and colleagues (Albin et al., 1989; Alexander et al., 1990) demonstrated the integration between the basal ganglia, thalamus, and cortex. They noted that specific cortical areas projected to separate areas within the basal ganglia, which in turn projected recurrently forming a closed-loop circuit to the original cortical areas through specific thalamic relay nuclei. The motor loop of this cortico-striato-pallido-thalamocortical circuit is most relevant to PD (Alexander et al., 1990). Within the motor loop there are direct (Striatum-Globus Pallidus interna (Gpi)) and indirect (Striatum-Globus pallidus externa (GPe)-Subthalamic nucleus (STN)-GPi) projections to the thalamus that have opposite functional effects and are driven by different dopaminergic receptors. The primarily excitatory D1 dopamine receptor is expressed in the direct pathway, while the inhibitory D2 subtype gives rise to the indirect pathway (Smith et al., 1998). Though both pathways project to the thalamus, the direct pathway uses GABA, which reduces activity in GPi/Substantia nigra pars reticulate (SNr), and thus limits inhibition of neurons in the thalamus, resulting in increasing thalamocortical output. The indirect pathway uses glutamate to excite the Gpi/SNr and thus has the opposite effect. In PD, the loss of dopamine in both the direct and the indirect pathways results in a net reduction of the inhibitory activity of Gpi and subsequent increased inhibition of the thalamocortical pathways causing hypokinetic movements (Delong, 1990).

Evidence from primate studies, operative recordings, and functional imaging has provided further details to this loop and increased our understanding of the very complex interaction among the various nuclei. It is now clear that the GPe does not simply project to STN but has reciprocal innervation from STN and direct projections GPi/SNr (Kopell et al., 2006). This loop within the loop as well as the direct cortical projections to both striatum and STN may be important in relaying sensory information into the basal ganglia and synchronizing oscillatory activity. Reciprocal corticothalamic projections between ventral intermediate (Vim) and motor cortex may also be involved.

As our understanding of anatomical interconnections increases, so does our understanding of the electrophysiology of the basal ganglia. The original paradigm was based on rate and helped classify movement disorders as hypo- or hyperkinetic. Data from the 1-methyl-4-phenyl-1,2,3,6-tetrahydropyridine (MPTP) monkey model of PD shows an increase in neuronal firing rates in STN and GPi (Filion and Tremblay, 1991; Bergman et al., 1994). Intraoperative MER in humans has confirmed these findings (Vitek et al., 1998). This theory does not, however, explain the so-called "Marsden and Obeso paradox", i.e., how do lesions in the thalamus improve the hypokinetic features of PD features while lesions in the Gpi improve hyperkinetic movements (Marsden and Obeso, 1994). Oscillatory patterns and synchronization of neuronal activity in PD are more common than in non-diseased states and thus may also play a role in the pathophysiology (Levy et al., 2002). Furthermore, abnormally wide receptive fields are also present in PD (Pessiglione et al., 2005).

Ablation and stimulating procedures seem to have different means of improving symptomatology, despite affecting similar volumes of tissue (McIntyre et al., 2004a). In MPTP monkeys, subthalamotomy reduces Gpi firing rates, but does not change oscillations (Wichmann et al., 1994). STN DBS leads to an increase in Gpi firing (Hashimoto et al., 2003). DBS may also work via increased synchronization by altering local field potentials (LFPs). LFPs, i.e., synchronized current oscillations, exist in low (13–20 Hz), high beta (20–35 Hz), gamma (70–100 Hz) and some very high frequency (70–300 Hz) rhythms. Beta frequency activity is abnormal in PD. In the off state or within minutes of turning off the DBS, the beta frequency dominates and inversely correlates with the akinesia (Alonso-Freuch et al., 2005; Brown, 2006). In the on state or with high frequency DBS, the higher frequencies in the gamma range return as dominant (Alonso-Freuch et al., 2005; Wingeier et al., 2006). The movement- and levodopa-dependent 300 Hz frequency may reflect normal processing in the basal ganglia and provides support of the excitatory mechanism of high frequency stimulation (Foffani et al., 2003).

The mechanism or mechanisms by which DBS modulates neuronal network function remains controversial. Most likely all of the above mechanisms are activated. The stimulation is from a high voltage cathode pulse. The cathode pulses preferentially stimulate axons resulting in both orthodromal and antidromal propagation. Unlike somata, axons can follow the high frequency firing rates and this activation of distal neural elements may lead to the reestablishment of appropriate synchronization that is believed to be the key element to the mechanism of DBS stimulation.

Imaging is also increasingly suggesting more excitatory evidence rather than inhibitory evidence (Davis et al., 1997; Grafton et al., 2006). A combination of both may be necessary but the excitatory effects appear predominant. Positron emission tomography (PET) of human patients with subthalamotomies reveals a decrease in pallidal and thalamic metabolism while those with STN DBS show increased metabolism (Hershey et al., 2003; Trost et al., 2003). Regardless of the local activity change, it is possible that both ablation and stimulation replace pathological neuronal activity at the cortical level with no activity or a highly regular pattern respectively; thus though lesions and stimulation have different means of removing disruptive rhythms, both produce identical clinical effects (McIntyre et al., 2004b; Grill et al., 2004).

Advances in our understanding of both the anatomy and the electrophysiology of the basal ganglia will improve the selection of anatomical targets for movement disorder surgery.

PATIENT SELECTION

Regardless of the target or of the modality of treatment selected, preoperative patient selection criteria are the same and are critical for successful outcomes. Successful treatment is dependent on careful selection by a multidisciplinary team comprised of a functional neurosurgeon, movement disorders neurologist, and neuropsychologist. Surgery is indicated in patients with PD who continue to have significant motor disability despite maximal medical treatment. The best candidates have idiopathic PD and have motor symptoms that are responsive to levodopa. Generally it is recommended that patients have had PD for greater than 5 years to ensure that their parkinsonism is not due to another disease, such as multisystem atrophy or other Parkinson plus syndromes which do not benefit from surgical procedures (Litvan et al., 1996; McKeith et al., 1996); in fact, the majority of reported stimulation patients have a mean duration of disease between 12 and 15 years (Limousin et al., 1998; Houeto et al., 2000). Studies of surgical treatment of PD patients with shorter disease duration are underway. Preliminary work suggests that patients with disease duration of 6–7 years also significantly benefit from DBS (Schupbach et al., 2007). The symptoms that are most apt to be ameliorated with surgery are tremor, motor fluctuations, dyskinesias, bradykinesias, and rigidity. Gait, postural instability, dysphagia, and dysarthria respond to a much lesser degree.

Preoperative screening of these patients is extensive and is generally completed by a movement disorders neurologist. The severity of disease is assessed via a battery of tests, including the Unified Parkinson's Disease Rating Scale (UPDRS) both on and of medication, the Core Assessment Program for Intracerebral Transplantation (CAPIT) protocol, Hoehn and Yahr staging, and timed tests of motor function. The goal of surgery is to improve symptoms to the level achieved by levodopa. In selecting surgical candidates, the greatest predictor is the preoperative response to levodopa; levodopa responsiveness is defined as at least >25–50% improvement in UPDRS motor scores compared to an off-medication state (Welter et al., 2002). Most surgical patients reported in the literature, however, include patients with a preoperative levodopa response of 40–70% (Lang et al., 2006). These patients generally had a UPDRS motor score greater than 40 in the medication off state and less than 30 in the on state (Lang et al., 2006).

Patients undergo MRI to assure that no structural causes of PD are present. MRI thus may be helpful in ruling out certain Parkinson plus syndromes, as well as hydrocephalus, tumors, and ischemic disease. An extensive battery of neuropsychological testing prior to surgery is also necessary to assess the cognitive status of patients. These tests include measures of overall cognition and depression as well as specific tests of memory, verbal function, and the effect of disease on daily living. Patients scoring less than 25 of 30 on mini-mental status evaluation (MMSE) may not be cooperative in the operating room and more importantly may not achieve any significant benefit from surgery due to their profound cognitive disability (Hobson and Meara, 1999). In light of the possibility for psychological effects of the surgery, great vigilance is given to monitoring patients for depression peri-operatively. Other confounding variables which are relative contraindications to surgery include advanced age (>75 years old in most series), systemic co-morbidities which preclude anesthesia, and significant white matter disease on MRI (Lang et al., 2006).

ABLATIVE PROCEDURES

Though in most cases stimulation has become the surgical treatment of choice for PD, as described, certain patients may benefit from lesioning. Lesioning may also provide a means of treatment in patients with systemic illnesses that predispose them to infection or erosion at the hardware sites (unpublished observations-Figure 10.1). Patients with lesions also do not require the vigorous post-operative

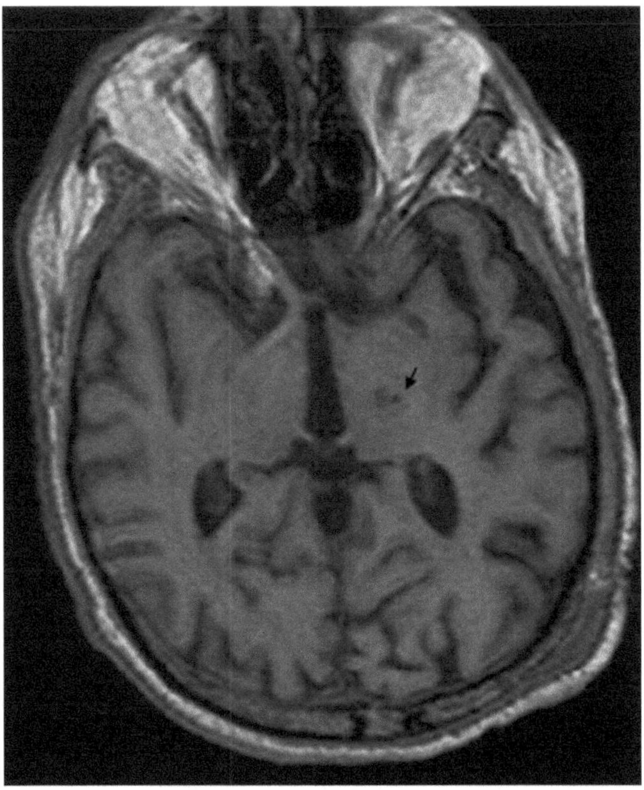

FIGURE 10.1 This axial T1-weighted MRI demonstrates a radiofrequency thalamotomy medial to a prior DBS lead. The lead relieved tremor but at high voltage caused corticospinal symptoms. When recurrent infection necessitated DBS removal (marked by arrow), lesioning was performed medial to the lead. The patient had complete relief of tremor.

programming sessions that DBS treatment often necessitates. Ablative procedures have been validated by three class I randomized trials. The first one examined pallidotomy versus medical therapy (Vitek et al., 2003) and found that patients undergoing lesioning did significantly better than those receiving medication alone at both 6 months and 2 years. Comparisons of lesioning with DBS, specifically pallidotomy versus GPi DBS and thalamotomy versus thalamic DBS, showed that beneficial effects were similar, but suggested a higher rate of adverse side effects in patients with ablative procedures (Merello et al., 1999; Schuurman et al., 2000). That is not to imply that death or severe neurological deficits cannot occur with DBS. Bilateral thalamotomy or pallidotomy has generally been discouraged due to the association with speech problems, dysphagia, and cognitive deficits.

During both ablative and stimulation surgery, patients undergo pre-operative imaging. MRI is performed on the day of surgery in cases where a stereotactic frame is used (Leksell series G™ (Elekta, Atlanta, Ga) or CRW-fn™ frame (Intega Radionics, Burlington, Ma)). Proper alignment of the frame with the patient's anatomy simplifies targeting adjustments, allows consistent angles of approach, and decreases error that can be introduced by correcting for malposition. In cases where a frameless system NexFrame (Medtronic, Minneapolis, MN) is used, the patient undergoes MRI and CT scan in the week prior to surgery after fidicual posts are placed.

It is important to note the MR images are not, however, always of sufficient quality to visualize targets and nuclear boundaries may be indistinct. Use of higher field strength (3T) may improve the reliability of this technique (Slavin et al., 2006). MR is also highly susceptible to image distortion. Indirect targeting may then be used and is based on identification of the anterior and posterior commissures (AC and PC, respectively) and the subsequent use of standard distances to approximate the location of a desired target (Andrade-Souza et al., 2005). The distance between AC and PC may vary significantly between individuals, which limits the accuracy of this technique (Van Buren and Maccubin, 1962; Brierley and Beck, 1959). Thus, targeting based on imaging alone is affected by a number of potential problems: the errors introduced by the use of radiological methods, the ability or lack thereof to appropriately visualize landmarks, and individual anatomic variations (Van Buren and Maccubin, 1962).

Human and mechanical imprecision may also be introduced during positioning of the stereotactic frame/frameless system and/or during targeting. Furthermore, intraoperative variables, such as mechanical loading effects on the frame from positioning or the headstage, or brain shift from CSF loss, may also affect accuracy. All of these factors introduce sources of error, which can be additive. Subsequently, there is a need to correct for those errors before the lead is placed. One means of quality assurance is intraoperative electrophysiological monitoring (Yeung et al., 1994). This monitoring also allows for further adjustment of the anatomical target selected on MRI imaging with electrophysiological data. Though the amount and type of monitoring that is necessary is heavily debated, most groups use some means of monitoring. Options include microelectrode recording (MER) and/or a form of stimulation; microstimulation is possible in lesioning surgery and both micro- and macrostimulation are possible in DBS surgery. Often a combination of these techniques is used. We routinely employ all three (Figure 10.2).

The MER detects and amplifies the activity of individual neurons (Sierens and Bakay, 2004). It allows for identification of the structural borders, eloquent structures and sensorimotor territory, localization of the somatopic arrangement, and outlines of the three-dimensional shape of the targeted nuclei. Micro- and macrostimulation are used to test for side effects and, therefore, to assure that lesions or stimulators are optimally placed. In lesion surgeries, once the target is adequately mapped with MER and the physiological target selected, the microelectrode is replaced with a lesioning probe. A lesion generator system is then used for ablation after a test lesion of 42°C for 60 s. This test lesion is reversible. During the test lesion, both improvements and adverse effects are noted. If no adverse effects are noted, an irreversible lesion is then made by heating the electrode tip to 70° for 60 s. The patient is carefully examined after the lesion and heating is repeated at 5° increments up to 85° as long as there are no adverse effects. The patterns and size of lesions are determined based on the intraoperative clinical responses (Vitek et al., 1998). Adverse

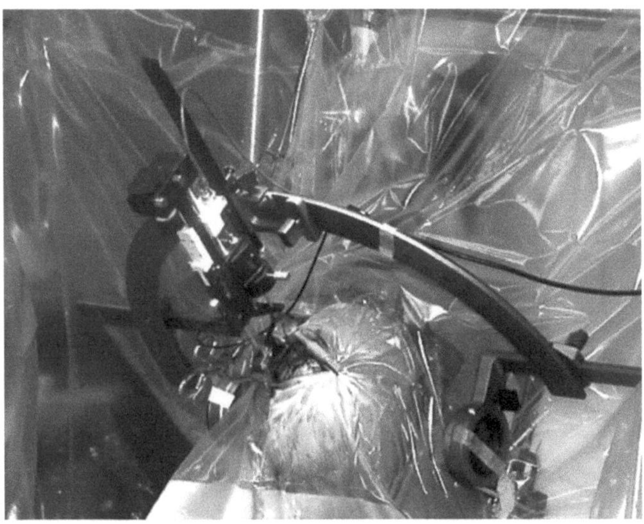

FIGURE 10.2 This photograph demonstrates the Leksell Series G frame set-up for electrophysiology with the FHC microdrive (Frederick Haer Corporation, Brunswick, ME). A clear plastic barrier separates the patient from the operative field with complete sterility behind the barrier and yet allows the surgeon to view direct interactions between the patient and the movement disorders neurologist.

effects may occur in cases where lesions are too large or suboptimally placed. Conversely, a lesion that is too small may result in the patient needing a second procedure.

Other complications are also possible, the most serious of which is intracranial hemorrhage (ICH). The incidence of ICH is approximately 2–3%; most, however, are asymptomatic, observed only on post-operative brain imaging (Terao et al., 2003). In approximately 1% of patients, symptomatic ICH may lead to a range of neurological sequelae ranging from mild confusion to a permanent neurological deficit to death. Up to 5% of patients may also report nausea, headaches, or transient post-operative confusion. Seizures are rare. Specific adverse effects can also occur depending on the area targeted (Deuschl et al., 2006b).

PALLIDOTOMY

The posteroventral GPi is preferentially targeted in surgical procedures, as this region is specifically related to the motor output of the basal ganglia. The patients that most often benefit from pallidotomy are those with asymmetric disease, as the procedure is generally performed unilaterally, due to adverse effects that may occur with bilateral procedures. In cases of patients with bilateral disease, the dominant side is first treated. Complete details of the procedure have been previously published (Vitek et al., 1998).

Briefly, a combination of direct and indirect targeting on MRI is generally used. A pallidotomy target of 2–3 mm anterior to middle cerebellar peduncle (MCP), 2–6 mm below the intercommisural line, and 18–21 mm off midline is chosen (Laitinen et al., 1992; Vitek et al., 1998). Intraoperative MER then allows the confirmation of targeting in the sensorimotor portion of the Gpi by recording neuronal activity in response to patient's active and passive movements during surgery (Figure 10.3). These kinesthetic responses may be blunted by Parkinson's medications and anesthetics (Levy et al., 2001; Starr et al., 2004); thus, surgery is performed in an off-medication state and with minimal if any anesthesia. Microstimulation is then used to assess the proximity of the microelectrode to optic tracts and the internal capsule, through assessments of visual perceptions and sensations of muscle contractions respectively. Once the optimal target is chosen, lesioning ensues. The patient is assessed carefully for side effects as well as beneficial effects. There are often immediate improvements in both bradykinesia and rigidity, and occasionally tremor. The induction of dyskinesias may occur and may be a good prognostic sign. Complications specific to the pallidotomy lesion itself include limb and facial weakness or partial visual field deficits (Vitek et al., 1998).

Most pallidotomy series report excellent relief of symptoms. Short-term reduction of contralateral dyskinesias and tremor by 80–90%, of rigidity by 50–60%, of bradykinesia by 30%, and of gait stability of 30% have been reported

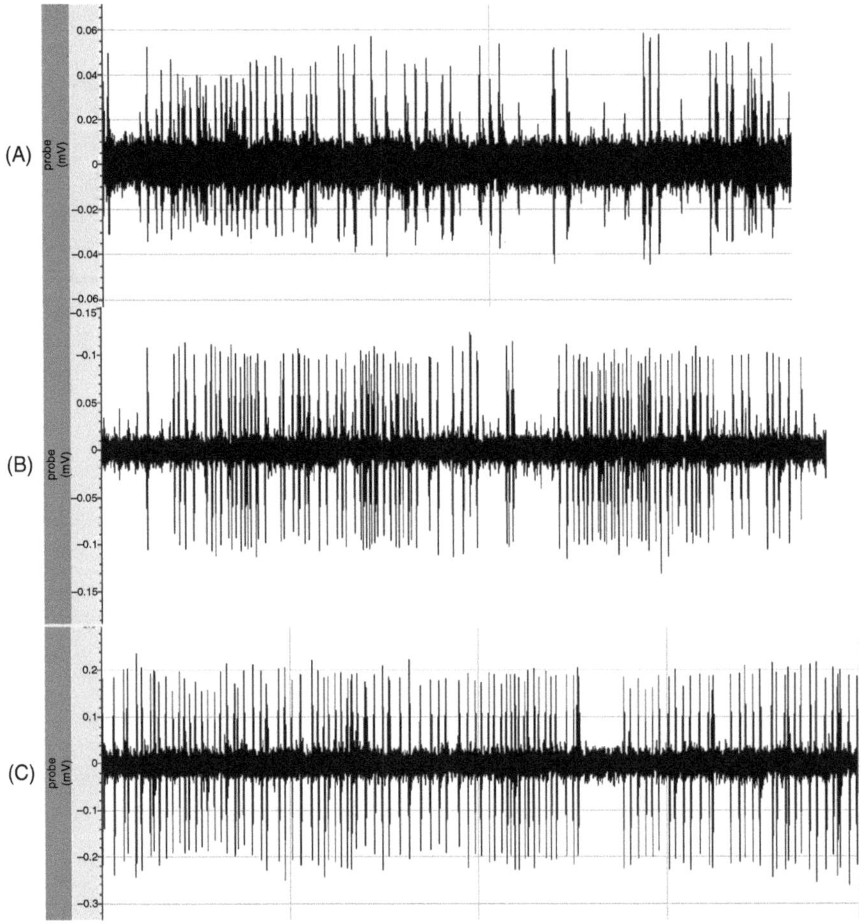

FIGURE 10.3 Electrophysiological recordings during placement of a GPi DBS are illustrated: (A) Insertional activity from putamen. (B) GPe cells characteristically are high frequency with irregular and intermittent pausing cells (C) A typical GPi neuron with high frequency and few, short pauses. All tracings are over a 2 s time period.

(Lang et al., 1997; Vitek et al., 2003). At long-term follow-up of 2–5 years, a significant benefit is still appreciated in all medication off symptoms, except gait (Vitek et al., 2003; Fine et al., 2000). Generally, the size of the lesion is thought to impact overall outcome while the location of the lesion within the Gpi appears to affect which symptoms are treated (RE Gross et al., 1999). The main limitations of pallidotomy are that it mostly ameliorates contralateral symptoms with only a minimal effect on unilateral symptoms and that its effects are irreversible.

THALAMOTOMY

Though in the 1990s, pallidotomy was generally considered the lesion of choice for patients with PD, there was a subset of PD patients that were benefited with thalamotomy. The Vim nucleus of the thalamus was found to be the most effective thalamic target for the treatment of tremor (Lozano, 2000). Successful tremor reduction was noted in 70–90% of patients (Tasker, 1990; Fox et al., 1991; Jankovic et al., 1995). PD patients who had asymmetric upper extremity tremor were candidates for thalamotomy and generally did well (Gray et al., 2002); however, because thalamotomy ineffectively treated other symptoms of PD and PD is a progressive disease, pallidotomy replaced thalamic surgery for the majority of patients.

For thalamic surgery, again a combination of direct and indirect targeting on MRI is generally used. The target is the hand area at the Vim and ventralis oralis posterior (Vop) border, which is approximately 6 mm anterior to PC, 10–10.5 lateral to the third ventricular wall, and in the AC–PC plane. In order to electrophysiologically find this border, it is helpful to use the ventralis caudalis (Vc) as a navigational landmark during MER. Microstimulation in the Vc results in a precise paresthesia will help to confirm the location, as opposed activation of the medial lemniscus which produces widespread paresthesia (hemiparesthesia or whole limb). Just anterior to Vc, cells are responsive to deep pressure (Bakay et al., 1992). Once the Vc is identified, a second MER is performed 2–4 mm anteriorly to target Vim. Once the Vim has been adequately mapped, an RF probe with a 1.1 mm diameter and 2–4 mm exposed tip (Integra Radionics) is inserted and a test lesion is made (Bakay et al., 1992). The effects on contralateral tremor, strength, coordination, and speech are noted. In PD, speech problems may preexist and care must be taken not to exacerbate them. An irreversible lesion is made by heating the electrode tip to 65° for 60 s (longer times do little to expand the size of the lesion). After the lesion, the patient is carefully examined and as long as there are no adverse effects the lesioning is repeated at 10 increments up to 85°. Complications specific to thalamotomy lesions include persistent paresthesias, gait difficulty, dysarthria, and cerebellar symptoms.

SUBTHALAMOTOMY

Subthalamotomy or campotomy was used during the 1960s as an alternative to thalamic surgery for the alleviation of tremor and rigidity (Spiegel et al., 1963). Because of the potential for development of hemiballism as a complication, it was largely abandoned for many years. The success of STN DBS has led to a small resurgence of interest in this procedure. There is some early recent evidence that subthalamotomy can be done safely, with similar beneficial effects to other surgeries with a low rate of transient involuntary movements (Su et al., 2003; Patel et al., 2003). At this point, however, it is not a common treatment option. Stimulation of the subthalamic nucleus is, however, considered in most centers to be the surgical treatment of choice for PD.

DEEP BRAIN STIMULATION

More than a century after the first report of human cortical stimulation was published (Bartholow, 1874), it was noted that stimulation resulted in clinical benefits similar to those achieved with surgical lesioning, but that these effects were also reversible (Benabid et al., 1991). In 1991, Benabid et al. combined implantable pacemaker technology with advances enabling chronically implanted deep brain electrodes and published the first reports of the successful use of DBS (Benabid et al., 1991, 1996). Since that time, DBS has become increasing used in the treatment of movement disorders. The first widespread use of DBS in the United States and Europe was for the treatment of tremor (Benabid et al., 1991, 1996). Thalamic targets were initially assessed followed by stimulation of the Gpi based on previous pallidotomy work. Over the last 10 years, DBS has emerged as a mainstay in the surgical treatment of movement disorders. A recent prospective randomized-pairs trial has shown bilateral STN DBS to be significantly more effective than medical management of patients with PD at 6-month follow-up (Deuschl et al., 2006a). Stimulation procedures have also been successful in patients with previous lesioning surgeries (Galvez Jimenez et al., 1998; Moligner et al., 2002).

In addition to complications that may arise due to lesioning procedures, DBS has a unique set of possible complications related to hardware. The hardware-related complication rate in one review of DBS was 8.4% per electrode-year (Oh et al., 2002). Over a 6-year period, 20 of 79 patients (25.3%) had hardware-related complications, with 40% of these occurring after 1 year. Thus patients are at risk for hardware failure throughout the lifetime of the device. Partial or complete removal of hardware may be necessary in 4.6% of patients (Voges et al., 2006). Batteries also need to be replaced, at an average of every 45 months (Bin-Mahfoodh et al., 2003). Having a DBS system in place also affects the daily living and future medical care of these patients. Electromagnetic interference occurs with some routine medical procedures including diathermy and more importantly MRI. Generally patients with DBS can undergo cranial MRI if the machine is set up in accordance to specific guidelines established by Medtronic. Body MRI is generally contraindicated; one patient with DBS who underwent an MRI of the lumbar spine developed hemiplegia immediately, presumably secondary to heating of the electrode (Henderson et al., 2005). Another DBS patient underwent diathermy for pain control after dental work and became vegetative, due to electrode heating (Nutt et al., 2001). Thus, though DBS offers significant long-term improvements in quality of life (Lyons and Pahwa, 2005) for many people, these patients must be vigilant in regards to future procedures.

PALLIDAL STIMULATION

The two most common sites for DBS are the Gpi and STN. Vim stimulation for PD has been largely abandoned, but may still be considered in select patients with unilateral tremor. GPi and STN stimulation have been shown not only to control tremor but also to treat other cardinal symptoms of PD, such as rigidity, bradykinesia, and l-dopa-induced motor complications such as dyskinesias and on–off fluctations (Volkmann et al., 1998, 2001; Krack et al., 2003; Deuschl et al., 2006a). Whether the GPi or the STN is a better target for stimulation is still debated (Anderson et al., 2005; Rodriguez-Oroz et al., 2005). The improvement in motor outcomes following STN DBS seem to be more consistent likely because of the smaller size of the STN and thus a lower potential for variability in targeting. Reductions in levo-dopa induced dyskinesias and bradykinesias also may be greater with STN DBS (Burchiel et al., 1999; Anderson et al., 2005; Rodriguez-Oroz et al., 2005). However, higher rates of cognitive and emotional side effects are seen after STN surgery (Anderson et al., 2005; Rodriguez-Oroz et al., 2005). Definitive comparison of safety and efficacy of the two targets awaits two randomized trials by NIH and the Veterans Administration.

The basic approach for stimulation procedures is similar to that for lesioning procedures. Patients are selected using the same criteria and pre-operative imaging is obtained. Surgical planning software is also used to simulate the electrode's trajectory and avoid blood vessels, sulci, eloquent structures, and when possible the ventricular cavity (Figure 10.4). Atlases may also be used in conjunction with MRI and planning software (Yelnik et al., 2003). Once the optimal target is selected and confirmed with electrophysiology, the macroelectrode is placed. (Model 3387 or 3389; Medtronic, Inc.; Minneapolis, MN). Macrostimulation is then performed through the DBS electrode and adverse effects are noted. If undesirable effects such as muscle contractions and persistent severe paresthesias are noted at voltages that may be used during programming, repositioning of the electrode may be required. Intraoperative fluoroscopy is also used to confirm placement. Once the DBS lead is optimally positioned and secured, the operation is completed and patients undergo a post-operative MRI to ensure accurate lead placement (Figure 10.5). Approximately 1 week later, patients return to have a second procedure to implant an internal pulse generator, i.e., the pacemaker of the DBS system. Initial programming is performed approximately 4 weeks after the initial surgery to allow for resolution of any "microlesion" effect, i.e., any transient clinical benefit resulting from the act of placing the electrode itself.

Improvements in medication off-state UPDRS motor scores following Gpi DBS ranges from approximately 30 to 55% (Ghika et al., 1998; Volkmann et al., 1998; Rodriguez-Oroz et al., 2005). It is important to note that these procedures are performed bilaterally, so that improvements are also bilateral as opposed to benefits seen with pallidotomy. Dyskinesias and tremor are lessened most significantly; rigidity and gait are also improved. Unilateral Gpi DBS has the most robust effects contralaterally, with less effect on the ipsilateral side (Gross et al., 1997). Specific adverse events related to Gpi stimulation and similar to those seen with pallidotomy and include paresthesias, muscle contractions, and involuntary visual phenomena (Deuschl et al., 2006b).

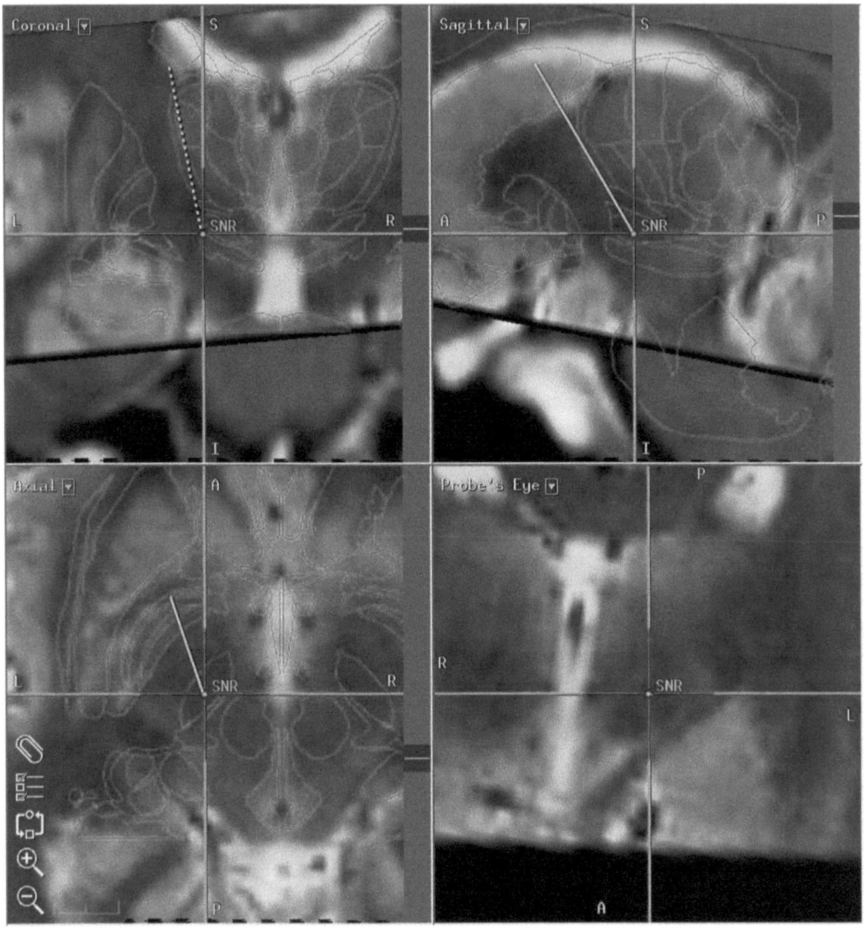

FIGURE 10.4 Surgical planning on the StealthStation™ using Framelink™ (Medtronic-Sofamor-Danek) allows multiplanar imaging of the planned trajectory. A Schaltenbrand and Wahren deformable atlas is overlaid onto the brain image. This shows a trajectory from an STN DBS placement. The area marked SNr correlates with the electrophysiology found at that point along the trajectory.

SUBTHALAMIC STIMULATION

Target selection for DBS was initially based on experiences with pallidotomy, but left clinicians and scientists wondering if another more ideal target may still exist for stimulation procedures. Investigation of the STN as a target was based on the nuclei's known role in modulation of ganglionic output. Successful high frequency stimulation in non-human primates allowed for its eventual clinical use (Bergman et al., 1990; Aziz et al., 1991). Prior to these studies, the STN was considered a potentially dangerous location. Due to the more uniform outcomes

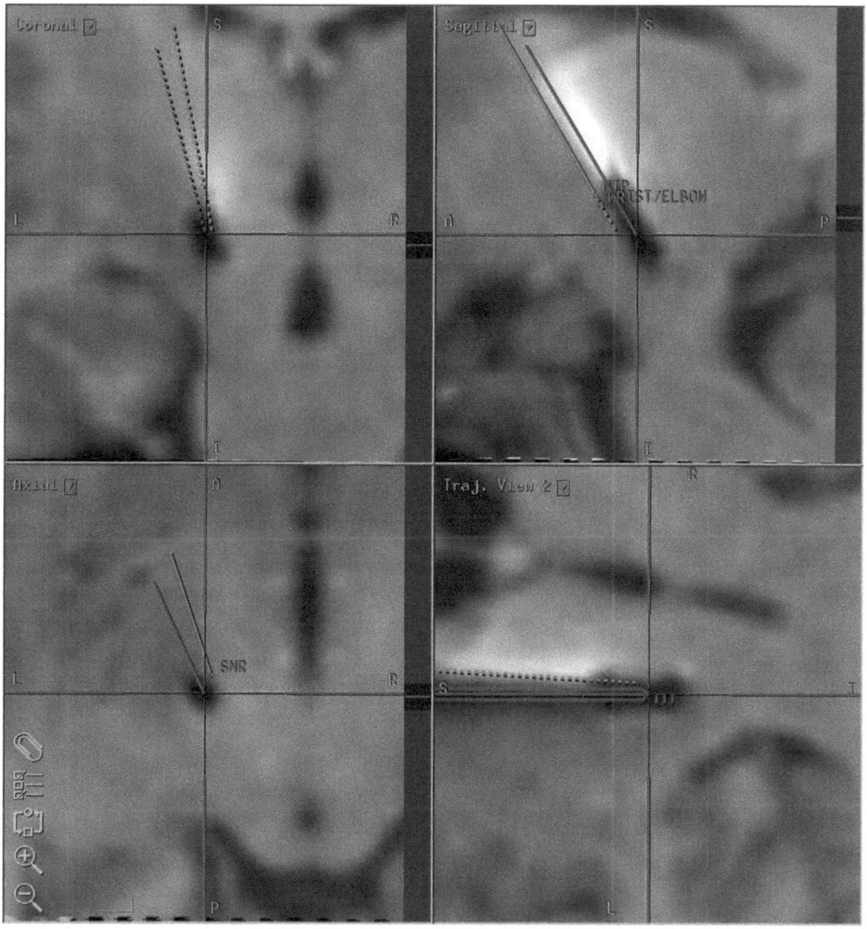

FIGURE 10.5 Post-operative image from the same case. The intraoperative electrophysiological data obtained during MER directed that the DBS electrode be placed more laterally and posteriorly than the original trajectory.

following STN DBS, presumability due to its smaller size and subsequent less potential for targeting variability, it is currently the preferred target for stimulation at most centers. UPDRS motor score improvement of 50–70% at 1-year follow-up have been generally reported (Kumar et al., 1998; Limousin et al., 1998; Fraix et al., 2000; Volkmann et al., 2001; Simuni et al., 2002). Like Gpi DBS, improvements are seen in tremor, levodopa-induced dyskinesias, rigidity, and bradykinesias. Five-year follow-up has shown that while improvements in tremor and rigidity were sustained, worsening of bradykinesia, speech, postural stability, and gait occurred (Krack et al., 2003).

Of note, improvements with unilateral STN DBS do not appear as pronounced as with bilateral stimulation; however, patients with unilateral disease and even some with symmetric disease have been noted to receive sufficient benefit and not require additional surgery (Germano et al., 2004; Slowinski et al., 2007). Unilateral procedures are shorter and often better tolerated in patients who are older or have less cognitive reserve. Specific adverse events related to STN stimulation are due to inadvertent stimulation of the oculomotor nerve, internal capsule, and medial lemniscus. Dyskinesias, paresthesias, diplopia, dystonia, and motor contractions may occur (Voges et al., 2006; Deuschl et al., 2006). In addition, cognitive and psychiatric complications have also been well documented. Smeding et al. (2006) reported that psychiatric complications occurred in 9% of STN DBS patients versus 3% of patients treated with medication alone. Patients with STN DBS also showed greater decline in neuropsychological testing for verbal fluency and memory (Funkiewiez et al., 2004; Smeding et al., 2006). Whether this decline impacts daily living has not been well established.

In an attempt to maximize clinical motor benefit and potentially reduce the neuropsychological effects of STN DBS, the sensorimotor region is generally targeted. Multiple studies suggest that the anterodorsal sector of the STN (especially laterally) or just dorsal within the zona incerta (Zi) is the most clinically effective site for stimulation (Saint Cyr et al., 2002; Starr, 2002; Voges et al., 2002). MRI resolution at 1.5 T does not allow differentiation of the two sites. The target is 3 mm posterior to MCP, 11–12 mm lateral from midline, and 4 mm below the AC–PC plane. Indirect localization based on the red nucleus as an internal landmark may also be beneficial (Zonenshayn et al., 2000; Cuny et al., 2002; Danish et al., 2006). A typical MER trajectory passes through the anterior thalamus, Zi, fields of Forel (H1 and H2), STN, and finally the SNr (Starr et al., 1998; Zonenshayn et al., 2000) (Figure 10.6). Micro- and macrostimulation can decrease rigidity or tremor and may induce dyskinesias. Internal pulse generator placement and programming is performed as with Gpi Stimulation.

THALAMIC STIMULATION

Thalamic stimulation for PD is currently rarely offered, but may be successful in a select subset of patients. Contralateral arm tremor is improved in 71–92% of patients at 3 months (Benabid et al., 1996) and in 74% at 1 year (Koller et al., 1997). Contralateral foot tremor also is improved in 50–90% of the patients at early follow-up (Ondo et al., 1998). Like thalamotomy, Vim stimulation may cause dysarthria, cerebellar ataxia, and paresthesias in up to 10% of patients (Benabid et al., 1996; Koller et al., 1997; Ondo et al., 1998). Bilateral procedures are associated with a significantly higher complication rate and are generally discouraged and/or performed in a staged fashion (Pahwa et al., 2006b). Stimulation of the thalamus may be efficacious through activation of the cerebellothalamocortical pathway rather than inhibition (Molnar et al., 2004, 2005), although the precise method of action remains controversial.

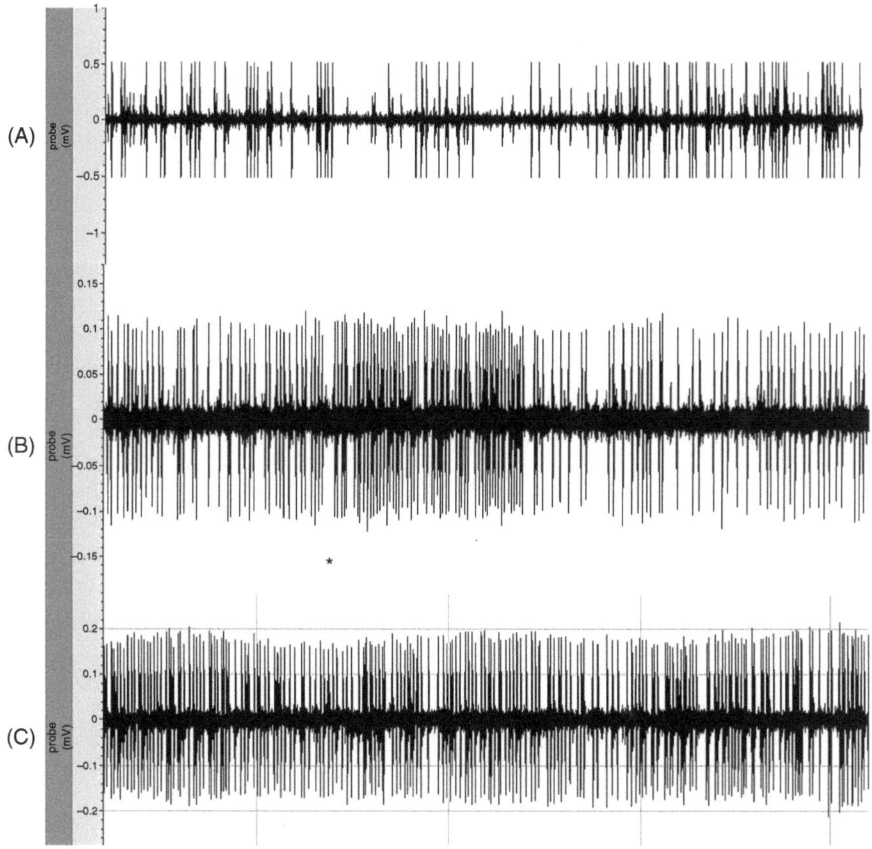

FIGURE 10.6 Electrophysiological recordings from a case of STN DBS: (A) A non-bursting thalamic cell. The anterior thalamus is low density and low amplitude. (B) Typical STN recordings are demonstrated with a kinesthetic STN cell that responds to wrist flexion with increased activity (starting at the asterisk). (C) SNr cells generally have high frequency regular firing rates. All tracings are over a 2 s time period.

CONCLUSION

Deep brain stimulation has provided a meaningful improvement in quality of life to many patients with PD and other movement disorders. The future of the field depends on further refinements in target and patient selection as well as in the device itself. Target selection has been based primarily on results that occurred previously with lesioning, some animal research, and lessons learned while treating other movement disorders. This method of trial and error has for the most part been successful, but does not always result in a clear consensus as to which target is best and whether different targets should be advocated for different symptoms. As our understanding of the anatomy and electrophysiology in PD improves, target selection may be altered. For example, gait disturbance and postural instability are only moderately improved with STN/Gpi stimulation. Recognition of modulatory effects on basal ganglia of the pedunculopontine nucleus (PPN) in animals (Jenkinson et al., 2004) led to bilateral PPN DBS electrodes in two patients who demonstrated significant improvement in gait and postural instability at low frequencies (20–25 Hz) (Plaha and Gill 2005).

Improvements in our understanding of stimulation technologies will also allow us to optimize treatments. As surgery becomes available to patients with less severe disease, device optimization is crucial to allow younger patients who may have the devices for decades to function relatively normally. Future leads must be more durable and cause less inflammatory reaction. They will need to easily better resist electromagnetic interference. Further understanding of electrode properties on brain parenchyma, through studies such as the Medtronic sponsored surveillance "Brain Autopsy Research program", is essential. In the near future, clinicians will have the ability to combine DBS systems with delivery systems technology, much as endovascular technologies have advanced with the advent of drug eluting stents. Ultimately, these systems may be augmented with other available technologies to further the growth of the fields of cellular and gene therapy in the treatment of PD, as DBS systems of the future may offer a delivery mechanism, and stimulation itself may result in more responsive cellular milieu (Lee et al., 2004; Iida et al., 2006).

REFERENCES

Albin RL, Young AB, and Penney JB (1989). The functional anatomy of basal ganglia disorders. *Trends Neurosci* 12: 366–375.

Alexander GE, Crutcher MD, and DeLong MR (1990). Basal ganglia-thalamocortical circuits: parallel substrates for motor, oculomotor, "prefrontal" and "limbic" functions. In Uylings HBM, Van Eden CG, De Bruin JPC, Corner MA, Feenstra MGP, eds. *Progress in Brain Research* Elsevier Science Publishers, New York. pp. 119–146.

Alonso-Freuch F et al. (2005). Slow oscillatory activity and levodopa-induced dyskinesias in Parkinson's disease. *Brain* 129: 1748–1757.

Anderson VC et al. (2005). Pallidal vs subthalamic nucleus deep brain stimulation in Parkinson disease. *Arch Neurol* 62: 554–560.

Andrade-Souza YM et al. (2005). Comparison of three methods of targeting the subthalamic nucleus for chronic stimulation in Parkinson's disease. *Neurosurgery* 56: 360–368.

Aziz TZ et al. (1991). Lesion of the subthalamic nucleus for the alleviation of 1-methyl-4-phenyl-1,2,3,6-tetrahydropyridine (MPTP)-induced parkinsonism in the primate. *Mov Disord* 6: 288–292.

Bakay RAE, Vitek JL, and Delong MR (1992). Thalamotomy for tremor. In S Rengachary, R Wilkins eds. *Neurosurgical Operative Atlas.* Vol 2, American Association of Neurological Surgeons, Park Ridge, IL. pp. 299–312.

Bartholow R (1874). Experimental investigations into the functions of the human brain. *Am J Med Sci* 67: 305–313.

Benabid AL et al. (1991). Long-term suppression of tremor by chronic stimulation of the ventral intermediate thalamic nucleus. *Lancet* 337: 403–406.

Benabid AL et al. (1996). Chronic electrical stimulation of the ventralis intermedius nucleus of the thalamus as a treatment of movement disorders. *J Neurosurg* 84: 203–214.

Bergman H, Wichmann T, and DeLong MR (1990). Reversal of experimental parkinsonism by lesions of the subthalamic nucleus. *Science* 249: 1436–1438.

Bergman H et al. (1994). The primate subthalamic nucleus. II. Neuronal activity in the MPTP model of parkinsonism. *J Neurophysiol* 72: 507–520.

Bin-Mahfoodh M et al. (2003). Longevity of batteries in internal pulse generators used for deep brain stimulation. *Stereotact Funct Neurosurg* 80: 56–60.

Brierley JB and Beck E (1959). The significance in human stereotactic brain surgery of individual variation in the diencephalon and globus pallidus. *J Neurol Neurosurg Psychiatry* 22: 287–298.

Brooks DJ (1998). The early diagnosis of Parkinson's disease. *Ann Neurol* 44: S10–S18.

Brown P (2006). Bad oscillations in Parkinson's disease. *J Neural Transm Suppl* 70: 27–30.

Burchiel KJ et al. (1999). Comparison of pallidal and subthalamic nucleus deep brain stimulation for advanced Parkinson's disease: results of a randomized, blinded pilot study. *Neurosurgery* 45:1375–1382.

Cuny E, Guehl D, Burbaud P et al. (2002). Lack of agreement between direct magnetic resonance imaging and statistical determination of a subthalamic target: the role of electrophysiological guidance. *J Neurosurg* 97: 591–597.

Danish SF et al. (2006). Conventional MRI is inadequate to delineate the relationship between the red nucleus and subthalamic nucleus in Parkinson's disease. *Stereotact Funct Neurosurg* 84: 12–18.

Davis KD et al. (1997). Globus pallidus stimulation activates the cortical motor system during alleviation of parkinsonian symptoms. *Nat Med* 3: 671–674.

DeLong MR (1990). Primate models of movement disorders of basal ganglia origin. *Trends Neurosci* 13: 281–285.

Deuschl G et al. (2006a). A randomized trial of deep-brain stimulation for Parkinson's disease. *N Engl J Med* 355: 896–908.

Deuschl G et al. (2006b). Deep brain stimulation: postoperative issues. *Mov Disord* 21: S219–S237.

Filion M and Tremblay L (1991). Abnormal spontaneous activity of globus pallidus neurons in monkeys with MPTP-induced parkinsonism. *Brain Res* 547: 142–151.

Fine J et al. (2000). Long-term follow-up of unilateral pallidotomy in advanced Parkinson's disease. *N Engl J Med* 342: 1708–1714.

Foffani G et al. (2003). 300-Hz subthalamic oscillations in Parkinson's disease. *Brain* 126: 2153–2163.

Fox MW, Ahlskog JE, and Kelly PJ (1991). Stereotactic ventrolateralis thalamotomy for medically refractory tremor in post-levodopa era Parkinson's disease patients. *J Neurosurg* 75: 723–730.

Fraix V et al. (2000). Effect of subthalamic nucleus stimulation on levodopa-induced dyskinesia in Parkinson's disease. *Neurology* 55, 1921–1923.

Funkiewiez A et al. (2004). Long term effects of bilateral subthalamic nucleus stimulation on cognitive function, mood, and behaviour in Parkinson's disease. *J Neurol Neurosurg Psychiatry* 75: 834–839.

Galvez Jimenez N et al. (1998). Pallidal stimulation in Parkinson's disease patients with a prior unilateral pallidotomy. *Can J Neurol Sci* 25: 300–305.

Germano IM et al. (2004). Unilateral stimulation of the subthalamic nucleus in Parkinson disease: a double-blind 12-month evaluation study. *J Neurosurg* 101: 36–42.

Ghika J et al. (1998). Efficiency and safety of bilateral contemporaneous pallidal stimulation (deep brain stimulation) in levodopa-responsive patients with Parkinson's disease with severe motor fluctuations: a 2-year follow-up review. *J Neurosurg* 89: 713–718.

Grafton ST et al. (2006). Normalizing motor-related brain activity: subthalamic nucleus stimulation in Parkinson disease. *Neurology* 66: 1192–1199.

Gray A et al. (2002). Quality of life outcomes following surgical treatment of Parkinson's disease. *Mov Disord* 17: 68–75.

Grill WM, Snyder AN, and Miocinovic S (2004). Deep brain stimulation creates an informational lesion of the stimulated nucleus. *Neuroreport* 15: 1137–1140.

Gross C et al. (1997). High-frequency stimulation of the globus pallidus internalis in Parkinson's disease: a study of seven cases. *J Neurosurg* 87: 491–498.
Gross C et al. (1999). From experimentation to the surgical treatment of Parkinson's disease: prelude or suite in basal ganglia research? *Prog Neurobiol* 59: 509–532.
Gross RE et al. (1999). Relationship of lesion location to clinical outcome following microelectrode-guided pallidotomy for Parkinson's disease. *Brain* 122: 405–416.
Hashimoto T et al. (2003). Stimulation of the Subthalamic Nucleus Changes the Firing Pattern of Pallidal Neurons. *J Neurosci* 23: 1916–1923.
Henderson JM, et al. (2005). Permanent neurological deficit related to magnetic resonance imaging in a patient with implanted deep brain stimulation electrodes for Parkinson's disease: case report. *Neurosurgery* 57: E1063.
Hershey T et al. (2003). Cortical and subcortical blood flow effects of subthalamic nucleus stimulation in PD. *Neurology* 61: 816–882.
Hobson P and Meara J (1999). The detection of dementia and cognitive impairment in a community population of elderly people with Parkinson's disease by use of the CAMCOG neuropsychological test. *Age Ageing* 28: 39–43.
Houeto JL et al. (2000). Subthalamic stimulation in Parkinson disease: a multidisciplinary approach. *Arch Neurol* 57: 461–465.
Iida Y et al. (2006). Transthoracic direct current shock facilitates intramyocardial transfection of naked plasmid DNA infused via coronary vessels in canines. *Gene Ther* 13: 906–916.
Jankovic J et al. (1995). Outcome after stereotactic thalamotomy for parkinsonian, essential, and other types of tremor. *Neurosurgery* 37: 680–687.
Jenkinson N et al. (2004). Pedunculopontine nucleus stimulation improves akinesia in a Parkinsonian monkey. *Neuroreport* 15: 2621–2624.
Koller W et al. (1997). High-frequency unilateral thalamic stimulation in the treatment of essential and parkinsonian tremor. *Ann Neurol* 42: 292–299.
Kopell BH et al. (2006). Anatomy and physiology of the basal ganglia: implications for deep brain stimulation for Parkinson's disease. *Mov Disord* 21: 238–246.
Krack P et al. (2003). Five-Year follow-up of bilateral stimulation of the subthalamic nucleus in advanced Parkinson's disease. *N Engl J Med* 349: 1925–1934.
Kumar R, et al. (1998). Double-blind evaluation of subthalamic nucleus deep brain stimulation in advanced Parkinson's disease. *Neurology* 51: 850–855.
Laitinen LV, Bergenheim AT, and Hariz MI (1992). Leksell's posteroventral pallidotomy in the treatment of Parkinson's disease. *J. Neurosurg* 76: 53–61.
Lang AE et al. (1997). Posteroventral medial pallidotomy in advanced Parkinson's disease. *N Engl J Med* 337: 1036–1042.
Lang AE et al. (2006). Deep brain stimulation: preoperative issues. *Mov Disord* 21: S171–S196.
Lee PY, Chesnoy S, and Huang L (2004). Electroporatic delivery of TGF-beta1 gene works synergistically with electric therapy to enhance diabetic wound healing in db/db mice. *J Invest Dermatol* 123: 791–798.
Levy R et al. (2001). Effects of apomorphine on subthalamic nucleus and globus pallidus internus neurons in patients with Parkinson's disease. *J Neurophysiol* 86: 249–260.
Levy R, Hutchison WD, Lozano AM, et al. (2002). Synchronized neuronal discharge in the basal ganglia of parkinsonian patients is limited to oscillatory activity. *J Neurosci* 22: 2855–2861.
LeWitt PA and Nyholm D (2004). New developments in levodopa therapy. *Neurology* 62: S9–16.
Limousin P et al. (1998). Electrical stimulation of the subthalamic nucleus in advanced Parkinson's disease. *N Engl J Med* 339: 1105–1111.
Litvan I et al. (1996). Natural history of progressive supranuclear palsy (Steele-Richardson-Olszewski syndrome) and clinical predictors of survival: a clinicopathological study. *J Neurol Neurosurg Psychiatry* 60: 615–620.
Lozano AM (2000). Vim thalamic stimulation for tremor. *Arch Med Res* 31: 266–269.
Lyons KE and Pahwa R (2005). Long-term benefits in quality of life provided by bilateral subthalamic stimulation in patients with Parkinson disease. *J Neurosurg* 103: 252–255.

Marsden CD and Obeso JA (1994). The functions of the basal ganglia and the paradox of stereotaxic surgery in Parkinson's disease. *Brain* 117: 877–897.
McIntyre CC et al. (2004a). Electric field and stimulating influence generated by deep brain stimulation of the subthalamic nucleus. *Clin Neurophysiol* 115: 589–595.
McIntyre CC et al. (2004b). Cellular effects of deep brain stimulation: model-based analysis of activation and inhibition. *J Neurophysiol* 91: 1457–1469.
McKeith IG et al. (1996). Consensus guidelines for the clinical and pathologic diagnosis of dementia with Lewy bodies (DLB): report of the consortium on DLB international workshop. *Neurology* 45: 1113–1124.
Merello M et al. (1999). Unilateral radiofrequency lesion versus electrostimulation of posteroventral pallidum: a prospective randomized comparison. *Mov Disord* 14: 50–56.
Moligner A et al. (2002). Subthalamic nucleus stimulation in patients with a prior pallidotomy. *J Neurosurg* 96: 660–665.
Molnar GF et al. (2004). Thalamic deep brain stimulation activates the cerebellothalamocortical pathway. *Neurology* 63: 907–909.
Molnar GF et al. (2005). Changes in cortical excitability with thalamic deep brain stimulation. *Neurology* 64: 1913–1919.
Nutt JG et al. (2001). DBS and diathermy interaction induces severe CNS damage. *Neurology* 56: 1384–1386.
Oh MY et al. (2002). Long-term hardware-related complications of deep brain stimulation. *Neurosurgery* 50: 1268–1274.
Ondo W et al. (1998). Unilateral thalamic deep brain stimulation for refractory essential tremor and Parkinson's disease tremor. *Neurology* 51: 1063–1069.
Pahwa R et al. (2006a). Practice parameter: treatment of Parkinson disease with motor fluctuations and dyskinesia (an evidence-based review): report of the Quality Standards Subcommittee of the American Academy of Neurology. *Neurology* 66: 983–995.
Pahwa R et al. (2006b). Long-term evaluation of deep brain stimulation of the thalamus. *J Neurosurg* 104: 506–512.
Patel NK et al. (2003). Unilateral subthalamotomy in the treatment of Parkinson's disease. *Brain* 126: 1136–1145.
Pessiglione M et al. (2005). Thalamic neuronal activity in dopamine-depleted primates: evidence for a loss of functional segregation within basal ganglia circuits. *J Neurosci* 25: 1523–1531.
Plaha P and Gill SS (2005). Bilateral deep brain stimulation of the pedunculopontine nucleus for Parkinson's disease. *Neuroreport* 16: 1883–1887.
Pollak P et al. (2002). Treatment results: Parkinson's disease. *Mov Disord* 17: S75–S83.
Pool J (1954). Psychosurgery in older people. *J Am Geriat Soc* 2: 456–465.
Rizzone M et al. (2001). Deep brain stimulation of the subthalamic nucleus in Parkinson's disease: effects of variation in stimulation parameters. *J Neurol Neurosurg Psychiatry* 71: 215–219.
Rodriguez-Oroz MC et al. (2005). Bilateral deep brain stimulation in Parkinson's disease: a multicentre study with 4 years follow-up. *Brain* 128: 2240–2249.
Saint-Cyr JA et al. (2002). Localization of clinically effective stimulating electrodes in the human subthalamic nucleus on magnetic resonance imaging. *J Neurosurg* 97: 1152–1166.
Schupbach WM et al. (2007). Neurosurgery at an earlier stage of Parkinson disease: a randomized, controlled trial. *Neurology* 68: 267–271.
Schuurman PR et al. (2000). A comparison of continuous thalamic stimulation and thalamotomy for suppression of severe tremor. *N Engl J Med* 342: 461–468.
Sierens D and Bakay RAE (2004). Is Microelectrode recording necessary? The case in favor. In Z Israel KJ, Burchiel eds. *Microelectrode Recording in Movement Disorder Surgery*. Thieme Medical Publishers, New York. pp. 186–196.
Simuni T et al. (2002). Bilateral stimulation of the subthalamic nucleus in patients with Parkinson disease: a study of efficacy and safety. *J Neurosurg* 96: 666–672.
Slavin KV et al. (2006). Direct visualization of the human subthalamic nucleus with 3T MR imaging. *Am J Neuroradiol* 27, 80–84.

Slowinski JL et al. (2007). Unilateral deep brain stimulation of the subthalamic nucleus for Parkinson disease. *J Neurosurg* 106: 626–632.

Smeding HM et al. (2006). Neuropsychological effects of bilateral STN stimulation in Parkinson disease: a controlled study. *Neurology* 66: 1830–1836.

Smith Y et al. (1998). Microcircuitry of the direct and indirect pathways of the basal ganglia. *Neuroscience* 86: 353–387.

Spiegel EA et al. (1963). Campotomy in various extrapyramidal disorders. *J Neurosurg* 20: 871–884.

Starr PA (2002). Placement of deep brain stimulators into the subthalamic nucleus or Globus pallidus internus: technical approach. *Stereotact Funct Neurosurg* 79: 118–145.

Starr PA, Vitek JL, and Bakay RAE (1998). Deep brain stimulation for movement disorders. *Neurosurg Clin N Amer* 9: 381–402.

Starr PA et al. (2004). Microelectrode-guided implantation of deep brain stimulators into the globus pallidus internus for dystonia: techniques, electrode locations, and outcomes. *Neurosurg Focus* 17 E4.

Su PC et al. (2003). Treatment of advanced Parkinson's disease by subthalamotomy: one-year results. *Mov Disord* 18: 531–538.

Tasker RR (1990). Thalamotomy. *Neurosurg Clin N Am* 1: 841–864.

Terao T et al. (2003). Hemorrhagic complication of stereotactic surgery in patients with movement disorders. *J Neurosurg* 98: 1241–1246.

Trost M et al. (2003). Evolving metabolic changes during the first postoperative year after subthalamotomy. *J Neurosurg* 99: 872–878.

Van Buren JM and Maccubbin DA (1962). An outline atlas of the human basal ganglia with estimation of anatomical variants. *J Neurosurg* 19: 811–839.

Vitek JL et al. (1998). Microelectrode-guided pallidotomy: technical approach and its application in medically intractable Parkinson's disease. *J Neurosurg* 88: 1027–1043.

Vitek JL et al. (2003). Randomized trial of pallidotomy versus medical therapy for Parkinson's disease. *Ann Neurol* 53: 558–569.

Voges J et al. (2002). Bilateral high-frequency stimulation in the subthalamic nucleus for the treatment of Parkinson disease: correlation of therapeutic effect with anatomical electrode position. *J Neurosurg* 96: 269–279.

Voges J et al. (2006). Deep-brain stimulation: long-term analysis of complications caused by hardware and surgery–experiences from a single centre. *J Neurol Neurosurg Psychiatry* 77: 868–872.

Volkmann J et al. (1998). Bilateral high-frequency stimulation of the internal globus pallidus in advanced Parkinson's disease. *Ann Neurol* 44: 953–961.

Volkmann J et al. (2001). Safety and efficacy of pallidal or subthalamic nucleus stimulation in advanced PD. *Neurology* 56: 548–551.

Volkmann J et al. (2002). Introduction to the programming of deep brain stimulators. *Mov Disord* 17: 181–187.

Welter ML et al. (2002). Clinical predictive factors of subthalamic stimulation in Parkinson's disease. *Brain* 125: 575–583.

Wichmann T, Bergman H, and DeLong MR (1994). The primate subthalamic nucleus. I. Functional properties in intact animals. *J Neurophysiol* 72: 494–506.

Wingeier B et al. (2006). Intra-operative STN DBS attenuates the prominent beta rhythm in the STN in Parkinson's disease. *Exp Neurol* 197 244–251.

Yelnik J et al. (2003). Localization of stimulating electrodes in patients with Parkinson disease by using a three-dimensional atlas-magnetic resonance imaging coregistration method. *J Neurosurg* 99: 89–99.

Yeung D et al. (1994). Systematic analysis of errors in target localization and treatment delivery in stereotactic radiosurgery (SRS). *Int J Radiat Oncol Biol Phys* 28: 493–498.

Zonenshayn M et al. (2000). Comparison of anatomic and neurophysiological methods for subthalamic nucleus targeting. *Neurosurgery* 47: 282–292.

11

GENE AND CELLULAR TRANSPLANTATION THERAPIES FOR HUNTINGTON'S DISEASE

SHILPA RAMASWAMY AND JEFFREY H. KORDOWER

Department of Neuroscience, Rush University Medical Center, Chicago, IL 60612

Huntington's disease (HD) is a genetic neurodegenerative disorder that is caused by a mutation in the IT15 gene on chromosome 4. Patients with HD suffer from a multitude of symptoms occurring in the cognitive, motor and personality realms. Unfortunately no efficient therapy exists that can tackle all of these symptoms while at the same time delaying or preventing cell death. Cell death in HD occurs predominantly in the projection neurons of the striatum, and it is this population of neurons that most therapies focus on protecting. However, cortical degeneration also plays a prominent role in the manifestation of deficits in higher order cognitive functions and must be targeted while designing a therapy. The discovery of toxin-induced and genetic models in both rodents and nonhuman primates has allowed for the comprehensive testing of therapies before they reach the clinic. Pharmaceutical therapies are currently the most commonly used to treat HD patients. These may be efficient at temporarily tackling the symptoms of HD but do not address the inevitable disease progression. There are a few neuroprotective therapies currently in clinical trials but are somewhat limited in their effectiveness. Cell replacement strategies are also in use but have been shelved in the past few years due to lack of proper funding. This review will discuss neuroprotective gene therapies and restorative cell transplantation therapies that are in use for HD research and therapy.

Copyright © 2008 Elsevier Inc.
All rights of reproduction in any form reserved.

INTRODUCTION

Huntington's disease is a genetic disorder inherited in an autosomal dominant manner. The disease is caused by a mutation manifested in the IT15 or *huntingtin* gene located on chromosome 4 (The Huntington's Disease Collaborative Research Group, 1993). The huntingtin protein encoded by the IT15 or huntingtin gene, is a large 3000-amino acid and 350-kDa protein. In exon 1 of the huntingtin gene is a glutamine tract encoded by an expanded polyglutamine (CAG) region. In normal individuals there are anywhere from 8 to 27 CAG repeats and this increases to over 35 repeats in HD patients (Andrew et al., 1993). Mutations in the 28–30 range are considered a permutation. This suggests that the carrier will not manifest with clinical symptoms but that the repeat can expand into a full, symptom causing mutation when transmitted to an offspring.

The devastating and incurable symptoms of HD include cognitive, motor and psychiatric disturbances. The cognitive symptoms often present years before the other signs and include deficits in executive functions, procedural memory and psychomotor skills (Heindel et al., 1988; Bylsma et al., 1990; Lange et al., 1995). These symptoms are very debilitating to the patients, often more so than the motor signs. Unfortunately, most therapies in use today do not endeavor to deal with this problem and focus mostly on treating the motor signs. The classic motor phenotype of HD is chorea – random, spontaneous and involuntary dance-like movements. Although HD is most recognized by the symptom of chorea, other motor signs like dysarthria, hyperreflexia and abnormal eye movements are common and may precede chorea (Penney, et al., 1990). Symptoms of bradykinesia, rigidity and dystonia that are common features of Parkinson's disease occur as the disease progresses and predominate the late stages of HD (Young et al., 1986). Patients also suffer from psychiatric symptoms like obsessive-compulsive disorder, interpersonal sensitivity, anxiety, paranoia and depression (Duff et al., 2007; Marshall et al., 2007). These disturbances are very common and can occur up to 10 years prior to the onset of motor signs. In his first description of the disease, George Huntington described HD patients as having a tendency to an insanity that often leads to suicide (Huntington, 2003).

The multitude of symptoms in HD makes it difficult to produce one therapy that can treat all the components. In the past, most treatments have focused on treating chorea without tackling the cognitive and personality changes. Recently however, researches have realized the importance of treating the less evident but just as if not more devastating cognitive and psychiatric symptoms.

CELL DEATH MECHANISMS

The cell death seen in HD directly contributes to all the debilitating symptoms in patients. Cell death occurs primarily in the medium-sized spiny neurons of the striatum that comprise approximately 95% of striatal neurons. These cells express γ-aminobutyric acid (GABA) as their neurotransmitter and specific subpopulations co-express either substance P or enkephalin. The two populations of medium spiny neurons show variable vulnerability in different stages of the disease (Reiner et al., 1988). In the early stages of the HD, neurons that co-express enkephalin and project to the globus pallidus via the indirect pathway are particularly vulnerable. The indirect pathway which is normally involved in the inhibition of voluntary movements (Albin et al., 1989; Alexander and Crutcher, 1990) is therefore disrupted, leading to the activation of cortical motor circuits that produce the hallmark hyperkinetic, choreiform movements seen in early HD. In later stages of the disease, death occurs in the neurons that co-express substance P and project to the globus pallidus via the direct circuit. Disruption of this pathway which is normally involved in the initiation of voluntary movements (Albin et al., 1989; Alexander and Crutcher, 1990) blocks the activation of the pre-motor and supplementary motor cortices producing hypokinetic symptoms. Populations of interneurones like the large cholinergic and medium aspiny neurons are spared in the diseased brain (Ferrante et al., 1987; Cicchetti et al., 2000).

Cortical degeneration is also a prominent feature of HD to which the decline in cognitive function has been attributed. Cell loss occurs prominently in layers V and VI of the cortex (Hedreen et al., 1991) in select association areas. Degeneration has been reported in the dorsal frontal cortex (Hedreen et al., 1991), the dorsolateral prefrontal cortex (Selemon et al., 2004) and in Broadman's area 9. In the past it was thought that cell death in the cortex is a result of Wallerian degeneration of axon terminals from the striatum. However, layer VI of the cortex projects not to the striatum but to the thalamus, claustrum, and other cortical areas, it is likely that cortical cell loss is a primary process. Therefore, it may be necessary to tackle cortical cell death separately in addition to treating striatal neuron loss.

While the exact cause of cell death in HD is unclear, many theories have been put forth to explain and possibly combat neurodegeneration in the brain. The predominant cause of cell death is thought to be a result of the mutant huntingtin protein which is neurotoxic. While the mechanism of mutant huntingtin-induced toxicity is controversial, several theories involving the formation of cytoplasmic inclusions have been put forth (Waelter et al., 2001). Huntingtin aggregates may be toxic because they can sequester proteins that are essential for cell viability and survival. Aggregates can recruit transcription factors (Perez et al., 1998), caspases (Sanchez et al., 1999) and protein kinases (Meriin et al., 2001). Huntingtin aggregates can also sequester CREB-binding protein, a major player in cell survival, and prevent its function (Nucifora et al., 2001). Thus by sequestering and inhibiting the function of otherwise viable proteins, mutant huntingtin aggregates can retard the efficient functioning of otherwise normal neurons. Mutant huntingtin has also been shown to have toxic properties independent of its ability to form aggregates (Saudou et al., 1998). These results indicate that some inherent properties of mutant huntingtin irrespective of its ability to remain soluble or form aggregates causes death to neurons.

Mitochondrial impairment has also been implicated as a mechanism of cell death in the HD brain. An energy deficit exists in brain cells caused by impaired glucose metabolism due to decreased mitochondrial ATP production. Several enzymes involved in the tricarboxylic acid (TCA) cycle and the electron transport chain are downregulated in brain. A decrease in enzymes like aconitase in the caudate, putamen and cortex (Tabrizi et al., 1999), and reduced complex II, III and IV activities in the caudate and putamen (Gu et al., 1996; Browne et al., 1997) leads to decreased ATP production. Glyceraldehyde-3-phosphate dehydrogenase (GAPDH), an enzyme involved in glucose break down, can interact strongly with mutant huntingtin (Burke et al., 1996) leading to enhanced inclusion formation and translocation of mutant huntingtin to the nucleus. Disruption of enzymes in the mitochondrial membrane can lead to leakage of electrons from the mitochondria. This produces reactive oxidative species (ROS) like the superoxide radical ($O_2^{-\bullet}$), hydrogen peroxide (H_2O_2) and the hydroxyl radical (OH^-) that over time cause damage to DNA, mitochondria and proteins. The cells of HD patients show signs of DNA damage as indicated by the high levels of 8-hydroxydeoxyguanosine (OH8dG) in the putamen (Browne et al., 1997).

Glutamate-induced excitotoxicity is another mechanism of degeneration proposed to play a role in HD pathogenesis. Reduced ATP production in HD leads to improperly functioning ionic pumps including the Na^+–K^+ ATPase. This pump is required to create and maintain electronic gradients across cellular membranes. Impairment of pump activity leads to failure of the membrane to repolarize after an action potential has fired. This prolonged membrane depolarization leads to expulsion of the Mg^{2+} that normally blocks *N*-methyl-D-aspartic acid (NMDA) receptors. Opening the NMDA receptor causes an influx of Ca^{2+} which results in free radical production and oxidative damage (Beal, 1992). The medium spiny neurons of the striatum contain relatively high numbers of NMDA receptors and therefore may be more vulnerable to excitotoxic cell death (Gardian and Vecsei, 2004).

ANIMAL MODELS

Historically animal models have been established using two of the three known causes of cell death in HD – mitochondrial impairment and excitotoxicity. Until the isolation of the HD gene in 1993 (The Huntington's Disease Collaborative Research Group, 1993), it was not possible to create genetic animals modeled after the HD mutation. Invertebrate models like *Caenorhabditis elegans* and *Drosophila melanogaster* are often used because they allow for rapid and high throughput testing of specific hypotheses and novel therapeutic strategies. The *C. elegans* model is created by expressing expanded polyglutamine repeats in the worm nervous system (Brignull et al., 2006). The mutation in *C. elegans* leads to the accumulation of huntingtin aggregates and overall decreased motility. The drosophila model is created by expressing expanded CAG repeats in the eye of the fly (Jackson et al., 1998). Although the eye develops normally, photoreceptors develop inclusions and subsequently degenerate. Since invertebrates have short life spans, large numbers of animals can be generated quickly, easily and inexpensively. However, thorough evaluation of disease processes and novel therapeutics ultimately require models with much more complexity.

Most animal models of HD fall into two broad categories – genetic and nongenetic. Nongenetic models typically induce cell death either by disruption of mitochondrial machinery or by excitotoxic mechanisms. Mitochondrial impairment has been shown to result in cell death in animal models following either 3-nitropropionic acid (3-NP) or malonic acid treatment. The 3-NP is a mitochondrial toxin that irreversibly inhibits succinate dehydrogenase thereby disrupting both the TCA and the electron transport chain (Alston et al., 1977; Coles et al., 1979). It was discovered in 1991 when accidentally ingested by people in China when they ate moldy sugarcane coated with the fungus *Arthrinium* spp. This caused neuronal death in the caudate and putamen accompanied by dystonia (Ludolph et al., 1991). When 3-NP is administered to rodents (Ludolph et al., 1991; Beal et al., 1993; Guyot et al., 1997; Blum et al., 2001) and nonhuman primates (Palfi et al., 2000), findings in humans can be replicated. 3-NP is administered systemically upon which it crosses the blood–brain barrier and causes cell death in the lateral striatum. Quinolinic acid (QA) and kainic acid (KA) are toxins that cause cell death by excitotoxicity. These amino acids bind to NMDA and non-NMDA receptors, respectively, on striatal neurons thereby inducing cell death. QA and KA have been used in both rodent and primate models of HD. Both QA and KA are incapable of crossing the blood–brain barrier and therefore have to be injected directly into the striatum.

Discovery of the HD gene and advances in molecular technology have recently allowed for the development of genetic mouse and rat models. These models attempt to capture the genetic and progressive nature of HD by introducing genes expressing the mutated htt protein into the rodent's germline. The transgenic models express a mutant human huntingtin gene in a random location in the mouse genome. The R6/1 and R6/2 transgenic mouse models (Mangiarini et al., 1996) express a mutant exon 1 of the human htt gene with 114 CAG repeats and 150 repeats, respectively. The N171-82Q transgenic mouse model expresses the first 171 amino acids of the htt protein bearing 82 CAG repeats (Schilling et al., 1999). Yeast artificial chromosome (YAC) transgenic mice are created by cloning an artificial yeast vector containing an expanded polyglutamine repeat into the mouse genome (Hodgson et al., 1999). A transgenic rat model also exists with 51 CAG repeats, created by a process similar to that used in transgenic mice (von Horsten et al., 2003). Rats, in general, tend to live longer than mice and have a more complex behavioral repertoire making the transgenic rat model an attractive candidate for carrying out thorough, long-term therapeutic studies.

Knock-in mouse models, considered to be the most accurate models for HD from a genetic standpoint, are constructed by replacing a portion of the mouse htt gene with a mutant human copy containing an expanded CAG region. Therefore, these mice only have two copies of the htt gene – a wild-type and a mutant allele both under control of the mouse htt promoter. The most commonly used knock-in mouse models are the HdhQ (Wheeler et al., 2000) and the CAG (Menalled et al., 2003) lines.

Due to the impracticality of creating a transgenic nonhuman primate model for HD, investigators have attempted to use viral vectors to over-express the human mutant htt gene in the normal striatum. This allows for the insertion of the mutant htt protein directly into specific cells of interest. In the absence of a transgenic primate model, this approach is an extremely valuable tool for mechanistic and therapeutic studies in species higher on the phylogenetic scale and more capable of higher order functional processing.

THERAPIES

Since HD is a genetic disease, familial history of the disease and genetic testing can be used to predict disease risk. Genetic testing for HD has been available since the 1980s (linkage analysis until 1993) (Koller and Davenport, 1984; Harper et al., 1990) and is used today to detect the presence or absence of the HD mutation in at-risk patients (Silber et al., 1998; Creighton et al., 2003; Ramos-Arroyo et al., 2005). Unfortunately, since no current therapy exists to protect patients from the devastating effects of HD, most patients decide against being tested for the aberrant gene. However, for the benefit of patients that do opt for genetic testing, neuroprotective strategies for treating patients prior to the onset of symptoms and cell death would be invaluable. Data being collected now by the Huntington Study Group (HSG) in their Prospective Huntington's at Risk Observational Study (PHAROS) and Neurobiological Predictors of Huntington's Disease (PREDICT-HD) study will allow researchers to gain knowledge on the early pathogenesis of the disease (Huntington Study Group, 2006). For the benefit of patients that only present to the clinical after the onset of debilitation symptoms, therapies that restore behavioral and cellular functionality must be investigated. This review chapter will discuss both gene and cell transplantation therapies currently under investigation in the laboratory and in clinical trials.

GENE THERAPY

Gene therapy is the process of correcting defective disease-causing genes. Most gene therapy studies involve using viral vectors to express a gene of interest in the host cell. Other methods include direct infusion of the molecule to the area of interest, genetically engineering cells that express the molecule and nonbiological capsules or spheres that release the molecule over time. The most commonly studied gene therapy methods for HD involve using neurotrophic factors for cell protection or RNA interference for mutant huntingtin downregulation.

Neurotrophic Factors

Neurotrophic factors are a group of proteins that play a critical role in the growth, survival and maintenance of neurons. For the treatment of neurodegenerative diseases like HD, neurotrophic factors or the genes expressing them are inserted into the abnormal cell in an attempt at rescuing cell death and dysfunction. Although neurotrophic factors do not correct the underlying cause of cell damage, in the case of HD a mutation in the huntingtin gene, they attempt to rescue or at least minimize some of the downstream effects of the genetic defect. Neurotrophic factors used in studies of HD fall into the following three families – (1) Neurotrophins, (2) Glial cell line-derived neurotrophic factor (GDNF) family of ligands (GFLs) and (3) Cytokines.

Neurotrophins

Nerve Growth Factor Nerve Growth Factor (NGF) was discovered in the early 1950s as a protein that supported the development and survival of peripheral neurons. In the central nervous system, NGF is expressed by cells both during and after development. NGF is expressed by the target cells of the cholinergic basal forebrain neurons. These cells are the pyramidal neurons and glial cells of the cortex. NGF is also expressed by the pyramidal neurons, dentate granule neurons and the interneurons of the hippocampal formation. The significance of NGF to HD therapy is that it is expressed by the cholinergic interneurons of the striatum. Interestingly, these interneurons are one of the few populations of striatal cells that are spared in HD patients. Although a direct connection has not been established between NGF expression and cell sparing in the striatum it may provide support for testing the effects of NGF on the susceptible medium spiny neurons.

One of the first studies to test the effects of NGF on striatal neuroprotection used genetically altered fibroblasts (Schumacher et al., 1991). The fibroblasts were designed to secrete NGF in the striatum and were implanted prior to a quinolinate lesion. The researches found that the group that received NGF-fibroblasts had a significantly smaller excitotoxic lesion compared to the group that received fibroblasts not secreting NGF. This smaller lesion volume correlated with an increased survival of striatal neurons. These same NGF-secreting fibroblasts had similar effects when transplanted in the corpus callosum ipsilateral to the lesion (Frim et al., 1993). However, effects of these NGF-secreting fibroblasts where limited to the site of implantation. Cells transplanted contralateral to a localized QA lesion caused only a slight and insignificant reduction in lesion area. This indicates that the effects of NGF using this method of implantation are limited to a small area immediately surrounding the cells. Additionally, while there was robust graft survival up to 18 days post-implantation, these cells had a significant reduction in NGF expression indicating that this method of NGF administration is very limited both spatially and temporally. Another method of NGF administration used neural stem cells (NSCs) retrovirally transduced to express NGF. This study transplanted the NGF-expressing stem cells into the striatum 1 week prior to a unilateral QA lesion in rats (Martinez-Serrano and Bjorklund, 1996) and examined the response one month later. The results obtained were very promising with the NGF-stem cell transplanted striatum showing a significant protection of GABAergic medium spiny neurons and the cholinergic interneurons. NGF even reduced significantly the immune response brought on by the excitotoxic lesion.

Transplantation of foreign materials like stem cells and fibroblasts inevitably causes a severe immune response requiring the administration of immunosuppressant. Upon termination of the immunosuppressant an immune response may be mounted and the transplant destroyed. Therefore, it may be valuable to explore other nonbiological methods of administration. Administering NGF via polymer microspheres has been examined with promising results (Gouhier et al., 2000). An imaging technique called single photo emission computerized tomography (SPECT) was used to specifically examine the effects of NGF on the medium spiny neurons that express the D2 receptor. A striatal QA lesion causes a reduced D2R signal in the SPECT which is rescued slightly by NGF. This study only showed moderate neuroprotective effects that may be attributed to the low levels of NGF released by the spheres. There was however a significantly decreased immune response in response to the QA lesion and no immune reaction in response to the microspheres. A follow-up to this study examined microspheres that released higher levels of NGF for a longer period of time. These microspheres were able to release NGF for up to 2.5 months after which they degraded (Menei et al., 2000). While this may seem a reasonably long period of time, this method would require multiple surgeries in human patients every 2–3 months.

Noninvasive methods of gene delivery are always more favorable to direct brain implantation. However, these methods are not always feasible in administering molecules like NGF that do not cross the blood–brain barrier. One way to circumvent this problem is by conjugating NGF to an antibody against a receptor in the endothelial cells of blood vessels. OX-26 is one such transferring receptor to which NGF has been conjugated and administered intravenously (Kordower et al., 1994). This method of administration selectively prevents the degeneration of the cholinergic interneurons in the rat striatum after a QA lesion. This selective neuroprotection may be a result of the fact that these are the only neurons in the striatum that express the NGF receptor TrkA. However, it is possible that higher amounts of OX-26 conjugated NGF may produce a significantly decreased immune response which may in turn provide neuroprotection.

PREMISE FOR THE USE OF NGF NGF is a neurotrophic factor which supports the growth and survival of several populations of neurons. In the striatum it is expressed by the cholinergic interneurons that are selectively spared in HD patients. It is possible that this effect can be transferred to the more susceptible GABAergic medium spiny neurons. Unfortunately, the methods used thus far in HD research to administer NGF to the lesioned striatum may not be ideal for neuroprotection. Most of the methods used thus far seem to significantly reduce the immune response to an excitotoxic lesion thereby decreasing cell death. A direct effect of NGF would require binding to the NGF receptor TrkA which is lacking in medium spiny neurons (Sofroniew et al., 2001).

NGF when used in a QA rat model of HD prevents the QA-mediated decrease in glutathione (Cruz-Aguado et al., 2000). Glutathione is an antioxidant that prevents damage to the cell from ROS. HD is a disorder in which there is significant mitochondrial impairment leading to oxidative stress. An increase in NGF-induced glutathione may assist in combating the downstream effects of mitochondrial damage. It is necessary to note that HD patients show an increase in glutathione in the striatum which does not combat cell death (Jakel and Maragos, 2000). Therefore, this increase by itself may not be sufficient to mediate neuroprotection. Cell death in several transgenic and toxin-induced HD models may be attributed to the free radical damage by nitric oxide. NGF has been shown to significantly decrease the levels of nitric oxide that are otherwise upregulated by a QA lesion.

Several previous studies have shown support for the use of NGF for the treatment of HD. However, in the past few years this avenue of research has been largely abandoned. Testing this neurotrophin in transgenic rodent model may provide better insight into the effectiveness of this molecule for HD therapy.

Brain-Derived Neurotrophic Factor Brain-derived neurotrophic factor (BDNF) is another molecule that belongs to the neurotrophin family and is more widely used in studies of HD than NGF. BDNF is produced by cortical neurons and is anterogradely transported to the striatum (Canals et al., 2001). Cortical BDNF supports the survival of striatal neurons and in the face of injury or insult to the striatum, BDNF expression in the cortex is increased for transport to the striatum. Mutant huntingtin protein has a direct effect on cellular BDNF transport. Wild-type huntingtin protein assists in the transport of vesicular BDNF along microtubules (Gauthier et al., 2004). In HD, a reduction in wild-type huntingtin and the expression of mutant huntingtin disrupts transport of BDNF from the cortex resulting in a loss of trophic support to striatal neurons. In healthy striatal neurons, wild-type huntingtin protein enhances the expression BDNF (Zuccato et al., 2001) and thus provided a constant support to the neurons. In the HD brain, mutant huntingtin reduces BDNF levels contributing to neurotoxicity. Serum BDNF levels are significantly lower in HD patients compared to age-matched controls (Ciammola et al., 2007). Additionally, patients with longer CAG repeat lengths had lower BDNF levels. BDNF levels were also lower in patients that had a longer disease duration indicating a decrease in BDNF with disease progression. Patients with lower BDNF levels also performed worse on the motor and cognitive components of the Unified Huntington's Disease Rating Scale (UHDRS). The BDNF receptor TrkB is also downregulated in the motor cortex and caudate of HD patients (Gines et al., 2006). Similar results are seen in several knock-in mouse models indicating a decrease in TrkB expressing as a result of mutant huntingtin.

Evidence from transgenic mouse models suggests that restoration of BDNF levels in striatal neurons will attenuate the cell death seen in HD. In a transgenic mouse model created by crossing the R6/1 HD transgenic model and mice with lower BDNF levels, there is a specific loss of enkephalinergic striatal neurons (Canals et al., 2004). Lower BDNF levels are also associated with an exaggerated motor deficit in these mice. These motor deficits may be directly related to the increased number of huntingtin aggregates in the substantia nigra and the decreased striatal dopamine content in these double transgenic mice (Pineda et al., 2005). Restoring the levels of BDNF in these mice was shown to rescue enkephalinergic cells but not substance P positive cells from death.

In a QA-induced rat model of HD, transplanted cells were genetically engineered to express BDNF and were grafted prior to a QA lesion (Perez-Navarro et al., 2000b). BDNF protected both substance P and enkephalin positive neurons from cell death indicating that it may be useful in early and late stages of the disease process. BDNF has also been administered to the lesioned rat striatum using an adenoviral vector gene delivery method (Bemelmans et al., 1999). Rats were administered vector- delivered BDNF 2 weeks prior to a QA lesion. One month after the lesion, BDNF-administered brains showed a significant protection of striatal GABAergic medium spiny neurons and a reduced lesion size. While these and other studies have shown significant neuroprotection in the striatum brought on by DNF, no study to date has examined the effects of BDNF on motor symptoms. Cell survival is an essential feature of any therapy but is of little significance in the absence of proper cell function.

Studies using BDNF have also not adequately examined cognitive recovery after a lesion or in genetic models. A study wishing to examine the effects of BDNF on memory formation looked at long-term potentiation (LTP) in the hippocampus in HDh111 knock-in mice (Lynch et al., 2007). LTP, a process of synaptic plasticity involved in memory formation, is disrupted in this knock-in mouse model. Administration of BDNF to hippocampal slices from this knock-in model restores LTP function to normal and stable levels. Although the effects of BDNF appear to benefit both striatal and hippocampal neurons, further studies in transgenic mice and nonhuman primates are warranted. In-depth studies on the effects of BDNF on motor and cognitive deficits should be undertaken.

PREMISE FOR THE USE OF BDNF BDNF is a trophic factor released by cortical neuron for the support of striatal neurons. Since striatal medium spiny neurons are the predominant populations of cells that die in HD, BDNF may have the capacity to protect these neurons. HD patients have reduced levels of BDNF both in the brain and in blood. The levels of BDNF are directly proportional to performance on motor and cognitive tasks. Studies have shown that the huntingtin protein is co-localized with BDNF in 99% of cortical pyramidal neurons and in 75% of the BDNF-containing striatal interneurons (Fusco et al., 2003). A localized QA lesion in the striatum causes a decrease in huntingtin protein express and a concurrent decrease in BDNF expression. This suggests that the huntingtin protein is essential to the production of BDNF in cortical cells and a disruption of the protein by QA contributes to cell death.

Similar to NGF, BDNF is also expressed by the interneurons of the striatum that are spared in HD (Fusco et al., 2003). Several therapies that are currently in clinical trials mediate their effects through an upregulation of BDNF. PN401, a uridine pro-drug, causes improvement in the rotorod task of motor function in the R6/2 and N171-82Q transgenic mouse models (Saydoff et al., 2006). PN401 also protected neurons in both the cortex and the striatum and reduced huntingtin aggregates in the striatum. In the N171-82Q model, PN401 restored the levels of BDNF in the cortex. Riluzole, an antiexcitotoxic substance used in clinical trials of HD (Seppi et al., 2001), has been shown to increase BDNF levels when applied to cultured neurons (Mizuta et al., 2001). Several antidepressants have been shown to be effective by increasing levels of BDNF in the brain. Cysteamine is part of a group of antidepressants that has been used in animal models of HD and works by increasing cortical BDNF levels (Borrell-Pages et al., 2006). Results of several other studies similar to the ones mentioned above indicate that replacement of BDNF that is depleted in the HD brain may help reduce cell death and behavioral symptoms associated with the disease.

Cytokine Family – Ciliary Neurotrophic Factor

Ciliary neurotrophic factor (CNTF) is a member of the interleukin-6 family of cytokines. CNTF is a differentiating cytokine that drives cells toward a predominantly astrocytic fate. In HD, CNTF is the most widely studied neurotrophic factor. CNTF is the first and currently the only trophic factor to enter clinical trails in HD. CNTF has trophic effects on striatal neurons as seen in both in vitro and in vivo studies. Some of the earliest studies administered CNTF to the brain by direct infusion of the protein using pumps. An infusion cannula was implanted directly into the striatum and recombinant CNTF was continuously infused using an osmotic pump (Anderson et al., 1996). This method of CNTF administration was efficient at significantly reducing cell death within the QA-lesioned striatum. A major pitfall of this method of administration is the need to constantly infuse CNTF into striatum. Additionally, large amounts of CNTF may be needed at one time to establish adequate diffusion throughout the striatum.

In an attempt to establish relatively long-term express, many studies use cell lines genetically engineered to express CNTF. In one such study a baby hamster kidney (BHK) cell line was transfected with a vector expressing CNTF (Emerich et al., 1996) and transplanted into the rat striatum. Twelve days later these rats received ipsilateral lesions using QA. CNTF-producing implants not only provided significant neuroprotection from the QA lesion but also improved motor behavior in the amphetamine-induced rotation test. Similar results were seen using the same cells in a 3-NP nonhuman primate model of HD (Emerich et al., 1997). When administered 1 week prior to the lesion there is a significant reduction in lesion area in the CNTF group vs. the control group – an approximately 380% decrease in lesion area in the caudate and 300% decrease in the putamen. This decrease in lesion area was directly attributable to a neuroprotection of different populations of striatal neurons including the GABAergic medium spiny neurons and the cholinergic and NADPH positive interneurons.

The previously mentioned studies examined the effects of CNTF when administered prior to the onset of cell death. It is also very important to examine the efficiency of any therapy to protect neurons after the onset of cell death. In a study using the neurotoxin 3-NP, nonhuman primates received 3-NP for 2 months prior to the transplantation of BHK cells expressing CNTF into the striatum (Mittoux et al., 2000). At this stage of the lesion, animals had begun to experience motor and cognitive deficits but did not exhibit an overt cell death detected by MRI. After transplantation of CNTF-expressing cells, 3-NP was continued for 3 months to mimic the progressive cell death seen in the brains of HD patients. The 3-NP-treated monkeys that received BHK-CNTF cells showed improvements in all tests and a complete restoration of motor and cognitive function 3 months after treatment despite ongoing toxicity from continuous 3-NP administration. There was also significant neuroprotection in most areas of the striatum and no significant difference in cell number compared to unlesioned controls.

Due to great success in animal models, this therapy was taken to clinical trials for HD. In a phase I clinical trial, polymer capsules containing BHK-CNTF cells were transplanted into the lateral ventricles of six subjects with early HD (Bloch et al., 2004). Capsules were removed and replaced every 6 months over 2 years. Evaluation on the UHDRS, TFC and Mattis Dementia Rating Scale revealed no significant improvements on any of these tests. In patients that had the most active capsules still secreting CNTF at the time of explantation, there was a positive electrophysiological recording indicating an improvement in intracerebral neural connections. The disappointing results in this study may have been due to the inadequacy of the CNTF delivery methods. Capsules were found not to release sufficient trophic factor following explantation and inadequate amounts of CNTF may have diffused into the striatum from the ventricle.

Alternative methods of CNTF administration using viral-mediated gene therapy are currently being explored. In one such study, an adenoviral vector was used to transfect cells in the striatum with CNTF (Mittoux et al., 2002). Adenoviral-CNTF was administered 10, 30 or 90 days prior to a 3-NP lesion in rats. At all three time points there was significant and comparable protection of striatal neurons. There was also protection of neurons in striatal target areas in the globus pallidus and the cortex. This indicates that vector delivery of

CNTF allows for long-term sustained expression of the protein in the striatum conferring neuroprotection upon the transfected cells.

Lentiviral vectors are also currently used in animal models of HD. Similar to adenoviral delivery, lentiviral vectors allow for a widespread and long-term expression of CNTF. In a study using YAC72 transgenic mice at 5 and 8 months of age, lenti-CNTF administration to the striatum produced results contradictory to those seen in lesion-induced models. The striatum of wild-type mice injected with lenti-CNTF showed a significant decrease in expression of dopamine and cyclic AMP-regulated phosphoprotein (DARPP-32), a marker of dopamine signaling present in the GABAergic medium spiny neurons (Zala et al., 2004). These results in wild-type mice precluded the drawing of any conclusions regarding the neuroprotective effects of CNTF in transgenic mice. There was also a decrease in the total number of neurons in the striatum of both wild-type and transgenic mice treated with lenti-CNTF. While the exact mechanisms of CNTF-mediated downregulation of DARPP-32 is unclear, there may be some association between long-term (9 months) expression of CNTF in cells containing the mutant huntingtin protein. An increase in the activation of astrocytes within the CNTF-treated striatum may also play a role. Side-effects in response to gene delivery methods can be common in some patients. In cases where complications arise, it may become necessary to halt gene expression. In order to examine gene therapy methods that can be closely monitored and turned off if necessary, researchers are studying the effects of a lentiviral vector with a tetracycline-regulated promoter (Regulier et al., 2002). Such a vector can be significantly downregulated by administering oral doxycycline. This ensures rapid termination of the therapy if adverse side-effects were to occur.

Premise for the Use of CNTF CNTF is upregulated in the striatum following a localized QA lesion (Haas et al., 2004). This indicates a neuroprotective role for CNTF immediately following injury. While CNTF has produced promising results in some models of HD, it is important to note that striatal neurons do not express the CNTF receptor. The method of CNTF-mediated neuroprotection is yet unclear but may involve an astrocytic response. In the spinal cord, CNTF administration activates astrocytes which in turn mediates motor neuron survival (Albrecht et al., 2002).

Glial Cell Line-Derived Neurotrophic Factor Family of Ligands

Members of the glial cell line-derived neurotrophic factor (GDNF) family of ligands (GFLs) include GDNF, neurturin (NTN), artemin and persephin. GDNF and NTN are two members of this family that have been extensively studied in HD. These two molecules have similar sequence homologies and exert comparable effects on different populations of neurons. Both GDNF and NTN have been used extensively in Parkinson's disease research and in clinical trials (Nutt et al., 2003; Dass et al., 2006) due to their trophic effects on midbrain dopaminergic neurons (Akerud et al., 1999). GFLs are also recognized for their important role in the growth, development and trophic support of striatal neurons. Treatment of GABAergic neurons in ventral mesencephalic cultures with GDNF or NTN promotes cell density and neurite outgrowth (Ducray et al., 2006). The GFLs – GDNF and NTN – signal via the GDNF receptor (GFR) complexes GFRα1 and GFRα2, respectively, and the receptor tyrosine kinase (c-ret) (Sariola and Saarma, 2003). Although GFRα2 is the preferred receptor for NTN, it can also signal via the GFRα1 receptor. This is promising for HD research because only GFRα1 expressed in the striatum along with c-ret (Perez-Navarro et al., 1999; Cho et al., 2004).

Glial Cell Line-Derived Neurotrophic Factor The GDNF gene therapy has been studied extensively in models of HD. In one study, GDNF was directly infused into rat brain. GDNF was administered as either a single bolus injection into the lateral ventricle or infused over 2 weeks (Araujo and Hilt, 1997). Thirty minutes after GDNF delivery, rats were given a localized unilateral QA lesion. The method of GDNF infusion did not determine the outcome of the study. Both methods showed partially nonselective neuroprotection of all populations of striatal projection neurons. There was also an improvement in rotational behavior in rats treated with GDNF. Previous studies administering GDNF via a bolus intracerebroventricular (i.c.v.) injection have shown that the effects of GDNF administration last for up to 14 days (Lapchak et al., 1997). Unfortunately, this would require multiple bi-monthly injections of GDNF to ensure sustained benefit. In a clinical trial for Parkinson's disease, GDNF was administered as a single bolus injection given i.c.v. (Patel and Gill, 2007). The phase I open-labeled trial produced significant behavioral and anatomical improvements prompting a phase II randomized control study. In this study there was no clinical benefit to the GDNF infusion and some reported side-effects. The failure of the phase II study was attributed to the technical differences between this and the phase I trial and inadequate diffusion of GDNF from the ventricle to the striatum.

Fibroblasts engineered to express GDNF, when transplanted in the rat striatum, protect the brain from excitotoxic injury (Perez-Navarro et al., 1999). A fibroblast cell line was designed to over express GDNF and was transplanted into the rat striatum. A localized QA lesion was performed 24 h later. This study reported that GDNF had selective neuroprotective effects on the different populations of striatal projection neurons. GDNF was only capable of protecting the GABAergic neurons that co-express dynorphin and tachykinin (i.e., the neurons that project to the substantia nigra). GDNF could not protect the neurons that expressed enkephalin and projected to the globus pallidus. These differential effects may be explained by the evidence that both the adult striatum and adult globus pallidus express GDNF. There is evidence to suggest that an injury to the striatum causes an increase in GDNF in an attempt to protect the striatonigral circuitry (Schmidt-Kastner et al., 1994). It is therefore possible that direct administration of GDNF to the striatum is only effective at protecting those neurons that project to the substantia nigra. In a similar mechanism, GDNF in the globus pallidus may play a target-derived role in protecting the enkephalinergic neurons of the striatum.

Due to the side-effects associated with fibroblast transplantation in the brain (Hoffman et al., 1993), alternative methods of cellular delivery of trophic factors must be explored. NSCs designed to secrete neurotrophic factors may be a comparatively safer method of gene delivery. NSC expressing transgenes for GDNF and the fluorescent marker Luciferase have been used in toxin models of HD (Pineda et al., 2007). The Luciferase tagging of these NSCs allows for in vivo tracking of the location and migration of grafts. GDNF-expressing NSCs were transplanted bilaterally in the striatum 1 day prior to a unilateral QA lesion. Mice receiving transplants showed a 190% increase in preserved striatal neurons compared to lesioned control mice. This neuroprotection correlated with an improvement in amphetamine-induced rotational behavior. Mice receiving GDNF-expressing NSCs showed a more than 50% decrease in net rotations compared to lesioned control mice. Both anatomical and behavioral improvements were only seen in the group that received NSCs expressing GDNF and not in the group transplanted with control NSCs. GDNF-expressing NSCs injected in the lesioned striatum proliferated at a rate 1500% faster than the same cells in the unlesioned hemisphere. This response is most likely the result of signals from the host striatum in response to the QA lesion. This indicates that the graft is capable of complete integration into the host striatum.

Viral vectors are often used in administering GDNF to the striatum in HD models. A study using the adeno-associated viral vector examined the neuroprotective effects of AAV–GDNF when administered 2 weeks prior to a systemic 3-NP lesion in rats (McBride et al., 2003). Rats underwent 4 weeks of behavioral testing on an ambulatory rating scale and platform balance test. The performance of 3-NP lesioned animals on both these behavior tasks deteriorated after toxin administration. The AAV–GDNF-treated rats however showed near normal performance. Histological evaluation showed that AAV–GDNF-treated rats had 70% more NeuN-immunoreactive neurons in the striatum compared to lesioned control rats. AAV–GDNF-treated rats also had significantly less CD45 staining in the striatum compared to lesioned control rats, demonstrating a reduction in a microglial response to toxicity.

As a follow-up to this study, the effects of AAV–GDNF were examined in the N171-82Q transgenic mouse model of HD (McBride et al., 2006). Adult N171-82Q transgenic mice were administered bilateral injections of AAV–GDNF prior to the onset of overt behavioral symptoms. Animals were tested on a rotorod test of motor coordination and a hind limb clasping test. Animals receiving AAV–GDNF performed significantly better than untreated transgenic mice at all time points. The onset of clasping behavior was delayed in the AAV–GDNF-treated mice. The behavioral benefits imposed by AAV–GDNF treatment are directly related to striatal neuroprotection. AAV–GDNF-treated mice had a significant protection of striatal neurons compared to untreated transgenic mice. Additionally, AAV–GDNF prevented striatal cell shrinkage as seen in the untreated transgenic mice.

A study using lentiviral delivery of GDNF used the R6/2 transgenic mouse model of HD (Popovic et al., 2005). The R6/2 mouse model has an earlier onset and more rapidly progressing phenotype compared to the N171-82Q model. Researchers in this study injected mice with lenti-GDNF at 4–5 weeks of age. At this time point some R6/2 mice already started to develop symptoms (Mangiarini et al., 1996). Mice receiving lenti-GDNF did not show improvements in any of the behavioral or anatomical measurements. This may be attributed to the post-symptomatic delivery lenti-GDNF as opposed to the presymptomative delivery of AAV–GDNF (McBride et al., 2006) and the use of the more aggressive R6/2 mouse model by Popovic and co-workers (2005) as opposed to the less aggressive N171-82 mouse model employed by McBride and colleagues (2006)

Although studies using GDNF have yielded promising results in animal models of HD, this neurotrophic factor is not available for use in clinical trials. AMGEN Inc. has the current patent on GDNF, and it is therefore unavailable for clinical use. Therefore, alternatives to GDNF must be explored for therapeutic use.

Neurturin Neurturin (NTN) acts in a similar manner to GDNF and has been shown to be effective in animal models of HD. While GDNF selectively protects the substance P neurons of the direct circuit in a QA rat model of HD (Perez-Navarro et al., 1999), NTN has effects on the other population of projection neurons.

In a QA model, NTN has been shown to selectively protect the striatal neurons of the indirect circuit (Perez-Navarro et al., 2000a). In early stages of HD, these enkephalinergic neurons that project to the external segment of the globus pallidus are the first to degenerate. The potential of NTN to prevent the death of these cells may be useful in treatment during early stages of the disease. In the QA rat model of HD, endogenous NTN is upregulated in the striatum in response to injury. This may indicate a neuroprotective role for NTN in the presence of excitotoxic cell death. At high-enough doses, exogenous administration of NTN protects striatal neurons from cell death in both QA and kainic acid models of HD (Perez-Navarro et al., 2000a; Gratacos et al., 2001).

Our group has determined that gene delivery of NTN protects striatal neurons from degeneration and rescues motor deficits in 3-NP-treated rats and N171-82Q transgenic HD mice (Ramaswamy et al., 2006). Transgenic mice receiving AAV–NTN at the age of 5 weeks, prior to the onset of symptoms, show significantly improved behavior on the rotorod test compared to untreated transgenic controls. This behavioral improvement is linked to a significant protection of neurons in the striatum. Our study indicates that, when administered prior to the onset of symptoms, AAV–NTN can protect striatal neurons from death and cause improvements in motor tasks. We are currently examining the effects of AAV–NTN when administered after the onset of symptoms.

RNA Interference

Isolation of the huntingtin gene and discovery of the mutation that causes HD has allowed not only the creation of animal models that mimic this genetic defect but also the establishment of treatments that target the underlying cause of the disease. A recently developed therapy that hopes to target the gene mutation is RNA interference (RNAi). This therapy attempts to use short interfering RNA (siRNA), short hairpin RNA (shRNA) or microRNA (miRNA) molecules to shut down the production of the mutant huntingtin protein. These short RNA molecules bind to the huntingtin protein mRNA and trigger a cascade of events that results in the degradation of the mRNA. This inhibits the translation of a large number of huntingtin mRNA, significantly downregulating huntingtin protein expression within cells. It is important to understand that both mutant and wild-type mRNA expressions will be downregulated by this process.

A study using shRNA in the N171-82Q transgenic mouse model showed a 50–55% decrease in the N171-82Q mRNA in the injected striatum and a complete elimination of mutant huntingtin-positive inclusions (Harper et al., 2005). There was also a rescue of motor deficits on the rotorod test. In the R6/2 transgenic mouse model, siRNAs against the R6/2 huntingtin mRNA reduced brain atrophy and neuronal inclusions (Wang et al., 2005). This study also saw a rescue of motor deficits (rotorod, clasping and open-field tests) and an increase in animal survival. The previously described studies examined the effects of RNAi in presymptomatic models. In a post-symptomatic study using shRNAs, nuclear inclusions were reduced in the striatum even after they had begun to form (Machida et al., 2006). Cellular phenotypes that are normally downregulated in this model (DARPP-32, enkephalin) were restored.

RNAi is a very promising strategy in the therapy of HD. In HD where the cause of the disease is completely genetic, effectively downregulating levels of the mutant huntingtin protein may stop the damage caused by the mutant protein in its tracks. The one major drawback of RNAi therapy in HD is that treatment downregulates the expression of both wild-type and mutant huntingtin protein. Ongoing studies are using RNAi to decrease expression of wild-type huntingtin and look for any adverse effects (Chen et al., 2005; Omi et al., 2005). Some researchers are also attempting to develop strategies of allele-specific targeting in which only mutant huntingtin expression would be downregulated.

CELL TRANSPLANTATION THERAPY

Although HD is a genetic disorder which can be detected prior to the onset of symptoms, most patients do not consult a physician until after they develop signs of the disease. At this stage, there likely is extensive and irreversible cell death in the striatum. In such cases, a neuroprotection therapy may not be ideal and cell replacement therapies should be considered. Cell transplantation came to the forefront of HD research after one of the first and most critical experiments conducted by Isacson and co-workers (Isacson et al., 1985). This group grafted cell from the rat fetal ganglionic eminence into the striatum of rats that had previously received bilateral ibotenic acid-induced excitotoxic lesions. Not only did the grafts survive and improve motor function, they also improved cognitive function (Isacson et al., 1986). This is one of the first demonstrations that restoring function solely to the striatum can improve higher order cognitive function. Since these initial landmark studies, there have been numerous examples in which fetal striatal grafts improve function in toxin-based rodent models of HD (Dunnett and Rosser, 2004).

For the purposes of the review paper we will discuss cell replacement therapies based on two different types of tissue for transplantation – fetal tissue/cells and adult NSCs.

Human Fetal Tissue Transplants

Several clinical trials using fetal embryonic tissue transplants into the striatum have been conducted with varying levels of success. An important factor in this therapy is the selection of appropriate tissue to obtain optimal therapeutic benefit. For HD therapy, isolation of tissue destined to a striatal fate is ideal for transplantation therapies. The Network of European CNS Transplantation and Restoration has determined that DARPP-32 positive-striatal neurons begin to develop at week 7 post-conception in the ganglionic eminence (Naimi et al., 1996). At 8.5 weeks post-conception the lateral ganglionic eminence separates from the medial ganglionic eminence. At this time all the medium spiny projection neurons of the striatum are isolated in the lateral ganglionic eminence, and the medial ganglionic eminence is devoid of striatal neurons destined to a GABAergic fate. Consequently, tissue ideal for striatal transplants will arise from the lateral ganglionic eminence although whole ganglion implants are also conducted.

Tissue from the lateral ganglionic eminence of embryonic day 16 (E16) rat fetuses has been transplanted into the striatum of QA-lesioned rats (Chen et al., 2002). Rats were lesioned unilaterally with QA and 1 month later transplanted with the fetal tissue. The QA lesion caused ipsilateral rotations in response to amphetamine administration and this behavior was reversed at the graft transplantation. Five months after transplantation, animals were anaesthetized and electrophysiological recordings were obtained from single cells in the striatum. When dopamine is applied to the intact striatum it inhibits striatal neuron activity. In the QA-lesioned striatum, a higher dose of dopamine is required to inhibit neuron activity. In the striatum transplanted with fetal striatal tissue, dopamine signaling was restored to levels seen in unlesioned controls.

Numerous successful studies using fetal tissue transplants in animal models of HD prompted clinical trials using fetal striatal grafts. There are currently four clinical trials in progress for HD using fetal tissue transplants. The NEST-UK study transplanted cell suspensions from whole ganglionic eminences of 9.5 to 12-week-old fetuses unilaterally into the striatum of early and mid-stage HD patients (n = 4) (Rosser et al., 2002). Each patient received two whole ganglionic eminences from one fetus transplanted into multiple sites in both the caudate and putamen. Patients were evaluated 6 months postoperatively to look for side-effects relating to the surgery. There was adequate graft survival in all patients indicating that this method of transplantation is feasible for future HD trials. No significant changes in motor, cognitive or psychological measurements were seen due to the short-term follow-up. Unimpressive results in this study could have been due to the short evaluation interval, unilateral injections or the use of whole ganglionic eminence. The main goal of this study was to establish the safety of this protocol. The researchers concluded that there were no significant adverse side-effects related to transplantation and consequently were proceeding to trials using bilateral transplants.

In another clinical trial conducted in Los Angeles, lateral ganglionic eminences were transplanted bilaterally into the striatum of three patients (Kopyov et al., 1998). Each patient received transplants from 5–8 donors and each gestational age of each fetus was determined by crown-to-rump length (CRL from –20 to 32 mm). As indicated by MRI, all three patients exhibited striatal atrophy prior to transplantation and the first patient also exhibited frontal atrophy. A deoxyglucose PET scan revealed striatal hypometabolism as is common in HD patients. At 12 months after surgery, there was an increase in T1 weighted signaling in the striatum compared to preoperative levels indicating graft survival and potential graft growth. Motor scores in the UHDRS improved significantly in all three patients, 6 months after transplantation. Cognitive symptoms also showed some improvement although there was variability between patients (Philpott et al., 1997). A few years later, two of the three patients died and came to autopsy. The first patient received two grafts in the right putamen, three in the left putamen and one in the left caudate. There was gradual progression of his disease and he died of pneumonia at age 54. The second patient had received eight grafts: three in each putamen and one in each caudate. Three months after transplantation she showed improved ambulation but a constant deterioration in speech. Between 9 months and 3 years she experienced increased falls that confined her to a wheelchair at 3 years. At 4 years she had to wear a protective helmet and started taking haloperidol. Like patient 1 she died of pneumonia at the age of 41 years. Upon examination of the brains, there was pathology that would be expected in HD patients. Frontal cortical atrophy, dilated lateral ventricles and severe bilateral atrophy of the caudate and putamen were seen in both patients. The brains also showed gliosis in the caudate and putamen, reduction in neuron number and ubiquitin-positive inclusions. All six grafts survived in patient 1, whereas seven out of the eight grafts survived in patient 2. Cell phenotype was analyzed in the graft and cells had adopted a striatal phenotype. Cells stained positive for calbindin and DARPP-32 which are markers of medium spiny projection neurons. This was expected as cells transplanted were from the lateral ganglionic eminence. Interestingly, some cells also stained positive for calretinin which is a marker for interneurons found in the striatum. This clinical trial showed that there is good graft survival and integration several years after transplantation.

There was also very minimal host immune response in spite of years without immunosuppressant.

A third study transplanted the whole ganglionic eminence from 7 to 9-week-old fetuses unilaterally into the striatum of five HD patients showing relatively early symptomology (Bachoud-Levi et al., 2000a). Patients had good graft survival, without overgrowth, but showed no improvements in motor or cognitive symptoms. These patients were scheduled for regrafting on the opposite side of the brain 1 year later (Bachoud-Levi et al., 2000b) and revaluated for cognitive and motor improvements. In three out of five patients there was stability in performances on executive function tasks, neurophysiological tests and chorea for 3 years compared to untreated controls. In two of these three patients, performance improved over the 3 years. In these three patients improvements in the conducted tests was correlated with an increase in striatal and cortical metabolisms, indicating a reconstruction of cortico-striatal circuits (Gaura et al., 2004). A follow-up of these patients indicated benefits to cognition and functionality 6 years post-implantation (Bachoud-Levi et al., 2006).

In a study by Hauser and colleagues (Hauser et al., 2002), seven patients received bilateral fetal transplants derived from the lateral ganglionic eminence (8–9 weeks post-conception). One patient died 18 months after transplantation from cardiovascular disease and his brain was evaluated for pathology (Freeman et al., 2000). Prior to death, this patient improved by 10 and 8 points on the UHDRS 12 and 15 months post-transplantation, respectively. Histological evaluation of the brain showed good graft survival and no mutant huntingtin labeling in the graft cells. This indicates that the graft tissue did not take on the pathology of the host cells over this short postoperative period. Host-derived dopaminergic fibers had grown into the graft. In the remaining patients, when evaluations were conducted after excluding one patient that suffered from a subdural hemorrhage, performance on the UHDRS was significantly improved 1 year after transplantation. There was a slowing in rate of yearly decline on the UHDRS and TFC tests after transplantation. There was no improvement on neuropsychological tests. These studies indicate that fetal tissue transplants are potentially viable options for treating HD and benefits can last for several years. However, practical issues related to tissue procurement remain a major obstacle for this approach.

Stem Cells

Issues with procurement of large numbers of fetal donors for transplantation have pushed transplantation research toward more modern donor tissue like stem cells. Important characteristics of stem cells for use in HD are the capability to differentiate into neurons, the capability to attain a GABAergic phenotype and the ability to re-establish lost circuitry. This section will describe transplantation studies using cells that are not derived from the fetal striatum.

Human fetal NSCs derived from the fetal cortex and treated with CNTF have been shown to attenuate the motor deficits associated with a 3-NP lesion in a rat model of HD (McBride et al., 2004). In a study conducted by our laboratory, the neuroanatomical and behavioral effects of human stem cell transplants placed into the striatum of QA-treated rats were examined. Rats received unilateral QA (200 nM/μl) injections into the striatum. One week later, rats were transplanted with stem cells derived from the human fetal cortex at E12. These cells were either pre-treated in culture media with CNTF or allowed to grow in culture media alone. Cortically derived neurospheres expanded in culture in the presence of CNTF which tend to differentiate into a GABAergic phenotype (Caldwell et al., 2001). CNTF pre-treatment of cultured neurospheres leads to a 40% increase in the number of GABAergic neurons compared to treatment with other neurotrophic factors. Cells of this phenotype are ideal for transplantation into models of HD. Rats transplanted with human stem cells performed significantly better over the 8 weeks of testing on the cylinder test as compared to lesioned rats treated with vehicle. Animals transplanted with CNTF-treated neurospheres also showed significant neuroprotection of the striatum compared to animals transplanted with untreated neurospheres. Stereological counts indicated that rats transplanted with CNTF-treated neurospheres had a 20% larger striatal volume compared to those receiving transplants of untreated neurospheres and a 27% larger striatal volume compared to rats injected with vehicle. Grafted cells were seen to migrate to projection areas of the striatum including the globus pallidus, entopeduncular nucleus and substantia nigra, pars reticulata. This study showed that stem cell transplants can improve motor performance when transplanted into an excitotoxically lesioned striatum. Additionally, if these cells are driven to a GABAergic phenotype prior to transplantation they can significantly protect the striatum from degeneration.

GDNF-expressing NSCs when transplanted into a QA-lesioned striatum protect neurons from degeneration and ameliorate motor deficits (Pineda et al., 2007). These NSCs also proliferated in response to the lesion. In an interesting study using the QA rat model, fetal cortical cells were administered intravenously through the tail vein (Lee et al., 2005). NSCs differentiated into neurons and glia and migrated to the cortex and preferentially to the lesioned striatum. Rats that received these transplants showed improved rotational behavior and had decreased striatal atrophy. This study indicates that peripheral administration of stem cells may be a viable and less intrusive method for treatment.

Alternative Transplantation Studies

Due to the limited availability of embryonic or fetal stem cells for therapy, many researchers are looking into alternative donor sources for transplantation. Such sources of stem cells are derived from umbilical cord blood, bone marrow and adult sources like the subventricular zone and dentate gyrus. In a QA-lesioned rat model of HD, rat bone marrow cells were injected bilaterally into the striatum (Lescaudron et al., 2003). Animals treated with bone marrow cells showed significant improvements in working memory performance compared to lesioned rats. However, there was no rescue from cell death in the striatum and less than 1% of transplanted cells expressed a neuronal phenotype. Preliminary results in a study using human umbilical cord cells showed that huntingtin transgenic mice receiving transplants had increased survival and decreased weight loss (Ende and Chen, 2001). These results are promising for the use of umbilical cord blood in HD, and further analysis of histological and symptomatic benefits should be conducted.

REFERENCES

Akerud P, Alberch J, Eketjall S et al. (1999). Differential effects of glial cell line-derived neurotrophic factor and neurturin on developing and adult substantia nigra dopaminergic neurons. *J Neurochem* 73(1): 70–78.

Albin RL, Young AB, and Penney JB (1989). The functional anatomy of basal ganglia disorders. *Trends Neurosci* 12(10): 366–375.

Albrecht PJ, Dahl JP, Stoltzfus OK et al. (2002). Ciliary neurotrophic factor activates spinal cord astrocytes, stimulating their production and release of fibroblast growth factor-2, to increase motor neuron survival. *Exp Neurol* 173(1): 46–62.

Alexander GE and Crutcher MD (1990). Functional architecture of basal ganglia circuits: neural substrates of parallel processing. *Trends Neurosci* 13(7): 266–271.

Alston TA, Mela L, and Bright HJ (1977). 3-Nitropropionate, the toxic substance of Indigofera, is a suicide inactivator of succinate dehydrogenase. *Proc Natl Acad Sci U S A* 74(9): 3767–3771.

Anderson KD, Panayotatos N, Corcoran TL et al. (1996). Ciliary neurotrophic factor protects striatal output neurons in an animal model of Huntington disease. *Proc Natl Acad Sci U S A* 93(14): 7346–7351.

Andrew SE, Goldberg YP, Kremer B et al. (1993). The relationship between trinucleotide (CAG) repeat length and clinical features of Huntington's disease. *Nat Genet* 4(4): 398–403.

Araujo DM and Hilt DC (1997). Glial cell line-derived neurotrophic factor attenuates the excitotoxin-induced behavioral and neurochemical deficits in a rodent model of Huntington's disease. *Neuroscience* 81(4): 1099–1110.

Bachoud-Levi A, Bourdet C, Brugieres P et al. (2000a). Safety and tolerability assessment of intrastriatal neural allografts in five patients with Huntington's disease. *Exp Neurol* 161(1): 194–202.

Bachoud-Levi AC, Remy P, Nguyen JP et al. (2000b). Motor and cognitive improvements in patients with Huntington's disease after neural transplantation. *Lancet* 356(9246): 1975–1979.

Bachoud-Levi AC, Gaura V, Brugieres P et al. (2006). Effect of fetal neural transplants in patients with Huntington's disease 6 years after surgery: a long-term follow-up study. *Lancet Neurol* 5(4): 303–309.

Beal MF (1992). Does impairment of energy metabolism result in excitotoxic neuronal death in neurodegenerative illnesses? *Ann Neurol* 31(2): 119–130.

Beal MF, Brouillet E, Jenkins BG et al. (1993). Neurochemical and histologic characterization of striatal excitotoxic lesions produced by the mitochondrial toxin 3-nitropropionic acid. *J Neurosci* 13(10): 4181–4192.

Bemelmans AP, Horellou P, Pradier L et al. (1999). Brain-derived neurotrophic factor-mediated protection of striatal neurons in an excitotoxic rat model of Huntington's disease, as demonstrated by adenoviral gene transfer. *Hum Gene Ther* 10(18): 2987–2997.

Bloch J, Bachoud-Levi AC, Deglon N et al. (2004). Neuroprotective gene therapy for Huntington's disease, using polymer-encapsulated cells engineered to secrete human ciliary neurotrophic factor: results of a phase I study. *Hum Gene Ther* 15(10): 968–975.

Blum D, Gall D, Cuvelier L et al. (2001). Topological analysis of striatal lesions induced by 3-nitropropionic acid in the Lewis rat. *Neuroreport* 12(8): 1769–1772.

Borrell-Pages M, Canals JM, Cordelieres FP et al. (2006). Cystamine and cysteamine increase brain levels of BDNF in Huntington disease via HSJ1b and transglutaminase. *J Clin Invest* 116(5): 1410–1424.
Brignull HR, Morley JF, Garcia SM et al. (2006). Modeling polyglutamine pathogenesis in *C. elegans*. *Methods Enzymol* 412: 256–282.
Browne SE, Bowling AC, MacGarvey U et al. (1997). Oxidative damage and metabolic dysfunction in Huntington's disease: selective vulnerability of the basal ganglia. *Ann Neurol* 41(5): 646–653.
Burke JR, Enghild JJ, Martin ME et al. (1996). Huntingtin and DRPLA proteins selectively interact with the enzyme GAPDH. *Nat Med* 2(3): 347–350.
Bylsma FW, Brandt J, and Strauss ME (1990). Aspects of procedural memory are differentially impaired in Huntington's disease. *Arch Clin Neuropsychol* 5(3): 287–297.
Caldwell MA, He X, Wilkie N et al. (2001). Growth factors regulate the survival and fate of cells derived from human neurospheres. *Nat Biotechnol* 19(5): 475–479.
Canals JM, Checa N, Marco S et al. (2001). Expression of brain-derived neurotrophic factor in cortical neurons is regulated by striatal target area. *J Neurosci* 21(1): 117–124.
Canals JM, Pineda JR, Torres-Peraza JF et al. (2004). Brain-derived neurotrophic factor regulates the onset and severity of motor dysfunction associated with enkephalinergic neuronal degeneration in Huntington's disease. *J Neurosci* 24(35): 7727–7739.
Chen GJ, Jeng CH, Lin SZ et al. (2002). Fetal striatal transplants restore electrophysiological sensitivity to dopamine in the lesioned striatum of rats with experimental Huntington's disease. *J Biomed Sci* 9(4): 303–310.
Chen ZJ, Kren BT, Wong PY et al. (2005). Sleeping beauty-mediated down-regulation of huntingtin expression by RNA interference. *Biochem Biophys Res Commun* 329(2): 646–652.
Cho J, Yarygina O, Oo TF et al. (2004). Glial cell line-derived neurotrophic factor receptor GFRalpha1 is expressed in the rat striatum during postnatal development. *Brain Res Mol Brain Res* 127(1–2): 96–104.
Ciammola A, Sassone J, Cannella M et al. (2007). Low brain-derived neurotrophic factor (BDNF) levels in serum of Huntington's disease patients. *Am J Med Genet B Neuropsychiatr Genet* 144(4): 574–577.
Cicchetti F, Prensa L, Wu Y et al. (2000). Chemical anatomy of striatal interneurons in normal individuals and in patients with Huntington's disease. *Brain Res Brain Res Rev* 34(1–2): 80–101.
Coles CJ, Edmondson DE, and Singer TP (1979). Inactivation of succinate dehydrogenase by 3-nitropropionate. *J Biol Chem* 254(12): 5161–5167.
Creighton S, Almqvist EW, MacGregor D et al. (2003). Predictive, pre-natal and diagnostic genetic testing for Huntington's disease: the experience in Canada from 1987 to 2000. *Clin Genet* 63(6): 462–475.
Cruz-Aguado R, Turner LF, Diaz CM et al. (2000). Nerve growth factor and striatal glutathione metabolism in a rat model of Huntington's disease. *Restor Neurol Neurosci* 17(4): 217–221.
Dass B, Kladis T, Chu Y et al. (2006). RET expression does not change with age in the substantia nigra pars compacta of rhesus monkeys. *Neurobiol Aging* 27(6): 857–861.
Ducray A, Krebs SH, Schaller B et al. (2006). GDNF family ligands display distinct action profiles on cultured GABAergic and serotonergic neurons of rat ventral mesencephalon. *Brain Res* 1069(1): 104–112.
Duff K, Paulsen JS, Beglinger LJ et al. (2007). Psychiatric symptoms in Huntington's disease before diagnosis: the Predict-HD Study. *Biol Psychiatry* 62(12): 1341–1346.
Dunnett SB and Rosser AE (2004). Cell therapy in Huntington's disease. *NeuroRx* 1(4): 394–405.
Emerich DF, Lindner MD, Winn SR et al. (1996). Implants of encapsulated human CNTF-producing fibroblasts prevent behavioral deficits and striatal degeneration in a rodent model of Huntington's disease. *J Neurosci* 16(16): 5168–5181.
Emerich DF, Winn SR, Hantraye PM et al. (1997). Protective effect of encapsulated cells producing neurotrophic factor CNTF in a monkey model of Huntington's disease. *Nature* 386(6623): 395–399.

Ende N and Chen R (2001). Human umbilical cord blood cells ameliorate Huntington's disease in transgenic mice. *J Med* 32(3–4): 231–240.

Ferrante RJ, Beal MF, Kowall NW et al. (1987). Sparing of acetylcholinesterase-containing striatal neurons in Huntington's disease. *Brain Res* 411(1): 162–166.

Freeman TB, Cicchetti F, Hauser RA et al. (2000). Transplanted fetal striatum in Huntington's disease: phenotypic development and lack of pathology. *Proc Natl Acad Sci U S A* 97(25): 13877–13882.

Frim DM, Short MP, Rosenberg WS et al. (1993). Local protective effects of nerve growth factor-secreting fibroblasts against excitotoxic lesions in the rat striatum. *J Neurosurg* 78(2): 267–273.

Fusco FR, Zuccato C, Tartari M et al. (2003). Co-localization of brain-derived neurotrophic factor (BDNF) and wild-type huntingtin in normal and quinolinic acid-lesioned rat brain. *Eur J Neurosci* 18(5): 1093–1102.

Gardian G and Vecsei L (2004). Huntington's disease: pathomechanism and therapeutic perspectives. *J Neural Transm* 111(10–11): 1485–1494.

Gaura V, Bachoud-Levi AC, Ribeiro MJ et al. (2004). Striatal neural grafting improves cortical metabolism in Huntington's disease patients. *Brain* 127(Pt 1): 65–72.

Gauthier LR, Charrin BC, Borrell-Pages M et al. (2004). Huntingtin controls neurotrophic support and survival of neurons by enhancing BDNF vesicular transport along microtubules. *Cell* 118(1): 127–138.

Gines S, Bosch M, Marco S et al. (2006). Reduced expression of the TrkB receptor in Huntington's disease mouse models and in human brain. *Eur J Neurosci* 23(3): 649–658.

Gouhier C, Chalon S, Venier-Julienne MC et al. (2000). Neuroprotection of nerve growth factor-loaded microspheres on the D2 dopaminergic receptor positive-striatal neurones in quinolinic acid-lesioned rats: a quantitative autoradiographic assessment with iodobenzamide. *Neurosci Lett* 288(1): 71–75.

Gratacos E, Perez-Navarro E, Tolosa E et al. (2001). Neuroprotection of striatal neurons against kainate excitotoxicity by neurotrophins and GDNF family members. *J Neurochem* 78(6): 1287–1296.

Gu M, Gash MT, Mann VM et al. (1996). Mitochondrial defect in Huntington's disease caudate nucleus. *Ann Neurol* 39(3): 385–389.

Guyot MC, Hantraye P, Dolan R et al. (1997). Quantifiable bradykinesia, gait abnormalities and Huntington's disease-like striatal lesions in rats chronically treated with 3-nitropropionic acid. *Neuroscience* 79(1): 45–56.

Haas SJ, Ahrens A, Petrov S et al. (2004). Quinolinic acid lesions of the caudate putamen in the rat lead to a local increase of ciliary neurotrophic factor. *J Anat* 204(4): 271–281.

Harper PS, Morris MJ, and Tyler A (1990). Genetic testing for Huntington's disease. *BMJ* 300(6732): 1089–1090.

Harper SQ, Staber PD, He X et al. (2005). RNA interference improves motor and neuropathological abnormalities in a Huntington's disease mouse model. *Proc Natl Acad Sci U S A* 102(16): 5820–5825.

Hauser RA, Sandberg PR, Freeman TB et al. (2002). Bilateral human fetal striatal transplantation in Huntington's disease. *Neurology* 58(11): 1704.

Hedreen JC, Peyser CE, Folstein SE et al. (1991). Neuronal loss in layers V and VI of cerebral cortex in Huntington's disease. *Neurosci Lett* 133(2): 257–261.

Heindel WC, Butters N, and Salmon DP (1988). Impaired learning of a motor skill in patients with Huntington's disease. *Behav Neurosci* 102(1): 141–147.

Hodgson JG, Agopyan N, Gutekunst CA et al. (1999). A YAC mouse model for Huntington's disease with full-length mutant huntingtin, cytoplasmic toxicity, and selective striatal neurodegeneration. *Neuron* 23(1): 181–192.

Hoffman D, Breakefield XO, Short MP et al. (1993). Transplantation of a polymer-encapsulated cell line genetically engineered to release NGF. *Exp Neurol* 122(1): 100–106.

Huntington G (2003). On chorea. George Huntington, M.D. *J Neuropsychiatry Clin Neurosci* 15(1): 109–112.

Huntington Study Group (2006). Huntington Study Group. *http://www.huntington-study-group.org*.

Isacson O, Brundin P, Gage FH et al. (1985). Neural grafting in a rat model of Huntington's disease: progressive neurochemical changes after neostriatal ibotenate lesions and striatal tissue grafting. *Neuroscience* 16(4): 799–817.

Isacson O, Dunnett SB, and Bjorklund A (1986). Graft-induced behavioral recovery in an animal model of Huntington disease. *Proc Natl Acad Sci U S A* 83(8): 2728–2732.

Jackson GR, Salecker I, Dong X et al. (1998). Polyglutamine-expanded human huntingtin transgenes induce degeneration of *Drosophila* photoreceptor neurons. *Neuron* 21(3): 633–642.

Jakel RJ and Maragos WF (2000). Neuronal cell death in Huntington's disease: a potential role for dopamine. *Trends Neurosci* 23(6): 239–245.

Koller WC and Davenport J (1984). Genetic testing in Huntington's disease. *Ann Neurol* 16(4): 511–513.

Kopyov OV, Jacques S, Lieberman A et al. (1998). Safety of intrastriatal neurotransplantation for Huntington's disease patients. *Exp Neurol* 149(1): 97–108.

Kordower JH, Charles V, Bayer R et al. (1994). Intravenous administration of a transferrin receptor antibody-nerve growth factor conjugate prevents the degeneration of cholinergic striatal neurons in a model of Huntington disease. *Proc Natl Acad Sci U S A* 91(19): 9077–9080.

Lange KW, Sahakian BJ, Quinn NP et al. (1995). Comparison of executive and visuospatial memory function in Huntington's disease and dementia of Alzheimer type matched for degree of dementia. *J Neurol Neurosurg Psychiatry* 58(5): 598–606.

Lapchak PA, Gash DM, Collins F et al. (1997). Pharmacological activities of glial cell line-derived neurotrophic factor (GDNF): preclinical development and application to the treatment of Parkinson's disease. *Exp Neurol* 145(2 Pt 1): 309–321.

Lee ST, Chu K, Park JE et al. (2005). Intravenous administration of human neural stem cells induces functional recovery in Huntington's disease rat model. *Neurosci Res* 52(3): 243–249.

Lescaudron L, Unni D, and Dunbar GL (2003). Autologous adult bone marrow stem cell transplantation in an animal model of Huntington's disease: behavioral and morphological outcomes. *Int J Neurosci* 113(7): 945–956.

Ludolph AC, He F, Spencer PS et al. (1991). 3-Nitropropionic acid-exogenous animal neurotoxin and possible human striatal toxin. *Can J Neurol Sci* 18(4): 492–498.

Lynch G, Kramar EA, Rex CS et al. (2007). Brain-derived neurotrophic factor restores synaptic plasticity in a knock-in mouse model of Huntington's disease. *J Neurosci* 27(16): 4424–4434.

Machida Y, Okada T, Kurosawa M et al. (2006). rAAV-mediated shRNA ameliorated neuropathology in Huntington disease model mouse. *Biochem Biophys Res Commun* 343(1): 190–197.

Mangiarini L, Sathasivam K, Seller M et al. (1996). Exon 1 of the HD gene with an expanded CAG repeat is sufficient to cause a progressive neurological phenotype in transgenic mice. *Cell* 87(3): 493–506.

Marshall J, White K, Weaver M et al. (2007). Specific psychiatric manifestations among preclinical Huntington disease mutation carriers. *Arch Neurol* 64(1): 116–121.

Martinez-Serrano A and Bjorklund A (1996). Protection of the neostriatum against excitotoxic damage by neurotrophin producing, genetically modified neural stem cells. *J Neurosci* 16(15): 4604–4616.

McBride JL, During MJ, Wuu J et al. (2003). Structural and functional neuroprotection in a rat model of Huntington's disease by viral gene transfer of GDNF. *Exp Neurol* 181(2): 213–223.

McBride JL, Behrstock SP, Chen EY et al. (2004). Human neural stem cell transplants improve motor function in a rat model of Huntington's disease. *J Comp Neurol* 475(2): 211–219.

McBride JL, Ramaswamy S, Gasmi M et al. (2006). Viral delivery of glial cell line-derived neurotrophic factor improves behavior and protects striatal neurons in a mouse model of Huntington's disease. *Proc Natl Acad Sci U S A* 103(24): 9345–9350.

Menalled LB, Sison JD, Dragatsis I et al. (2003). Time course of early motor and neuropathological anomalies in a knock-in mouse model of Huntington's disease with 140 CAG repeats. *J Comp Neurol* 465(1): 11–26.

Menei P, Pean JM, Nerriere-Daguin V et al. (2000). Intracerebral implantation of NGF-releasing biodegradable microspheres protects striatum against excitotoxic damage. *Exp Neurol* 161(1): 259–272.

Meriin AB, Mabuchi K, Gabai VL et al. (2001). Intracellular aggregation of polypeptides with expanded polyglutamine domain is stimulated by stress-activated kinase MEKK1. *J Cell Biol* 153(4): 851–864.

Mittoux V, Joseph JM, Conde F et al. (2000). Restoration of cognitive and motor functions by ciliary neurotrophic factor in a primate model of Huntington's disease. *Hum Gene Ther* 11(8): 1177–1187.

Mittoux V, Ouary S, Monville C et al. (2002). Corticostriatopallidal neuroprotection by adenovirus-mediated ciliary neurotrophic factor gene transfer in a rat model of progressive striatal degeneration. *J Neurosci* 22(11): 4478–4486.

Mizuta I, Ohta M, Ohta K et al. (2001). Riluzole stimulates nerve growth factor, brain-derived neurotrophic factor and glial cell line-derived neurotrophic factor synthesis in cultured mouse astrocytes. *Neurosci Lett* 310(2–3): 117–120.

Naimi S, Jeny R, Hantraye P et al. (1996). Ontogeny of human striatal DARPP-32 neurons in fetuses and following xenografting to the adult rat brain. *Exp Neurol* 137(1): 15–25.

Nucifora FC, Jr, Sasaki M, Peters MF et al. (2001). Interference by huntingtin and atrophin-1 with cbp-mediated transcription leading to cellular toxicity. *Science* 291(5512): 2423–2428.

Nutt JG, Burchiel KJ, Comella CL et al. (2003). Randomized, double-blind trial of glial cell line-derived neurotrophic factor (GDNF) in PD. *Neurology* 60(1): 69–73.

Omi K, Hachiya NS, Tokunaga K et al. (2005). siRNA-mediated inhibition of endogenous Huntington disease gene expression induces an aberrant configuration of the ER network in vitro. *Biochem Biophys Res Commun* 338(2): 1229–1235.

Palfi S, Leventhal L, Goetz CG et al. (2000). Delayed onset of progressive dystonia following subacute 3-nitropropionic acid treatment in *Cebus apella* monkeys. *Mov Disord* 15(3): 524–530.

Patel NK and Gill SS (2007). GDNF delivery for Parkinson's disease. *Acta Neurochir Suppl* 97(Pt 2): 135–154.

Penney JB, Jr, Young AB, Shoulson I et al. (1990). Huntington's disease in Venezuela: 7 years of follow-up on symptomatic and asymptomatic individuals. *Mov Disord* 5(2): 93–99.

Perez MK, Paulson HL, Pendse SJ et al. (1998). Recruitment and the role of nuclear localization in polyglutamine-mediated aggregation. *J Cell Biol* 143(6): 1457–1470.

Perez-Navarro E, Arenas E, Marco S et al. (1999). Intrastriatal grafting of a GDNF-producing cell line protects striatonigral neurons from quinolinic acid excitotoxicity in vivo. *Eur J Neurosci* 11(1): 241–249.

Perez-Navarro E, Akerud P, Marco S et al. (2000a). Neurturin protects striatal projection neurons but not interneurons in a rat model of Huntington's disease. *Neuroscience* 98(1): 89–96.

Perez-Navarro E, Canudas AM, Akerund P et al. (2000b). Brain-derived neurotrophic factor, neurotrophin-3, and neurotrophin-4/5 prevent the death of striatal projection neurons in a rodent model of Huntington's disease. *J Neurochem* 75(5): 2190–2199.

Philpott LM, Kopyov OV, Lee AJ et al. (1997). Neuropsychological functioning following fetal striatal transplantation in Huntington's chorea: three case presentations. *Cell Transplant* 6(3): 203–212.

Pineda JR, Canals JM, Bosch M et al. (2005). Brain-derived neurotrophic factor modulates dopaminergic deficits in a transgenic mouse model of Huntington's disease. *J Neurochem* 93(5): 1057–1068.

Pineda JR, Rubio N, Akerud P et al. (2007). Neuroprotection by GDNF-secreting stem cells in a Huntington's disease model: optical neuroimage tracking of brain-grafted cells. *Gene Ther* 14(2): 118–128.

Popovic N, Maingay M, Kirik D et al. (2005). Lentiviral gene delivery of GDNF into the striatum of R6/2 Huntington mice fails to attenuate behavioral and neuropathological changes. *Exp Neurol* 193(1): 65–74.

Ramaswamy S, McBride JL, Zhou L et al. (2006) AAV-Neurturin (Cere-120) therapy in a transgenic mouse model of Huntington's disease (SFN Abstract). *Society for Neuroscience* 472.11., Abstract.

Ramos-Arroyo MA, Moreno S, and Valiente A (2005). Incidence and mutation rates of Huntington's disease in Spain: experience of 9 years of direct genetic testing. *J Neurol Neurosurg Psychiatry* 76(3): 337–342.

Regulier E, Pereira dA, Sommer B et al. (2002). Dose-dependent neuroprotective effect of ciliary neurotrophic factor delivered via tetracycline-regulated lentiviral vectors in the quinolinic acid rat model of Huntington's disease. *Hum Gene Ther* 13(16): 1981–1990.

Reiner A, Albin RL, Anderson KD et al. (1988). Differential loss of striatal projection neurons in Huntington disease. *Proc Natl Acad Sci U S A* 85(15): 5733–5737.

Rosser AE, Barker RA, Harrower T et al. (2002). Unilateral transplantation of human primary fetal tissue in four patients with Huntington's disease: NEST-UK safety report ISRCTN no 36485475. *J Neurol Neurosurg Psychiatry* 73(6): 678–685.

Sanchez I, Xu CJ, Juo P et al. (1999). Caspase-8 is required for cell death induced by expanded polyglutamine repeats. *Neuron* 22(3): 623–633.

Sariola H and Saarma M (2003). Novel functions and signalling pathways for GDNF. *J Cell Sci*, 116(Pt 19): 3855–3862.

Saudou F, Finkbeiner S, Devys D et al. (1998). Huntingtin acts in the nucleus to induce apoptosis but death does not correlate with the formation of intranuclear inclusions. *Cell* 95(1): 55–66.

Saydoff JA, Garcia RA, Browne SE et al. (2006). Oral uridine pro-drug PN401 is neuroprotective in the R6/2 and N171-82Q mouse models of Huntington's disease. *Neurobiol Dis* 24(3): 455–465.

Schilling G, Becher MW, Sharp AH et al. (1999). Intranuclear inclusions and neuritic aggregates in transgenic mice expressing a mutant N-terminal fragment of huntingtin. *Hum Mol Genet* 8(3): 397–407.

Schmidt-Kastner R, Tomac A, Hoffer B et al. (1994). Glial cell-line derived neurotrophic factor (GDNF) mRNA upregulation in striatum and cortical areas after pilocarpine-induced status epilepticus in rats. *Brain Res Mol Brain Res* 26(1–2): 325–330.

Schumacher JM, Short MP, Hyman BT et al. (1991). Intracerebral implantation of nerve growth factor-producing fibroblasts protects striatum against neurotoxic levels of excitatory amino acids. *Neuroscience* 45(3): 561–570.

Selemon LD, Rajkowska G, and Goldman-Rakic PS (2004). Evidence for progression in frontal cortical pathology in late-stage Huntington's disease. *J Comp Neurol* 468(2): 190–204.

Seppi K, Mueller J, Bodner T et al. (2001). Riluzole in Huntington's disease (HD): an open label study with one year follow up. *J Neurol* 248(10): 866–869.

Silber E, Kromberg J, Temlett JA et al. (1998). Huntington's disease confirmed by genetic testing in five African families. *Mov Disord* 13(4): 726–730.

Sofroniew MV, Howe CL, and Mobley WC (2001). Nerve growth factor signaling, neuroprotection, and neural repair. *Annu Rev Neurosci* 24: 1217–1281.

Tabrizi SJ, Cleeter MW, Xuereb J et al. (1999). Biochemical abnormalities and excitotoxicity in Huntington's disease brain. *Ann Neurol* 45(1): 25–32.

The Huntington's Disease Collaborative Research Group (1993). A novel gene containing a trinucleotide repeat that is expanded and unstable on Huntington's disease chromosomes. *Cell* 72(6): 971–983.

von Horsten S, Schmitt I, Nguyen HP et al. (2003). Transgenic rat model of Huntington's disease. *Hum Mol Genet* 12(6): 617–624.

Waelter S, Boeddrich A, Lurz R et al. (2001). Accumulation of mutant huntingtin fragments in aggresome-like inclusion bodies as a result of insufficient protein degradation. *Mol Biol Cell* 12(5): 1393–1407.

Wang YL, Liu W, Wada E et al. (2005). Clinico-pathological rescue of a model mouse of Huntington's disease by siRNA. *Neurosci Res* 53(3): 241–249.

Wheeler VC, White JK, Gutekunst CA et al. (2000). Long glutamine tracts cause nuclear localization of a novel form of huntingtin in medium spiny striatal neurons in HdhQ92 and HdhQ111 knock-in mice. *Hum Mol Genet* 9(4): 503–513.

Young AB, Shoulson I, Penney JB et al. (1986). Huntington's disease in Venezuela: neurologic features and functional decline. *Neurology* 36(2): 244–249.

Zala D, Bensadoun JC, Pereira dA et al. (2004). Long-term lentiviral-mediated expression of ciliary neurotrophic factor in the striatum of Huntington's disease transgenic mice. *Exp Neurol* 185(1): 26–35.

Zuccato C, Ciammola A, Rigamonti D et al. (2001). Loss of huntingtin-mediated BDNF gene transcription in Huntington's disease. *Science* 293(5529): 493–498.

PART

IV

Novel Therapies for Alzheimer's Disease

12

ANTI-AMYLOID-β IMMUNOTHERAPY AS A TREATMENT FOR ALZHEIMER'S DISEASE

JOHN R. CIRRITO[*,†,§,**] AND DAVID M. HOLTZMAN[†,‡,§,**]

[*]Department of Psychiatry,
[†]Neurology,
[‡]Molecular Biology and Pharmacology,
[§]Hope Center for Neurological Disorders,
[**]Alzheimer's Disease Research Center, Washington University School of Medicine, St. Louis, MO 63110

ABSTRACT

Accumulation of the amyloid-β (Aβ) peptide in different aggregated conformations within the brain is central to the pathogenesis of Alzheimer's disease (AD). In brains of AD patients, amyloid plaques are surrounded by swollen, dystrophic neurites. Particular aggregated species of Aβ are also toxic to neurons and are proposed to inhibit synaptic transmission. In order to prevent or reverse Aβ-related pathology, several therapeutic strategies are aimed at reducing brain Aβ levels. For instance, active immunization with Aβ enables the body to augment the clearance or neutralization of Aβ, thus preventing Aβ from having a toxic effect. Both active and passive immunization strategies have been demonstrated to significantly reduce the amount of Aβ plaques and the associated neuritic dystrophy within the brain as well as reverse behavioral deficits in animal models of disease. It appears that several distinct mechanisms likely account for the reduction in Aβ levels. In 2002, a clinical trial in humans using active immunization against the Aβ peptide was halted early due to adverse side effects, including death, in a subset of individuals. Importantly however, there was a hint that there may be cognitive benefits as well as reduced amyloid pathology in response to immunization. Immunization

Copyright © 2008 Elsevier Inc.
All rights of reproduction in any form reserved.

strategies are currently being designed to target-specific aspects of the immune response in order to minimize or prevent adverse side effects while still reducing Aβ levels and ameliorating disease symptoms.

INTRODUCTION

Amyloid plaques and neurofibrillary tangles are hallmark pathologies that characterize Alzheimer's disease (AD). Amyloid plaques consist primarily of a 40–42 amino acid peptide called amyloid-β (Aβ) that is aggregated in fibrils that contain a high β-sheet structure. Plaques become insoluble and deposit within the brain extracellular space. Amyloid plaques are typically associated with swollen, dystrophic neurites, astrogliosis, and activated microglia which together comprise a neuritic plaque. In contrast to amyloid plaques, neurofibrillary tangles accumulate intracellularly within neurons. Aβ is normally produced by neurons within the brain and secreted into the brain extracellular space where during the pathogenesis of AD it changes conformation, becomes insoluble, and deposits as plaques. Thus far, Aβ does not have a known, physiologic function, though a growing literature has demonstrated that under certain experimental conditions, Aβ can modulate synaptic transmission. The role Aβ plays in normal synaptic function or in the setting of the disease state is still unresolved.

Tangles are comprised of an abnormally hyperphosphorylated microtubule-associated protein, tau, that when it aggregates, winds into paired helical filaments both within the nerve cell body and within neurites. While both pathologies are associated with brain injury in AD and in part contribute to signs and symptoms of the disease, evidence strongly suggests that accumulation of Aβ in some form is the initiating and propagating event necessary for AD pathogenesis. Genetic mutations that lead to familial forms of AD affect either Aβ production or Aβ fibrillogenesis (Selkoe, 2001). Down Syndrome, or trisomy 21, is associated with a variety of phenotypes in different organ systems. All individuals with Down Syndrome develop early onset AD pathology with Aβ beginning in the teenage years or earlier and dementia onset usually beginning in the fifth decade (Head and Lott, 2004). Interestingly, the amyloid precursor protein (*APP*) gene is located on chromosome 21 and this results in increased *APP* gene dose and a resulting 50% or more increase in Aβ production in affected individuals. It is of note that there is at least one known case with the phenotype of Down Syndrome that results from partial triplication of chromosome 21 that does not include *APP*. Although this case exhibited many of the typical symptoms and signs of Down Syndrome such as mental retardation, Aβ pathology did not develop (Prasher et al., 1998). In contrast, mutations in tau can cause disease; however, they do not lead to Aβ deposition and AD (Tsuboi, 2006). Mouse models have been generated that exhibit both amyloid plaques and neurofibrillary tangles (Oddo et al., 2003a). Interestingly, the presence of amyloid plaques precedes the formation of neurofibrillary tangle pathology, suggesting that the presence of amyloid plaques may exacerbate tangle formation. Because Aβ plays a key role in disease pathogenesis and progression, many therapeutic strategies are aimed at lowering Aβ levels.

In order to study particular disease processes, many researchers have relied on animal models of AD. The most common models overexpress mutated forms of human APP, which is cleaved to produce Aβ. These mutations, in humans, lead to early onset, familial forms of AD. While there are a wide-variety of APP-based animal models available, one common factor among almost all of them is the age-dependent accumulation of Aβ in plaques within the brain (Dodart et al., 2002a). For instance, in the PDAPP mouse model, one of the first AD mouse models that was generated, there is overexpression of a mutated form of APP that causes a greater percentage of Aβ1-42 to be produced than normal (Games et al., 1995). Aβ1-42 is more prone to aggregate than Aβ peptides of other lengths. By increasing the relative amount of Aβ1-42 that is produced, it is more likely that an initial aggregate or seed will form. Since Aβ fibrillization appears to proceed via a process of seeded polymerization, this likely initiates the process that ultimately leads to AD pathology. PDAPP mice develop Aβ plaques throughout the hippocampus and in certain cortical regions by 12 months of age. The macroscopic structure of these plaques looks remarkably similar to plaques found within the human AD brain. PDAPP mice also exhibit plaque-dependent disruptions in learning and memory (Dodart et al., 1999). APP transgenic mouse models do not recapitulate all aspects of AD, however. For instance, most of these models do not have marked cell death except in the vicinity of plaques and none of the singly transgenic animals for APP contain neurofibrillary tangles, the other hallmark in AD pathology. This may be due to the fact that murine tau and human tau have a different primary sequence. Despite some limitations to these animal models, they are particularly useful for studying specific aspects of the disease such as Aβ aggregation, production, and clearance mechanisms.

During the pathogenesis of AD, Aβ converts from a soluble, random coil conformation into a high β-sheet, insoluble conformation which deposits with the extracellular space of the brain. What precipitates this conversion is unknown; however, in vitro and in vivo experiments strongly suggest that elevated Aβ1-42 levels can lead to early onset of plaque formation. Consequently, in an attempt to delay or prevent AD, therapeutic strategies are being developed to lower Aβ1-42 levels or lower the ratio of Aβ42/40 in order to stave off plaque formation. These tactics include preventing Aβ formation using β- and γ-secretase inhibitors and γ-secretase modulators which interfere with the cleavage of APP to produce Aβ (Xia, 2003). Alternatively, immunization or vaccination strategies are being developed to enhance clearance of Aβ from the brain. The goal of these strategies is to reduce or prevent the pathology and events that directly contribute to disease pathogenesis and progression. This is in contrast to current clinically approved therapies for AD such as cholinesterase inhibitors and an NMDA receptor antagonist, which appear to predominantly target disease symptoms.

ACTIVE IMMUNIZATION

When a microbe enters the body, the immune system normally recognizes it as foreign and develops an immune response to eliminate the foreign agent from the body. This process involves the production of antibodies and activation of immune cells, which can take several days, by which time it is possible that the microbe has already caused an infection or disease. Vaccination, or active immunization, is used to boost the immune system's reaction before a person comes into contact with a particular microbe. A non-infectious agent that resembles the actual microbe is injected into a person who thereby generates an immune response. Later, when the body sees the real microbe, the body immediately recognizes it as foreign and hosts a rapid and strong immune response, thereby preventing the microbe from being harmful. The first successful vaccination was to fight smallpox virus in the late 18th century (Jenner, 1798). People were infected with cowpox virus, which is molecularly similar to smallpox virus but is much less harmful. This primed the body's immune response so that when immunized people later contracted the smallpox virus, the body was able to rapidly develop a strong immune response to rid the body of the virus before it could cause disease. Vaccination against many pathogens are routinely used world-wide to boost the immune response prior to infection.

The same general vaccination strategy was utilized in 1999 to rid APP transgenic mice of Aβ-containing plaques. Schenk and colleagues at Elan Pharmaceuticals actively immunized APP transgenic mice with aggregated forms of synthetic human Aβ42 peptide (Schenk et al., 1999). This primed the immune system to recognize Aβ as foreign, thereby enabling the immune system to clear normal, endogenously produced Aβ. Mice were chronically immunized prior to the onset of AD-like pathology. At 1 year of age, immunized mice contained significantly fewer amyloid plaques compared to age-matched, sham-treated mice. Astrogliosis was also reduced in Aβ42-treated mice. Additionally, immunization of older mice that already contained amyloid plaques appeared to prevent additional plaques from forming and may have reduced the number of pre-existing plaques as well. Immunization was much more effective at reducing AD-like pathology when treatment was begun prior to onset of plaque formation. Mice developed a humoral and cellular immune response following immunization with Aβ meaning that anti-Aβ antibodies were produced and a particular type of T-lymphocyte response also took place. In addition, activated microglia were detected within the brain. Interestingly, the brains of immunized mice contained punctate, Aβ-immunoreactive cells, presumably microglia, whereas saline-treated mice did not contain similar cells. Based on this evidence, the authors suggested that immunization somehow leads to activation of microglial cells which were then able to phagocytose amyloid plaques directly.

Since the publication of this method, numerous groups have observed that active immunization can prevent or reduce plaque formation in APP transgenic mouse models (for review, see Holtzman et al., 2002). Successful approaches have addressed the mode by which the body is immunized, including the traditional cutaneous administration, as well as intranasal (Lemere et al., 2000) and transcutaneous (Nikolic et al., 2007) administrations of the immunogen. The adjuvant used to enhance the immune response has also been screened to tailor the immune response toward specific goals. Freud's adjuvant, which was used in the initial studies of Aβ immunization (Schenk et al., 1999), promotes a cytotoxic T_h1 immune response. In order to drive higher humoral immune response, alum was also used in active immunization studies of APP transgenic mouse models to effectively reduce Aβ plaque load (Asuni et al., 2006). In a screen comparing several adjuvants, Cribbs and colleagues found that the type of adjuvant used was a critical factor in determining how the immune system responded to Aβ immunization (Cribbs et al., 2003). Depending on the goals of the immunization, the adjuvant can be chosen to elicit a more T_h1 cytotoxic response; as is the case with Freud's and QS21 adjuvants, or a more T_h2 humoral response, as is the case with alum.

Vaccination immunogens that have been tested include various conformations of Aβ ranging from very large β-sheet aggregates (Schenk et al., 1999) to small molecular weight, oligomeric forms of Aβ (Zhou et al., 2005). Peptide fragments of Aβ have also been used to tailor the immune response. For instance, APP transgenic mice that were immunized with a multimeric Aβ1-15 peptide caused a humoral immune response without activating a cellular response (Maier et al., 2006; Seabrook et al., 2007). This immunogen was specifically chosen to generate a B-cell mediated immune response. The T-cell epitope appears to exist between amino acids 15–42 of Aβ (Cribbs et al., 2003; Monsonego et al., 2003), whereas the B-cell epitope exists between amino acids 1–15 (McLaurin et al., 2002; Cribbs et al., 2003). Altered structures of the immunogen itself, such as 16 Aβ1-15 peptides radiating from a branched lysine core, have also designed specifically to enhance the immune reaction and withstand degradation. In total, active immunization using a variety of methodologies has been shown to lower Aβ levels within the body, as well as reduce amyloid plaques and associated pathologies in mouse models of disease.

PASSIVE IMMUNIZATION

Passive immunization involves the direct administration of antibodies, thereby bypassing the need for the body to react to a foreign microbe. Administered antibodies bind to their target, thus neutralizing the microbe before it can do harm. This bypasses the need for the body to generate any active processes. This approach is particularly useful in cases when (1) an active immune response must be avoided, (2) a very specific and controllable immune response is required, or (3) an individual cannot host a strong enough immune reaction. Additionally,

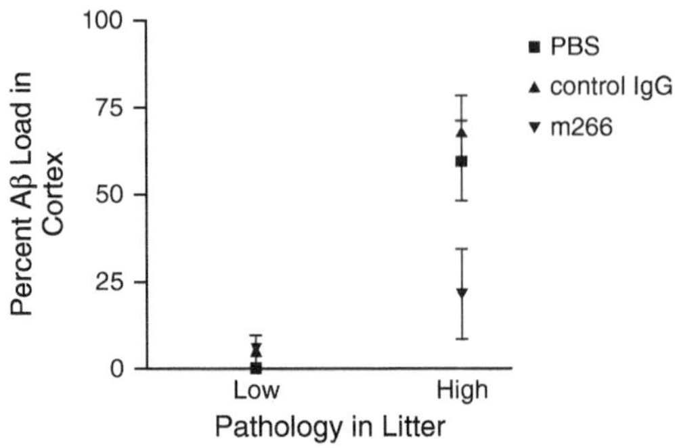

FIGURE 12.1 Chronic passive immunization with an anti-Aβ antibody reduces Aβ deposition in a mouse model of AD. There was a significant effect following chronic immunization with the mouse-anti-Aβ antibody, m266, in reducing Aβ burden in PDAPP mice when using the parental line as a covariant. When comparing m266 to controls in all groups (high- and low- pathology litters), $p = 0.0082$. When comparing m266 to controls in only the high-pathology litters, $p = 0.00025$. Reprinted with permission from DeMattos et al. (2001). *Proc Natl Acad Sci USA* 98: 8850–8855.

passive immunization can be stopped at any time whereas active immunization lasts much longer. Soon after the first active Aβ immunization study was published, Bard and colleagues and DeMattos and colleagues demonstrated that chronic, intraperitoneal administration of certain anti-Aβ monoclonal antibodies also reduced plaques within the brains of APP transgenic mice (Figure 12.1) (Bard et al., 2000; DeMattos et al., 2001). In the DeMattos study, starting at an age prior to plaque formation, mice were treated bi-weekly for 2 months with an antibody, called m266, that recognizes the central domain of Aβ. Littermates were treated with saline or a control IgG. At 9 months of age, mice that received the anti-Aβ antibody had significantly reduced plaque burden within the hippocampus and cortex compared to both controls. Immunized mice also had elevated levels of Aβ within the plasma and CSF; however, all of the plasma Aβ was bound to the anti-Aβ antibody. Because m266 does not bind fibrillar forms of Aβ, the authors suggested that one of the mechanisms of action of this particular antibody was to sequester Aβ within the periphery, thereby preventing it from forming Aβ plaques within the brain. Similar to active immunization, passive immunization was more effective at reducing plaque burden when begun prior to the onset of Aβ deposition (Bard et al., 2003; Wilcock et al., 2004b; Hartman et al., 2005).

A wide-variety of anti-Aβ antibodies have been utilized in passive immunization of APP transgenic mice to reduce plaque burden. One advantage of passive immunization is the ability to choose an antibody with very specific properties. For instance, the passive immunization trial by DeMattos and colleagues utilized an antibody that bound soluble Aβ very well but had very low affinity for fibrillar, deposited forms of the peptide (DeMattos et al., 2001). Antibodies specific to various regions of Aβ, including to the N-terminus of the peptide and Aβ40 and Aβ42, have also successfully reduced plaque burden (Wilcock et al., 2004b; Asami-Odaka et al., 2005; Levites et al., 2006b). Monoclonal antibody F(ab)2 fragments, which are able to bind Aβ, but are unable to activate microglia, also reduce plaques in APP transgenic mice (Tamura et al., 2005). Similar to peripheral administration, intracranial administration of antibodies is also capable of clearing amyloid plaques. When anti-Aβ antibodies were applied directly to the cortical surface, there was a local removal of Aβ plaques that extended approximately 100–200 μm deep within 3 days (Bacskai et al., 2001). This response was accompanied by marked microglial activation. Following intracranial administration, the microglial response appears to aid in plaque removal; however, because F(ab)2 fragments can also substantially reduce plaque burden, there must be other mechanisms that also play a role (Bacskai et al., 2002).

Following active immunization the body continually produces an immune response that includes anti-Aβ antibody production to maintain low Aβ levels. In passive immunization, however, anti-Aβ antibodies must be administered on a regular basis (approximately every few weeks) in order to maintain the ability to reduce Aβ levels. Several groups have utilized gene therapy to express single-chain variable fragments (scFvs) of anti-Aβ antibodies within the brain. ScFvs encode a portion of the antibody gene that binds to the target, but lacks the coding regions for the constant portions of the antibody. This approach is similar to expressing F(ab)2 fragments genetically. Using this methodology, the body can continually produce specific antibodies without the need for repetitive administrations or boosters. By packing scFvs into adeno-associated viruses (AAV), researchers were able to express an anti-Aβ scFv within neurons of APP transgenic mice (Fukuchi et al., 2006). One year after infection, an anti-Aβ scFv was still detectable within the brain. Neuronal cell bodies and the extracellular space stained positive for the scFv, suggesting that the antibody fragment was produced by neurons and then secreted. ScFv-treated mice had significantly reduced Aβ burden within the hippocampus compared to saline-treated mice. Levites and colleagues also demonstrated that anti-Aβ scFvs can reduce plaque burden (Levites et al., 2006a). CRND8 APP transgenic mice were injected on postnatal day zero with an AAV-scFv that recognized all species of Aβ (pan-Aβ) or one that was specific for Aβ40 or Aβ42. All three of these scFvs displayed reduced Aβ plaque burden as well as reduced total brain Aβ levels as assessed biochemically. Expression of anti-Aβ scFvs via a gene delivery approach provides an opportunity to passively immunize with specific antibodies without the need to continually administer antibodies.

Interestingly, while some antibodies are more effective at lowering plaque burden than others, the consensus is that anti-Aβ antibodies with a broad array of properties and that target many different types of Aβ can ameliorate plaque burden in mouse models of disease. While experiments such as those listed above have been performed in animal models, the lessons learned provide a framework for future immunization trials in humans, as well as provide important insights into mechanisms governing Aβ immunization.

MECHANISMS OF PLAQUE REDUCTION FOLLOWING IMMUNIZATION

While it is clear that immunization against Aβ can effectively reduce plaque burden or neutralize the effects of toxic forms of Aβ, there appear to be three mechanisms that are likely to be involved in such a response: (1) cell-mediated phagocytosis of Aβ; (2) peripheral sequestration of Aβ; and (3) direct binding and dissolution of plaques (Figure 12.2). One mechanism of plaque removal following active or passive immunization that is likely to occur under certain conditions is that microglia, and possibly macrophages, phagocytose Aβ either in a soluble form or fibrillar form. In either active or passive immunization, antibodies that bind to Aβ within plaques can mediate the cellular uptake and clearance of Aβ. This mechanism was highlighted in the initial publication of active immunization and was supported by the fact that immunized mice appeared to have Aβ-immunoreactive cells, whereas control-treated mice did not

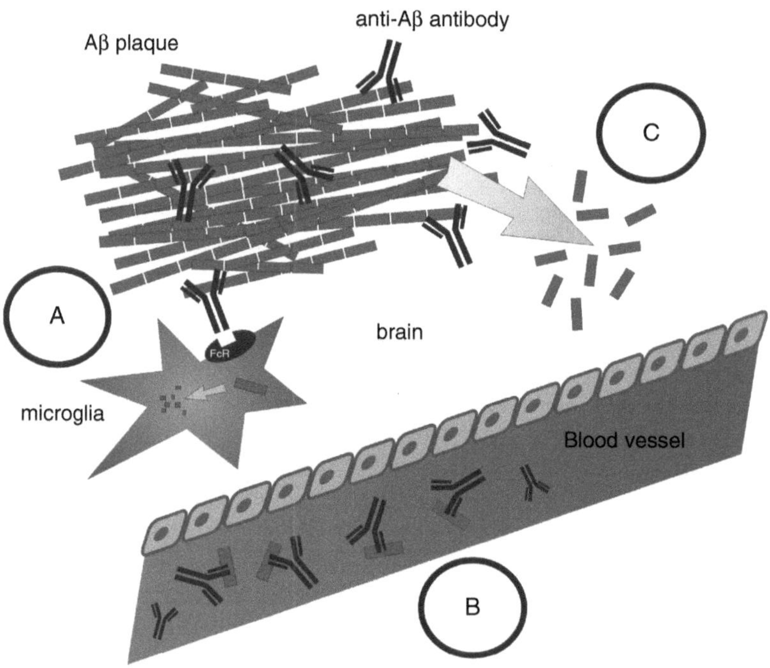

FIGURE 12.2 Mechanisms of immunization-mediated clearance of Aβ. (A) Opsonization mechanism. Anti-Aβ antibodies cross the blood–brain barrier and bind to Aβ with amyloid plaques. Microglia recognize the antibody via the Fc receptor and phagocytose the amyloid plaque. (B) Peripheral sink mechanism. Antibodies within the blood bind to Aβ, thus preventing it crossing into the brain or increasing the net flux of Aβ from brain into blood. (C) Direct dissolution mechanism. Antibodies bind to amyloid plaque within the brain, thereby changing the conformation of the Aβ aggregate and facilitating Aβ re-entry into the soluble phase. (see Plate 5)

(Schenk et al., 1999). Other groups have also demonstrated increased numbers of activated microglia surrounding plaques of immunized mice using both active and passive immunization strategies (Bard et al., 2000; Wilcock et al., 2004b).

Opsonization, or cell-mediated clearance, of Aβ plaques cannot be the only mechanism of Aβ clearance, however. Microglia-mediated phagocytosis in response to immunization would require that anti-Aβ antibodies be bound to the plaque; certain anti-Aβ antibodies can bind to Aβ in plaques and mediate microglial movement and clearance of plaques. The variable coding regions of the antibody binds to the target while the constant regions of the antibody bind to Fc receptors on the microglia to induce phagocytosis. In situations where the Fc-mediated response is reduced or prevented, there can still be effective reductions in Aβ pathology. F(ab)2 fragments retain the ability to bind to their target, but lack the Fc portion of the antibody, thus preventing the antibody from binding to Fc receptors on microglia. APP transgenic mice that were passively immunized intraperitoneally with monoclonal F(ab)2 fragments still had a robust reduction in plaque burden despite the inability to activate microglia (Tamura et al., 2005). Direct administration of anti-Aβ F(ab)2 fragments into the brain also lowered plaque load (Bacskai et al., 2002). Similarly, Das and colleagues demonstrated that APP transgenic mice bred to mice lacking the Fc receptor still had a significant reduction in plaque burden following active immunization (Das et al., 2003). Consequently, while microglial-mediated phagocytosis of Aβ is capable of reducing peptide levels, it is unlikely to be the only mechanism at work in the setting of Aβ immunization.

A second mechanism mediating plaque reduction is the sequestration of Aβ outside the brain. DeMattos and colleagues cited that following passive immunization with the monoclonal antibody, m266, there was not a cellular reaction in the brain that appeared responsible for the decreased plaque load (DeMattos et al., 2001). First, in their paradigm, antibody penetrance across the blood–brain barrier was likely too low to substantially bind to plaques. Second, m266 has a high affinity for soluble Aβ, but does not bind to deposited, insoluble Aβ making it unlikely that the antibody could mediate a microglial phagocytic response even if it did enter the brain. Following immunization with m266, plasma Aβ levels increased 1000-fold over baseline levels. These findings led to the "peripheral sink hypothesis"; that anti-Aβ antibodies sequester Aβ within the blood, thus lowering brain Aβ levels and preventing plaque formation. Consistent with this hypothesis, non-antibody agents such as gelsolin and GM1, both of which are Aβ-binding proteins, are also able to reduce plaque burden in an APP mouse model (Matsuoka et al., 2003). Neither of these proteins will initiate an immune response nor will they cross the blood–brain barrier in any substantial amount, suggesting that binding Aβ within the periphery alone is sufficient to reduce brain Aβ deposition.

The third mechanisms by which immunization could modulate plaque burden is via direct dissolution of plaques caused by antibody binding. Anti-Aβ antibodies bound to Aβ can alter the conformation of Aβ, thus breaking apart existing plaques, preventing new plaques from forming, or preventing the local toxic effects of Aβ. In vitro experiments with synthetic Aβ fibrils demonstrate that anti-Aβ antibodies can dissociate fibrils in a test tube (Solomon et al., 1996). Antibodies with epitopes near the N-terminus (amino acids 3–6 of Aβ) were especially effective at inhibiting fibril formation, though experiments with central domain antibodies also can markedly inhibit fibril formation (Legleiter et al., 2004).

The presence of any one of these mechanisms does not preclude the possibility that other mechanisms are also at play. In fact, it is likely that several mechanisms may be at work simultaneously. David Morgan and colleagues have taken an approach of injecting anti-Aβ antibodies directly into the brain (Wilcock et al., 2003). In this paradigm, over 24 h, diffuse plaques are cleared in a microglia-independent manner, which is consistent with the dissolution hypothesis. Compact plaques were also cleared by antibody injections; however, these plaques took approximately 2–3 days to be removed. Slower clearance of compact plaques required microglial activation and was inhibited if F(ab)2 fragments were injected or if the immune response was inhibited with anti-inflammatory agents such as dexamethasone. Systemic application of monoclonal antibodies can (1) cause a rapid elevation of plasma Aβ levels consistent with the peripheral sink hypothesis and (2) simultaneously decorate plaques with the same antibodies (Wilcock et al., 2004b). The antibodies bound to plaques increase the number of activated microglia. The three mechanisms employed for Aβ clearance following immunization are not mutually exclusive, but instead likely work synergistically to accomplish the task of reducing plaque burden. A wealth of data has been generated in animal models to elucidate how the immune response can be tailored to create a desired reaction to Aβ.

PATHOLOGIC CONSEQUENCES OF Aβ IMMUNIZATION

Prevention of Aβ plaque formation and possible removal of existing plaques are likely direct effects of immunization on AD-like pathologies in APP transgenic mice. Interestingly, Aβ immunization can also improve other pathologies such as dystrophic neurites and neurofibrillary tangles. Given that Aβ can be associated with these structures, but is not directly involved in their framework, the amelioration of these other pathologies is likely indirect.

Recent advances using in vivo imaging of the brain have provided a unique perspective of the dynamic nature of amyloid plaques and the surrounding tissue. These techniques utilize cranial windows placed on the skull of living mice. Anesthetized mice can then be placed on multi-photon microscope to image fluorescent molecules in the living brain. A variety of tools are available to assist in these endeavors. Fluorescent dyes, such as methoxy-X04, bind to amyloid plaques in vivo and provide a measure of plaque size in the living brain (Brendza et al., 2005). Administration of antibodies under these cranial windows

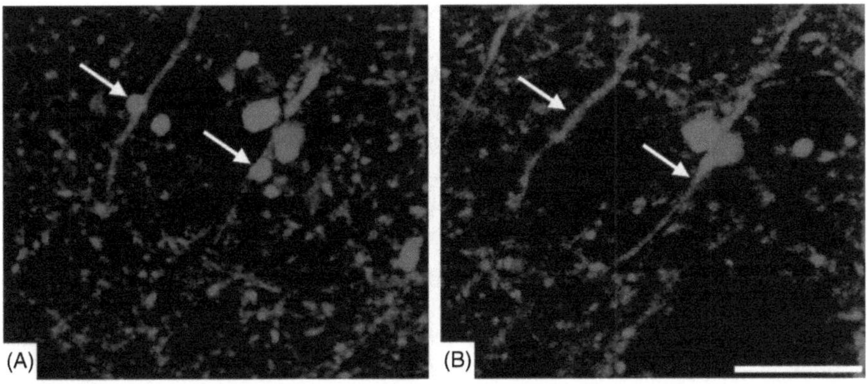

FIGURE 12.3 Morphological recovery of amyloid-associated neuritic dystrophy in a *PDAPP;Thy1:YFP* double-transgenic mouse after anti-Aβ antibody (10D5) treatment. PDAPP transgenic mice overexpressing human APP were bred to mice expressing YFP within a subset of their cortical neurons and processes. The two panels show the same population of YFP-labeled neurites, which are associated with a neuritic plaque imaged on the initial day of surgery (A) and 72 h later (B). The antibody was administered directly to the surface of the brain during the cranial-window surgical procedure on day 0. The arrows in the day 0 image (A) point to 2 enlarged dystrophic neurites, which are absent in the day 3 image (B). The associated amyloid is not shown. Scale bar: 20 μm. Reprinted with permission from Brendza et al. (2005). *J Clin Invest* 115: 428–433. (see Plate 6)

reduced local plaque burden, as well as rapidly reversed abnormal geometries of dystrophic neurites over the course of days or weeks (Lombardo et al., 2003). Brendza and colleagues bred APP-transgenic mice to mice that express yellow fluorescent protein (YFP) within a subset of their cortical neurons (Brendza et al., 2005). The YFP entirely fills the cell body and the processes of these neurons, making them visible during in vivo imaging. Brendza found that when anti-Aβ antibodies were applied directly to the brain, a significant number of dystrophic neurites recovered to morphologically normal appearances over 3 days (Figure 12.3). While the functional status of the recovered neurites remains unknown, the fact that some of the neuronal processes reverted to a normal appearance is promising that immunization may improve aspects of AD-like pathology that were perhaps thought to be irreversible. The reversible damage to neuronal processes suggests that the damaged neurites either retain the ability to repair themselves or that Aβ exerts continual damage to surrounding neurites, which improves when the deleterious species of Aβ is removed. Buttini and colleagues found that chronic active and passive immunization prevented loss of synaptic structures in APP transgenic mice (Buttini et al., 2005). Using in vivo imaging, Prada and colleagues demonstrated that passive immunization with anti-Aβ antibodies directly to the brain surface can also clear vascular amyloid deposits over the span of several weeks (Prada et al., 2007).

In AD it appears that Aβ in some form, as either plaques or oligomers, contributes to disease pathogenesis (Selkoe, 2001). Triple transgenic mice (3xTg) that express mutated forms of APP, presenilin-1, and tau develop both amyloid and tau pathologies, suggest that amyloid pathology precedes the formation of tau pathology (Oddo et al., 2003a). In addition to the two pathological hallmarks of AD, these mice also display deficits in synaptic plasticity (Oddo et al., 2003b) as well as in learning and memory (Billings et al., 2005). In order to determine if reductions in amyloid load could modulate tau pathology, an anti-Aβ antibody was administered directly to the brains of these mice (Oddo et al., 2004). As expected, passive immunization reduced the number of plaques within the cortex of the 3xTg mice. Importantly, tau pathology was also decreased in these mice following Aβ immunization. This provided strong evidence that Aβ directly contributed to tau pathology and that reductions in Aβ levels could reverse accumulations in tau. The same group also demonstrated that active immunization and passive, systemic immunization had similar affects in reducing both Aβ and tau pathologies (Oddo et al., 2006).

Together, these studies suggest that Aβ immunization does not have the limited effect of just reducing plaque burden, but may potentially reverse other pathologies associated to AD as well.

BEHAVIORAL CONSEQUENCES OF Aβ IMMUNIZATION

In addition to reducing plaque load, Aβ immunization has functional consequences as well. Immunization has been shown to improve learning and memory in mouse models of AD, as well as prevent or even reverse other pathologies besides amyloid plaques. In some cases this appears to be directly related to reductions in Aβ plaques; however, in other cases there appear to be consequences of immunization that are related to Aβ but are possibly independent of plaques.

To determine if immunization could reduce plaque load and provide improvement in learning and memory, APP and presenilin-1 double transgenic mice were actively immunized with Aβ at 7 months of age when plaque deposition is already present but prior to the onset of disruptions in memory (Morgan et al., 2000). At 11 months of age, mice were tested on the radial arm water maze as a measure of spatial working memory. Mice that received a control immunization displayed typical deficits in spatial memory, whereas mice immunized with Aβ could ultimately perform as well as non-transgenic mice. This suggested that immunization not only reduced plaque burden, but could also significantly improve behavioral performance as well. Analogous findings were published at the same time using similar tests in a different mouse model of AD (Janus et al., 2000). In addition, Hartman and colleagues found that chronic, passive immunization reduced plaque burden in aged PDAPP transgenic mice and improved performance on the Morris Water Maze test of spatial memory (Figure 12.4) (Hartman et al., 2005). In this paradigm, immunized mice also had improvements in synaptic plasticity. Acute hippocampal brain slices generated from immunized mice had marked improvement in long-term potentiation compared to control,

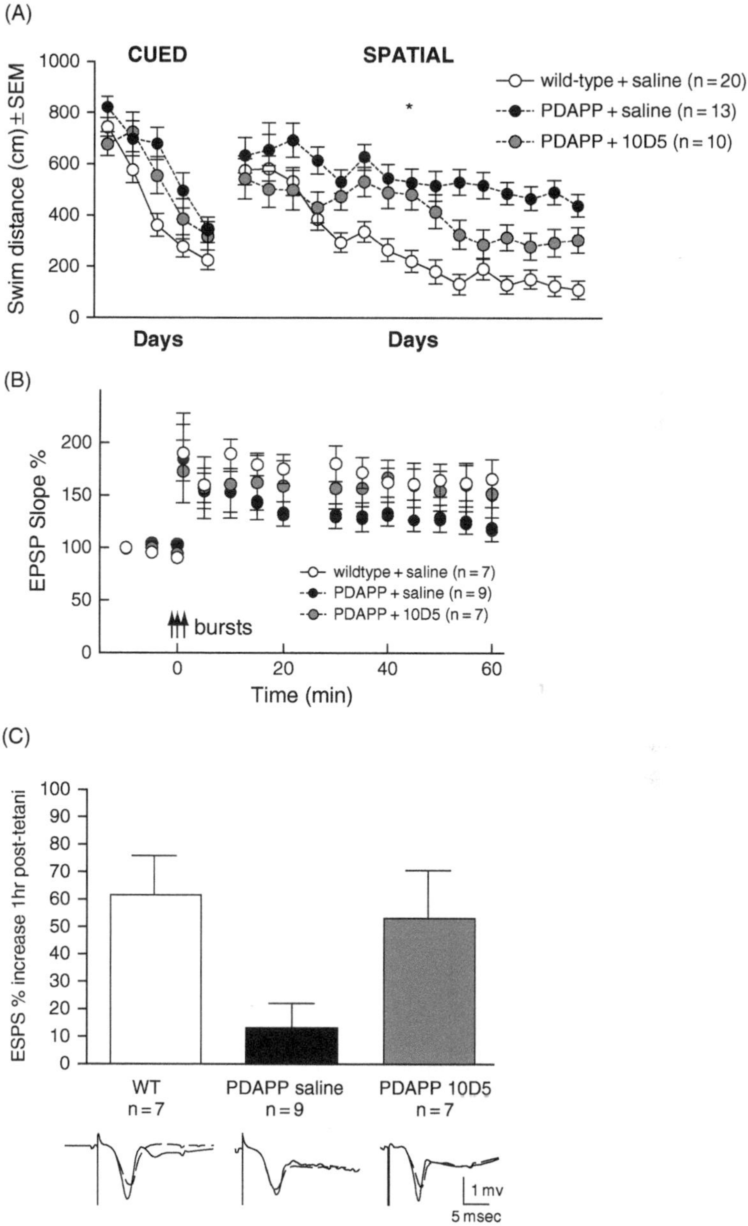

FIGURE 12.4 Chronic passive immunization improves learning and memory and long-term potentiation. (A) Treatment with anti-Aβ antibody, m10D5, improved the spatial learning performance of aged PDAPP mice. Each point represents the average of four daily trials. Significant main effects of age are indicated by* Treatment with m10D5 improved hippocampal LTP in aged PDAPP mice. (B) Six series of tetanic bursts consisting of six trains of six pulses were delivered at time 0 (arrows). At 60 min after tetani, PDAPP plus saline mice exhibited impaired LTP compared with PDAPP plus 10D5 and WT mice. (C) The bars represent levels of LTP 60 min after the induction of tetani. Traces depict EPSPs before (dotted traces) and 60 min after (solid traces) tetanic bursts. Error bars represent SEM. Reprinted with permission from Hartman et al. (2005) *J Neurosci* 25: 6213–6220.

saline-treated PDAPP mice. This finding is consistent with results by Butini and colleagues that demonstrated that Aβ immunization prevents changes in synaptic structures (Buttini et al., 2005). This conclusion suggested that in certain instances immunization could reverse disruptions in learning and memory.

While chronic, active immunization can ameliorate plaque burden and improve behavior, there are several reports of acute effects of immunization on behavior that appear independent of plaque burden. Dodart and colleagues gave passive immunization with the monoclonal anti-Aβ antibody m266 to 22-month-old PDAPP mice that already contained Aβ plaques (Dodart et al., 2002b). Mice were administered the antibody intraperitoneally once per week for 6 weeks. This significantly improved their performance on the object recognition task compared to control-treated mice. In fact, 22-month-old APP transgenic mice that were immunized performed nearly identical to 8-month-old wild-type mice. Interestingly, however, there was no significant change in plaque burden in immunized mice; despite improvements in memory there was still ample plaque burden in the immunized mice. This suggested that plaques may not be the only species of Aβ responsible for disruptions in learning and memory. Dodart went on to demonstrate that a single administration of an anti-Aβ antibody 24 h prior to a memory task could also significantly improve performance. Again, over this short time period it is unlikely that the antibody would have any substantial effect on Aβ plaque load. This result strongly suggested that while immunization could improve learning and memory in APP transgenic mice, the effect may be due to neutralization of a pool of Aβ independent of plaques, such as a pool of soluble Aβ. Kotilinek and colleagues found that passive immunization of Tg2576 APP transgenic mice with an N-terminal anti-Aβ antibody provided similar rapid improvements in performance in the Morris Water maze test (Kotilinek et al., 2002). These effects also appeared to be independent of any changes in plaque load. Together, these findings challenged the overwhelming hypothesis that Aβ within plaques was the sole mediator of cognitive impairment in APP transgenic mice and possibly humans. These results implied the possibility of a soluble pool of Aβ, perhaps in equilibrium with plaques, that could modulate behavior and could be affected over the range of hours to days.

It has been demonstrated that soluble, oligomeric species of Aβ can be toxic to neurons, as well as modulate synaptic transmission (Walsh et al., 1999). Synthetic oligomeric species inhibit synaptic transmission in dissociated neuronal cultures (Hartley et al., 1999) and inhibit LTP in acute brain slices (Wang et al., 2002). Cell-derived, oligomeric species appear even more potent at inhibiting synaptic plasticity in acute brain slices (Townsend et al., 2006) as well as in vivo (Walsh et al., 2002). Given their soluble nature, it seemed plausible that rapid effects of immunization could be mediated by neutralization of Aβ oligomers. Lee and colleagues passively immunized APP transgenic mice with a monoclonal antibody that is specific for oligomeric forms of Aβ, NAB61 (Lee et al., 2006). This antibody does not recognize monomeric forms of Aβ and only binds to a subset of amyloid plaques in brain tissue. Tg2576 APP transgenic mice of

17- to 19-months old were immunized four times over a 2-week period. Mice at this age already contain a substantial number of Aβ plaques. The immunized mice had no measurable change in Aβ levels within the brain; however, they had significant improvement in spatial memory as tested by the Morris Water Maze. This discovery strongly suggested that neutralization of Aβ oligomers was responsible for at least a portion of the improvements in learning and memory following immunization.

Both active and passive Aβ immunizations reduce Aβ plaque burden, ameliorate other types of AD-like pathology, and improve learning and memory. Immunization studies in APP transgenic animal models were the first practical tests of the Aβ hypothesis, that Aβ was a major causative factor ultimately leading to cognitive deficits in AD. By specifically reducing Aβ, many of the downstream aspects of disease were prevented or reversed, namely the dystrophic neurites and cognitive deficits.

NEGATIVE CONSEQUENCES OF IMMUNIZATION

While Aβ immunization reduces AD-like pathologies via several mechanisms, there are other consequences of immunization that may not be as beneficial. Active immunization can induce both humoral and cellular immune responses depending on the immunization protocol, while passive immunization induces events downstream of antibody binding to a target. While some of these effects are necessary to reduce Aβ pathology, not all are required and evidence suggests some of these processes can actually be harmful.

Aβ accumulates in two different locations in the CNS: in parenchymal Aβ plaques in the setting of AD and in meningeal and cerebral blood vessels in form of cerebral amyloid angiopathy (CAA) that occurs both in AD and sometimes independently of AD (Attems, 2005). CAA consists primarily of fibrillar conformations of Aβ40 deposited in the wall of arterioles in the brain. These Aβ structures weaken the blood vessel wall, thereby leading to poor function and increased risk of hemorrhage and infarcts. While AD and CAA are considered separate diseases, 30–90% of patients with AD also contain CAA to differing degrees (Jellinger and Attems, 2005). Additionally, because Aβ is the key component of the pathology in both AD and CAA, it is important to consider how treatment of one pathology may affect the other. While most APP mouse models develop Aβ plaques, not all models develop CAA. Therefore, when assessing how immunization affects CAA it is important to consider if the model that is immunized also accumulates Aβ surrounding blood vessels.

Pfeifer and colleagues demonstrated that passive immunization with an N-terminal-specific, anti-Aβ antibody caused a twofold increase in incidence of microvascular hemorrhage in an APP model that develops substantial CAA (Pfeifer et al., 2002). Similarly, Wilcock and colleagues also found that passively immunized mice had greater incidence of microhemorrhage in a different

APP transgenic mouse model (Wilcock et al., 2004a). Despite the hemorrhages, however, immunized mice had reductions in Aβ plaques and improvements in learning and memory. Racke and colleagues demonstrated that the incidence and severity of microhemorrhage depended on the type of antibody used for immunization (Racke et al., 2005). N-terminal anti-Aβ antibodies with a high affinity to fibrillar forms of Aβ increased microhemorrhage, whereas an anti-Aβ antibody that did not bind fibrillar Aβ had no effect on hemorrhage. Essentially, passive immunization led to deleterious events only if the injected antibodies could bind to the fibrillar Aβ associated with the CAA in blood vessels.

Passive immunization may exacerbate cerebral microhemorrhage via two mechanisms: (1) destabilization of the vessel wall structure by immunization-dependent removal of amyloid or (2) activating an antibody-dependent immune response that leads to compromised and weak blood vessels. Burbach and colleagues took an ultrastructural approach to assess the effect of passive immunization on cerebral microhemorrhage (Burbach et al., 2007). Beginning at 20 months of age, APP23 transgenic mice were passively immunized for 5 weeks with a N-terminal anti-Aβ antibody. At this age, these mice contained ample Aβ plaques as well as substantial CAA. At the end of the immunization, animals were sacrificed and their brains were processed for ultrastructural analysis using electron microscopy. The authors were particularly interested in the difference between bleeding vessels in control-treated mice and bleeding vessels in immunized mice in order to obtain information as to why and how immunization could exacerbate hemorrhage. However, the tissue surrounding microhemorrhages in immunized and control-treated mice looked very similar. There was no sign of structural destabilization, nor was there sign of a differential immune response in immunized mice. The number of macrophages near the site of hemorrhage was the same in immunized and control mice. In fact, the only difference between bleeding and non-bleeding vessels that contained CAA was the number of macrophages. While this study could not discount the possibility of subtle structural changes induced by immunization, the overall conclusion was that passive immunization does not cause an overt change in vessel structure or immune response that would contribute to greater hemorrhage. Therefore, the root cause of the increased number and severity of hemorrhages following immunization is still unknown. Given that certain types of immunization seem to be more deleterious than others, it is reasonable, and appears plausible, to design an immunization strategy that favors amyloid plaque reduction without adversely affecting CAA.

Aβ IMMUNIZATION TRIALS IN HUMANS

Given that Aβ immunization remarkably prevents, and in some cases reverses, Aβ pathology in mouse models, it is a promising, potential therapy to treat and possibly cure AD in humans. In 2000, the companies Elan and Wyeth initiated a phase I double-blind, safety and immunogenicity clinical trial in the United Kingdom to test the safety of active immunization with fibrillar, synthetic Aβ42, AN1792. Eighty subjects were enrolled at four sites with 64 subjects receiving up to four injections over 24 weeks and 16 subjects receiving placebo. Fifty-one of these subjects enrolled in the extension phase of the study and received up to four more doses over an additional 48 weeks. More than one-quarter of subjects experienced adverse side effects possibly linked with the treatment, including four serious adverse effects that were non-fatal. Five deaths occurred during the study; however, none were thought to be treatment-related at the time. In one of the fatalities, meningoencephalitis was ultimately found at autopsy (Nicoll et al., 2003). Almost half of the study participants had a positive anti-AN1792 antibody titer at some point during the 72-week study (Hock et al., 2002, 2003). The antibody response was correlated with higher dose and longer duration of AN1792. Though the study was only designed to test safety, one of the four measures of efficacy suggested a small improvement in cognition following immunization. The phase I clinical trial began in April 2000 and terminated in June 2002.

Given the initial safety results and preclinical data, a phase II clinical trial was initiated in October 2001. Of the 372 patients enrolled in the study, 298 patients received active immunization with AN1792. In January 2002, the study was suspended after 6% of patients developed meningoencephalitis (Orgogozo et al., 2003). Eventually, 18 patients developed encephalitis, of which 12 patients recovered and 6 patients were left with residual deficits (Schenk et al., 2004). Using a small sample size, Hock and colleagues originally reported that the patients that mounted an immune response to AN1792 showed slower rate of cognitive decline (Hock et al., 2003). The significant correlation between immune response and cognitive improvement was lost, however, when the sample size was increased (Gilman et al., 2005). In the later study, there was some statistically significant improvement on a neuropsychiatric test battery, particularly on certain memory tasks, that was linked with the degree of antibody response. Thus, in mild-to-moderate AD, there were small improvements in cognition in subjects that mounted an antibody response following active immunization with AN1792.

Pathologic analysis of several immunized patients has provided important insights into the effectiveness of the immunization clinical trial. One year after the trial was halted, three subjects had come to autopsy and their brains assessed for pathologic evidence of Aβ pathology. In all three individuals, there were signs of Aβ clearance in some brain regions (Nicoll et al., 2003; Ferrer et al., 2004; Masliah et al., 2005). Two of these patients experienced encephalitis during the clinical trial and showed signs of T-cell infiltration into the brain (Nicoll et al., 2003; Ferrer et al., 2004). It is likely that the encephalitis was caused by the T-cell response because there was no correlation between antibody titers and development of encephalitis (Ferrer et al., 2004). It has been speculated that the encephalitis was at least partially caused by the presence of CAA in some subjects (Gandy and Walker, 2004). The third subject that has come to autopsy displayed no neuritic or amyloid plaques within the frontal lobe, though there were signs of Aβ immunoreactivity within macrophages (Masliah et al., 2005). This suggests that there was amyloid present at one time, which was subsequently cleared from the brain. There was no apparent reduction in CAA or neurofibrillary tangles, however. In mouse models, it was demonstrated that Aβ immunization only reduced tau pathology if mice were immunized early in the progression of pathology, whereas there was no reduction in tau pathology if immunized late in the progression (Oddo et al., 2004). This suggests that it may still be possible to clear amyloid plaques and neurofibrillary tangles if subjects are immunized early in disease progression.

While the first Aβ active immunization clinical trial was halted early due to fatal side effects in some individuals, the information derived from the trial has provided important insights into the immunization strategy and has provided hope that, if designed correctly, a subsequent trial may be capable of reversing Aβ pathology as well as cognitive deficits associated with AD. Immunization strategies with animal models suggest that an active immunization protocol could be developed that would initiate a humoral immune response but lack a cell-mediated response which was likely responsible for the encephalitis. In April 2005, a passive immunization phase II clinical trial was initiated in the United States by Elan Pharmaceuticals and Wyeth Pharmaceuticals using an anti-Aβ antibody, AAB-001, to treat mild and moderate AD patients (ClinicalTrials.gov, 2005). At the time of this publication, no results from that trial have been published.

There are several advantages to a passive immunization strategy compared to an active immunization strategy. First, the dose of antibody can be tightly controlled. In the active immunization clinical trial, there was a wide range of immune responses as measured by the antibody titers following immunization. In fact, almost 50% of subjects did not mount an immune response at all. Since the antibody is administered directly in passive immunization, the ability or inability of an individual to mount an immune response is not a factor and treatment can be better controlled. Second, passive immunization can target a specific type of Aβ, given that the antibody can be prescreened for certain properties. Third and most importantly, if individuals begin showing serious side effects following passive immunization, the therapy can be stopped prior to serious or fatal consequences, which is not necessarily feasible with active immunization. A major disadvantage of passive immunization, however, is the potential cost of therapy since individuals may need to be treated for decades with an anti-Aβ antibody. Regardless, the cost of passive immunization therapy would still be substantially cheaper than institutionalization and could vastly improve the quality of life of individuals.

FUTURE GOALS

Though the initial Aβ immunization clinical trial proved to have fatal consequences in a limited number of individuals, the general strategy appears to have effectively treated some aspects of disease; Aβ plaques were reduced in the several individuals that have come to autopsy and at least some measures of cognition were improved. This provides a tremendous amount of hope that AD might be treatable with an immunization strategy if the methods are improved to increase safety. By combining data from the human clinical trial with pre-clinical animal model data, there is a plethora of information of how that could be possible. Based on current data, the ideal human immunization would likely mount a humoral, antibody response without activating T-cells, which may avoid encephalitis. Such a response would allow for microglial-mediated clearance of Aβ plaques, peripheral sequestration of Aβ, and direct dissolution of plaques. Antibodies could also bind a pool of soluble Aβ that may contribute to cognitive deficits as well. While a passive immunization strategy could achieve these results, it is still conceivable that an active immunization strategy could be designed to meet these criteria as well.

In the future, anti-Aβ therapies appear promising to treat the underlying cause of AD. Findings from animal models and humans strongly suggest that such strategies can reduce or prevent AD-related pathologies as well as improve cognition. While immunization is one promising strategy, other anti-Aβ approaches could be effective as well. Many pharmaceutical companies are also targeting β-secretase and γ-secretase (Xia, 2003). Both of these targets would reduce Aβ production and in mouse models have also been shown to reduce Aβ deposition (Best et al., 2007; Hussain et al., 2007). While research in the last several decades has provided exciting insights into AD mechanisms and potential therapeutic avenues, it is seems very likely that research in the coming decade may finally provide a treatment that targets the cause of the disease and will hopefully lead to a prevention or cure.

REFERENCES

Asami-Odaka A, Obayashi-Adachi Y, Matsumoto Y, Takahashi H, Fukumoto H, Horiguchi T, Suzuki N, and Shoji M (2005). Passive immunization of the Abeta42(43) C-terminal-specific antibody BC05 in a mouse model of Alzheimer's disease. *Neurodegener Dis* 2: 36–43.

Asuni AA, Boutajangout A, Scholtzova H, Knudsen E, Li YS, Quartermain D, Frangione B, Wisniewski T, and Sigurdsson EM (2006). Vaccination of Alzheimer's model mice with Abeta derivative in alum adjuvant reduces Abeta burden without microhemorrhages. *Eur J Neurosci* 24: 2530–2542.

Attems J (2005). Sporadic cerebral amyloid angiopathy: pathology, clinical implications, and possible pathomechanisms. *Acta Neuropathol (Berl)* 110: 345–359.

Bacskai BJ, Kajdasz ST, Christie RH, Carter C, Games D, Seubert P, Schenk D, and Hyman BT (2001). Imaging of amyloid-beta deposits in brains of living mice permits direct observation of clearance of plaques with immunotherapy. *Nat Med* 7: 369–372.

Bacskai BJ, Kajdasz ST, McLellan ME, Games D, Seubert P, Schenk D, and Hyman BT (2002). Non-Fc-mediated mechanisms are involved in clearance of amyloid-beta in vivo by immunotherapy. *J Neurosci* 22: 7873–7878.

Bard F, Cannon C, Barbour R, Burke RL, Games D, Grajeda H, Guido T, Hu K, Huang J, Johnson-Wood K, Khan K, Kholodenko D, Lee M, Lieberburg I, Motter R, Nguyen M, Soriano F, Vasquez

N, Weiss K, Welch B, Seubert P, Schenk D, and Yednock T (2000). Peripherally administered antibodies against amyloid beta-peptide enter the central nervous system and reduce pathology in a mouse model of Alzheimer disease. *Nat Med* 6: 916–919.

Bard F, Barbour R, Cannon C, Carretto R, Fox M, Games D, Guido T, Hoenow K, Hu K, Johnson-Wood K, Khan K, Kholodenko D, Lee C, Lee M, Motter R, Nguyen M, Reed A, Schenk D, Tang P, Vasquez N, Seubert P, and Yednock T (2003). Epitope and isotype specificities of antibodies to beta -amyloid peptide for protection against Alzheimer's disease-like neuropathology. *Proc Natl Acad Sci U S A* 100: 2023–2028.

Best JD, Smith DW, Reilly MA, O'Donnell R, Lewis HD, Ellis S, Wilkie N, Rosahl TW, Laroque PA, Boussiquet-Leroux C, Churcher I, Atack JR, Harrison T, and Shearman MS (2007). The novel gamma secretase inhibitor N-[cis-4-[(4-chlorophenyl)sulfonyl]-4-(2,5-difluorophenyl)cyclohexyl]-1,1, 1-trifluoromethanesulfonamide (MRK-560) reduces amyloid plaque deposition without evidence of notch-related pathology in the Tg2576 mouse. *J Pharmacol Exp Ther* 320: 552–558.

Billings LM, Oddo S, Green KN, McGaugh JL, and LaFerla FM (2005). Intraneuronal Abeta causes the onset of early Alzheimer's disease-related cognitive deficits in transgenic mice. *Neuron* 45: 675–688.

Brendza RP, Bacskai BJ, Cirrito JR, Simmons KA, Skoch JM, Klunk WE, Mathis CA, Bales KR, Paul SM, Hyman BT, and Holtzman DM (2005). Anti-Abeta antibody treatment promotes the rapid recovery of amyloid-associated neuritic dystrophy in PDAPP transgenic mice. *J Clin Invest* 115: 428–433.

Burbach GJ, Vlachos A, Ghebremedhin E, Del Turco D, Coomaraswamy J, Staufenbiel M, Jucker M, and Deller T (2007). Vessel ultrastructure in APP23 transgenic mice after passive anti-Abeta immunotherapy and subsequent intracerebral hemorrhage. *Neurobiol Aging* 28: 202–212.

Buttini M, Masliah E, Barbour R, Grajeda H, Motter R, Johnson-Wood K, Khan K, Seubert P, Freedman S, Schenk D, and Games D (2005). Beta-amyloid immunotherapy prevents synaptic degeneration in a mouse model of Alzheimer's disease. *J Neurosci* 25: 9096–9101.

ClinicalTrials.gov (2005) [database online]. AAB-001 in Patients with Mild to Moderate Alzheimer's disease. US National Institutes of Health.

Cribbs DH, Ghochikyan A, Vasilevko V, Tran M, Petrushina I, Sadzikava N, Babikyan D, Kesslak P, Kieber-Emmons T, Cotman CW, and Agadjanyan MG (2003). Adjuvant-dependent modulation of Th1 and Th2 responses to immunization with beta-amyloid. *Int Immunol* 15: 505–514.

Das P, Howard V, Loosbrock N, Dickson D, Murphy MP, and Golde TE (2003). Amyloid-beta immunization effectively reduces amyloid deposition in FcRgamma-/- knock-out mice. *J Neurosci* 23: 8532–8538.

DeMattos RB, Bales KR, Cummins DJ, Dodart JC, Paul SM, and Holtzman DM (2001). Peripheral anti-A beta antibody alters CNS and plasma A beta clearance and decreases brain A beta burden in a mouse model of Alzheimer's disease. *Proc Natl Acad Sci U S A* 98: 8850–8855.

Dodart JC, Meziane H, Mathis C, Bales KR, Paul SM, and Ungerer A (1999). Behavioral disturbances in transgenic mice overexpressing the V717F beta-amyloid precursor protein. *Behav Neurosci* 113: 982–990.

Dodart JC, Mathis C, Bales KR, and Paul SM (2002a). Does my mouse have Alzheimer's disease? *Genes Brain Behav* 1: 142–155.

Dodart JC, Bales KR, Gannon KS, Greene SJ, DeMattos RB, Mathis C, DeLong CA, Wu S, Wu X, Holtzman DM, and Paul SM (2002b). Immunization reverses memory deficits without reducing brain Abeta burden in Alzheimer's disease model. *Nat Neurosci* 5: 452–457.

Ferrer I, Boada Rovira M, Sanchez Guerra ML, Rey MJ, and Costa-Jussa F (2004). Neuropathology and pathogenesis of encephalitis following amyloid-beta immunization in Alzheimer's disease. *Brain Pathol* 14: 11–20.

Fukuchi K, Tahara K, Kim HD, Maxwell JA, Lewis TL, Accavitti-Loper MA, Kim H, Ponnazhagan S, and Lalonde R (2006). Anti-Abeta single-chain antibody delivery via adeno-associated virus for treatment of Alzheimer's disease. *Neurobiol Dis* 23: 502–511.

Games D, Adams D, Alessandrini R, Barbour R, Berthelette P, Blackwell C, Carr T, Clemens J, Donaldson T, and Gillespie F, et al. (1995). Alzheimer-type neuropathology in transgenic mice overexpressing V717F beta-amyloid precursor protein. *Nature* 373: 523–527.

Gandy S and Walker L (2004). Toward modeling hemorrhagic and encephalitic complications of Alzheimer amyloid-beta vaccination in nonhuman primates. *Curr Opin Immunol* 16: 607–615.

Gilman S, Koller M, Black RS, Jenkins L, Griffith SG, Fox NC, Eisner L, Kirby L, Rovira MB, Forette F, and Orgogozo JM (2005). Clinical effects of Abeta immunization (AN1792) in patients with AD in an interrupted trial. *Neurology* 64: 1553–1562.

Hartley DM, Walsh DM, Ye CP, Diehl T, Vasquez S, Vassilev PM, Teplow DB, and Selkoe DJ (1999). Protofibrillar intermediates of amyloid beta-protein induce acute electrophysiological changes and progressive neurotoxicity in cortical neurons. *J Neurosci* 19: 8876–8884.

Hartman RE, Izumi Y, Bales KR, Paul SM, Wozniak DF, and Holtzman DM (2005). Treatment with an amyloid-beta antibody ameliorates plaque load, learning deficits, and hippocampal long-term potentiation in a mouse model of Alzheimer's disease. *J Neurosci* 25: 6213–6220.

Head E and Lott IT (2004). Down syndrome and beta-amyloid deposition. *Curr Opin Neurol* 17: 95–100.

Hock C, Konietzko U, Papassotiropoulos A, Wollmer A, Streffer J, von Rotz RC, Davey G, Moritz E, and Nitsch RM (2002). Generation of antibodies specific for beta-amyloid by vaccination of patients with Alzheimer disease. *Nat Med* 8: 1270–1275.

Hock C, Konietzko U, Streffer JR, Tracy J, Signorell A, Muller-Tillmanns B, Lemke U, Henke K, Moritz E, Garcia E, Wollmer MA, Umbricht D, de Quervain DJ, Hofmann M, Maddalena A, Papassotiropoulos A, and Nitsch RM (2003). Antibodies against beta-amyloid slow cognitive decline in Alzheimer's disease. *Neuron* 38: 547–554.

Holtzman DM, Bales KR, Paul SM, and DeMattos RB (2002). Abeta immunization and anti-Abeta antibodies: potential therapies for the prevention and treatment of Alzheimer's disease. *Adv Drug Deliv Rev* 54: 1603–1613.

Hussain I, Hawkins J, Harrison D, Hille C, Wayne G, Cutler L, Buck T, Walter D, Demont E, Howes C, Naylor A, Jeffrey P, Gonzalez MI, Dingwall C, Michel A, Redshaw S, and Davis JB (2007). Oral administration of a potent and selective non-peptidic BACE-1 inhibitor decreases beta-cleavage of amyloid precursor protein and amyloid-beta production in vivo. *J Neurochem* 100: 802–809.

Janus C, Pearson J, McLaurin J, Mathews PM, Jiang Y, Schmidt SD, Chishti MA, Horne P, Heslin D, French J, Mount HT, Nixon RA, Mercken M, Bergeron C, Fraser PE, St George-Hyslop P, and Westaway D (2000). A beta peptide immunization reduces behavioural impairment and plaques in a model of Alzheimer's disease. *Nature* 408: 979–982.

Jellinger KA and Attems J (2005). Prevalence and pathogenic role of cerebrovascular lesions in Alzheimer disease. *J Neurol Sci* 229–230: 37–41.

Jenner E (1798). An inquiry into the causes and effects of the Variolae Vaccine, a disease discovered in some of the western counties of England particularly Gloucestershire, and known by the name of cow pox. Sampson Low, London.

Kotilinek LA, Bacskai B, Westerman M, Kawarabayashi T, Younkin L, Hyman BT, Younkin S, and Ashe KH (2002). Reversible memory loss in a mouse transgenic model of Alzheimer's disease. *J Neurosci* 22: 6331–6335.

Lee EB, Leng LZ, Zhang B, Kwong L, Trojanowski JQ, Abel T, and Lee VM (2006). Targeting amyloid-beta peptide (Abeta) oligomers by passive immunization with a conformation-selective monoclonal antibody improves learning and memory in Abeta precursor protein (APP) transgenic mice. *J Biol Chem* 281: 4292–4299.

Legleiter J, Czilli DL, Gitter B, DeMattos RB, Holtzman DM, and Kowalewski T (2004). Effect of different anti-Abeta antibodies on Abeta fibrillogenesis as assessed by atomic force microscopy. *J Mol Biol* 335: 997–1006.

Lemere CA, Maron R, Spooner ET, Grenfell TJ, Mori C, Desai R, Hancock WW, Weiner HL, and Selkoe DJ (2000). Nasal A beta treatment induces anti-A beta antibody production and decreases cerebral amyloid burden in PD-APP mice. *Ann N Y Acad Sci* 920: 328–331.

Levites Y, Jansen K, Smithson LA, Dakin R, Holloway VM, Das P, and Golde TE (2006a). Intracranial adeno-associated virus-mediated delivery of anti-pan amyloid beta, amyloid beta40, and amyloid beta42 single-chain variable fragments attenuates plaque pathology in amyloid precursor protein mice. *J Neurosci* 26: 11923–11928.

Levites Y, Das P, Price RW, Rochette MJ, Kostura LA, McGowan EM, Murphy MP, and Golde TE (2006b). Anti-Abeta42- and anti-Abeta40-specific mAbs attenuate amyloid deposition in an Alzheimer disease mouse model. *J Clin Invest* 116: 193–201.

Lombardo JA, Stern EA, McLellan ME, Kajdasz ST, Hickey GA, Bacskai BJ, and Hyman BT (2003). Amyloid-beta antibody treatment leads to rapid normalization of plaque-induced neuritic alterations. *J Neurosci* 23: 10879–10883.

Maier M, Seabrook TJ, Lazo ND, Jiang L, Das P, Janus C, and Lemere CA (2006). Short amyloid-beta (Abeta) immunogens reduce cerebral Abeta load and learning deficits in an Alzheimer's disease mouse model in the absence of an Abeta-specific cellular immune response. *J Neurosci* 26: 4717–4728.

Masliah E, Hansen L, Adame A, Crews L, Bard F, Lee C, Seubert P, Games D, Kirby L, and Schenk D (2005). Abeta vaccination effects on plaque pathology in the absence of encephalitis in Alzheimer disease.*Neurology* 64: 129–131.

Matsuoka Y, Saito M, LaFrancois J, Saito M, Gaynor K, Olm V, Wang L, Casey E, Lu Y, Shiratori C, Lemere C, and Duff K (2003). Novel therapeutic approach for the treatment of Alzheimer's disease by peripheral administration of agents with an affinity to beta-amyloid. *J Neurosci* 23: 29–33.

McLaurin J, Cecal R, Kierstead ME, Tian X, Phinney AL, Manea M, French JE, Lambermon MH, Darabie AA, Brown ME, Janus C, Chishti MA, Horne P, Westaway D, Fraser PE, Mount HT, Przybylski M, and St George-Hyslop P (2002). Therapeutically effective antibodies against amyloid-beta peptide target amyloid-beta residues 4–10 and inhibit cytotoxicity and fibrillogenesis. *Nat Med* 8: 1263–1269.

Monsonego A, Zota V, Karni A, Krieger JI, Bar-Or A, Bitan G, Budson AE, Sperling R, Selkoe DJ, and Weiner HL (2003). Increased T cell reactivity to amyloid beta protein in older humans and patients with Alzheimer disease. *J Clin Invest* 112: 415–422.

Morgan D, Diamond DM, Gottschall PE, Ugen KE, Dickey C, Hardy J, Duff K, Jantzen P, DiCarlo G, Wilcock D, Connor K, Hatcher J, Hope C, Gordon M, and Arendash GW (2000). A beta peptide vaccination prevents memory loss in an animal model of Alzheimer's disease. *Nature* 408: 982–985.

Nicoll JA, Wilkinson D, Holmes C, Steart P, Markham H, and Weller RO (2003). Neuropathology of human Alzheimer disease after immunization with amyloid-beta peptide: a case report. *Nat Med* 9: 448–452.

Nikolic WV, Bai Y, Obregon D, Hou H, Mori T, Zeng J, Ehrhart J, Shytle RD, Giunta B, Morgan D, Town T, and Tan J (2007). Transcutaneous beta-amyloid immunization reduces cerebral beta-amyloid deposits without T cell infiltration and microhemorrhage. *Proc Natl Acad Sci U S A* 104: 2507–2512.

Oddo S, Caccamo A, Kitazawa M, Tseng BP, and LaFerla FM (2003a). Amyloid deposition precedes tangle formation in a triple transgenic model of Alzheimer's disease. *Neurobiol Aging* 24: 1063–1070.

Oddo S, Caccamo A, Shepherd JD, Murphy MP, Golde TE, Kayed R, Metherate R, Mattson MP, Akbari Y, and LaFerla FM (2003b). Triple-transgenic model of Alzheimer's disease with plaques and tangles: intracellular Abeta and synaptic dysfunction. *Neuron* 39: 409–421.

Oddo S, Billings L, Kesslak JP, Cribbs DH, and LaFerla FM (2004). Abeta immunotherapy leads to clearance of early, but not late, hyperphosphorylated tau aggregates via the proteasome. *Neuron* 43: 321–332.

Oddo S, Vasilevko V, Caccamo A, Kitazawa M, Cribbs DH, and LaFerla FM (2006). Reduction of soluble Abeta and tau, but not soluble Abeta alone, ameliorates cognitive decline in transgenic mice with plaques and tangles. *J Biol Chem* 281: 39413–39423.

Orgogozo JM, Gilman S, Dartigues JF, Laurent B, Puel M, Kirby LC, Jouanny P, Dubois B, Eisner L, Flitman S, Michel BF, Boada M, Frank A, and Hock C (2003). Subacute meningoencephalitis in a subset of patients with AD after Abeta42 immunization. *Neurology* 61: 46–54.

Pfeifer LA, White LR, Ross GW, Petrovitch H, and Launer LJ (2002). Cerebral amyloid angiopathy and cognitive function: the HAAS autopsy study. *Neurology* 58: 1629–1634.

Prada CM, Garcia-Alloza M, Betensky RA, Zhang-Nunes SX, Greenberg SM, Bacskai BJ, and Frosch MP (2007). Antibody-mediated clearance of amyloid-beta peptide from cerebral amyloid angiopathy revealed by quantitative in vivo imaging. *J Neurosci* 27: 1973–1980.

Prasher VP, Farrer MJ, Kessling AM, Fisher EM, West RJ, Barber PC, and Butler AC (1998). Molecular mapping of Alzheimer-type dementia in Down's syndrome. *Ann Neurol* 43: 380–383.

Racke MM, Boone LI, Hepburn DL, Parsadainian M, Bryan MT, Ness DK, Piroozi KS, Jordan WH, Brown DD, Hoffman WP, Holtzman DM, Bales KR, Gitter BD, May PC, Paul SM, and DeMattos RB (2005). Exacerbation of cerebral amyloid angiopathy-associated microhemorrhage in amyloid precursor protein transgenic mice by immunotherapy is dependent on antibody recognition of deposited forms of amyloid beta. *J Neurosci* 25: 629–636.

Schenk D, Barbour R, Dunn W, Gordon G, Grajeda H, Guido T, Hu K, Huang J, Johnson-Wood K, Khan K, Kholodenko D, Lee M, Liao Z, Lieberburg I, Motter R, Mutter L, Soriano F, Shopp G, Vasquez N, Vandevert C, Walker S, Wogulis M, Yednock T, Games D, and Seubert P (1999). Immunization with amyloid-beta attenuates Alzheimer-disease-like pathology in the PDAPP mouse. *Nature* 400: 173–177.

Schenk D, Hagen M, and Seubert P (2004). Current progress in beta-amyloid immunotherapy. *Curr Opin Immunol* 16: 599–606.

Seabrook TJ, Thomas K, Jiang L, Bloom J, Spooner E, Maier M, Bitan G, and Lemere CA (2007). Dendrimeric Abeta1-15 is an effective immunogen in wildtype and APP-tg mice. *Neurobiol Aging* 28(6): 813–823.

Selkoe DJ (2001). Alzheimer's disease: genes, proteins, and therapy. *Physiol Rev* 81: 741–766.

Solomon B, Koppel R, Hanan E, and Katzav T (1996). Monoclonal antibodies inhibit in vitro fibrillar aggregation of the Alzheimer beta-amyloid peptide. *Proc Natl Acad Sci U S A* 93: 452–455.

Tamura Y, Hamajima K, Matsui K, Yanoma S, Narita M, Tajima N, Xin KQ, Klinman D, and Okuda K (2005). The F(ab)'2 fragment of an Abeta-specific monoclonal antibody reduces Abeta deposits in the brain. *Neurobiol Dis* 20: 541–549.

Townsend M, Shankar GM, Mehta T, Walsh DM, and Selkoe DJ (2006). Effects of secreted oligomers of amyloid beta-protein on hippocampal synaptic plasticity: a potent role for trimers. *J Physiol* 572: 477–492.

Tsuboi Y (2006). Neuropathology of familial tauopathy. *Neuropathology* 26: 471–474.

Walsh DM, Hartley DM, Kusumoto Y, Fezoui Y, Condron MM, Lomakin A, Benedek GB, Selkoe DJ, and Teplow DB (1999). Amyloid beta-protein fibrillogenesis. Structure and biological activity of protofibrillar intermediates. *J Biol Chem* 274: 25945–25952.

Walsh DM, Klyubin I, Fadeeva JV, Cullen WK, Anwyl R, Wolfe MS, Rowan MJ, and Selkoe DJ (2002). Naturally secreted oligomers of amyloid beta protein potently inhibit hippocampal long-term potentiation in vivo. *Nature* 416: 535–539.

Wang HW, Pasternak JF, Kuo H, Ristic H, Lambert MP, Chromy B, Viola KL, Klein WL, Stine WB, Krafft GA, and Trommer BL (2002). Soluble oligomers of beta amyloid (1-42) inhibit long-term potentiation but not long-term depression in rat dentate gyrus. *Brain Res* 924: 133–140.

Wilcock DM, DiCarlo G, Henderson D, Jackson J, Clarke K, Ugen KE, Gordon MN, and Morgan D (2003). Intracranially administered anti-Abeta antibodies reduce beta-amyloid deposition by mechanisms both independent of and associated with microglial activation. *J Neurosci* 23: 3745–3751.

Wilcock DM, Rojiani A, Rosenthal A, Subbarao S, Freeman MJ, Gordon MN, and Morgan D (2004a). Passive immunotherapy against Abeta in aged APP-transgenic mice reverses cognitive deficits and depletes parenchymal amyloid deposits in spite of increased vascular amyloid and microhemorrhage. *J Neuroinflammation* 1: 24.

Wilcock DM, Rojiani A, Rosenthal A, Levkowitz G, Subbarao S, Alamed J, Wilson D, Wilson N, Freeman MJ, Gordon MN, and Morgan D (2004b). Passive amyloid immunotherapy clears amyloid and transiently activates microglia in a transgenic mouse model of amyloid deposition. *J Neurosci* 24: 6144–6151.

Xia W (2003). Amyloid inhibitors and Alzheimer's disease. *Curr Opin Investig Drugs* 4: 55–59.

Zhou J, Fonseca MI, Kayed R, Hernandez I, Webster SD, Yazan O, Cribbs DH, Glabe CG, and Tenner AJ (2005). Novel Abeta peptide immunogens modulate plaque pathology and inflammation in a murine model of Alzheimer's disease. *J Neuroinflammation* 2: 28.

PART V

Novel Therapies for Spinal Cord Injury and White Matter Demyelination

13

Axonal Plasticity and Regeneration in the Injured Spinal Cord

M.H. Tuszynski[*,†] and P. Lu[*,†]

*Department of Neurosciences, University of California-San Diego, La Jolla, CA 92093-0626

†Veterans Administration Medical Center, San Diego, CA 92165

ABSTRACT

The field of spinal cord injury research has made remarkable progress in the last 20 years, with elucidation of several mechanisms that underlie axonal growth failure in the adult CNS. This progress in turn has led to the identification of several potential means of enhancing axonal plasticity and regeneration in the adult CNS. This chapter will review progress in the field of spinal cord injury regeneration research, with a particular emphasis on the influences of growth factors and combinatorial therapies in maximizing the growth capacity of injured axonal systems.

INTRODUCTION

Remarkable progress has been made in the last 20 years in identifying mechanisms underlying the failure of axonal regeneration in the central nervous system (CNS). Following the identification of these underlying mechanisms, distinct therapeutic approaches have been tested in in vivo models of spinal cord injury to enhance central axonal structural plasticity, including the therapeutic administration of neurotrophic factors. Several experimental treatments enhance axonal growth, and some of these are beginning clinical trials. Nonetheless, the practical challenges in eliciting a sufficient degree of axonal sprouting or regeneration to realistically impact the large and complex spinal cord of the human remain considerable. Recent studies indicate that experimental approaches to combine therapies targeting distinct neural mechanisms can achieve superior

Copyright © 2008 Elsevier Inc.
All rights of reproduction in any form reserved.

axonal growth compared to therapies applied in isolation. The optimization of combinatorial therapies may provide the most promising hope for translation of therapies to human clinical trials.

CNS REGENERATION FAILURE VS. PNS REGENERATION SUCCESS

Whereas the adult mammalian spinal cord fails to regenerate, the crushed peripheral nerve often successfully regenerates. Several mechanisms contribute to the success of peripheral axonal regeneration, including the following: (1) Physical bridges form at sites of peripheral nerve injury that fill the lesion cavity, re-establishing a permissive physical matrix to support axonal attachment and growth (Williams et al., 1983; Chernousov and Carey, 2000; Dubovy, 2004). This matrix includes collagen, fibronectin, laminin, Schwann, cells and fibroblasts. (2) Nervous system growth factors, or neurotrophic factors, are secreted in appropriate spatial and temporal gradients to support axonal regeneration after peripheral nerve injury (Longo et al., 1983; Terenghi, 1999; Boyd and Gordon, 2003). Schwann cells produce many of these growth factors in the local injured and regenerating milieu (Assouline et al., 1987; Johnson et al., 1988; Meyer et al., 1992; Bhatheja and Field, 2006). (3) A set of genes is activated in the nucleus of the damaged peripheral neuron that supports axonal regeneration, including GAP-43, CAP-23, ß-tubulin, and others (Tetzlaff et al., 1991; Jenkins et al., 1993; Kury et al., 2001; Navarro et al., 2007). (4) Molecules of two classes have been identified that actively suppress axonal sprouting and regeneration in the CNS: molecules associated with central myelin, including nogo (Schnell and Schwab, 1990; Buchli and Schwab, 2005), myelin-associated glycoprotein (MAG) (McKerracher et al., 1994), oligodendrocyte-myelin glycoprotein (OMgp) (Wang et al., 2002b), semaphorins (Moreau-Fauvarque et al., 2003), and netrin, and molecules present in the extracellular matrix (ECM), particularly the chondroitin sulfate proteoglycan (CSPG) molecule NG2 (Jones et al., 2002; Silver and Miller, 2004; Fawcett, 2006).

Experimental efforts to enhance regeneration of the spinal cord have individually targeted many of the preceding mechanisms. These experimental approaches include placement of molecular, cellular, or "synthetic" bridges in the lesion cavity (Lakatos and Franklin, 2002; Novikova et al., 2003; Stokols et al., 2006); stimulation of the injured spinal cord with growth factors (Tuszynski, 2002); "conditioning" of neurons to activate intrinsic genetic programs and proteins related to an active growth state (Neumann and Woolf, 1999; Neumann et al., 2002; Qiu et al., 2002; Rossi et al., 2007); and efforts to neutralize myelin- or ECM-related inhibitors (Schnell and Schwab, 1990; Bradbury et al., 2002; Silver and Miller, 2004; Fawcett, 2006). In many cases, experiments combined two of these approaches by placing cellular transplants in a lesion cavity while simultaneously, for example, administering neurotrophic factors (Lu et al., 2004) or raising cyclic AMP (cAMP) levels (Nikulina et al., 2004; Pearse et al., 2004). Several studies report enhancement of axonal growth and, in some cases, incremental improvement in functional outcome. However, no properly conducted and controlled study to date has demonstrated very extensive structural or functional recovery after SCI, and in no case has convincing improvement of plasticity and regeneration been demonstrated in a larger animal model.

Indeed, there are substantial challenges in attempting to achieve recovery of function after spinal cord injury. In excess of 3 million axons project rostrally and caudally through the human spinal cord. The normal patterning of axonal projections during development is established by an exquisitely orchestrated set of mechanisms that occur both intrinsically within the neuron, and extrinsically in the environment surrounding the growing axon. Within the developing neuron, genes are sequentially activated that begin the process of axonal elongation, cytoskeletal stabilization, and receptor expression on the tips of growth cones that sense substrates and diffusible signals in the extracellular environment. The environment in turn influences the extending axon with diffusible molecules such as netrins, semaphorins, ephrins, growth factors, and extracellular matrix molecules that contribute to attraction or repelling of growth. The array of intrinsic and extrinsic processes that control axon growth is precisely timed, and perturbation of single contributors to axonal elongation in the developing nervous system can lead to mistargeting, axonal withdrawal or neuron death.

Clearly, axonal elongation during development depends on "combinations" of events. Both the developing axon and its environment utilize a number of disparate mechanisms to orchestrate a series of events that result in growth that is rapid, accurate and precise. If substantial axonal growth is to be achieved in the injured adult CNS, it seems reasonable to expect that multiple mechanisms will also need to be addressed. The manipulation of any single or pair of factors may be sufficient to detect axonal growth in a rodent model of SCI; however, the probability of achieving growth of sufficient numbers of axons, and over distances of at least several centimeters in the larger primate spinal cord, will likely require convergence of treatments targeting multiple cellular and extracellular mechanisms. Simple, one-therapy solutions seem unlikely to address this complex problem. Yet we have learned a great deal from individual approaches to enhancing the plasticity and regeneration of the injured spinal cord. We will review some of these approaches now.

INDIVIDUAL APPROACHES TO ENHANCING SPINAL CORD PLASTICITY AND REGENERATION

GROWTH FACTORS

Several years ago we and others examined the ability of injured adult spinal cord axons to respond to growth factors (Schnell et al., 1994; Tuszynski et al., 1994, 1996; Xu et al., 1995; Bregman et al., 1997; Grill et al., 1997; Kobayashi et al.,

1997; Ye and Houle, 1997; Jackeman et al., 1998; Blesch et al., 1999; Bradbury et al., 1999; Liu et al., 1999; Lu et al., 2001; Ruitenberg et al., 2003). Our group introduced nerve growth factor (NGF) into the intact and partially lesioned spinal cord, using techniques of gene delivery. Suspensions of autologous fibroblasts genetically modified to secrete NGF were injected into either the central gray matter of the nonlesioned thoracic spinal cord, or were embedded into collagen matrices that were then injected into dorsally hemisected spinal cord lesion cavities of adult rats (Tuszynski et al., 1994, 1996). We found similar patterns of axonal responsiveness to NGF in both the nonlesioned and lesioned spinal cord: dorsal root ganglion (DRG) nociceptive axons and cerulospinal axons densely penetrated NGF-secreting cell grafts placed in the adult spinal cord (Figure 13.1). This finding established the fact that adult axons retain sensitivity to growth factors, that patterns of growth factor sensitivity after injury in adulthood recapitulate developmental patterns of growth factor sensitivity, and that axons extend in

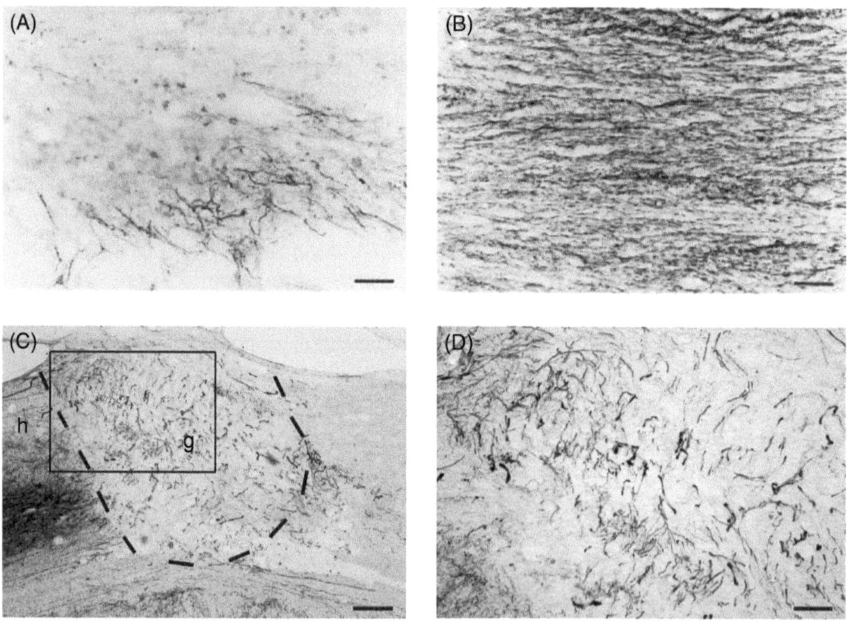

FIGURE 13.1 Spinal cord axon growth is induced by grafts of NGF- or BDNF-secreting cell grafts. (A) Neurofilament immunolabel shows modest axon penetration into control fibroblast graft, whereas (B) NGF-secreting graft is densely penetrated by axons 3 months post-injury (see Table 13.1 for list of responding axons). (C) BDA-labeled reticulospinal axons also extensively penetrate a BDNF-secreting bone marrow stromal cell graft (outlined by dashed lines) placed into site of mid-cervical injury. g, graft; h, host; graft shown 3 months post-injury. (D) Higher magnification of boxed areas from panel C, demonstrating BDA-labeled reticulospinal axons in BDNF-expressing bone marrow stromal cell graft. Scale bar, A–B, 15 μm; C, 210 μm; D, 80 μm.

large numbers when presented with a permissive growth environment containing growth factors. In this set of studies, a "bridge" to support axonal penetration was provided by the cellular graft in the lesion cavity. These results generally supported much earlier findings of Aguayo and colleagues indicating that central axons can extend for essentially unlimited distances in the permissive environment of the peripheral nerve "bridge" (David and Aguayo, 1981). Indeed, we now know that peripheral nerve bridges contain substantial quantities of NGF, BDNF, neurotrophin-3 (NT-3), ciliary neurotrophic factor (CNTF) and glial cell line-derived neurotrophic factor (GDNF) secreted by Schwann cells, which likely stimulate axon growth.

Over subsequent years, a number of groups proceeded to define the spectrum of axonal sensitivity to various growth factors in the injured spinal cord (Table 13.1). Of note, BDNF supports the growth of a number of neuronal populations that modify the activity of spinal cord motor neurons, including

TABLE 13.1 Sensitivity of spinal cord axons to growth factors

Growth factors	Injured axons	References
NGF	nociceptive spinal axons	Tuszynski et al., 1994, 1996
		Ramer et al., 2000
	cerulospinal axons	Tuszynski et al., 1994, 1996
BDNF	rubrospinal axons	Kobayashi et al., 1997
		Ye and Houle, 1997
		Liu et al., 1999
	raphespinal axons	Bregman et al., 1997
		Menei et al., 1998
	coerulospinal axons	Menei et al., 1998
	reticulospinal axons	Ye and Houle, 1997
		Jin et al., 2002
	vestibulospinal	Jin et al., 2002
	local motor axon	Lu et al., 2001
	local sensory axon (CGRP)	Lu et al., 2001
NT-3	corticospinal axons	Schnell et al., 1994
		Grill et al., 1997
	dorsal column sensory axons	Bradbury et al., 1999
NT-4/5	local motor axons	Blesch et al., 2004
	cerulospinal axons	Blesch et al., 2004
	reticulospinal axons	Blesch et al., 2004
	propriospinal axons	Blesch et al., 2004
GDNF	local motor axons	Blesch et al., 2003
	propriospinal axons	Blesch et al., 2003
	dorsal column sensory axons	Blesch et al., 2003
	nociceptive spinal axons	Ramer et al., 2003

cerulospinal, rubrospinal, raphaepsinal and reticulospinal axons (Kobayashi et al., 1997; Ye and Houle, 1997; Jackeman et al., 1998; Menei et al., 1998; Liu et al., 1999; Lu et al., 2001, 2005; Jin et al., 2002; Ruitenberg et al., 2003). Significantly, NT-3 enhances the growth of the most important upper motor neuron population in primates, corticospinal axons (Schnell et al., 1994; Grill et al., 1997; Blits et al., 2003). NT-3 also enhances the growth of ascending dorsal column sensory axons, a model system of great utility in studying both mechanisms and empirical features of axonal elongation in the adult spinal cord (Bradbury et al., 1999; Ramer et al., 2000; Lu et al., 2003; Taylor et al., 2006) (Table 13.1). These growth factors emerge as the most frequent subjects of therapeutic analysis in models of SCI because of their ability to influence the growth of motor axons, although other growth factors, including insulin-like growth factor-1 (IGF-1), the fibroblast growth factor (FGF) family, and others, remain of considerable interest. NGF and GDNF, on the other hand, by promoting growth of nociceptive spinal axons, may risk worsening dysfunctional axonal sprouting after SCI and worsening pain. This point highlights the fact that axonal growth is not a uniformly advantageous phenomenon: aberrant growth of either sensory or motor systems could have deleterious anatomical and functional consequences.

BDNF Effects on Axonal Growth

In a series of studies, several groups reported that BDNF could enhance the growth of supraspinal motor axons into permissive growth milieus placed at sites of SCI. Kobayashi and colleagues (1997) reported that BDNF promotes the growth of rubrospinal axons into a peripheral nerve graft placed at a site of cervical SCI, and prevents degeneration of the red nucleus cell body in the brainstem. Further, delayed delivery of BDNF to the red nucleus by one year in this model continued to demonstrate neurotrophic actions, reversing atrophy of neurons in the red nucleus even after this extended time point (Kwon et al., 2002). Liu and colleagues (1999) also reported that rubrospinal axons regenerated into cellular grafts of autologous fibroblasts genetically modified to secrete BDNF in a mid-cervical spinal cord lesion site. They further noted improvement in use of the forepaw on the side affected by the lesion, possibly as a result of local sprouting stimulated by BDNF. Reticulospinal axons, which modulate activity of motor neurons in the spinal cord, also respond to BDNF (Jin et al., 2002) (Figure 13.1). Lu and colleagues (2001) reported that BDNF did not influence the growth of corticospinal axons in the spinal cord, even though BDNF prevents the death of corticospinal neuronal cell bodies when applied to the cortex, most likely because the BDNF receptor trkB was not trafficked from the cortical soma to the spinal axon. Others described protective effects of BDNF on cortical neurons (Giehl and Tetzlaff, 1996). In none of these papers was axonal growth beyond the BDNF cell graft or peripheral nerve bridge demonstrated, however. During development, neurotrophic factors classically promote the growth of their responsive axons along chemotropic gradients, with growth maximal at the point of peak growth factor concentration. This appears to be the case in the adult injured spinal cord also: axons readily penetrate BDNF-secreting grafts, but have not been reliably observed to emerge from the grafts again.

NT-3 Effects on Axonal Growth

A related but slightly distinct phenomenon has been observed in assessing responsiveness of corticospinal axons to NT-3. In 1994, Schnell and colleagues reported that a single injection of NT-3 above a spinal cord lesion site, when combined with administration of the Nogo neutralizing body IN-1, significantly enhanced the distance that a corticospinal axon extends below a mid-thoracic spinal dorsal hemisection lesion site (Schnell et al., 1994). Corticospinal axons observed below the lesion site in this experiment might have originated either from: (1) dorsally lesioned corticospinal axons that regenerated through spared host gray matter underlying the dorsal hemisection lesion, or (2) from ventral corticospinal axons, spared by the dorsal hemisection lesion that subsequently sprouted below the lesion site in response to IN-1 antibody treatment. The single injection of NT-3 protein increased corticospinal axon sprouting above the lesion, and may have slightly increased the density of axons detected below in the lesion. In 1997, Grill and colleagues implanted autologous fibroblasts genetically modified to secrete NT-3 in a mid-thoracic spinal cord dorsal hemisection lesion site (Grill et al., 1997). Interestingly, corticospinal axons did not penetrate the NT-3 secreting cell graft, but did exhibit enhanced sprouting in host gray matter underlying the graft in the lesion site. Sprouting of CST axons occurred in gray matter above, underlying, and 1 mm caudal to the NT-3 graft/lesion site. Rats treated with NT-3 showed partial recovery of locomotion on a horizontal ladder (Grill et al., 1997). This report confirmed that NT-3 can influence injured adult corticospinal axons, and also revealed that corticospinal axons will not penetrate a substrate other than adult gray matter. To the present day, 10 years later, we and to our knowledge others have not clearly succeeded in promoting growth of injured adult corticospinal axons into a substrate placed into the site of an adult spinal cord lesion.

Sensory axons also respond to NT-3 (Bradbury et al., 1999; Ramer et al., 2000; Lu et al., 2003; Taylor et al., 2006). Unlike corticospinal tract axons, however, dorsal column sensory axons successfully penetrate a cell graft placed in a spinal cord lesion site, and their growth into a graft is significantly enhanced by NT-3 (Lu et al., 2003). Like BDNF-responsive descending motor axons, however, ascending dorsal column sensory axons do not emerge from grafts placed in lesion sites to bridge beyond the lesion cavity.

In summary, studies over the past decade clearly demonstrated that injured adult axons retain sensitivity to various growth factors. While mechanistically informative, these findings were of little practical value: axons must grow beyond the lesion site to have potential for facilitating functional recovery. The observation that corticospinal axons sprout through host gray matter surrounding a dorsal hemisection lesion site was also informative but of limited value in a practical sense: gray matter is usually entirely destroyed at sites of SCI in humans, due its greater vulnerability to ischemia, inflammation, and secondary injury. Another approach was required to support axonal bridging beyond the lesion, a point that will be explored in a discussion of combination therapies below.

PHARMACOLOGICAL ENHANCEMENT OF AXONAL GROWTH: cAMP AND INOSINE

While growth factor administration is one means of stimulating the injured CNS axon, the spinal cord injury literature has also identified several other means of stimulating the injured axon. Among these are increases in cAMP levels in injured neurons, a topic explored in detail in Chapter 17. For several years it has been known that injury to the peripheral process of a dorsal root ganglion axon enhances the growth of the central axon of the same neuron in the spinal cord (McQuarrie et al., 1977; Neumann et al., 1999). This "preconditioning" enhancement of central axonal growth was subsequently shown to depend in part on elevated levels of cyclic-AMP in neurons (see Chapter 17). When cAMP-modulating compounds are administered in models of SCI, several investigators have demonstrated increased axonal growth (Neumann et al., 2002; Qiu et al., 2002; Lu et al., 2004; Nikulina et al., 2004; Pearse et al., 2004). The effects of cAMP modulation have now been shown to extend beyond dorsal root ganglion sensory axons, and to include descending motor systems arising from the brainstem (Lu et al., 2004; Nikulina et al., 2004; Pearse et al., 2004). Efforts are continuing to identify specific molecular mechanisms underlying preconditioning effects and to apply them singly, or in combination with other approaches, in the injured spinal cord (below).

Benowitz and colleagues (see Chapter 1) reported another means of augmenting axonal growth in the injured CNS: administration of the purine analog inosine (Benowitz et al., 1999, 2002). When applied to a spinal cord hemisection lesion, inosine reportedly enhanced sprouting of nonlesioned corticospinal axons across the spinal cord midline. The mechanism of action is related to upregulation of growth-associated proteins in cells including GAP-43, alpha tubulin, L1, and others. The potential value of inosine in models of central axonal injury and regeneration continues to be studied.

DEGRADING THE EXTRACELLULAR MATRIX

While one strategy for enhancing neural repair after spinal cord injury is to stimulate injured axons and neurons, another strategy for promoting axonal growth is to neutralize or degrade inhibitors in the CNS environment. At least two general classes of inhibitors to axonal regeneration in the CNS have been identified: inhibitory extracellular matrix (ECM) molecules (see Chapters 3 and 14), and inhibitory proteins associated with adult myelin (see Chapters 15 and 16).

Inhibitory extracellular matrix molecules are deposited at sites of SCI over several weeks and may persist indefinitely, consisting primarily of chondroitin sulfate proteoglycans (Chapters 3 and 14; Jones et al., 2002, 2003). The inhibitory extracellular matrix can be degraded either enzymatically using chondroitinase, or by administration of pharmacological compounds (Chapters 3 and 14; Bradbury et al., 2002), resulting in enhanced growth of injured axons as described in detail in other chapters. Recent studies have combined degradation of the inhibitory extracellular matrix with other treatments, and are addressed in "Degrading the Extracellular Matrix."

NEUTRALIZING MYELIN AND INACTIVATING RHO

Chapters 15 and 16 examine this topic in greater detail. That adult myelin inhibits axonal growth in experimental systems has been clear for an extended period (Schwab and Caroni, 1988). In fact, a number of proteins that inhibit axonal growth have now been discovered that are associated with adult CNS myelin, including Nogo, myelin-associated glycoprotein (MAG), oligodendrocyte-myelin glycoprotein (OMgp), semaphorins, ephrins, and netrin.

As reviewed in other chapters, the most extensively studied myelin inhibitor has been nogo. Efforts to neutralize inhibitory influences began in 1990 using antibodies to nogo (Schnell and Schwab, 1990). The receptor to which nogo, MAG and OMgp bind, termed Nogo-66, was identified in 2001 (Fournier et al., 2001) and was subsequently found to require association with the neurotrophin receptor p75 to signal inhibition (Wang et al., 2002a). Nogo-related axonal inhibition is also mediated by receptor binding to TROY (Park et al, 2005; Shao et al., 2005) and LINGO-1 (Mi et al., 2004). Nogo/nogo-66 receptor activation converges on intracellular mechanisms including Rho (Yamashita and Tohyama, 2003) to mediate inhibition of axon growth. Collectively, this impressive body of work identified several potential therapeutic targets for intervention to enhance CNS plasticity and regeneration: reduction or blockade of Nogo protein; blockade of binding to Nogo-66, p75, TROY or LINGO-1; or blockade of Rho activation (Winton et al., 2002).

When applied experimentally to in vivo models of SCI, enhanced axonal regeneration following various forms of nogo neutralization was reported (Schnell and Schwab, 1990; GrandPre, 2002). However, conflicting reports were also published, indicating that deletion of components of the nogo gene in transgenic mice did not enhance axonal regeneration, that p75 knockout mice did not exhibit enhanced regeneration, and that axonal labeling in some studies may have led to misinterpretation of experimental findings (Song et al., 2004; Zheng et al., 2005; Steward et al., 2007). While some uncertainty on these points remains, neutralization of nogo signaling has entered early stage clinical trials for acute SCI, using nogo antibodies. Similarly, clinical trials have begun in acute SCI using C3-like proteins that inactivate Rho (Dergham et al., 2002;

Winton et al., 2002), an approach that could more broadly target inhibition mediated by nogo, MAG, and OMgp simultaneously.

COMBINATORIAL THERAPIES SUPPORT AXONAL BRIDGING BEYOND SPINAL CORD LESION SITES

The preceding sections summarize findings indicating that a variety of strategies can enhance the plasticity and regeneration of injured CNS axons. Almost all of the cited studies utilized mouse or rodent models, however, making predictive statements of benefit for human SCI difficult. For example, while it is believed that we can model the human mechanism of spinal cord injury relatively well in animals using spinal contusion devices, the fact remains that human injury most frequently results from fracture-dislocation of the spine (this is rarely used as a model of injury in rodents), and that spinal cord compression may persist in humans for substantial time periods before an opportunity for surgical decompression occurs. These distinctions in mechanism of injury in human vs. experimental injury may generate important differences in the resulting basic pathophysiology of acute injury. Second, the physical size of the human spinal cord is vastly different from that of rodents. Distances over which axonal regeneration may result in functional recovery in rodents could be ineffective in humans. Third, the structure and functional anatomy of motor systems differs in rodents and humans. For example, forelimb use in rats is heavily influenced by the rubrospinal projection, but this system is vestigial in humans. Instead, the corticospinal projection is functionally critical for voluntary hand and leg use in humans. Further, the corticospinal projection in humans is in a different physical location of the spinal cord compared to rodents, and has distinct terminal inputs to intermediate and ventral gray matter compared to rodents, where projections are primarily in dorsal gray matter.

Thus, one must approach the translation of positive findings from rodent models to humans from a realistic perspective: while a single therapeutic manipulation in rodents may enhance anatomical plasticity and improve functional outcomes, it seems likely a priori that combination strategies to enhance plasticity and regeneration are more likely to be required to impact the spinal cord that is of greater size and of more complex function in advanced primates.

While the concept that combination therapies are likely to exhibit greater potency in eliciting axonal growth than single therapies has been stated for some time, recently this concept was subjected to more rigorous experimental analysis (Fischer et al., 2004; Lu et al., 2004; Nikulina et al., 2004; Pearse et al., 2004). Testing the hypothesis that combinatorial therapies exert greater effects on regeneration than single therapies is not a simple matter, however. One must control for each variable, and if there are more than two combinations to the therapy, then each potential combination must be tested. For example, if three treatments are tested, there must be seven test groups; if four potential combinations of treatments are tested, 15 groups must be tested. Assuming that detection of significant differences among groups requires a sample size of 10–12 animals per group, testing three combinations would require 70–80 subjects, and testing 4 combinations would require 150–170 subjects. These experiments become impractical. Yet proper controls are necessary, and a strategy we have taken is to test the most important controls in an initial round of experiments, returning to test additional control groups as indicated by ongoing experimental results.

Recently we reported that a combination of autologous cell bridges, growth factors, and cAMP-mediated stimulation of the injured neuronal soma can achieve axonal regeneration beyond a mid-cervical spinal cord lesion site (Lu et al., 2004). Cell bridges were placed in the lesion site to provide a substrate for the attachment and growth of regenerating axons; growth factor gradients were provided within and *beyond* the lesion site to test the hypothesis that axons would regenerate beyond the graft in the presence of a trophic gradient; and cAMP was injected into the region of the injured neuronal soma (the dorsal root ganglion) to upregulate expression of regeneration-associated genes and to enhance the ability of the injured axon to extend. The full combination of the cell bridge, trophic factor gradients, and cAMP-mediated stimulation of the neuronal soma achieved axonal bridging well beyond the lesion and up to the distal point of growth factor administration in the spinal cord, 1.5 mm beyond the lesion (Lu et al., 2004) (Figure 13.2). Any other two-component treatment combination failed to elicit bridging beyond the immediate host/graft interface. Thus, with an incomplete but logical set of control groups, one can conclude that the full combination treatment is more potent than pairwise therapies in this experiment.

Two other reports published in 2004 also suggested that combination therapies can generate more extensive anatomical or behavioral plasticity than individual therapies (Nikulina et al., 2004; Pearse et al., 2004). Interestingly, both of these studies, as well as our own, included cAMP augmentation of the neuronal growth state as a component of a combination therapy to improve outcomes in models of SCI. Augmentation of the neuronal growth state remains a major focus of several current efforts in SCI research, including attempts to discover novel genes related to enhanced central plasticity and regeneration (Costigan et al., 2002). Other combination therapies have been examined and report greater potency than individual treatments, including the combination of chondroitinase therapy with cell therapy (Fouad et al., 2005; Kim et al., 2006), chondroitinase with peripheral nerve grafts in the spinal cord (Houle et al., 2006), and chondriotinase with growth factors (Massey et al., 2007).

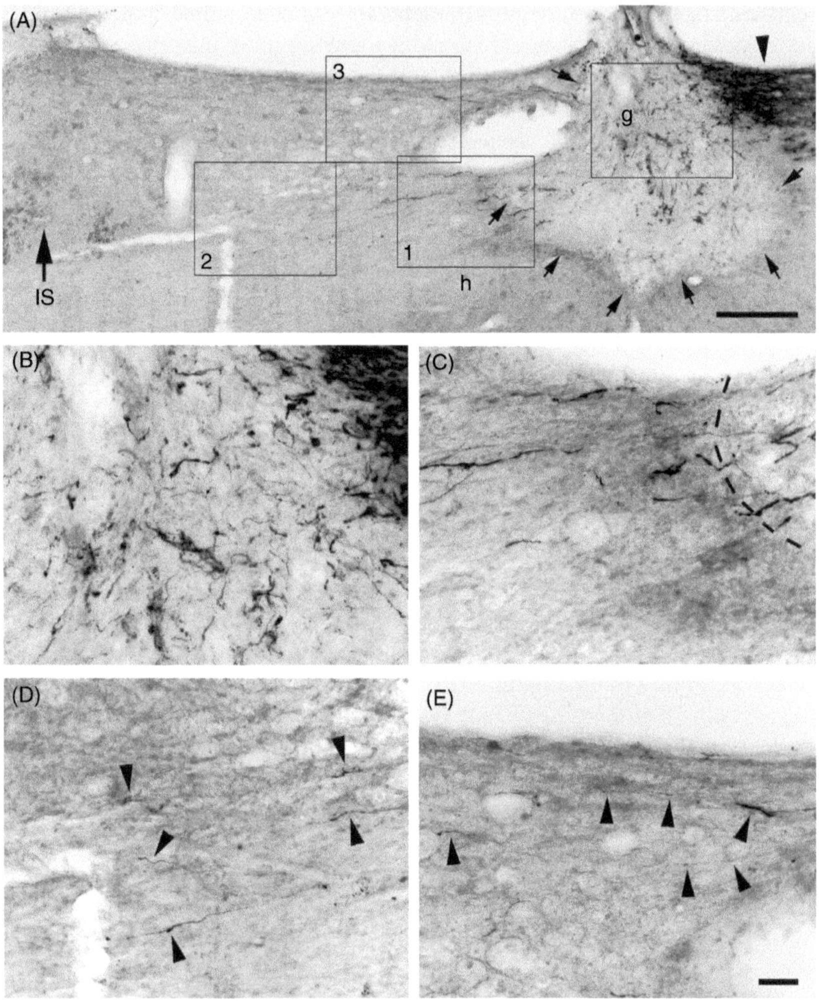

FIGURE 13.2 Sensory axons regenerate beyond spinal cord lesion sites after combined administration of intraganglionic cAMP and axonal application of NT-3. (A) Lower magnification view of sagittal section of spinal cord illustrating CTB-labeled dorsal column sensory axons approaching lesion site (arrowhead on upper right), MSC graft in lesion cavity (g; arrows indicate host/graft interface), and region rostral to lesion site (left side of figure). Large arrow and IS indicate rostral Injection Site of NT-3. (B) Boxed area of graft: CTB-labeled sensory axons penetrate graft in lesion site. (C) Higher magnification of box 1 from panel A, demonstrating crossing of CTB-labeled axons from graft into host white matter beyond the graft. Dashed lines indicate host/graft interface. This crossing occurs at a point well away from dorsal or ventral lesion regions, reducing likelihood that axons were spared by lesion. Lesion completeness was confirmed by failure to observe CTB-labeled sensory axons in medulla in lesioned subjects. (D) Higher magnification of box 2 from panel A, demonstrating several varicose CTB-labeled axons extending 0.5–0.7 mm beyond lesion site (arrowheads). (E) Higher magnification of box 3 from panel A, showing additional axons extending under the dorsal aspect of the spinal cord beyond the lesion site. Scale bars =280 μm (A), 44 μm (B–E). (From Lu et al., *J Neurosci* 2004; 24: 6406).

CONCLUSION

Great progress has been made over the last two decades in defining mechanisms that contribute to limitations in CNS axonal plasticity and regeneration after injury. Correspondingly, these insights have identified therapeutic directions. While collectively it is clear that several independent approaches can enhance the growth of CNS axons after injury, testing of combination therapies generally reveals greater effects than individual approaches. The experimental and practical hurdles to identifying optimal combination therapies are not trivial, but must be overcome to generate approaches that may ultimately benefit the human condition.

ACKNOWLEDGMENTS

Supported by the NIH (NS09881, NS42291), the Veterans Administration, the Canadian Spinal Research Organization, and the Adelson Medical Foundation.

REFERENCES

Assouline JG, Bosch P, Lim R et al. (1987). Rat astrocytes and Schwann cells in culture synthesize nerve growth factor-like neurite-promoting factors. *Brain Res* 428: 103–118.

Benowitz LI, Goldberg DE, Madsen JR et al. (1999). Inosine stimulates extensive axon collateral growth in the rat corticospinal tract after injury. *Proc Natl Acad Sci U S A* 96, 13486–13490.

Benowitz LI, Goldberg DE, and Irwin N (2002). Inosine stimulates axon growth in vitro and in the adult CNS. *Prog Brain Res* 137: 389–399.

Bhatheja K and Field J (2006). Schwann cells: origins and role in axonal maintenance and regeneration. *Int J Biochem Cell Biol* 38: 1995–1999.

Blesch A, Uy HS, Grill RJ et al. (1999). Leukemia inhibitory factor augments neurotrophin expression and corticospinal axon growth after adult CNS injury. *J Neurosci* 19: 3556–3566.

Blesch A and Tuszynski MH (2003). Cellular GDNF delivery promotes growth of motor and dorsal column sensory axons after partial and complete spinal cord transections and induces remyelination. *J Comp Neurol* 467: 403–417.

Blesch A, Yang H, Weidner Net al. (2004). Axonal responses to cellularly delivered NT-4/5 after spinal cord injury. *Mol Cell Neurosci* 27: 190–201.

Blits B, Oudega M, Boer GJ et al. (2003). Adeno-associated viral vector-mediated neurotrophin gene transfer in the injured adult rat spinal cord improves hind-limb function. *Neuroscience* 118: 271–281.

Boyd JG and Gordon T (2003). Neurotrophic factors and their receptors in axonal regeneration and functional recovery after peripheral nerve injury. *Mol Neurobiol* 27: 277–324.

Bradbury EJ, Khemani S, King VR et al. (1999). NT-3 promotes growth of lesioned adult rat sensory axons ascending in the dorsal columns of the spinal cord. *Eur J Neurosci* 11: 3873–3883.

Bradbury EJ, Moon LD, Popat RJ et al. (2002). Chondroitinase ABC promotes functional recovery after spinal cord injury. *Nature* 416: 636–640.

Bregman BS, McAtee M, Dai HN et al. (1997). Neurotrophic factors increase axonal growth after spinal cord injury and transplantation in the adult rat. *Exp Neurol* 148: 475–494.

Buchli AD and Schwab ME (2005). Inhibition of Nogo: a key strategy to increase regeneration, plasticity and functional recovery of the lesioned central nervous system. *Ann Med* 37: 556–567.

Chernousov MA and Carey DJ (2000). Schwann cell extracellular matrix molecules and their receptors. *Histol Histopathol* 15: 593–601.

Costigan M, Befort K, Karchewski L et al. (2002). Replicate high-density rat genome oligonucleotide microarrays reveal hundreds of regulated genes in the dorsal root ganglion after peripheral nerve injury. *BMC Neurosci* 3: 16.

David S and Aguayo AJ (1981). Axonal elongation into peripheral nervous system "bridges" after central nervous system injury in adult rats. *Science* 214: 931–933.

Dergham P, Ellezam B, Essagian C et al. (2002). Rho signaling pathway targeted to promote spinal cord repair. *J Neurosci* 22: 6570–6577.

Dubovy P (2004). Schwann cells and endoneurial extracellular matrix molecules as potential cues for sorting of regenerated axons: a review. *Anat Sci Int* 79: 198–208.

Fawcett JW (2006). Overcoming inhibition in the damaged spinal cord. *J Neurotrauma* 23: 371–383.

Fischer D, Petkova V, Thanos S et al. (2004). Switching mature retinal ganglion cells to a robust growth state in vivo: gene expression and synergy with RhoA inactivation. *J Neurosci* 24: 8726–8740.

Fouad K, Schnell L, Bunge MB, Schwab ME., Liebscher T, and Pearse DD (2005). Combining Schwann cell bridges and olfactory-ensheathing glia grafts with chondroitinase promotes locomotor recovery after complete transection of the spinal cord. *J Neurosci* 25: 1169–1178.

Fournier AE, GrandPre T, and Strittmatter SM (2001). Identification of a receptor mediating Nogo-66 inhibition of axonal regeneration. *Nature* 409: 341–346.

Giehl KM and Tetzlaff W (1996). BDNF and NT-3, but not NGF, prevent axotomy-induced death of rat corticospinal neurons in vivo. *Eur J Neurosci* 8: 1167–1175.

GrandPre T, Li S, and Strittmatter SM (2002). Nogo-66 receptor antagonist peptide promotes axonal regeneration. *Nature* 417: 547–551.

Grill R, Murai K, Blesch A et al. (1997). Cellular delivery of neurotrophin-3 promotes corticospinal axonal growth and partial functional recovery after spinal cord injury. *J Neurosci* 17: 5560–5572.

Houle JD, Tom VJ, Mayes D et al. (2006). Combining an autologous peripheral nervous system "bridge" and matrix modification by chondroitinase allows robust, functional regeneration beyond a hemisection lesion of the adult rat spinal cord. *J Neurosci* 26: 7405–7415.

Jackeman LB, Wei P, Guan Z et al. (1998). Brain-derived neurotrophic factor stimulates hindlimb stepping and sprouting of cholinergic fibers after spinal cord injury. *Exp Neurol* 154: 170–184.

Jenkins R, McMahon SB, Bond AB et al. (1993). Expression of c-Jun as a response to dorsal root and peripheral nerve section in damaged and adjacent intact primary sensory neurons in the rat. *Eur J Neurosci* 5: 751–759.

Jin Y, Fischer I, Tessler A et al. (2002). Transplants of fibroblasts genetically modified to express BDNF promote axonal regeneration from supraspinal neurons following chronic spinal cord injury. *Exp Neurol* 177: 265–275.

Johnson EM, Jr, Taniuchi M, and DiStefano PS (1988). Expression and possible function of nerve growth factor receptors on Schwann cells. *Trends Neurosci* 11: 299–304.

Jones LL, Yamaguchi Y, Stallcup WB et al. (2002). NG2 is a major chondroitin sulfate proteoglycan produced after spinal cord injury and is expressed by macrophages and oligodendrocyte progenitors. *J Neurosci* 22: 2792–2803.

Jones LL, Sajed D, and Tuszynski MH (2003). Axonal regeneration through regions of chondroitin sulfate proteoglycan deposition after spinal cord injury: a balance of permissiveness and inhibition. *J Neurosci* 23: 9276–9288.

Kim BG, Dai HN, Lynskey JV et al. (2006). Degradation of chondroitin sulfate proteoglycans potentiates transplant-mediated axonal remodeling and functional recovery after spinal cord injury in adult rats. *J Comp Neurol* 497: 182–198.

Kobayashi NR, Fan DP, Giehl KM et al. (1997). BDNF and NT-4/5 prevent atrophy of rat rubrospinal neurons after cervical axotomy, stimulate GAP-43 and Talpha1-tubulin mRNA expression, and promote axonal regeneration. *J Neurosci* 17: 9583–9595.

Kury P, Stoll G, and Muller HW (2001). Molecular mechanisms of cellular interactions in peripheral nerve regeneration. *Curr Opin Neurol* 14: 635–639.

Kwon BK, Liu J, Messerer C et al. (2002). Survival and regeneration of rubrospinal neurons 1 year after spinal cord injury. *Proc Natl Acad Sci U S A* 99: 3246–3251.

Lakatos A and Franklin RJ (2002). Transplant mediated repair of the central nervous system: an imminent solution? *Curr Opin Neurol* 15: 701–705.

Liu Y, Kim D, Himes BT et al. (1999). Transplants of fibroblasts genetically modified to express BDNF promote regeneration of adult rat rubrospinal axons and recovery of forelimb function. *J Neurosci* 19: 4370–4387.

Longo FM, Skaper SD, Manthorpe M et al. (1983). Temporal changes of neuronotrophic activities accumulating in vivo within nerve regeneration chambers. *Exp Neurol* 81: 756–769.

Lu P, Blesch A, and Tuszynski MH (2001). Neurotrophism without neurotropism: BDNF promotes survival but not growth of lesioned corticospinal neurons. *J Comp Neurol* 436: 456–470.

Lu P., Jones LL, Snyder EY et al. MH (2003). Neural stem cells constitutively secrete neurotrophic factors and promote extensive host axonal growth after spinal cord injury. *Exp Neurol* 181: 115–129.

Lu P, Yang H, Jones LL et al. (2004). Combinatorial therapy with neurotrophins and cAMP promotes axonal regeneration beyond sites of spinal cord injury. *J Neurosci* 24: 6402–6409.

Lu P, Jones LL, and Tuszynski MH (2005). BDNF-expressing marrow stromal cells support extensive axonal growth at sites of spinal cord injury. *Exp Neurol* 191: 344–360.

Massey JM, Amps J, Viapiano MS et al. (2007). Increased chondroitin sulfate proteoglycan expression in denervated brainstem targets following spinal cord injury creates a barrier to axonal regeneration overcome by chondroitinase ABC and neurotrophin-3. *Exp Neurol* (Epub ahead of print)

McKerracher L, David S, Jackson DL et al. (1994). Identification of myelin-associated glycoprotein as a major myelin-derived inhibitor of neurite growth. *Neuron* 13: 805–811.

McQuarrie IG, Grafstein B, and Gershon MD (1977). Axonal regeneration in the rat sciatic nerve: effect of a conditioning lesion and of dbcAMP. *Brain Res* 132: 443–453.

Menei P, Montero-Menei C, Whittemore SR et al. (1998). Schwann cells genetically modified to secrete human BDNF promote enhanced axonal regrowth across transected adult rat spinal cord. *Eur J Neurosci* 10: 607–621.

Meyer M, Matsuoka I, Wetmore C et al. (1992). Enhanced synthesis of brain-derived neurotrophic factor in the lesioned peripheral nerve: different mechanisms are responsible for the regulation of BDNF and NGF mRNA. *J Cell Biol* 119: 45–54.

Mi S, Lee X, Shao Z et al. (2004). LINGO-1 is a component of the Nogo-66 receptor/p75 signaling complex. *Nat Neurosci* 7: 221–228.

Moreau-Fauvarque C, Kumanogoh A, Camand E et al. (2003). The transmembrane semaphorin Sema4D/CD100, an inhibitor of axonal growth, is expressed on oligodendrocytes and upregulated after CNS lesion. *J Neurosci* 23: 9229–9239.

Navarro X, Vivo M, and Valero-Cabre A (2007). Neural plasticity after peripheral nerve injury and regeneration. *Prog Neurobiol* 82: 163–201.

Neumann S and Woolf CJ (1999). Regeneration of dorsal column fibers into and beyond the lesion site following adult spinal cord injury. *Neuron* 23: 83 91.

Neumann S, Bradke F, Tessier-Lavigne M et al. (2002). Regeneration of sensory axons within the injured spinal cord induced by intraganglionic cAMP elevation. *Neuron* 34: 885–893.

Nikulina E, Tidwell JL, Dai HN et al.(2004). The phosphodiesterase inhibitor rolipram delivered after a spinal cord lesion promotes axonal regeneration and functional recovery. *Proc Natl Acad Sci U S A* 101: 8786–8790.

Novikova LN, Novikov LN, and Kellerth JO (2003). Biopolymers and biodegradable smart implants for tissue regeneration after spinal cord injury. *Curr Opin Neurol* 16: 711–715.

Park JB, Yiu G, Kaneko S et al. (2005). A TNF receptor family member, TROY, is a coreceptor with Nogo receptor in mediating the inhibitory activity of myelin inhibitors. *Neuron* 45: 345–351.

Pearse DD, Pereira FC, Marcillo AE et al. (2004). cAMP and Schwann cells promote axonal growth and functional recovery after spinal cord injury. *Nat Med* 10: 610–616.

Qiu J, Cai D, Dai H et al. (2002). Spinal axon regeneration induced by elevation of cyclic AMP. *Neuron* 34: 895–903.

Ramer MS, Priestley JV, and McMahon SB (2000). Functional regeneration of sensory axons into the adult spinal cord. *Nature* 403: 312–316.

Ramer MS, Bradbury EJ, Michael GJ et al. (2003). Glial cell line-derived neurotrophic factor increases calcitonin gene-related peptide immunoreactivity in sensory and motoneurons in vivo. *Eur J Neurosci* 18: 2713–2721.

Rossi F, Gianola S, and Corvetti L (2007). Regulation of intrinsic neuronal properties for axon growth and regeneration. *Prog Neurobiol* 81: 1–28.

Ruitenberg MJ, Plant GW, Hamers FP et al. (2003). Ex vivo adenoviral vector-mediated neurotrophin gene transfer to olfactory ensheathing glia: effects on rubrospinal tract regeneration, lesion size, and functional recovery after implantation in the injured rat spinal cord. *J Neurosci* 23: 7045–7058.

Schnell L and Schwab ME (1990). Axonal regeneration in the rat spinal cord produced by an antibody against myelin-associated neurite growth inhibitors. *Nature* 343: 269–272.

Schnell L, Schneider R, Kolbeck R et al. (1994). Neurotrophin-3 enhances sprouting of corticospinal tract during development and after adult spinal cord lesion. *Nature* 367: 170–173.

Schwab ME and Caroni P (1988). Oligodendrocytes and CNS myelin are nonpermissive substrates for neurite growth and fibroblast spreading in vitro. *J Neurosci* 8: 2381–2393.

Shao Z, Browning JL, Lee X et al. (2005). TAJ/TROY, an orphan TNF receptor family member, binds Nogo-66 receptor 1 and regulates axonal regeneration. *Neuron* 45: 353–359.

Silver J and Miller JH (2004). Regeneration beyond the glial scar. *Nat Rev Neurosci* 5: 146–156.

Song XY, Zhong JH, Wang X et al. (2004). Suppression of p75NTR does not promote regeneration of injured spinal cord in mice. *J Neurosci* 24: 542–546.

Steward O, Zheng B, Banos K et al. (2007). Response to: Kim et al., "axon regeneration in young adult mice lacking Nogo-A/B." Neuron 38, 187–199. *Neuron* 54: 191–195.

Stokols S, Sakamoto J, Breckon C et al. (2006). Templated agarose scaffolds support linear axonal regeneration. *Tissue Eng* 12: 2777–2787.

Taylor L, Jones L, Tuszynski MH et al. (2006). Neurotrophin-3 gradients established by lentiviral gene delivery promote short-distance axonal bridging beyond cellular grafts in the injured spinal cord. *J Neurosci* 26: 9713–9721.

Terenghi G (1999). Peripheral nerve regeneration and neurotrophic factors. *J Anat* 194 (Pt 1): 1–14.

Tetzlaff W, Alexander SW, Miller FD et al. (1991). Response of facial and rubrospinal neurons to axotomy: changes in mRNA expression for cytoskeletal proteins and GAP-43. *J Neurosci* 11: 2528–2544.

Tuszynski MH, Peterson DA, Ray J et al. (1994). Fibroblasts genetically modified to produce nerve growth factor induce robust neuritic ingrowth after grafting to the spinal cord. *Exp Neurol* 126: 1–14.

Tuszynski MH, Gabriel K, Gage FH et al. (1996). Nerve growth factor delivery by gene transfer induces differential outgrowth of sensory, motor, and noradrenergic neurites after adult spinal cord injury. *Exp Neurol* 137: 157–173.

Tuszynski MH (2002). Growth-factor gene therapy for neurodegenerative disorders. *Lancet Neurol* 1: 51–57.

Wang KC, Kim JA, Sivasankaran R et al. (2002a). P75 interacts with the Nogo receptor as a co-receptor for Nogo, MAG and OMgp. *Nature* 420: 74–78.

Wang KC, Koprivica V, Kim JA et al. (2002b). Oligodendrocyte-myelin glycoprotein is a Nogo receptor ligand that inhibits neurite outgrowth. *Nature* 417: 941–944.

Williams LR, Longo FM, Powell HC et al. (1983). Spatial-temporal progress of peripheral nerve regeneration within a silicone chamber: parameters for a bioassay. *J. Comp Neurol* 218: 460–470.

Winton MJ, Dubreuil CI, Lasko D et al. (2002). Characterization of new cell permeable C3-like proteins that inactivate Rho and stimulate neurite outgrowth on inhibitory substrates. *J Biol Chem* 277: 32820–32829.

Xu XM, Guenard V, Kleitman N et al. (1995). A combination of BDNF and NT-3 promotes supraspinal axonal regeneration into Schwann cell grafts in adult rat thoracic spinal cord. *Exp Neurol* 134: 261–272.

Yamashita T and Tohyama M (2003). The p75 receptor acts as a displacement factor that releases Rho from Rho-GDI. *Nat Neurosci* 6: 461–467.

Ye JH and Houle JD (1997). Treatment of the chronically injured spinal cord with neurotrophic factors can promote axonal regeneration from supraspinal neurons. *Exp Neurol* 143: 70–81.

Zheng B, Atwal J, Ho C et al. (2005). Genetic deletion of the Nogo receptor does not reduce neurite inhibition in vitro or promote corticospinal tract regeneration in vivo. *Proc Natl Acad Sci U S A* 102: 1205–1210.

14

Modifying the Extracellular Matrix as a Treatment to Improve Functional Recovery after Spinal Cord Injury

James W. Fawcett

Cambridge University Centre for Brain Repair, Robinson Way, Cambridge CB2 2PY, UK

Abstract

The CNS extracellular matrix has many functions relevant to spinal cord repair. The main axon growth inhibitory activity in glial scar tissue comes from chondroitin sulphate proteoglycans (CSPGs), and is mainly due to their sulphated glycosaminoglycan chains. Digestion with chondroitinase removes these chains and promotes axon regeneration. Sialidiase also promotes axon regeneration, probably by affecting responses to MAG. The CSPGs are also involved in the termination of plasticity at the end of critical periods, probably through their association with perineuronal nets which form around neurons at this time. Digestion with chondroitinase can re-activate plasticity in the adult CNS. Tenascin-R is present in perineuronal nets, and animals lacking it have increased plasticity. The formation of perineuronal nets is triggered by the production of a link protein, which may be a future therapeutic target.

Copyright © 2008 Elsevier Inc.
All rights of reproduction in any form reserved.

THE CNS EXTRACELLULAR MATRIX

There is an abundant extracellular matrix (ECM) in the mammalian central nervous system (CNS), making up approximately 20% of its volume. The extent of the matrix was only appreciated recently, because the aldehyde fixatives used for electron microscopy dissolve many matrix components, and in earlier studies hypotonic fixatives were used that expanded the cells to occupy the vacated space. The ECM is a complex structure that fulfills several roles. It comprises the molecules that form the structure of the matrix, but also present in the matrix compartment are a huge range of molecules that have been shed from the cell surface or secreted by neurons and glia. Most of these shed or secreted molecules bind to the matrix to some extent, mainly to the negatively charged glycosaminoglycan (GAG) chains of the chondroitin sulfate proteoglycans (CSPGs) and heparan sulfate proteoglycans (HSPGs). For many molecules, binding to GAGs influences their localization with the CNS extracellular compartment, but may also be necessary for their action on their receptors. The ECM can therefore have effects through the direct actions of matrix molecules, through presenting diffusible and other molecules to receptors, through modulating the binding of cell surface molecules with diffusible, shed, or matrix-associated molecules and through its structural properties (Gallagher, 2006).

The backbone of the CNS ECM is the repeating disaccharide polymer hyaluronan, produced through the action of hyaluronan synthase enzymes situated under the cell surface. This disaccharide polymer is unsulfated, unlike the GAGs of HSPGs and CSPGs. The CSPGs consist of a core protein to which is attached one or more chondroitin sulfate (CS) GAG chains. The majority of the CSPGs belong to the lectican family (neurocan, aggrecan, brevican, versican) which has a link protein module in their protein core, enabling them to bind to hyaluronan. In addition, there are CSPGs that do not belong to this family, phosphacan, NG2 and neuroglycan C. Phosphacan is the secreted form of the transmembrane phosphatase RPTP beta/zeta. NG2 is membrane-attached, found on oligodendrocyte precursors but also shed into the matrix, and neuroglycan C is present on some neurons. Several of the CSPGs bind to tenascin, the main glycoprotein of the CNS ECM, and laminin is also a binding partner. Three link proteins are found in the ECM, Crtl1, Bral2 and Bral1 which act to stabilize the binding of the CPSGs to hyaluronan. All of these molecules bind together to form a complex interlinked and stable structure, giving the ECM its structural properties (Galtrey and Fawcett, 2007). The degree to which the CNS matrix is condensed and organized varies. There are highly condensed and stable matrix structures not unlike cartilage known as perineuronal nets (PNNs), which form late in postnatal development around some neurons, and there are similar structures at nodes of Ranvier (Celio et al., 1998). Much of the CNS ECM, however, is a rather diffuse structure which can be dissolved away in simple physiological buffers. The condensed matrix of PNNs is formed around neurons between 14 and 30 days in rats, probably around 2–5 years in humans. In the parts of the

CNS so far examined, all the neurons surrounded by these structures express a hyaluronan synthase enzyme and either express a link protein, or are contacted by terminals of a neuron that produces link protein (Oohashi et al., 2002; Bekku et al., 2003; Carulli et al., 2005a,b, 2007). The PNNs can be demolished by digestion with hyaluronidase, which also releases CSPGs, particularly neurocan, from CNS tissue, showing that hyaluronan is the backbone of these structures. The CSPG and tenascin components of PNNs are produced both by the neurons and by surrounding glia. At the moment of PNN formation during development, or when formation is triggered in the visual cortex by light exposure after dark rearing, neurons begin to produce link protein (Carulli et al., 2007). The link protein Ctrl1 is therefore probably the initiator of formation of these organized matrix structures, and preliminary evidence suggests that mice lacking Ctrl1 in their CNS lack PNNs around the dendrites of their neurons.

There are two families of cell surface attached HSPGs, the transmembrane syndecans and the GPI-linked glypicans. These are found on the surface of both neurons and glia in the CNS. In addition, there is perlecan in basement membranes. The HSPGs are widely expressed during all stages of development and in the adult (Yamaguchi, 2001).

There are two members of the tenascin family, tenascin-C and tenascin-R. Tenascin-C is the predominant form during development and is upregulated again after injury. Tenascin-R is found particularly in the PNNs in the adult CNS. Tenascin-R and link protein are involved in the stabilization of these structures, because in knockout animals the PNNs are not fully formed (Bruckner et al., 2000; Joester and Faissner, 2001). Because many of the CSPGs bind to tenascin it is an important component of the interlocking set of molecules that together construct the CNS matrix. Also found in the ECM are the glycoproteins, thrombospondin and osteopontin. These molecules interact with cell surface integrins, but their interactions with other ECM molecules are less well studied than those of laminin and tenascins.

Attached to the cell surface is the adhesion molecule N-CAM. This may be modified by the addition of polysialic acid chains, producing bulky matrix around these cells which can prevent the interaction of adhesion molecules with neighboring cell surfaces (Bruses and Rutishauser, 2001).

The structures of the CNS matrix bind to many other types of molecule. It is probably true that almost all the molecules released by or cleaved from cells interact with the ECM to some degree, and particularly with the GAG chains of the CSPGs and HSPGs. These GAG chains are repeating disaccharides that are modified by sulfation and epimerization. Sulfation can occur at several positions on the disaccharide residues; each form of sulfation being performed by one or more sulfotransferase enzymes found in the Golgi. There are also extracellular sulfatases that can modify the HSPGs after they have reached the cell surface (Esko and Selleck, 2002; Uchimura et al., 2006). By sulfating the GAG chains in different positions and patterns and by epimerization, charge patterns can be produced that endow the GAG chains with binding sites that can be of high specificity and affinity. The binding of molecules to these GAG epitopes can localize particular molecules to particular places, and is essential for the function of many molecules (Yamaguchi, 2001). Members of the fibroblast growth factor (FGF) family, for instance, bind to HS GAG where those GAGs are sulfated in such a way as to provide a binding site. Some FGFs have little biological activity in the presence of their receptor unless there is an HS GAG present which can form a ternary complex between FGF, FGF receptor and GAG (Allen and Rapraeger, 2003). Some axon guidance molecules, for instance slits and netrins, rely on interactions with GAGs for their normal developmental effects (Kantor et al., 2004; Hussain et al., 2006). The pattern of sulfation of GAGs, and the subsequent modification of HS GAGs by sulfatases, is therefore a major determinant of the function of the CNS ECM. It is therefore of interest that the sulfation pattern of both CS and HS GAGs changes during development, changes after injury, and is different between the organized matrix of PNNs and the diffuse soluble matrix (Kitagawa et al., 1997; Properzi et al., 2005; Deepa et al., 2006). It is possible to extract matrix enriched in PNN components, and the GAG from this form of matrix has different binding properties to that of GAG from the diffused matrix compartment. For instance, brain-derived neurotrophic factor (BDNF) binds with high affinity to PNN GAG but with low affinity to GAG from soluble matrix.

THE EXTRACELLULAR MATRIX IN AXON REGENERATION

Interest in the role of the ECM in axon regeneration came from observations showing that various matrix components, particularly tenascin-C and CSPGs, are upregulated in regions of CNS damage (Brodkey et al., 1995; Silver and Miller, 2004). Tenascin-C is anti-adhesive to many forms of neuron in vitro, and inhibits axon growth from many neurons, although it promotes axon growth from some embryonic neuronal types (Meiners et al., 1999; Rigato et al., 2002). There is a large alternatively spliced domain in the fibronectin repeats of tenascin, in which there are probably two domains that can promote axon growth, one via F3/F11/contactin, one probably via a beta1 integrin (Rigato et al., 2002). One of these domains, the FN-D domain is upregulated in the injured CNS, but it is not clear whether CNS axons possess an integrin that can interact with it. Interest in the CSPGs as possible inhibitory molecules in the injured CNS came from three main sources. First, these molecules in vitro are generally inhibitory to the growth of most but not all types of axon (Grumet et al., 1996). Second, the molecules are upregulated in all sites of CNS injury. Third, comparing astrocyte cell lines that are inhibitory or permissive for axon growth showed that the inhibitory cell lines produce inhibitory ECM, with CSPGs being mainly responsible for this inhibition (Fidler et al., 1999; Carulli et al., 2005a,b), and fourth, preventing the sulfation of CSPGs produced by astrocytes removed much of their inhibitory activity on axon regeneration (Smith-Thomas et al., 1995). Where the CNS is injured there is a change in expression of several CSPGs. Neurocan, versican, NG2, brevican and phosphacan are upregulated, produced mainly by glial cells. Neurocan and phosphacan are produced by astrocytes and oligodendrocyte precursors, versican V2 and NG2 by oligodendrocyte precursors, brevican by astrocytes (Morgenstern et al., 2002; Jones et al., 2002). The role of individual CSPGs in producing inhibition of axon regeneration is not clear, since it is probable that they can compensate for one another's function to some extent. The role of NG2 has been examined in a knockout animal and using an NG2 antibody with somewhat contradictory results, the knockout showing no regeneration after spinal cord injury and infusion of the antibody producing increased regeneration (Tan et al., 2006). The inhibitory properties of CSPGs depend largely on their GAG chains, as demonstrated by the reduction of inhibition after chondroitinase treatment, and the GAGs must be sulfated in order to be inhibitory. The composition of the CS GAGs changes somewhat after injury. There is a large overall increase in the quantity of GAG in regions of injury, and an increase in the proportion of GAG that is 6-sulfated. However, the GAG-sulfation pattern does not revert to the embryonic pattern in which most GAG is 6-sulfated and the most of the GAG in the injured CNS remains 4-sulfated (Properzi et al., 2005). The most effective strategy for promoting axon regeneration has been to use a treatment that targets all the CSPGs by removing their GAG chains, the enzyme chondroitinase, as described later.

THE EXTRACELLULAR MATRIX IN PLASTICITY

Plasticity is rather a ill-defined term that encompasses the sum of axon sprouting, terminal sprouting, new synapse formation and changes in synaptic strength. Although the term is imprecise, it is widely used in the field of CNS repair, because following interventions or injuries to the nervous system there is a spontaneous return of function, the precise mechanism of which is not fully understood. However, it is probably mainly due to plasticity, through a combination of all the above mechanisms. There are really no situations in which the full set of structural events that underlie behavioural recovery after injury have been defined, hence the continued usefulness of the term "plasticity". There is evidence that the properties of the ECM can affect all the types of change described above.

A characteristic of CNS plasticity is the existence of critical periods. These are windows shortly after birth during which there is a high level of plasticity. They were originally described in the visual system, particularly around ocular dominance plasticity which is seen in young mammals (rats up to 35 days, humans up to 5–7 years). There is now evidence for critical periods for plasticity in most other part of the CNS, including the spinal cord where it terminates rather earlier (Buonomano and Merzenich, 1998). There is a general observation that young animals and humans during critical periods can compensate more completely for CNS damage than adults, known as the Kennard principle, and it is assumed that this is due to greater levels of plasticity. High levels of plasticity, however, can be maladaptive. This may be seen in young children after traumatic brain injury, where the comprehensive reorganization can lead to cognitive decline (Kolb and Whishaw, 1998; Montour-Proulx et al., 2004). It is probable that increased plasticity, particularly if it is restricted to chosen regions of the CNS, could be of considerable benefit following CNS damage. Several of the treatments that have been shown to produce behavioural improvements after spinal cord injury, including NogoA-blocking antibodies, inosine and chondroitinase, promote both axon regeneration and plasticity, and it is probable that much of their effect is through promoting plasticity (Schwab, 2002; Rhodes and Fawcett, 2004).

There are several changes in the CNS ECM around the time that critical periods end. Most visibly the formation of PNNs around neuronal cell bodies and their dendrites occurs at this time, and there are also changes in the sulfation pattern of the CS GAGs with a decrease in 6-sulfated GAG and an increase in the 4-sulfated form (Properzi et al., 2005). Also with development there is a general decrease in the amount of N-CAM that is polysialated, although levels remain higher in regions such as the hippocampus and olfactory bulb where higher levels of plasticity remain (Bruses and Rutishauser, 2001).

How might these matrix changes affect CNS plasticity? At present the mechanisms are not well understood. There is much evidence that integrins play an important part in controlling the motility of dendritic spines, in the insertion of glutamate receptors into synapses and on synaptic strength (Rohrbough et al., 2000). The binding partners of these integrins are not defined, but there is evidence for a role of astrocyte-produced thrombospondin in synaptogenesis presumably via interaction with integrins, and tenascin and laminin are also possible integrin partners that are present (Christopherson et al., 2005). There is an extensive literature on the inhibition of integrin–laminin interactions by CSPGs in the promotion of axon growth, so it is likely that CSPGs around synapses could affect integrin function and thereby synaptic behavior. BDNF plays a key part in the control of long-term changes in synaptic strength and in the closure of critical periods (Mandolesi et al., 2005). The fact that BDNF binds with high affinity to the CS glycans of PNNs may be a link to matrix changes, critical periods and plasticity. The HSPGs, syndecans and glypicans, are another set of molecules that have functions in controlling plasticity. Syndecans are concentrated in dendritic spines, associated with ephrin B2, and levels of syndecan affect the number of spines (Yamaguchi, 2002). It is possible that CSPGs could affect the binding of syndecan with its partners. Tenasin-R is concentrated in PNNs, expressing the HNK-1 sugar epitope during early development, and this also affects synaptic strength (Dityatev and Schachner, 2006). Finally, there are many other trophic factors and diffusible molecules with potential activity on neurites and synapses that may bind to CSPGs and HSPGs in the matrix and therefore affect plasticity in different ways that may be controlled by the binding properties of the GAGs, which in turn are controlled by the patterns of sulfation on these long sugar chains.

There is a large literature addressing these issues, and readers wishing to know more are referred to two recent reviews (Hensch, 2004; Dityatev and Schachner, 2006).

PHYSIOLOGICAL MODIFICATION OF THE EXTRACELLULAR MATRIX

The matrix is modified in various ways during normal life, and some of these modifications are probably part of the mechanism that controls plasticity. The CSPGs, particularly neurocan and aggrecan, are secreted as large single polypeptide chains. However, in the adult CNS they are found mainly in forms that result from their cleavage by extracellular proteases. Thus neurocan has a single cleavage site resulting in the division of the molecule into neurocan-N and neurocan-C, and the majority of neurocan in the adult CNS is in this form. Neurocan-N contains the domain that associates with hyuloronan and is therefore the form found in PNNs. After injury, most of the upregulated neurocan is in the uncleaved form, indicating that proteolytic activity in the developing glial scar is not great (Asher et al., 2000). The physiological significance of the two forms of neurocan is not known. Cell surface CSPGs such as NG2 and RPTP-beta can be cleaved from the cell surface by local proteases and then appear as soluble CSPGs in the ECM, but again the significance of this is not understood. The proteolytic modification that has received the greatest attention is that initiated by plasticity-inducing changes, for instance those that initiate ocular dominance plasticity. An early event in this form of plasticity is the release of tissue plasminogen activator (TPA) from the relevant regions of the cortex. In the absence of TPA release, the plastic changes are absent or attenuated (Samson and Medcalf, 2006). Exactly what the TPA does is not well understood. The assumption is that it activates other extracellular proteases, leading in turn to modification of molecules in the ECM. Many of the CSPGs and other matrix molecules are very sensitive to proteolytic action, but it is not known which molecules have to be modified in order for cortical connectivity to change. There is also evidence that cleavage of pro-BDNF into its active form is involved (Pang et al., 2004).

TREATMENTS TO MODIFY THE EXTRACELLULAR MATRIX

The matrix being as complicated as it is, there are clearly many ways in which it can be modified, many outwith the scope of this chapter, in which we highlight the effects of enzymatic modification. The enzymes that have been used to modify CNS matrix are chondroitinase, which digests the CS–GAGs into disaccharides, heparinase and heparitinase which do the same for HS–GAGs, hyaluronidase which digests hyaluronan into disaccharides, sialidase which digests sialic acid chains, including the glycan-binding determinant on the inhibitory myelin molecule myelin-associated glycoprotein (MAG) and neuraminidase which removes the polysialic acid chains from N-CAM. Sulfatases have also been used to remove sulfates from HS–GAGs, with important effects on CNS development and tumor biology (Uchimura et al., 2006). From the perspective of CNS repair, the most interesting results have come from the use of chondroitinase. There are in fact several bacterial chondroitinases, labeled ABC, AC, AB, depending on their substrate specificity. (A, B and C refer to the sulfation and epimerization modifications of the GAG chains.) Most experiments have used chondroitinase ABC (ChABC), which digest all the types of CS–GAGs. The enzyme is produced by the bacterium *Proteus vulgaris*, and the preparation used for most experiments is extracted from it. ChABC in addition to its action on CS–GAGs, also has activity on hyaluronan, so that CNS regions injected with the enzyme show no binding of the hyaluronan-binding protein probe for a week or more afterwards. The digestion of CS–GAGs in tissue samples is partial, even if small pieces of tissue are digested in vitro. This presumably reflects the fact that CSPGs may be associated with large, stable aggregates of matrix, which presumably exclude the enzyme. The overall action of the enzyme is therefore complex. In addition to digestion of the CS–GAG chains, some CSPG core proteins, particularly neurocan, are now released from their binding to the ECM, and are therefore absent from regions in which chondroitinase has been injected, and may take several weeks to return (Galtrey et al., 2007). On the other hand, other CSPGs including aggrecan are not removed from their stable association with PNNs. After chondroitinase injection there is greatly increased general staining of the digested region with antibodies that recognize the short four-sugar adaptor stubs that are left attached to the protein cores, indicating that there is a considerable amount of CSPG core protein remaining. This stub staining remains for over a month in the brain, rather less in the spinal cord, showing that the digested proteins are turned over very slowly (Bruckner et al., 1998). Hyaluronidase digestion also releases CSPG core proteins from their attachment to the ECM, in some cases more effectively than ChABC. Thus following hyaluronidase injection, almost all the CSPGs associated with PNNs are absent from spinal cord tissue. Hyaluronidase also has some activity against CS–GAGs.

MATRIX MODIFICATION AND AXON REGENERATION

The rationale for using enzyme treatments to promote axon regeneration comes from experiments showing that CSPGs are the main axon growth inhibitory molecules produced by astrocytes, and the main inhibitory molecules of the glial scar (Fawcett and Asher, 1999; Grimpe and Silver, 2002). Individual purified CSPGs, including neurocan, aggrecan, NG2, brevican, versican and phosphacan have been shown to inhibit the growth of some axons (Margolis and Margolis,

1997). The degree of inhibition is both context and axon dependent. Thus a high concentration of laminin will overcome the inhibitory effects of CSPGs, and some embryonic axons will grow in the presence of CPSGs, particularly phosphacan/DSD-1 while growth of axons from other forms of neuron is inhibited (Rigato et al., 2002). Next came the demonstration that axon growth in three-dimensional astrocyte tissues in vitro is inhibited, although again embryonic axons are able to grow in this environment (Fawcett et al., 1989). The demonstration that the ECM of astrocytes is responsible for their inhibitory activity came from the study of clonal astrocyte cell lines, some permissive for axon regeneration, some inhibitory. It was shown that inhibition by these cells correlated with the inhibitory nature of their ECM, that this inhibition was produced by CSPGs and that inhibition due to astrocytes could be removed either by digestion with chondroitinase or by preventing the sulfation of the CSPGs by treatment with sodium chlorate (Smith-Thomas et al., 1994). Alongside these in vitro observations were demonstrations that CSPGs are greatly upregulated in regions of CNS damage and around other forms of CNS lesion such as amyloid plaques (DeWitt et al., 1994; McKeon et al., 1995; Hoke and Silver, 1996). The CSPGs in the injured brain were shown to be produced by astrocytes, oligodendrocyte precursors and meningeal cells. The logic, therefore, was that axon regeneration after injury fails within the environment of the glial scar, and that this scar tissue contains high levels of CPSG, whose inhibitory properties in relevant in vitro models are dependent on the CS–GAG chains. The first successful attempt to promote axon regeneration using enzymes used the nigro-striatal pathway. This was chosen because its axons are known to have a high regenerative potential, and are readily visualized. Chondroitinase was injected as a bolus on alternate days, the rationale being that successful digestion of CS will require a threshold level of enzyme, but sustained digestion is not required because of the slow turnover of matrix molecules. In these experiments, around 2000 out of a total of 40 000 axons regenerated through the glial scar to return to the striatum (Moon et al., 2001). This experiment has since been repeated, with similar results, and with the observation that chondroitinase treatment has an effect on the formation of a reactive astrocyte barrier to axon regeneration (Li et al., 2007). This success led to an experiment in which lesions of the dorsal columns of the spinal cord were treated, again using bolus infusions through a cannula overlying the lesion site. Regeneration of both ascending sensory and descending corticospinal axons was seen for distances of 7 mm or more, although the number of axons was modest (Bradbury et al., 2002). Again these experiments have been repeated with similar results (Caggiano et al., 2005). The enzyme has been particularly successful at inhibitory boundaries. Where Schwann cells and astrocytes interface, there is almost no mixing of the cell types, leading to a sudden change in glial environment. Axons are not able to cross these boundaries from a Schwann cell to an astrocyte environment, tending to stop and remain in the very attractive Schwann cell environment. It is this failure of axons to exit from Schwann cell or peripheral nerve grafts that makes them unsuitable for the repair of spinal cord injuries. Chondroitinase has now been applied to the ends of these grafts, with the result that axons are now able to exit the grafts back into the spinal cord, with return of spinal cord function (Steinmetz et al., 2005; Houle et al., 2006). Most importantly, this return of function was abolished when the nerve grafts were re-lesioned, showing unequivocally that these regenerated axons were responsible for the return of function. This is the first spinal cord repair experiment with a re-lesion component, and therefore the first in which it is clear that the regenerating axons rather than spinal cord plasticity is responsible for functional recovery. A similar barrier to regeneration is seen at the root entry zones. At the dorsal roots, sensory axons pass from a Schwann cell to an astrocyte environment, and, following damage to the dorsal root, axons would have to make the same transition. Similarly, patients may have the spinal roots avulsed from the cord as a result of trauma. Surgical attempts to re-implant the roots are not very successful because few, if any sensory axons are able to regenerate from the re-implanted dorsal root into the cord. In both these models, ChABC infusion has been successful in allowing axons to cross these Schwann cell/astrocyte boundaries to re-enter CNS territory (Grimpe et al., 2005; Yang et al., 2006). There are various other matrix-modifying enzymes that degrade the various forms of glycan chain. Some of these have overlapping activities. Thus ChABC also has some activity against hyaluronan, while testicular hyaluronidase has chondroitinase activity at some pH levels. However, testicular hyaluronidase treatment was not effective at promoting axon regeneration in the CNS (Moon et al., 2003). Sialidase was used in a recent experiment to promote regeneration of sensory axons into the spinal cord from an avulsed dorsal root (Yang et al., 2006). The rationale for the use of the enzyme was that it might affect the inhibitory activity of MAG. There are many actions of sialic acid structures in axonogenesis and development that are outwith the scope of this chapter. Heparinase and heparitinase degrade the glycan chains of HSPGs. In general, HSPGs promote axon growth through a variety of mechanisms, including the presentation of trophic factors and the signaling actions of syndecans. Digestion of these molecules would therefore probably not be helpful to CNS repair. However, there might be rationale for the use of heparitinase in some situations, because some inhibitory axon guidance molecules, particularly slit, bind to HSPGs and are presented by them, and slits are upregulated in the injured CNS (Wehrle et al., 2005; Hussain et al., 2006).

MATRIX MODIFICATION AND PLASTICITY

The thought that enzymatic treatment might promote plasticity came from experiments in which chondroitinase was used to treat partial lesions of the spinal cord at the C4 level. The treatment led to some axon regeneration, and also to impressive return of function (Bradbury et al., 2002). That return of function began within 1–2 weeks after chondroitinase treatment, while improvements that were entirely due to axon regeneration would be expected not to start for at least 3 weeks after injury. The conclusion was that some of the return of function caused by chondroitinase treatment was due to an effect on promoting plasticity. Since then, several further demonstrations have been made of the plasticity-promoting effect of chondroitinase. In the early postnatal visual system, there is ocular dominance plasticity. This is a consequence of the projections from the two eyes competing for connection space on the visual cortex. When one eye is disadvantaged by eyelid closure or a similar manipulation, the other eye gains increased cortical space, and becomes the dominant input to most cortical neurons. As with many forms of plasticity, ocular dominance plasticity has a critical period, after which it is much diminished. In rats, the critical period closes at around 35 days, in humans at around 6 years of age. Chondroitinase was injected around the visual cortex of adult rats which no longer possessed ocular dominance plasticity, and in which one eye was closed. Chondroitinase-treated animals showed a degree of plasticity that was almost as great as that seen during the critical period (Pizzorusso et al., 2002). In humans, ocular dominance plasticity can have adverse consequences when one eye is misaligned in a squint, or when one eye has a lens opacity. The affected eye almost completely loses the ability to influence cortical activity and produce vision, a condition known as amblyopia. If the eye defect is corrected before 5 years old, the eye can regain its influence on the cortex, but the older the patient the less it is possible to correct amblyopia. Pizzorusso et al. (2006) created a rat model of amblyopia by suturing closed one eye beyond the end of the critical period and were then able to make it possible for the rats to regain function of the eye when it was re-opened by treatment of the cortex with chondroitinase. In the medulla, the sensory input from the spinal cord terminates in a topographic manner in the sensory nuclei. After a dorsal column injury at C8, the cuneate nucleus is approximately four-fifths denervated. In normal adult animals, the remaining projection expands slightly into the vacated space, but after chondroitinase treatment the projection expands to a much greater extent (Massey et al., 2006). In another model of spinal cord plasticity, the median and ulnar nerves were repaired or crossed over to produce inaccurate regeneration of axons. In adult animals, the CNS is unable to compensate for the inaccurate sensory and muscle connections in the periphery and the result is poor function, as occurs in humans after peripheral nerve repair. In the rat, this was shown by poor skilled forepaw function and grip strength after nerve repair or crossover. Injection of chondroitinase into the spinal cord improved both skilled paw function and grip strength, suggesting that plasticity within the spinal cord had allowed compensation for the inaccurate peripheral connections (Galtrey et al., 2007). Chondroitinase injection into the injured spinal cord promotes sprouting of several descending and sensory pathways (Barritt et al., 2006).

How might chondroitinase treatment promote CNS plasticity? The answer is not fully worked out, but the possible way in which the ECM controls plasticity is becoming clearer. The CSPGs in the CNS exist in three main compartments.

There is a large amount of free-floating ECM, which can be extracted by gentle dissociation in saline or buffer solutions. This form of matrix is plentiful throughout development and in the adult CNS. Some CSPGs are associated with cell membranes, and can be shed from there into the free-floating compartment. The third compartment consists of condensed and very stable matrix, which can only be extracted in solutions such as 6M Urea, or can be made soluble by digestion of hyaluronan using hyaluronidase. This form of matrix is mostly found around some classes of neurons in the form of PNNs (Deepa et al., 2006). These PNNs are condensations of matrix consisting of CSPGs, hyaluronan, tenascin-R and link proteins. They surround the cell bodies and dendrites of neurons, encasing the synapses in dense matrix (Zaremba et al., 1990; Hockfield et al., 1990; Lander et al., 1997; Matthews et al., 2002; Carulli et al., 2005a,b, 2007). The PNNs appear in the CNS at the time that critical periods end, and are digested by chondroitinase. Because they are found in a place where they could influence new synapse formation, because their formation is controlled by neuronal electrical activity, and because they appear at the right time and contain large amounts of CSPGs, it has been assumed that the PNNs might play a part in the control of plasticity and the ending of critical periods (Zaremba et al., 1989; Kalb and Hockfield, 1994; Lander et al., 1997; Pizzorusso et al., 2002). This has yet to be proven. However, it appears that the formation of PNNs is triggered by the production of link proteins by neurons, and a brain link protein knock-out animal is now available which will allow this question to be addressed, since it lacks PNNs on the dendrites in the CNS. There are other changes to CSPGs during development. The sulfation pattern of the GAG chains changes from a predominance of 6-sulfation during embryogenesis to 4-sulfation postnatally (Kitagawa et al., 1997; Properzi et al., 2005). The form of versican changes from the V0 and V1 forms which are found in embryos to the V2 form postnatally. However, these changes happen rather sooner in development than the ending of critical periods. At present, it is not known how the ECM might affect plasticity. The processes that underlie plasticity are axonal sprouting, new synapse formation and changes in synaptic strength. It is not at all clear which of these processes are promoted by chondroitinase digestion, nor, in most cases, is it clear which of these processes are involved in the functional changes that are ascribed to plasticity. Several matrix components can affect short-term synaptic strength, particularly tenascin [for review see (Dityatev and Schachner, 2006)]. The CSPGs have been shown to inhibit axon growth, and therefore presumably sprouting. It is possible that they also have direct effects on synapses and synapse formation. Many of the actions of proteoglycans are not through direct binding of the GAGs to a receptor, but rather due to their ability to bind and present other active molecules. The CS–GAGs in the ECM compartments of the CNS have different sulfation patterns, and would therefore be expected to have different binding specificities. They might, therefore, act to bind and present particular molecules to structures in contact with PNNs. We have recently shown that BDNF, which has many effects on synaptogenesis and plasticity, binds with greater affinity to PNN glycans than it does to the glycans of the free-floating or membrane-associated CSPGs. It is likely that other molecules associate specifically with PNNs in this way, and this could be one way in which they act. The CSPGs can also affect the interaction of integrins with matrix glycoproteins. Integrins are involved in long-term synaptic changes. They probably interact with glycoproteins such as tenascin-R, thrombospondin and laminins which are found in the ECM around synapses (Rohrbough et al., 2000). It is possible that the CSPGs in the matrix affect plastic changes through their ability to influence such interactions.

HOW MIGHT MODIFICATION OF THE EXTRACELLULAR MATRIX FORM PART OF A TREATMENT FOR SPINAL CORD INJURY?

It is generally accepted that an optimum therapy for spinal cord injury will involve more than one treatment. For instance, for the promotion of axon regeneration it makes sense in principle to combine a treatment that promotes the intrinsic ability of axons to grow with one that makes their environment more permissive. Modification of the ECM should be helpful in several ways. First, the glial scar that forms around injury sites or around the sites of implantation of glial grafts contains very high levels of inhibitory CSPG, and digestion of this material with chondroitinase should be helpful for the promotion of axon regeneration in this area. Second, matrix modification with chondroitinase promotes plasticity in the CNS through mechanisms that are as yet poorly understood. It will probably be useful to use chondroitinase to promote plasticity in the cord in the majority of patients that retain some functional axons passing through the injury site. Matrix modification may also be useful for allowing better integration of glial, stem or neuronal cells grafted into the brain or spinal cord. The question of how, when and where to apply the treatment remains to be worked out. In most of the experiments using chondroitinase, it has been applied acutely at or near the time of injury. Recent experiments from our laboratory indicate that treatment can be delayed at least for a week. However, the experiment in which chondroitinase was injected into the uninjured cord in order to promote plastic changes to compensate for inaccurate peripheral nerve regeneration suggests that the enzyme might be efficacious at almost any time, including long after injury (Galtrey et al., 2007). The site of treatment needs to be worked out. Clearly, for promoting axon regeneration in the glial scar application to the injury site is necessary. However, if the aim is to promote plasticity, a much more widespread treatment of the cord is indicated. The method of applying the treatment is problematic. When applied at a fresh injury site, the tissue is rather porous with few diffusion barriers, and the effect of the enzyme can be fairly widespread. However, enzyme infused intrathecally has to penetrate the meninges away from the injury site, which restricts penetration. Penetration of the enzyme is better than for most proteins, since by digesting the CSPGs in the ECM the enzyme removes a major diffusion barrier, but penetration is still limited. Infusion of chondroitinase intrathecally or ICV in rats produces widespread digestion at the injury site, and also produces digestion within the white matter of the uninjured part of the cord (Caggiano et al., 2005). However, grey matter digestion is absent or limited. Injection directly into the parenchyma of the cord on either side of the injury has been the method of administration in some recent experiments, and this certainly produces functional recovery. In human patients, there would be an opportunity to make direct injections at the time of spinal decompression or stabilization surgery, which is frequently performed after spinal cord injury. Chondroitinase, being a bacterial enzyme, is immunogenic. This will limit the number of treatments that can be given. However, our recent work has shown that a single injection can lower the glycan level for 10 days or more, and that the enzyme remains active at the lesion site for at least 3 days, so the number of injections required may not be large.

FUTURE PERSPECTIVES

A considerable amount of work still needs to be done to work out the best way to use chondroitinase for spinal cord repair. For the future it may also be possible to produce a less immunogenic enzyme by modification of groups that provoke the immune system. Recently a version that is more heat stable has been produced, which will be a useful development. Viral vectors will be produced that allow the enzyme to be expressed by spinal cord cells. However, expression of bacterial proteins in mammalian cells is accompanied by difficulties of coding sequence and glycosylation. The main development for the future, however, may come from a thorough understanding of how the ECM affects axon growth and plasticity, leading to new treatments with greater activity and specificity.

REFERENCES

Allen, BL and Rapraeger, AC (2003). Spatial and temporal expression of heparan sulfate in mouse development regulates FGF and FGF receptor assembly. *J Cell Biol* 163: 637–648.

Asher RA, Fidler PS, Morgenstern DA et al. (2000). Neurocan is upregulated in injured brain and in cytokine-treated astrocytes. *J Neurosci* 20: 2427–2438.

Barritt AW, Davies M, Marchand F et al. (2006). Chondroitinase ABC promotes sprouting of intact and injured spinal systems after spinal cord injury. *J Neurosci* 26: 10856–10867.

Bekku Y, Su WD, Hirakawa S et al. (2003). Molecular cloning of Bral2, a novel brain-specific link protein, and immunohistochemical colocalization with brevican in perineuronal nets. *Mol Cell Neurosci* 24: 148–159.

Bradbury EJ, Moon LDF, Popat RJ et al. (2002). Chondroitinase ABC promotes axon regeneration and functional recovery following spinal cord injury. *Nature* 416: 636–640.

Brodkey JA, Laywell ED, Obrien TF et al. (1995). Focal brain injury and up-regulation of a developmentally- regulated extracellular-matrix protein. *J Neurosurg* 82: 106–112.

Bruckner G, Bringmann A, Hartig W et al. (1998). Acute and long-lasting changes in extracellular-matrix chondroitin-sulphate proteoglycans induced by injection of chondroitinase ABC in the adult rat brain. *Exp Brain Res* 121: 300–310.

Bruckner G, Grosche J, Schmidt S et al. (2000). Postnatal development of perineuronal nets in wild-type mice and in a mutant deficient in tenascin-R. *J Comp Neurol* 428: 616–629.

Bruses JL and Rutishauser U (2001). Roles, regulation, and mechanism of polysialic acid function during neural development. *Biochimie* 83: 635–643.

Buonomano DV and Merzenich MM (1998). Cortical plasticity: from synapses to maps. *Annu Rev Neurosci* 21: 149–186.

Caggiano AO, Zimber MP, Ganguly A et al. (2005). Chondroitinase ABCI improves locomotion and bladder function following contusion injury of the rat spinal cord. *J Neurotrauma* 22: 226–239.

Carulli D, Deepa SS, and Fawcett JW. (2007). Upregulation of aggrecan, link protein 1 and hyaluronan synthases during formation of perineuronal nets in the rat cerebellum. *J Comp Neurol* 501: 83–94.

Carulli D, Laabs T, Geller HM et al. (2005a). Chondroitin sulfate proteoglycans in neural development and regeneration. *Curr Opin Neurobiol* 15: 116–120.

Carulli D, Rhodes KE, Brown DJ et al. (2005b). The composition of perineuronal nets in the adult rat cerebellum and the cellular origin of their components. *J Comp Neurol* 494: 559–577.

Celio MR, Spreafico R, De Biasi S et al. (1998). Perineuronal nets: past and present. *Trends Neurosci* 21: 510–514.

Christopherson KS, Ullian EM, Stokes CC et al. (2005). Thrombospondins are astrocyte-secreted proteins that promote CNS synaptogenesis. *Cell* 120: 421–433.

Deepa SS, Carulli D, Galtrey C et al. (2006). Composition of perineuronal net extracellular matrix in rat brain: a different disaccharide composition for the net-associated proteoglycans. *J Biol Chem* 281(26): 17789–17800.

DeWitt DA, Richey PL, Praprotnik D et al. (1994). Chondroitin sulfate proteoglycans are a common component of neuronal inclusions and astrocytic reaction in neurodegenerative diseases. *Brain Res* 656: 205–209.

Dityatev A and Schachner M (2006). The extracellular matrix and synapses. *Cell Tissue Res* 326: 647–654.

Esko JD and Selleck SB (2002). Order out of chaos: assembly of ligand binding sites in heparan sulfate. *Annu Rev Biochem* 71: 435–471.

Fawcett JW and Asher RA (1999). The glial scar and CNS repair. *Brain Res Bull* 49: 377–391.

Fawcett JW, Housden E, Smith-Thomas L et al. (1989). The growth of axons in three dimensional astrocyte cultures. *Dev Biol* 135: 449–458.

Fidler PS, Schuette K, Asher RA et al. (1999). Comparing astrocytic cells lines that are inhibitory or permissive for axon growth: the major axon-inhibitory proteoglycan is NG2. *J Neurosci* 19: 8778–8788.

Gallagher JT (2006). Multiprotein signalling complexes: regional assembly on heparan sulphate. *Biochem Soc Trans* 34: 438–441.

Galtrey CM and Fawcett JW (2007). The role of chondroitin sulfate proteoglycans in regeneration and plasticity in the central nervous system. *Brain Res Rev* 54: 1–18.

Galtrey CM, Asher RA, Nothias F et al. (2007). Promoting plasticity in the spinal cord with chondroitinase improves functional recovery after peripheral nerve repair. *Brain* 130[Pt 4]: 926–939.

Grimpe B and Silver J (2002). The extracellular matrix in axon regeneration. *Prog Brain Res* 137: 333–349.

Grimpe B, Pressman Y, Lupa MD et al. (2005). The role of proteoglycans in Schwann cell/astrocyte interactions and in regeneration failure at PNS/CNS interfaces. *Mol Cell Neurosci* 28: 18–29.

Grumet M, Friedlander DR, and Sakurai T (1996). Functions of brain chondroitin sulfate proteoglycans during development - interactions with adhesion molecules. *Perspect Dev Neurobiol* 3: 319–330.

Hensch TK (2004). Critical period regulation. *Annu Rev Neurosci* 27: 549–579.

Hockfield S, Kalb RG, Zaremba S et al. (1990). Expression of neural proteoglycans correlates with the acquisition of mature neuronal properties in the mammalian brain. *Cold Spring Harb Symp Quant Biol* 55: 505–513.

Hoke A and Silver J (1996). Proteoglycans and other repulsive molecules in glial boundaries during development and regeneration of the nervous system. *Prog Brain Res* 108: 149–163.

Houle JD, Tom VJ, Mayes D et al. (2006). Combining an autologous peripheral nervous system "bridge" and matrix modification by chondroitinase allows robust, functional regeneration beyond a hemisection lesion of the adult rat spinal cord. *J Neurosci* 26: 7405–7415.

Hussain SA, Piper M, Fukuhara N et al. (2006). A molecular mechanism for the heparan sulfate dependence of slit-robo signaling. *J Biol Chem* 281:39693–39698.

Joester A and Faissner A (2001). The structure and function of tenascins in the nervous system. Matrix Biol 20: 13–22.

Jones LL, Yamaguchi Y, Stallcup WB et al. (2002). NG2 is a major chondroitin sulfate proteoglycan produced after spinal cord injury and is expressed by macrophages and oligodendrocyte progenitors. *J Neurosci* 22: 2792–2803.

Kalb RG and Hockfield S (1994). Electrical activity in the neuromuscular unit can influence the molecular development of motor neurons. *Dev Biol* 162: 539–548.

Kantor DB, Chivatakarn O, Peer KL et al. (2004). Semaphorin 5A is a bifunctional axon guidance cue regulated by heparan and chondroitin sulfate proteoglycans. *Neuron* 44: 961–975.

Kitagawa H, Tsutsumi K, Tone Y et al. (1997). Developmental regulation of the sulfation profile of chondroitin sulfate chains in the chicken embryo brain. *J Biol Chem* 272: 31377–31381.

Kolb B and Whishaw IQ (1998). Brain plasticity and behavior. *Annu Rev Psychol* 49: 43–64.

Lander C, Kind P, Maleski M et al. (1997). A family of activity-dependent neuronal cell-surface chondroitin sulfate proteoglycans in cat visual cortex. *J Neurosci* 17: 1928–1939.

Li HP, Homma A, Sango K et al. (2007). Regeneration of nigrostriatal dopaminergic axons by degradation of chondroitin sulfate is accompanied by elimination of the fibrotic scar and glia limitans in the lesion site. *J Neurosci Res*85(3): 536–547.

Mandolesi G, Menna E, Harauzov A et al. (2005). A role for retinal brain-derived neurotrophic factor in ocular dominance plasticity. *Curr Biol* 15: 2119–2124.

Margolis RU and Margolis RK (1997). Chondroitin sulfate proteoglycans as mediators of axon growth and pathfinding. *Cell Tissue Res* 290: 343–348.

Massey JM, Hubscher CH, Wagoner MR et al. (2006). Chondroitinase ABC digestion of the perineuronal net promotes functional collateral sprouting in the cuneate nucleus after cervical spinal cord injury. *J Neurosci* 26: 4406–4414.

Matthews RT, Kelly GM, Zerillo CA et al. (2002). Aggrecan glycoforms contribute to the molecular heterogeneity of perineuronal nets. *J Neurosci* 22: 7536–7547.

McKeon RJ, Höke A, and Silver J. (1995). Injury-induced proteoglycans inhibit the potential for laminin-mediated axon growth on astrocytic scars. *Exp Neurol* 136: 32–43.

Meiners S, Mercado MLT, Nur-e-Kamal MSA et al. (1999). Tenascin-C contains domains that independently regulate neurite outgrowth and neurite guidance. *J Neurosci* 19: 8443–8453.

Montour-Proulx I, Braun CM, Daigneault S et al. (2004). Predictors of intellectual function after a unilateral cortical lesion: study of 635 patients from infancy to adulthood. *J Child Neurol* 19: 935–943.

Moon LDF, Asher RA, Rhodes KE et al. (2001). Regeneration of CNS axons back to their original target following treatment of adult rat brain with chondroitinase ABC. *Nat Neurosci* 4: 465–466.

Moon LDF, Asher RA, and Fawcett JW (2003). Limited increase in sprouting of severed CNS axons following treatment of adult rat brain with hyaluronidase. *J Neurosci Res* 71: 23–37.

Morgenstern DA, Asher RA, and Fawcett JW (2002). Chondroitin sulphate proteoglycans in the CNS injury response. *Prog Brain Res* 137: ch22.

Oohashi T, Hirakawa S, Bekku Y et al. (2002). Bral1, a brain-specific link protein, colocalizing with the versican V2 isoform at the nodes of Ranvier in developing and adult mouse central nervous systems. *Mol Cell Neurosci* 19: 43–57.

Pang PT, Teng HK, Zaitsev E et al. (2004). Cleavage of proBDNF by tPA/plasmin is essential for long-term hippocampal plasticity. *Science* 306: 487–491.

Pizzorusso T, Medini P, Berardi N et al. (2002). Reactivation of ocular dominance plasticity in the adult visual cortex with chondroitinase ABC. *Science* 298: 1248–1251.

Pizzorusso T, Medini P, Landi S et al. (2006). Structural and functional recovery from early monocular deprivation in adult rats. *Proc Natl Acad Sci U S A* 103: 8517–8522.

Properzi F, Carulli D, Asher RA et al. (2005). Chondroitin 6-sulphate synthesis is up-regulated in injured CNS, induced by injury-related cytokines and enhanced in axon-growth inhibitory glia. *Eur J Neurosci* 21: 378–390.

Rhodes KE and Fawcett JW (2004). Chondroitin sulphate proteoglycans: preventing plasticity or protecting the CNS? *J Anat* 204: 33–48.

Rigato F, Garwood J, Calco V et al. (2002). Tenascin-C promotes neurite outgrowth of embryonic hippocampal neurons through the alternatively spliced fibronectin type III BD domains via activation of the cell adhesion molecule F3/contactin. *J Neurosci* 22: 6596–6609.

Rohrbough J, Grotewiel MS, Davis RL et al. (2000). Integrin-mediated regulation of synaptic morphology, transmission, and plasticity. *J Neurosci* 20: 6868–6878.

Samson AL and Medcalf RL (2006). Tissue-type plasminogen activator: a multifaceted modulator of neurotransmission and synaptic plasticity. *Neuron* 50: 673–678.

Schwab ME (2002). Increasing plasticity and functional recovery of the lesioned spinal cord. *Prog Brain Res* 137: 351–359.

Silver J and Miller JH (2004). Regeneration beyond the glial scar. *Nat Rev Neurosci* 5: 146–156.

Smith-Thomas L, Fok-Seang J, Stevens J et al. (1994). An inhibitor of neurite outgrowth produced by astrocytes. *J Cell Sci* 107: 1687–1695.

Smith-Thomas L, Stevens J, Fok-Seang J et al. (1995). Increased axon regeneration in astrocytes grown in the presence of proteoglycan synthesis inhibitors. *J Cell Sci* 108: 1307–1315.

Steinmetz MP, Horn KP, Tom VJ et al. (2005). Chronic enhancement of the intrinsic growth capacity of sensory neurons combined with the degradation of inhibitory proteoglycans allows functional regeneration of sensory axons through the dorsal root entry zone in the mammalian spinal cord. *J Neurosci* 25: 8066–8076.

Tan AM, Colletti M, Rorai AT et al. (2006). Antibodies against the NG2 proteoglycan promote the regeneration of sensory axons within the dorsal columns of the spinal cord. *J Neurosci* 26: 4729–4739.

Uchimura K, Morimoto-Tomita M, Bistrup A et al. (2006). HSulf-2, an extracellular endoglucosamine-6-sulfatase, selectively mobilizes heparin-bound growth factors and chemokines: effects on VEGF, FGF-1, and SDF-1. *BMC Biochem* 7: 2.

Wehrle R, Camand E, Chedotal A et al. (2005). Expression of netrin-1, slit-1 and slit-3 but not of slit-2 after cerebellar and spinal cord lesions. *Eur J Neurosci* 22: 2134–2144.

Yamaguchi Y (2001). Heparan sulfate proteoglycans in the nervous system: their diverse roles in neurogenesis, axon guidance, and synaptogenesis. *Semin Cell Dev Biol* 12: 99–106.

Yamaguchi Y (2002). Glycobiology of the synapse: the role of glycans in the formation, maturation, and modulation of synapses. *Biochim Biophys Acta* 1573: 369–376.

Yang LJ, Lorenzini I, Vajn K et al. (2006). Sialidase enhances spinal axon outgrowth in vivo. *Proc Natl Acad Sci U S A* 103: 11057–11062.

Zaremba S, Guimaraes A, Kalb RG et al. (1989). Characterization of an activity-dependent, neuronal surface proteoglycan identified with monoclonal antibody Cat-301. *Neuron* 2: 1207–1219.

Zaremba S, Naegele JR, Barnstable CJ et al. (1990). Neuronal subsets express multiple high-molecular-weight cell-surface glycoconjugates defined by monoclonal antibodies Cat-301 and VC1.1. *J Neurosci* 10: 2985–2995.

15

MYELIN NEUTRALIZATION FOR SPINAL CORD INJURY AND STROKE

A.R. WALMSLEY, D. BOURIKAS, AND A.K. MIR

Neuroscience Research, Novartis Institutes for Biomedical Research, 4056 Basel, Switzerland

ABSTRACT

Injuries to the central nervous system (CNS) in humans are associated with a low degree of neurological recovery and, in most cases, life-long debilitation. This, however, is not due to an intrinsic inability of CNS axons to regenerate; rather the environment of the CNS is highly inhibitory to axonal regeneration and plasticity. In recent years, great advances have been made in our understanding of the molecular components contributing to the inhibitory environment of the CNS. The majority of these have been identified as membrane proteins present in CNS myelin, such as myelin-associated glycoprotein, Nogo-A and oligodendrocyte-myelin glycoprotein, that inhibit neurite outgrowth by binding to the multimeric Nogo-66 receptor (NgR1) complex on the neuron and activating the small GTPase RhoA. Accordingly, blockade of either Nogo-A, the NgR receptor complex or RhoA using biological-based inhibitors has been shown to promote neurite outgrowth on CNS myelin substrates in vitro and, more importantly, enhance axonal regeneration and plasticity in vivo leading to improved functional recovery in animal models of acute CNS injury. Based on these promising findings, clinical trials have now been instigated in acute spinal cord injury using a function-blocking anti-Nogo-A antibody and a RhoA antagonist, respectively.

INTRODUCTION

Functional recovery following injury to the central nervous system (CNS) in humans, such as spinal cord injury (SCI) and stroke, is exceptionally poor,

Copyright © 2008 Elsevier Inc.
All rights of reproduction in any form reserved.

leaving the affected individual with neurological deficits which are life-long and debilitating (Schwab, 2004). This is in stark contrast to injuries in the peripheral nervous system (PNS) where complete regain of function is typically observed. Early anatomical observations of spinal cord lesions in cats showed that transected CNS axons failed to elongate but rather retracted away from the lesion area, leading to the assumption that adult CNS neurons had an inadequate regenerative capacity (Ramon y Cajal, 1928). This was later disproved by the finding that transected CNS axons could regenerate over long distances in a transplanted PNS environment, and, reciprocally, transected PNS neurons

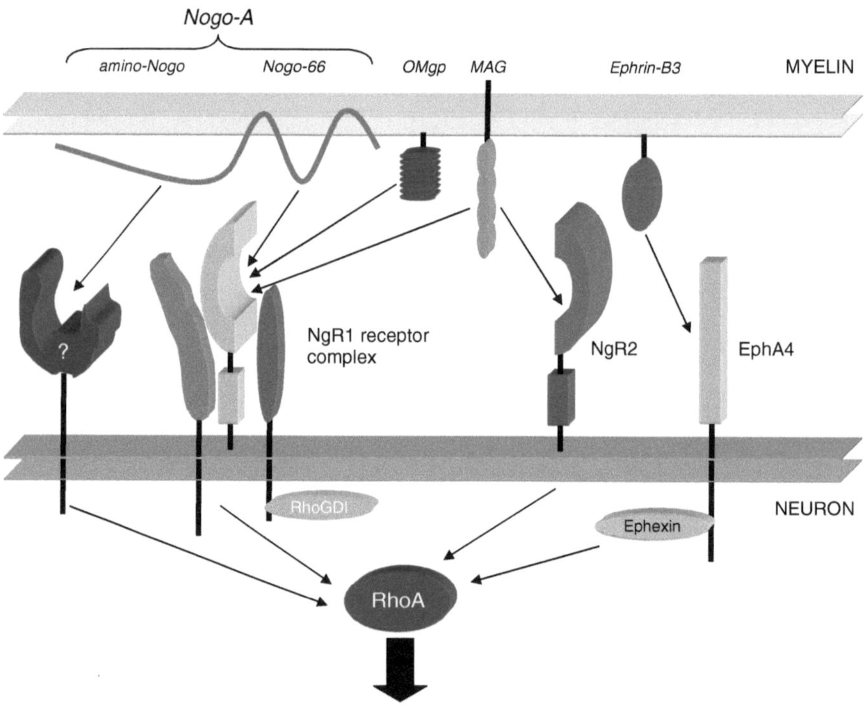

FIGURE 15.1 Myelin-associated inhibitors of neurite outgrowth. The CNS myelin membrane proteins MAG, OMgp and the Nogo-66 domain of Nogo-A bind to overlapping sites on the neuronal GPI-anchored receptor NgR1 (yellow) which signals to the inside of the cell via the transmembrane proteins LINGO-1 (purple) and $p75^{NTR}$ or TROY/TAJ (orange). MAG also binds to the NgR1 homologue, NgR2, the transmembrane co-receptors of which remain to be identified. Apart from Nogo-66, Nogo-A has additional inhibitory regions in the amino-Nogo-A domain which signal through an unknown receptor complex (brown) independent of NgR1. All these ligands activate the small GTPase RhoA in the neuron leading to actin rigidification and neurite outgrowth inhibition. MAG and Nogo-66 appear to activate RhoA by enhancing the binding of the $p75^{NTR}$ intracellular domain to RhoGDI which precludes its association with RhoA allowing the activation of RhoA by GEFs. Myelin-associated ephrin-B3 activates neuronal EphA4 receptors and mediates neurite outgrowth inhibition possibly via the activation of RhoA by the GEF ephexin. (see Plate 7)

failed to regenerate in a transplanted CNS environment (Aguayo et al., 1978; Richardson et al., 1980; David and Aguayo, 1981). In vitro studies showed that neurite outgrowth from CNS neurons was robust on CNS gray matter, PNS myelin, astrocytes and Schwann cells but was inhibited on CNS myelin and oligodendrocytes, the myelin-producing cells of the CNS (Schwab and Thoenen, 1985; Carbonetto et al., 1987; Schwab and Caroni, 1988; Savio and Schwab, 1989). Furthermore, the regenerative and plastic capacity of the CNS is relatively high during development but is strongly suppressed at the onset of myelination (Bates and Stelzner, 1993; Bregman et al., 1993; Saunders et al., 1998). Indeed, this suppression can be circumvented by ablating oligodendrocytes during development using anti-mitotic agents or X-irradiation (Savio and Schwab, 1990; Keirstead et al., 1992). Cumulatively, these findings led to the conclusion that CNS myelin contained factors that were inhibitory to neurite outgrowth in vitro and most likely axonal regeneration and neuroplasticity in vivo. In recent years, several of these factors have been identified as myelin membrane proteins, namely myelin-associated glycoprotein (MAG), Nogo-A, oligodendrocyte-myelin glycoprotein (OMgp) and ephrin-B3, and in some cases their role in axonal regeneration and plasticity has now been firmly established in vivo (Figure 15.1).

MYELIN-ASSOCIATED INHIBITORS OF NEURITE OUTGROWTH

NOGO-A

Nogo-A (a.k.a. reticulon 4A) is mainly expressed in oligodendrocytes and is localized on the inner and outer membranes of the myelin sheath (Chen et al., 2000; GrandPre et al., 2000; Prinjha et al., 2000; Huber et al., 2002). Nogo-A has an N-terminal domain (amino-Nogo-A) followed by a reticulon homology domain containing a 66-amino acid-residue extracellular loop (Nogo-66) bounded by two putative transmembrane domains. Although amino-Nogo-A is cytoplasmically orientated in the endoplasmic reticulum, this topology appears to be inverted at the cell surface of oligodendrocytes where amino-Nogo-A faces the extracellular space (Oertle et al., 2003; Voeltz et al., 2006). As an isolated domain, Nogo-66 has the ability to induce growth cone collapse and inhibit neurite outgrowth via a multimeric neuronal receptor complex containing a ligand-binding subunit known as the Nogo-66 receptor (NgR1) (Fournier et al., 2001) (Figure 15.1). The amino-Nogo-A domain, on the other hand, not only inhibits neurite outgrowth but also inhibits the adherence and spreading of non-neuronal cells via a receptor complex that does not appear to require NgR1 (Niederost et al., 2002; Oertle et al, 2003; Schweigreiter et al., 2004) (Figure 15.1). Indeed, several antibodies with epitopes within the amino-Nogo-A domain suppress the neurite outgrowth inhibitory activity of Nogo-A, oligodendrocytes and CNS myelin in vitro and promote axonal regeneration and plasticity in vivo (Caroni and Schwab, 1988; Schnell and Schwab, 1990; Bregman et al., 1995; Chen et al., 2000; Merkler et al., 2001; Wiessner et al., 2003; Liebscher et al., 2005; Markus et al., 2005; Freund et al., 2006).

MAG

MAG (a.k.a. Siglec-4) is a single transmembrane sialic acid-binding lectin expressed by oligodendrocytes and Schwann cells and is primarily located in the paranodal loops and periaxonal membranes of myelin (Kelm et al., 1994; Domeniconi and Filbin, 2005). The effect of MAG on neurite outgrowth is dependent on the age of the neuron: neurite outgrowth is enhanced by MAG in neonatal and early postnatal neurons and is inhibited by MAG in later postnatal and adult neurons, an effect that can be blocked with antibodies against the MAG ectodomain (Johnson et al., 1989; McKerracher et al., 1994; Mukhopadhyay et al., 1994). As a sialic acid binding protein, MAG binds to neuronal surface gangliosides, in particular GT1b and GD1a, and, correspondingly, its neurite outgrowth inhibitory activity can be suppressed either by blocking complex ganglioside biosynthesis, by function-blocking antibodies against GT1b or GD1a or by synthetic sialoglycans (Yang et al., 1996; Vinson et al., 2001; Vyas et al., 2002, 2005). Despite this, the sialic acid-binding activity of MAG does not appear to be necessary for its neurite outgrowth activity and this may be explained by the recent finding that MAG can also bind in a sialic acid-independent manner to neuronal receptor complexes containing the ligand-binding subunits NgR1 or its homologue NgR2 (Tang et al., 1997; Domeniconi et al., 2002; Liu et al., 2002; Venkatesh et al., 2005) (Figure 15.1).

OMGP

OMgp is a glycosylphosphatidylinositol (GPI)-anchored glycoprotein containing eight tandem leucine-rich repeats (LRRs) and C-terminal serine–threonine repeats and is expressed in the CNS on mature oligodendrocytes and a subset of neurons with expression peaking in the CNS at the later stages of myelination (Vourc'h and Andres, 2004). Purified or recombinant OMgp has been shown to induce growth cone collapse and inhibit neurite outgrowth in vitro via the NgR1 receptor complex (Kottis et al., 2002; Wang et al., 2002a) (Figure 15.1). However, the sole role of OMgp in the restriction of axonal regeneration and plasticity in vivo has yet to be reported.

EPHRIN-B3

Ephrin-B3 acts as a repulsive factor in the guidance of corticospinal tract (CST) axons during development with expression continuing into adulthood on oligodendrocytes (Flanagan and Vanderhaeghen, 1998). Consistent with its repulsive activity during development, a soluble form of the ephrin-B3 ectodomain has been shown to inhibit neurite outgrowth by activating EphA4 receptors on postnatal CNS neurons (Benson et al., 2005) (Figure 15.1). Furthermore, axonal regeneration and functional recovery is augmented by the genetic ablation of EphA4 in mice, although here the primary effect appears to be the suppression of astrogliosis and glial scar formation (Goldshmit et al., 2004).

RECEPTOR COMPLEXES INVOLVED IN NEURITE OUTGROWTH INHIBITION

NgR1 is expressed on CNS neurons and comprises an N-terminal ligand-binding domain (LBD) containing eight tandem LRRs flanked by N- and C-terminal subdomains (LRRNT and LRRCT, respectively), a C-terminal domain that is required for the interaction of the receptor with transmembrane co-receptors and a GPI anchor (Fournier et al., 2001; Wang et al., 2002b; Barton et al., 2003). The LBD forms a banana-shaped structure with Nogo-66, MAG and OMgp binding to overlapping sites in the middle of the concave face of the molecule (He et al., 2003; Lauren et al., 2007) (Figure 15.1). Although the Nogo-66 domain of Nogo-A is an absolute requirement for the activation of NgR1, the NgR1 LBD can also bind to extracellularly orientated regions of Nogo-A immediately N- and C-terminal to Nogo-66 leading to an augmented inhibitory activity of the Nogo-66 domain (Hu et al., 2005; Lauren et al., 2007). Like many other GPI-linked proteins, NgR1 is localized in cholesterol and sphingolipid-enriched detergent-insoluble membrane microdomains, also known as lipid rafts, which have been implicated in a number of biological processes ranging from cell signaling to the sorting of protein and lipids (Fournier et al., 2002; Pignot et al., 2003; Yu et al., 2004). Although lipid raft integrity is clearly essential for the signaling of NgR1 ligands (Yu et al., 2004), this can still occur, albeit with lesser efficiency, when NgR1 is re-located outside rafts (Fournier et al., 2002).

As NgR1 lacks a cytoplasmic signaling domain, signal transduction requires the formation of a complex between NgR1 and the LRR-containing transmembrane protein LINGO-1 (*L*RR and *I*g domain-containing, *N*o*go* receptor-interacting protein a.k.a LRRN6A or LERN1) in concert with either p75^{NTR} or TROY/TAJ, both members of the tumor necrosis factor receptor family (Wang et al., 2002b; Wong et al., 2002; Yamashita et al., 2002; Mi et al., 2004; Park et al., 2005; Shao et al., 2005) (Figure 15.1). Binding of Nogo-66, MAG or OMgp to the NgR1 receptor complex leads to an activation of RhoA resulting in actin rigidification and growth cone collapse (Niederost et al., 2002; Yamashita et al., 2002; Mi et al., 2004; Schweigreiter et al., 2004). The activation of RhoA appears to occur by the binding of the intracellular domain (ICD) of p75^{NTR} to Rho-guanine diphosphate dissociation inhibitor (Rho-GDI), which maintains RhoA in an inactive GDP-bound state, and this interaction displaces RhoA from Rho-GDI allowing its subsequent activation by guanine-nucleotide exchange factors (GEFs) (Yamashita et al., 2003) (Figure 15.1). Furthermore, it has been shown that RhoA activation requires the p75^{NTR} ICD to be liberated from the protein by gamma secretase-mediated intramembrane proteolysis induced by the NgR1 ligand MAG (Domeniconi et al., 2005). However, a contrasting report has shown that neurotrophins also enhance the intramembrane proteolysis of p75^{NTR} leading to a reduction in the activation of Rho-A and enhanced neurite outgrowth on CNS myelin substrates (Logan et al., 2006).

Binding of ligands to the NgR1 receptor complex also trans-activates the kinase domain of the epidermal growth factor receptor (EGFR) leading to auto-phosphorylation of the receptor and activation of the extracellular signal regulated kinase pathway (Koprivica et al., 2005). In this respect, EGFR kinase inhibitors and over-expression of a kinase-dead form of EGFR suppresses the inhibitory activity of CNS myelin in vitro. This trans-activation of EGFR requires Ca^{2+} ions, the intracellular level of which is elevated by myelin-associated inhibitors such as MAG by yet unknown mechanisms (Bandtlow et al., 1993; Wong et al., 2002). That the cognate ligands of EGFR, such as EGF, induce receptor auto-phosphorylation but fail to inhibit neurite outgrowth suggests that EGFR-mediated signaling is necessary, but not sufficient, for myelin-mediated neurite outgrowth inhibition (Koprivica et al., 2005).

NgR2, a homologue of NgR1, also binds to MAG but not to Nogo-66 or OMgp (Venkatesh et al., 2005) (Figure 15.1). Unlike for NgR, the MAG binding site on NgR2 extends outside the LRRNT/LRR/LRRCT domain and requires regions within the C-terminal domain. Being a GPI-anchored protein and thus lacking an ICD, NgR2 must, like NgR, associate with an as yet unknown signal transducing co-receptor(s).

DOWNSTREAM EFFECTORS

RHOA

Rho-A is a member of the Ras homology family of small guanosine triphosphatases (GTPases) that is involved in the rigidification of the actin cytoskeleton (Fukata et al., 2003). Irreversible inhibition of RhoA by ADP-ribosylation of its effector domain with *Clostridium botulinum* C3 transferase or inhibition of its downstream effector, Rho-associated serine-threonine kinase (ROCK), with the low molecular weight compound Y27632 suppresses the activity of many CNS neurite outgrowth inhibitors (Lehmann et al., 1999; Vinson et al., 2001; Dergham et al., 2002; Niederost et al., 2002). To date, RhoA/ROCK activation in neuronal cells has been shown to be critically required for the neurite outgrowth inhibitory activity not only of NgR1 ligands, but also the amino-Nogo-A domain of Nogo-A and the chondriotin sulphate proteoglycans (CSPGs) of the astroglial scar, the receptors of which are distinct from NgR1 and remain to be elucidated (Niederost et al., 2002; Mi et al., 2004; Schweigreiter et al., 2004) (Figure 15.1). Hence, RhoA represents a pivotal node in the signaling pathways of neurite outgrowth inhibitors.

CALCIUM AND PROTEIN KINASE C

NI-35, most likely a proteolytic fragment of the amino-Nogo-A domain of Nogo-A, has been shown to elevate intracellular Ca^{2+} within the neuronal growth cone immediately prior to its collapse and retraction (Bandtlow et al., 1993). Indeed, this collapse of the growth cone could be prevented by inhibiting Ca^{2+} release from caffeine-sensitive intracellular calcium stores using Dantrolene. Other myelin-associated neurite outgrowth inhibitors such as MAG and Nogo-66, along with the CSPGs of the glial scar, also elevate intracellular Ca^{2+} levels in neurons and activate protein kinase C (PKC) (Wong et al., 2002; Hasegawa et al., 2004; Sivasankaran et al., 2004). Accordingly, the neurite outgrowth inhibitory activity of NgR1 ligands, CNS myelin and CSPGs is attenuated when PKC is inhibited with either small molecular weight compounds or by overexpression of a dominant-negative mutant (Sivasankaran et al., 2004). Although one study reported that PKC inhibition prevents RhoA activation by neurite outgrowth inhibitors, an independent study showed that the activation of Rho-A did not require PKC activity (Hasegawa et al., 2004; Sivasankaran et al., 2004). Whether RhoA is upstream, downstream or independent of PKC activation remains to be confirmed.

BLOCKING THE NEURITE OUTGROWTH INHIBITORY ACTIVITY OF MYELIN IN VIVO

ANTIBODY NEUTRALIZATION OF NOGO-A

Several antibodies with disparate epitopes in the amino-Nogo-A domain of Nogo-A have been demonstrated to improve axonal regeneration/neuroplasticity and functional recovery in animal models of acute CNS injury (Caroni and Schwab, 1988; Schnell and Schwab, 1990; Bregman et al., 1995; Chen et al., 2000; Merkler et al., 2001; Wiessner et al., 2003; Liebscher et al., 2005; Markus et al., 2005; Freund et al., 2006). Due to their large size, Nogo-A antibodies most likely block the binding of all the inhibitory domains of Nogo-A to their cognate receptors (Figure 15.2). In addition, Nogo-A antibody treatment has been shown to downregulate Nogo-A in neurons and oligodendrocytes in vitro and in vivo possibly by enhancing the degradation of the protein due to the internalization of the antibody: Nogo-A complex into lysosomal structures (Weinmann et al., 2006) (Figure 15.2). To circumvent the impermeability of the blood–brain barrier to antibodies, Nogo-A antibodies have been administered in acute CNS injury models either via implantation into the CNS parenchyma of hybridoma cells secreting the antibody or by continuous infusion of the antibody into the cerebrospinal fluid from an implanted osmotic mini-pump via the intrathecal or intracerebroventricular routes.

In rats with an incomplete transection of the thoracic spinal cord, intrathecal treatment with Nogo-A antibodies from the time of the lesion enhances the

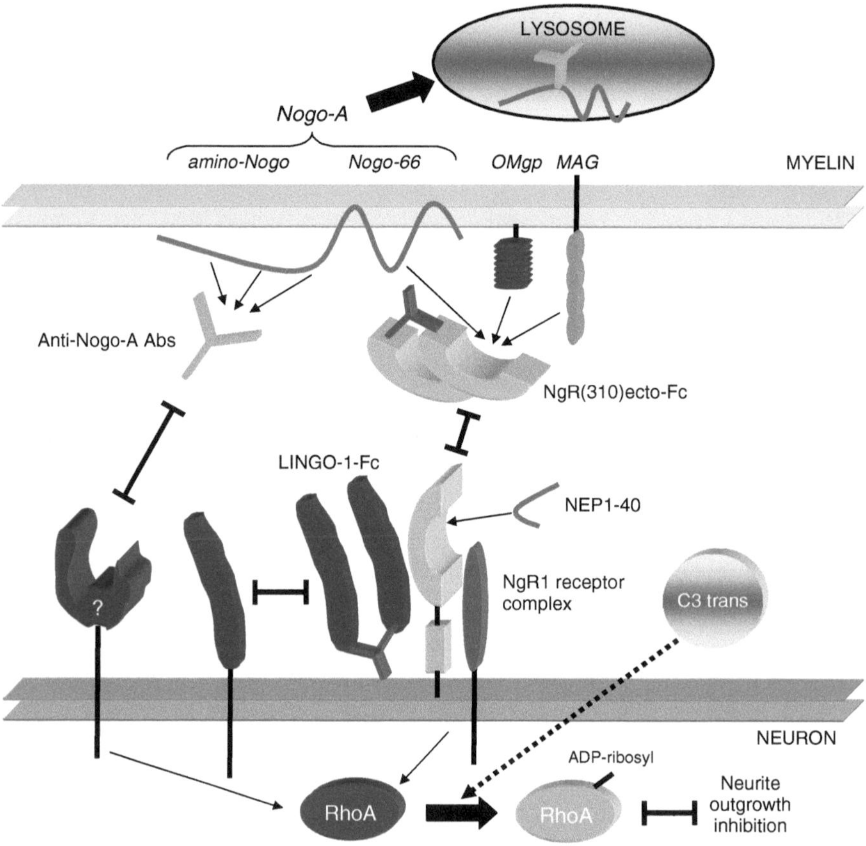

FIGURE 15.2 Approaches to block myelin-mediated neurite outgrowth inhibition Anti-Nogo-A antibodies to disparate epitopes in the amino-Nogo-A region of Nogo-A are likely to sterically hinder the binding of Nogo-A to its cognate neuronal receptors. The anti-Nogo-A antibody: Nogo-A complex is also internalized into lysosomal structures leading to a downregulation of the Nogo-A protein in myelin. NgR(310)ecto-Fc is a soluble form of the ligand-binding domain of NgR1 that binds NgR1 ligands but cannot signal as it lacks the C-terminal domain necessary to form an interaction with the transmembrane co-receptors of the complex. Similarly, LINGO-1-Fc is a soluble form of the LINGO-1 ectodomain that binds to NgR1 but cannot signal as it lacks an intracellular domain. The NEP1-40 peptide comprises the first 40 amino acids of Nogo-66 that binds to NgR but does not activate the receptor hence blocking its activation by Nogo-66. NEP1-40, however, fails to block the activation of NgR1 by MAG, possibly because the binding sites for MAG and Nogo-66 on the receptor are overlapping but distinct. The exoenzyme C3 transferase irreversibly inactivates RhoA by ADP-ribosylation of the effector domain and has now been engineered to yield membrane permeable variants. (see Plate 8)

sprouting of transected CST fibers rostral to the lesion, the CST being one of the main descending tracts involved in locomotor function, and increases the number of CST fibers extending up to several millimeters caudal to the lesion, although these fibers account for only a small percentage of the number prior to the lesion (Schnell and Schwab, 1990; Bregman et al., 1995; Merkler et al., 2001; Liebscher et al., 2005). This enhanced axonal regeneration is paralleled by improvements in functional recovery as demonstrated using a variety of sensorimotor function readouts that include the Basso, Beattie and Bresnahan (BBB) locomotor rating scale (Basso et al., 1995), stride length of the animals walking on a treadmill, performance in crossing a narrowing beam and limb usage during swimming. In addition, Nogo-A antibody treatment reduces the time taken by SCI rats to regain the ability to void their bladder autonomously (Liebscher et al., 2005). Using electromyography, Nogo-A antibody treatment has been shown to improve step-cycle duration and rhythmicity of muscle activity in animals walking on a treadmill (Merkler et al., 2001). More importantly, treatment with Nogo-A antibodies has no effect on the perception of SCI rats to heat and mechanical stimuli, hence arguing against the formation of aberrant neuronal connections (Merkler et al., 2001; Liebscher et al., 2005). Although Nogo-A antibody treatment induces axonal sprouting in both non-injured and injured animals, in non-injured animals the newly formed fibers are transient and disappear over time, whereas in injured animals they persist, possibly because they form functionally relevant connections (Bareyre et al., 2002). Along with CST fibers, Nogo-A antibody treatment has also been shown to enhance the regeneration of raphespinal and coeruleospinal fibers (Bregman et al., 1995).

In non-human primates, the effects of Nogo-A antibody treatment has been recently assessed in adult macaque monkeys with an incomplete transection of the cervical spinal cord (Freund et al., 2006). Manual dexterity of the monkeys is monitored using a modified Brinkman board test scoring the number of food pellets retrieved from horizontal and vertical slots in the board over a certain time period, a task which is ablated immediately after the lesion as it requires the monkeys to precisely grip the food pellet using their thumb and index finger. Intrathecal treatment with Nogo-A antibodies results in a rapid and almost complete recovery in the ability to perform this task irrespective of the size of the lesion whereas treatment with a control antibody gives a slower rate of recovery, the degree of which inversely correlates with the lesion size. At the anatomical level, Nogo-A antibody treatment reduces the number of transected axons rostral to the lesion bearing terminal retraction bulbs and, caudal to the lesion, increases the number and length of CST axonal arbors which frequently bear swellings indicative of synaptic contacts (Freund et al., 2006, 2007). Although no nociceptive tests were performed, the treated monkeys co-operated well with the experimenter and did not lose weight or exhibit signs of pain when handled. This demonstrates that the neutralization of Nogo-A in higher primates, where the organization of the CST is very similar to that of humans, enhances the rate and degree of recovery in manual dexterity and, importantly, does not lead to any overt neurological defects. In this regard, a human anti-Nogo-A antibody has now entered clinical trials in acute SCI patients.

As for SCI, Nogo-A antibody treatment has shown significant beneficial effects in rat models of acute brain cortical injury. Following the ablation of one hemi-cortex by either permanent middle cerebral artery occlusion (pMCAO) or photothrombosis, Nogo-A antibody treatment significantly improves the ability of rats to use their affected forepaw in retrieving food pellets (Papadopoulos et al., 2002; Wiessner et al., 2003; Markus et al., 2005; Seymour et al., 2005). This correlates with an increase at the anatomical level in the number of contralesional corticofugal fibers forming collateral sprouts that traverse the midline and appear to innervate ipsilesional targets. That these collateral sprouts are forming functional ipsilesional connections is indicated by the finding that Nogo-A antibody treatment increases the number of movements of the affected forelimb following microstimulation of the contralesional hemi-cortex (Emerick et al., 2003). Functional recovery occurs even when treatment is instigated 1 week after injury and, more importantly with regards to stroke in humans, has been demonstrated in aged rats (Markus et al., 2005; Seymour et al., 2005). Apart from enhanced collateral sprouting, Nogo-A antibody treatment also increases dendritic sprouting of pyramidal neurons following unilateral pMCAO, a process which may contribute to the formation of novel connections (Papadopoulos et al., 2006). Whereas this effect is sustained in lesioned animals, Nogo-A antibody treatment of non-lesioned animals leads to a transient and fully reversible dendritic sprouting of pyramidal neurons. Nogo-A antibody treatment has no effect on the infarct volume, consistent with the effect being neuroregenerative rather then neuroprotective (Papadopoulos et al., 2002; Wiessner et al., 2003).

INHIBITION OF NGR AND LINGO-1 WITH BIOLOGICAL ANTAGONISTS

The NEP1-40 peptide comprises the first 40 amino acid residues of Nogo-66 and acts as an NgR1 antagonist in that it can bind to the LBD of NgR1 and competes with Nogo-66 for binding to the LBD but fails to activate the receptor (GrandPre et al., 2002) (Figure 15.2). However, although MAG and Nogo-66 have overlapping binding sites on NgR1, NEP1-40 does not inhibit the activity of MAG (Liu et al., 2002). Another inhibitor of NgR1 signaling, NgR(310)ecto-Fc, is a soluble form of the LBD of NgR that binds all the ligands of NgR1 but cannot signal as it lacks the C-terminal domain required to interact with the transmembrane co-receptors of the NgR1 complex (Fournier et al., 2002; Barton et al., 2003) (Figure 15.2). Intrathecal treatment of rats with NEP1-40 or NgR(310)ecto-Fc after incomplete transection of the thoracic spinal cord enhances the sprouting of transected CST fibers rostral to the lesion, increases the number of CST fibers caudal to the lesion and improves functional recovery as assessed by sensorimotor tests such as the BBB locomotor score (GrandPre et al., 2002; Li et al., 2004). Many of the regenerated fibers caudal to the lesion bear varicosities containing the synaptic marker synaptophysin. NEP1-40 and NgR(310)ecto-Fc treatments also enhance the regeneration of raphespinal fibers and the subsequent ablation of these fibers with the neurotoxin DHT results in a partial loss of regained function, highlighting the contribution of this fiber tract to recovery. Interestingly, NEP1-40 is also effective when administered subcutaneously even when the treatment is instigated one week after the lesion, suggesting that in SCI rats the blood–brain barrier is permeable to the peptide for a prolonged time period and that axonal regeneration remains amenable to enhancement 1 week after injury (Li and Strittmatter, 2003). Similar improvements in axonal regeneration and functional recovery have been reported in transgenic mice expressing the NgR1 LBD under the control of the astrocyte-specific glial fibrillary acidic protein promoter (Li et al., 2005).

Intracerbroventricular treatment of rats with NgR(310)ecto-Fc after unilateral pMACO has no effect on cortical infarct volume but does result in a more rapid and improved functional recovery as assessed in the food pellet retrieval and rotarod tests (Lee et al., 2004). Similar to Nogo-A antibody treatment, NgR(310)ecto-Fc treatment increases the number of contralesional corticofugal fibers sprouting across the midline of the cervical spinal cord and this is likely to contribute to the improvement in functional recovery.

A soluble form of the LINGO-1 ectodomain, LINGO-1-Fc, blocks the binding of full length LINGO-1 to the NgR1 receptor complex and, as it lacks a cytoplasmic domain, prevents signaling (Mi et al., 2004) (Figure 15.2). Intrathecal treatment of rats with LINGO-1-Fc following incomplete spinal cord transection enhances the regeneration of CST and rubrospinal fibers and improves functional recovery as assessed by the BBB locomotor score and forelimb usage in a vertical exploration test (Ji et al., 2006). LINGO-Fc treatment also suppresses RhoA activation in the lesion area and reduces the apoptosis of oligodendrocytes and transected neurons.

INHIBITION OF RHOA WITH C3 TRANSFERASE

Blocking RhoA in vivo is challenging due to the intracellular location of the target and the large size of its inhibitor, the bacterially derived exoenzyme C3 transferase. Despite this, axonal regeneration has been demonstrated with C3 transferase treatment following optic nerve crush and spinal cord transection, possibly because the enzyme can reach its target due to the mechanical disruption of the axonal plasma membrane by the injury (Lehmann et al., 1999; Dergham et al., 2002) (Figure 15.2). Importantly, C3 transferase treatment enhances CST axonal regeneration and improves functional recovery following incomplete transection of the spinal cord in rats (Dergham et al., 2002). C3 transferase has consequently been engineered to yield a membrane-permeable form, Clethrin©, which is currently in Phase I/II trials in acute SCI patients.

PKC INHIBITORS

Intrathecal treatment with the PKC inhibitor Gö6976 following incomplete spinal cord transection in rats improves the regeneration of ascending dorsal column axons but, unexpectedly, has no effect on the regeneration of descending CST axons that are mainly responsible for hindlimb motor function (Sivasankaran et al., 2004). Whether the beneficial effects on the regeneration of ascending tracts are sufficient to improve functional recovery has yet to be reported.

CONCLUSION

In recent years, there has been a dramatic expansion in our knowledge of the inhibitory proteins, receptors and signaling pathways involved in the restriction of axonal regeneration and plasticity within the CNS. Blocking Nogo-A with antibodies or the NgR1 receptor complex with biological antagonists facilitates functional recovery in animal models of acute SCI and brain injury. Even RhoA, the mutual downstream effector of neurite outgrowth inhibitors, has proven tractable using deactivating enzymes engineered to be membrane permeable. The question as to whether these approaches will be sufficient as mono-therapies to significantly improve recovery in SCI patients will hopefully be provided from ongoing clinical trials.

Though the majority of neurite outgrowth inhibitors are myelin membrane proteins, equally potent inhibitors have also been identified in the astroglial scar such as CSPGs and semaphorin3A (Niederost et al., 1999; Schmalfeldt et al., 2000; Bradbury et al., 2002; De Winter et al, 2002; Kaneko et al., 2006). Indeed, a recent study has highlighted the beneficial effects of a low molecular weight inhibitor of semaphorin3A in rats with a complete transection of the spinal cord (Kaneko et al., 2006). Apart from neurite outgrowth inhibitors, the astroglial scar represents a physical barrier to axonal regeneration across the lesion. The finding that astrogliosis and subsequent scar formation is diminished in mice lacking EphA4 further widens the scope of potential therapeutic strategies to improve functional recovery following SCI (Goldshmit et al., 2004).

Alongside their roles in CNS injury, it is now emerging that Nogo-A and NgR may also be involved in a wide spectrum of CNS diseases. In the case of Alzheimer's disease (AD), NgR and the reticulons, of which Nogo-A is a member, have been shown to inhibit the generation of pathogenic Abeta peptides by their interaction with the amyloid precursor protein (APP) and beta-site APP cleaving enzyme, respectively (He et al., 2004; Murayama et al., 2006; Park et al., 2006a). Indeed, peripheral administration of NgR(310)ecto-Fc reduced Abeta plaque load and improved spatial memory in a mouse model of AD (Park et al., 2006b). Furthermore, active and passive immunization against a fragment of amino-Nogo-A suppresses disease in a mouse model of multiple sclerosis (Karnezis et al., 2004). In patients with amyloid lateral sclerosis (ALS), Nogo-A is ectopically expressed in the muscle and the level of expression correlates with the severity of the disease (Dupuis et al., 2002; Jokic et al., 2005). Interestingly, genetic ablation of Nogo-A has been shown to significantly increase survival time in a mouse model of ALS (Jokic et al., 2006). These findings thus considerably expand the therapeutic potential of approaches blocking Nogo-A and NgR.

REFERENCES

Aguayo AJ, Dickson R, Trecarten J et al. (1978). Ensheathment and myelination of regenerating PNS fibres by transplanted optic nerve glia. *Neurosci Lett* 9: 97–104.

Bandtlow CE, Schmidt MF, Hassinger TD et al. (1993). Role of intracellular calcium in NI-35-evoked collapse of neuronal growth cones. *Science* 259: 80–83.

Bareyre FM, Haudenschild B, and Schwab ME (2002). Long-lasting sprouting and gene expression changes induced by the monoclonal antibody IN-1 in the adult spinal cord. *J Neurosci* 22: 7097–7110.

Barton WA, Liu BP, Tzvetkova D et al. (2003). Structure and axon outgrowth inhibitor binding of the Nogo-66 receptor and related proteins. *EMBO J* 22: 3291–3302.

Basso DM, Beattie MS, Bresnahan JC (1995). A sensitive and reliable locomotor rating scale for open field testing in rats. *J Neurotrauma* 12: 1–21.

Bates CA and Stelzner DJ (1993). Extension and regeneration of corticospinal axons after early spinal injury and the maintenance of corticospinal topography. *Exp Neurol* 123: 106–117.

Benson MD, Romero MI, Lush ME et al. (2005). Ephrin-B3 is a myelin-based inhibitor of neurite outgrowth. *Proc Natl Acad Sci U S A* 102: 10694–10699.

Bradbury EJ, Moon LD, Popat RJ et al. (2002). Chondroitinase ABC promotes functional recovery after spinal cord injury. *Nature* 416: 636–640.

Bregman BS, Kunkel-Bagden E, Reier PJ et al. (1993). Recovery of function after spinal cord injury: mechanisms underlying transplant-mediated recovery of function differ after spinal cord injury in newborn and adult rats. *Exp Neurol* 123: 3–16.

Bregman BS, Kunkel-Bagden E, Schnell L et al. (1995). Recovery from spinal cord injury mediated by antibodies to neurite growth inhibitors. *Nature* 378: 498–501.

Carbonetto S, Evans D, and Cochard P (1987). Nerve fiber growth in culture on tissue substrata from central and peripheral nervous systems. *J Neurosci* 7: 610–620.

Caroni P and Schwab ME (1988). Antibody against myelin-associated inhibitor of neurite growth neutralizes nonpermissive substrate properties of CNS white matter. *Neuron* 1: 85–96.

Chen MS, Huber AB, van der Haar ME et al. (2000). Nogo-A is a myelin-associated neurite outgrowth inhibitor and an antigen for monoclonal antibody IN-1. *Nature* 403: 434–439.

David S and Aguayo AJ (1981). Axonal elongation into peripheral nervous system "bridges" after central nervous system injury in adult rats. *Science* 214: 931–933.

De Winter F, Holtmaat AJ, and Verhaagen J (2002). Neuropilin and class 3 semaphorins in nervous system regeneration. *Adv Exp Med Biol* 515: 115–139.

Dergham P, Ellezam B, Essagian C et al. (2002). Rho signaling pathway targeted to promote spinal cord repair. *J Neurosci* 22: 6570–6577.

Domeniconi M, Cao Z, Spencer T et al. (2002). Myelin-associated glycoprotein interacts with the Nogo66 receptor to inhibit neurite outgrowth. *Neuron* 35: 283–290.

Domeniconi M and Filbin MT (2003).Overcoming inhibitors in myelin to promote axonal regeneration. *J Neurol Sci* 233: 43–47.

Domeniconi M, Zampieri N, Spencer T et al. (2005). MAG induces regulated intramembrane proteolysis of the p75 neurotrophin receptor to inhibit neurite outgrowth. *Neuron* 46: 849–855.

Dupuis L, Gonzalez de Aguilar JL, di Scala F et al. (2002). Nogo provides a molecular marker for diagnosis of amyotrophic lateral sclerosis. *Neurobiol Dis* 10: 358–365.

Emerick AJ, Neafsey EJ, Schwab ME et al. (2003). Functional reorganization of the motor cortex in adult rats after cortical lesion and treatment with monoclonal antibody IN-1. *J Neurosci* 23: 4826–4830.

Flanagan JG and Vanderhaeghen P (1998). The ephrins and Eph receptors in neural development. *Annu Rev Neurosci* 21: 309–345.

Fournier AE, GrandPre T, and Strittmatter SM (2001). Identification of a receptor mediating Nogo-66 inhibition of axonal regeneration. *Nature* 409: 341–346.

Fournier AE, Gould GC, Liu BP et al. (2002). Truncated soluble Nogo receptor binds Nogo-66 and blocks inhibition of axon growth by myelin. *J Neurosci* 22: 8876–8883.

Freund P, Schmidlin E, Wannier T et al. (2006). Anti-Nogo-A treatment enhances corticospinal tract sprouting and functional recovery after unilateral cervical lesion in adult primates. *Nat Med* 12: 790–792.

Freund P, Wannier T, Schmidlin E et al. (2007). Anti-Nogo-A antibody treatment enhances sprouting of corticospinal axons rostral to a unilateral cervical spinal cord lesion in adult macaque monkey. *J Comp Neurol* 502: 644–659.

Fukata M, Nakagawa M, and Kaibuchi K (2003). Roles of Rho-family GTPases in cell polarisation and directional migration. *Curr Opin Cell Biol* 15: 590–597.

Goldshmit Y, Galea MP, Wise G et al. (2004). Axonal regeneration and lack of astrocytic gliosis in EphA4-deficient mice. *J Neurosci* 24: 10064–10073.

GrandPre T, Nakamura F, Vartanian T et al. (2000). Identification of the Nogo inhibitor of axon regeneration as a Reticulon protein. *Nature* 403: 439–444.

GrandPre T, Li S, and Strittmatter SM (2002). Nogo-66 receptor antagonist peptide promotes axonal regeneration. *Nature* 417: 547–551.

Hasegawa Y, Fujitani M, Hata K et al. (2004). Promotion of axon regeneration by myelin-associated glycoprotein and Nogo through divergent signals downstream of Gi/G. *J Neurosci* 24: 6826–6832.

He XL, Bazan JF, McDermott G et al. (2003). Structure of the Nogo receptor ectodomain: a recognition module implicated in myelin inhibition. *Neuron* 38: 177–185.

He W, Lu Y, Qahwash I et al. (2004). Reticulon family members modulate BACE1 activity and amyloid-beta peptide generation. *Nat Med* 10: 959–965.

Hu F, Liu BP, Budel S et al. (2005). Nogo-A interacts with the Nogo-66 receptor through multiple sites to create an isoform-selective subnanomolar agonist. *J Neurosci* 25: 5298–5304.

Huber AB, Weinmann O, Brosamle C et al. (2002). Patterns of Nogo mRNA and protein expression in the developing and adult rat and after CNS lesions. *J Neurosci* 22: 3553–3567.

Ji B, Li M, Wu WT et al. (2006). LINGO-1 antagonist promotes functional recovery and axonal sprouting after spinal cord injury. *Mol Cell Neurosci* 33: 311–320.

Johnson PW, Abramow-Newerly W, Seilheimer B et al. (1989). Recombinant myelin-associated glycoprotein confers neural adhesion and neurite outgrowth function. *Neuron* 3: 377–385.

Jokic N, Gonzalez de Aguilar JL, Pradat PF et al. (2005). Nogo expression in muscle correlates with amyotrophic lateral sclerosis severity. *Ann Neurol* 57: 553–556.

Jokic N, Gonzalez de Aguilar JL, Dimou L, Lin S, Fergani A, Ruegg MA, Schwab ME, Dupuis L, and Loeffler JP (2006). The neurite outgrowth inhibitor Nogo-A promotes denervation in an amyotrophic lateral sclerosis model. *EMBO Rep* 7: 1162–1167.

Kaneko S, Iwanami A, Nakamura M et al. (2006). A selective Sema3A inhibitor enhances regenerative responses and functional recovery of the injured spinal cord. *Nat Med* 12: 1380–1389.

Karnezis T, Mandemakers W, McQualter JL et al. (2004). The neurite outgrowth inhibitor Nogo A is involved in autoimmune-mediated demyelination. *Nat Neurosci* 7: 736–744.

Keirstead HS, Hasan SJ, Muir GD et al.(1992). Suppression of the onset of myelination extends the permissive period for the functional repair of embryonic spinal cord. *Proc Natl Acad Sci U S A* 89: 11664–11668.

Kelm S, Pelz A, Schauer R et al. (1994). Sialoadhesin, myelin-associated glycoprotein and CD22 define a new family of sialic acid-dependent adhesion molecules of the immunoglobulin superfamily. *Curr Biol* 4: 965–972.

Koprivica V, Cho KS, Park JB et al. (2005). EGFR activation mediates inhibition of axon regeneration by myelin and chondroitin sulfate proteoglycans. *Science* 310: 106–110.

Kottis V, Thibault P, Mikol D et al. (2002). Oligodendrocyte-myelin glycoprotein (OMgp) is an inhibitor of neurite outgrowth. *J Neurochem* 82: 1566–1569.

Lauren J, Hu F, Chin J et al. (2007). Characterization of myelin ligand complexes with neuronal Nogo-66 receptor family members. *J Biol Chem* 282: 5715–5725.

Lee JK, Kim JE, Sivula M et al. (2004). Nogo receptor antagonism promotes stroke recovery by enhancing axonal plasticity. *J Neurosci* 24: 6209–6217.

Lehmann M, Fournier A, Selles-Navarro I et al. (1999). Inactivation of Rho signaling pathway promotes CNS axon regeneration. *J Neurosci* 19: 7537–7547.

Li S and Strittmatter SM (2003). Delayed systemic Nogo-66 receptor antagonist promotes recovery from spinal cord injury. *J Neurosci* 23: 4219–4227.

Li S, Liu BP, Budel S et al. (2004). Blockade of Nogo-66, myelin-associated glycoprotein, and oligodendrocyte myelin glycoprotein by soluble Nogo-66 receptor promotes axonal sprouting and recovery after spinal injury. *J Neurosci* 24: 10511–10520.

Li S, Kim JE, Budel S et al. (2005). Transgenic inhibition of Nogo-66 receptor function allows axonal sprouting and improved locomotion after spinal injury. *Mol Cell Neurosci* 29: 26–39.

Liebscher T, Schnell L, Schnell D et al. (2005). Nogo-A antibody improves regeneration and locomotion of spinal cord-injured rats. *Ann Neurol* 58: 706–719.

Liu BP, Fournier A, GrandPre T et al. (2002). Myelin-associated glycoprotein as a functional ligand for the Nogo-66 receptor. *Science* 297: 1190–1193.

Logan A, Ahmed Z, Baird A et al. (2006). Neurotrophic factor synergy is required for neuronal survival and disinhibited axon regeneration after CNS injury. *Brain* 129: 490–502.

Markus TM, Tsai SY, Bollnow MR et al. (2005). Recovery and brain reorganization after stroke in adult and aged rats. *Ann Neurol* 58: 950–953.

McKerracher L, David S, Jackson DL et al. (1994). Identification of myelin-associated glycoprotein as a major myelin-derived inhibitor of neurite growth. *Neuron* 13: 805–811.

Merkler D, Metz GA, Raineteau O et al. (2001). Locomotor recovery in spinal cord-injured rats treated with an antibody neutralizing the myelin-associated neurite growth inhibitor Nogo-A. *J Neurosci* 21: 3665–3673.

Mi S, Lee X, Shao Z et al. (2004). LINGO-1 is a component of the Nogo-66 receptor/p75 signaling complex. *Nat Neurosci* 7: 221–228.

Mukhopadhyay G, Doherty P, Walsh FS et al. (1994). A novel role for myelin-associated glycoprotein as an inhibitor of axonal regeneration. *Neuron* 13: 757–767.

Murayama KS, Kametani F, Saito S et al. (2006). Reticulons RTN3 and RTN4-B/C interact with BACE1 and inhibit its ability to produce amyloid beta-protein. *Eur J Neurosci* 24: 1237–1244.

Niederost BP, Zimmermann DR, Schwab ME et al. (1999). Bovine CNS myelin contains neurite growth-inhibitory activity associated with chondroitin sulfate proteoglycans. *J Neurosci* 19: 8979–8989.

Niederost B, Oertle T, Fritsche J et al. (2002). Nogo-A and myelin-associated glycoprotein mediate neurite growth inhibition by antagonistic regulation of RhoA and Rac1. *J Neurosci* 22: 10368–10376.

Oertle T, van der Haar ME, Bandtlow CE et al. (2003). Nogo-A inhibits neurite outgrowth and cell spreading with three discrete regions. *J Neurosci* 23: 5393–5406.

Papadopoulos CM, Tsai SY, Alsbiei T et al. (2002). Functional recovery and neuroanatomical plasticity following middle cerebral artery occlusion and IN-1 antibody treatment in the adult rat. *Ann Neurol* 51: 433–441.

Papadopoulos CM, Tsai SY, Cheatwood JL et al. (2006). Dendritic plasticity in the adult rat following middle cerebral artery occlusion and nogo-A neutralization. *Cereb Cortex* 16: 529–536.

Park JB, Yiu G, Kaneko S et al. (2005). A TNF receptor family member, TROY, is a coreceptor with Nogo receptor in mediating the inhibitory activity of myelin inhibitors. *Neuron* 45: 345–351.

Park JH, Gimbel DA, GrandPre T et al. (2006a). Alzheimer precursor protein interaction with the Nogo-66 receptor reduces amyloid-beta plaque deposition. *J Neurosci* 26: 1386–1395.

Park JH, Widi GA, Gimbel DA et al. (2006b). Subcutaneous Nogo receptor removes brain amyloid-beta and improves spatial memory in Alzheimer's transgenic mice. *J Neurosci* 26: 13279–13286.

Pignot V, Hein AE, Barske C et al. (2003). Characterization of two novel proteins, NgRH1 and NgRH2, structurally and biochemically homologous to the Nogo-66 receptor. *J Neurochem* 85: 717–728.

Prinjha R, Moore SE, Vinson M et al. (2000). Inhibitor of neurite outgrowth in humans. *Nature* 403: 383–384.

Ramon y Cajal S (1928). *Degeneration and Regeneration of the Nervous System.* R. May, Trans. Reprint: Hafner, New York, 1968.

Richardson PM, McGuinness UM, and Aguayo AJ (1980). Axons from CNS neurons regenerate into PNS grafts. *Nature* 284: 264–265.

Saunders NR, Kitchener P, Knott GW et al. (1998). Development of walking, swimming and neuronal connections after complete spinal cord transection in the neonatal opossum, Monodelphis domestica. *J Neurosci* 18: 339–355.

Savio T and Schwab ME (1989). CNS white matter, but not gray matter, is nonpermissive for neuronal cell adhesion and fiber outgrowth. J. Neurosci. 9: 1126–1133.

Savio T and Schwab ME (1990). Lesioned corticospinal tract axons regenerate in myelin-free rat spinal cord. *Proc Natl Acad Sci U S A* 87: 4130–4133.

Schmalfeldt M, Bandtlow CE, Dours-Zimmermann MT et al. (2000). Brain derived versican V2 is a potent inhibitor of axonal growth. *J Cell Sci* 113: 807–816.

Schnell L and Schwab ME (1990). Axonal regeneration in the rat spinal cord produced by an antibody against myelin-associated neurite growth inhibitors. *Nature* 343: 269–272.

Schwab ME and Thoenen H (1985). Dissociated neurons regenerate into sciatic but not optic nerve explants in culture irrespective of neurotrophic factors. *J Neurosci* 5: 2415–2423.

Schwab ME and Caroni P (1988). Oligodendrocytes and CNS myelin are nonpermissive substrates for neurite growth and fibroblast spreading in vitro. *J Neurosci* 8: 2381–2393.

Schwab ME (2004). Nogo and axon regeneration. *Curr Opin Neurobiol* 14: 118–124.

Schweigreiter R, Walmsley AR, Niederost B et al. (2004). Versican V2 and the central inhibitory domain of Nogo-A inhibit neurite growth via p75NTR/NgR-independent pathways that converge at RhoA. *Mol Cell Neurosci* 27: 163–174.

Seymour AB, Andrews EM, Tsai SY et al. (2005). Delayed treatment with monoclonal antibody IN-1 1 week after stroke results in recovery of function and corticorubral plasticity in adult rats. *J Cereb Blood Flow Metab* 25: 1366–1375.

Shao Z, Browning JL, Lee X et al. (2005). TAJ/TROY, an orphan TNF receptor family member, binds Nogo-66 receptor 1 and regulates axonal regeneration. *Neuron* 45: 353–359.

Sivasankaran R, Pei J, Wang KC et al. (2004). PKC mediates inhibitory effects of myelin and chondroitin sulfate proteoglycans on axonal regeneration. *Nat Neurosci* 7: 261–268.

Tang S, Shen YJ, DeBellard ME et al. (1997). Myelin-associated glycoprotein interacts with neurons via a sialic acid binding site at ARG118 and a distinct neurite inhibition site. *J Cell Biol* 138: 1355–1366.

Venkatesh K, Chivatakarn O, Lee H et al. (2005). The Nogo-66 receptor homolog NgR2 is a sialic acid-dependent receptor selective for myelin-associated glycoprotein. *J Neurosci* 25: 808–822.

Vinson M, Strijbos PJ, Rowles A et al. (2001). Myelin-associated glycoprotein interacts with ganglioside GT1b. A mechanism for neurite outgrowth inhibition. *J Biol Chem* 276: 20280–20285.

Voeltz GK, Prinz WA, Shibata Y et al. (2006). A class of membrane proteins shaping the tubular endoplasmic reticulum. *Cell* 124: 573–586.

Vourc'h P and Andres C (2004). Oligodendrocyte myelin glycoprotein (OMgp): evolution, structure and function. *Brain Res Brain Res Rev* 45: 115–124.

Vyas AA, Patel HV, Fromholt SE et al. (2002). Gangliosides are functional nerve cell ligands for myelin-associated glycoprotein (MAG), an inhibitor of nerve regeneration. *Proc Natl Acad Sci U S A* 99: 8412–8417.

Vyas AA, Blixt O, Paulson JC et al. (2005). Potent glycan inhibitors of myelin-associated glycoprotein enhance axon outgrowth in vitro. *J Biol Chem* 280: 16305–16310.

Wang KC, Koprivica V, Kim JA et al. (2002a). Oligodendrocyte-myelin glycoprotein is a Nogo receptor ligand that inhibits neurite outgrowth. *Nature* 417: 941–944.

Wang KC, Kim JA, Sivasankaran R et al. (2002b). P75 interacts with the Nogo receptor as a co-receptor for Nogo, MAG and OMgp. *Nature* 420: 74–78.

Weinmann O, Schnell L, Ghosh A et al. (2006). Intrathecally infused antibodies against Nogo-A penetrate the rat CNS and downregulate the endogenous neurite growth inhibitor Nogo-A. *Mol Cell Neurosci* 32: 161–173.

Wiessner C, Bareyre FM, Allegrini PR et al. (2003). Anti-Nogo-A antibody infusion 24 hours after experimental stroke improved behavioral outcome and corticospinal plasticity in normotensive and spontaneously hypertensive rats. *J Cereb Blood Flow Metab* 23: 154–165.
Wong ST, Henley JR, Kanning KC et al. (2002). A p75(NTR) and Nogo receptor complex mediates repulsive signaling by myelin-associated glycoprotein. *Nat Neurosci* 5: 1302–1308.
Yamashita T, Higuchi H, and Tohyama M (2002). The p75 receptor transduces the signal from myelin-associated glycoprotein to Rho. *J Cell Biol* 157: 565–570.
Yamashita T and Tohyama M (2003). The p75 receptor acts as a displacement factor that releases Rho from Rho-GDI. *Nat Neurosci* 6: 461–467.
Yang LJ, Zeller CB, Shaper NL et al. (1996). Gangliosides are neuronal ligands for myelin-associated glycoprotein. *Proc Natl Acad Sci U S A* 93: 814–818.
Yu W, Guo W, and Feng L (2004). Segregation of Nogo66 receptors into lipid rafts in rat brain and inhibition of Nogo66 signaling by cholesterol depletion. *FEBS Lett* 577: 87–92.

16

Strategies to Inhibit Signaling Through Nogo Receptor 1 for Spinal Cord Injury and Stroke

Jane K. Relton* and Paul H. Weinreb#

*Department of Neuropharmacology and
#Department of Drug Discovery, Biogen Idec, 14 Cambridge Center, Cambridge, MA 02142

Can we make the CNS more permissive for axonal regeneration after injury? Inhibitory components of myelin Nogo, oligodendrocyte myelin glycoprotein (OMgp) and myelin-associated glycoprotein (MAG) play an important role in preventing axonal outgrowth in the CNS and their effects are mediated by Nogo receptor 1 (NgR1). Over recent years much effort has focused on developing therapeutics targeting the NgR1 pathway with the intent of promoting axonal regeneration after spinal cord injury and stroke thus promoting neurological recovery. This chapter describes the biological rationale for NgR1 inhibition, the approaches that have been taken to develop therapeutics to block this pathway, and experimental evidence demonstrating the efficacy of NgR1 inhibition in animal models of spinal cord injury and stroke. We discuss the inherent challenges of developing an inhibitor of this pathway from an experimental and clinical perspective.

Copyright © 2008 Elsevier Inc.
All rights of reproduction in any form reserved.

INTRODUCTION

Traumatic injury to the spinal cord results in loss of motor and sensory function due to primary mechanical injury to ascending and/or descending axons and secondary injury (Baptiste and Fehlings, 2006; Fleming et al., 2006). The incidence of spinal cord injury in the US is approximately 4/100 000. However the prevalent population (i.e., the number of people alive today who have suffered a spinal cord injury) is much greater (225/296 000) since the average age of spinal cord injuries is approximately 38 years and patients have a life expectancy not dissimilar to life expectancy for those with no spinal cord injury (www.spinalcord.uab.edu). In light of these statistics, the potential benefit for a regenerative therapy in the prevalent spinal cord injured population is great.

Stroke, or cerebral ischemia, occurs as a result of loss of blood flow to the brain due to embolic or thrombotic vascular occlusion, or hemorrhage (Dirnagl et al., 1999). With an incidence rate of approximately 250–400/100 000 and a mortality rate of 30%, it represents the third leading cause of death in the developed world and those who survive usually sustain neurological deficits in motor and sensory function. Stroke is a heterogeneous disease due to complex vascular pathology and numerous clinical trials have failed to develop acute treatments to halt the neurodegeneration that occurs as a result of stroke (Savitz & Fisher, 2007). Significant spontaneous recovery can occur after stroke due to plasticity and current thinking is that this plasticity could be pharmacologically enhanced using pro-regenerative therapies (Carmichael, 2006).

Components of myelin constitute a major barrier to axonal outgrowth and regeneration. Over recent years the specific molecular entities responsible for axonal outgrowth inhibition have been identified (Filbin, 2003; Yiu and He, 2003) and characterized; Nogo (Chen et al., 2000; GrandPré et al., 2000; Prinjha et al., 2000), Myelin associated protein (MAG) (McKerracher et al., 1994; Mukhopadhyay et al., 1994) and oligodendrocyte glycoprotein (OMgp) (Wang et al., 2002a). These ligands bind to and mediate their biological actions through the Nogo-66 receptor (NgR) (Fournier et al., 2001; Domeniconi et al., 2002; Liu et al., 2002; Wang et al., 2002b) also termed NgR1. NgR1 is expressed on neurons throughout the CNS and different levels of expression may reflect differential susceptibility to growth inhibitory signals and the capacity of axons to regenerate (Huber et al., 2002; Hunt et al., 2002a).

Recombinant forms of myelin inhibitory proteins, fragments thereof or fractions isolated from myelin itself can induce growth cone collapse and limit process outgrowth and arborization in cell-based neurite outgrowth systems (McKerracher et al., 1994; Mukhopadhyay et al., 1994; Li et al., 1996; Chen et al., 2000; GrandPré et al., 2000; Prinjha et al., 2000; Domeniconi et al., 2002; Wang et al., 2002a). However, in vivo the relative roles of these proteins have been more difficult to define, given the mixed conclusions drawn from genetically modified mice; Bartsch et al., reported some time ago that MAG null mice do not exhibit enhanced regeneration (Bartsch et al., 1995) and null mutations in Nogo A and NogoA/B/C resulted in some regeneration of corticospinal axons in some but not all instances (Kim et al., 2003; Simonen et al., 2003; Zheng et al., 2003; Cafferty and Strittmatter, 2006). The debate continues as scientists put forward arguments for strain differences, surgical methodology, axonal pathway heterogeneity, and the nature of the mutant allele as plausible reasons for the differences in observations (Woolf, 2003; Dimou et al., 2006; Cafferty et al., 2007; Steward et al., 2007). Amidst the controversy of knockout studies, however, what does hold true is the demonstration that pharmacological inhibition of NgR1 signaling appears to be a tractable approach to promoting axonal outgrowth and regeneration and/or plasticity after acute CNS trauma induced by stroke or spinal cord injury (Lee et al., 2003; Kastin & Pan 2005).

Several approaches have been taken to inhibit signaling through NgR1 (reviewed by Teng & Tang, 2005). Namely, anti-Nogo antibodies, blockade of the NgR itself using a peptide antagonist (NEP1-40) (GrandPré et al., 2002; Li & Strittmatter, 2003) or an Fc-conjugated NgR ecto-domain (Li and Strittmatter, 2003; Ji et al., 2005; Wang et al., 2006), inhibition of NgR1 co-receptors LINGO-1 (Ji et al., 2006), or $p75^{NTR}$ (Song et al., 2004; Zheng et al., 2005) or inhibitors of NgR1 signaling mediators such as Rho kinase, the intracellular signaling molecule utilized by the NgR1 complex (Dergham et al., 2002; Dubreuil et al., 2003; Fournier et al., 2003). Each of these are described in more detail below.

NOGO RECEPTOR 1: ITS CO-RECEPTORS, LIGANDS, AND SIGNALLING PATHWAYS

The Nogo receptor NgR1 is a 473-amino acid leucine-rich repeat (LRR) protein that contains eight LRRs flanked by N- and C-terminal disulfide-containing "cap" domains, a C-terminal stalk region, and a glycophosphatidylinositol (GPI) anchorage site (Fournier et al., 2001). Through systematic deletion of various domains, it was shown that all eight of the LRR domains, as well as the cap regions, are required for Nogo binding (Fournier et al., 2002). The C-terminal stalk region, while not required for Nogo binding, is required for inhibitory signaling.

Since NgR1 itself is a GPI-linked protein, intracellular signaling through NgR1 is presumed to work through one or more co-receptors. Several components of the NgR1 signaling complex have been identified in recent years. These include the neurotrophin receptor p75 (Wang et al., 2002b), the LRR-containing protein LINGO-1 (Mi et al., 2004), and TAJ/TROY, a TNF family receptor (Park et al., 2005; Shao et al., 2005) (see Figure 16.1).

Rho A is an intracellular regulator of the cytoskeleton (Hall, 1998) and mediates the axonal outgrowth inhibitory effects of myelin (Neideröst et al., 2002; Winton et al., 2002). RhoA activation correlates with the inability of axons to elongate, and RhoA inactivation results in neurite outgrowth (Lehmann et al., 1999; Neideröst et al., 2002). Ligation of NgR1 results in RhoA inactivation (reviewed by Hunt et al., 2002b)

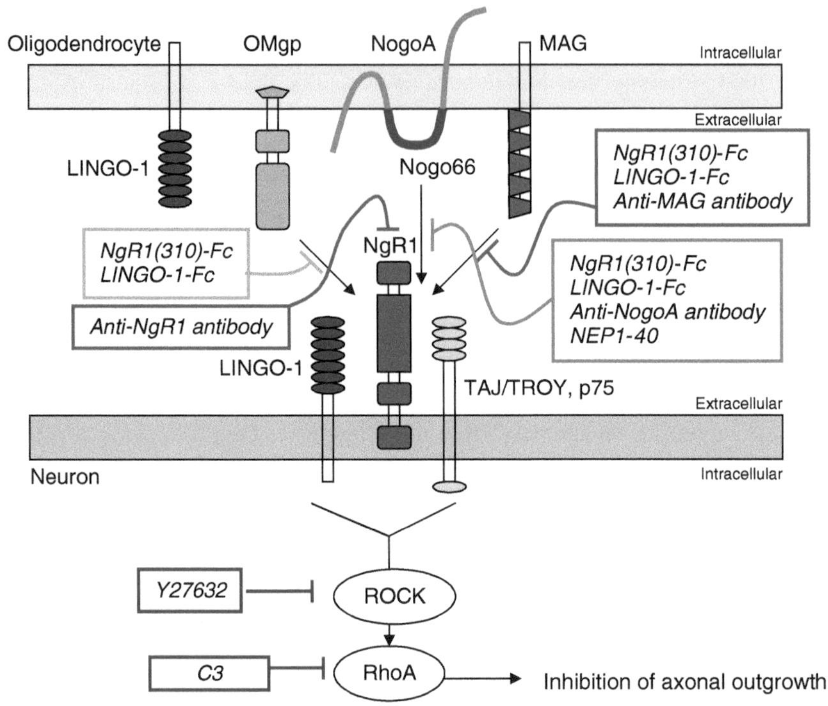

FIGURE 16.1 Schematic representation of NgR1 and its co-receptors and signaling mechanisms. The three characterized myelin ligands for NgR1 are Nogo-A, myelin-associated glycoprotein (MAG) and oligodendrocyte myelin glycoprotein (OMgp). The known co-receptors for NgR1 are p75, TAJ/TROY and LINGO-1. The signaling molecules are ROCK and RhoA. NgR1 and its co-receptors are expressed on neurons and its ligands are expressed on oligodendrocytes. The leucine rich repeats are required for interaction with Nogo-66, MAG and OMgp. The amino terminal of NogoA can be resent either in the cytosol or in the extracellular space. Shown in boxes are the potential sites of intervention and known inhibitors of signaling through NgR1. (see Plate 9)

STRATEGIES TO BLOCK SIGNALING THROUGH NOGO RECEPTOR 1

Inhibitors of the NgR1 pathway fall into several broad categories. The first approach is to block the interaction of myelin inhibitory proteins with NgR1, using either decoy receptors, soluble ligand-derived peptides, or antibodies directed against either the myelin constituents or the receptor (Hunt 2002b; Schwab, 2004). A second approach has been to target the signaling pathways downstream of NgR1 (Yiu and He, 2003). Thirdly, since NgR1 requires co-receptors to signal several components of the NgR1 signaling complex have been targeted. In this section, we discuss several classes of molecules which are being explored as therapeutic approaches.

RECOMBINANT PROTEINS

Soluble NOGO Receptors

One approach to disrupting NgR1 signaling is to use a "decoy" version of the receptor itself. One advantage of this approach is that, by blocking all myelin-inhibitory ligands, it should be less susceptible to the apparent redundancy of myelin inhibitory ligands. A recombinant NgR1 ectodomain comprising the first 310 amino acid residues of the receptor (NgR1(310) ecto) was shown to overcome the outgrowth inhibitory effects of myelin ligands in embryonic chick dorsal root ganglion cells (Fournier et al., 2002). Truncation of the recombinant protein at residue 310 retained ligand-binding activity, while eliminating NgR1 signaling. Subsequently, the protein was expressed as a fusion with an immunoglobulin Fc domain, in order to facilitate expression and improve in vivo stability (Li et al., 2004) This decoy receptor (NgR1(310)-Fc) binds to all three myelin ligands and promotes neurite outgrowth in a variety of cell-based neurite outgrowth assays (McGee and Strittmatter, 2003).

Co-receptors

A soluble p75-Fc protein blocked the interaction of NgR with p75 and reduced myelin inhibition on Nogo66, OMgp, and MAG (Wang et al., 2002b). Dominant negative LINGO-1 constructs and a LINGO-1-Fc fusion protein also showed diminished responses to myelin components (Mi et al., 2004). Similarly, dominant negative forms of TAJ/TROY or p75 or neurons from TAJ/TROY or p75 knockouts have reduced levels of RhoA activation and enhanced neurite outgrowth in the presence of myelin inhibitors (Wang et al., 2002b; Mi et al., 2004; Park et al., 2005; Shao et al., 2005; Zheng et al., 2005). Together these data suggest that blockade of NgR1–co-receptor interactions may present alternative approaches to overcoming myelin inhibition, through disruption of the cell-surface receptor complexes mediating the transduction of inhibitory signals.

ANTIBODIES

Anti-NOGO Antibodies

The specific protein components of the neurite outgrowth inhibitory activity found in CNS myelin were determined using a series of fractionation experiments (Caroni and Schwab, 1988). Two proteins with apparent masses of 35 and 250 kDa by SDS-PAGE were identified and named NI-35 and NI-250. Subsequent purification and sequence analysis of NI-250 led to the cloning of the full-length mammalian gene that we now refer to as Nogo. A monoclonal antibody raised against rat NI-250 (mAb IN-1) neutralized the growth inhibitory properties of both NI-35 and NI-250 as well as those of myelin itself (Caroni and Schwab 1988; Spillmann et al., 1998).

Anti-NgR1 Antibodies

A complementary alternative to the IN-1 approach is to use monoclonal antibodies to block the receptor NgR1. One such antibody, designated 7E11, has been described (Li et al., 2004a). The epitope for this antibody, which blocks the interaction of Nogo66 with NgR1, was mapped to a linear epitope within LRR3. It is likely that a single NgR1 antibody would have the ability to block multiple ligands, through either direct blockade of the ligand-binding site(s) or through steric hindrance, since there appears to be significant overlap among the NgR1 binding sites for each of the three known ligands (see "PEPTIDE/SMALL MOLECULE INHIBITORS") and the area affected by steric hindrance of a single IgG would span several of the NgR1 LRR repeats (Schimmele and Plückthun, 2005).

PEPTIDE/SMALL MOLECULE INHIBITORS

Peptides

As an alternative to recombinant proteins, synthetic peptides have been employed as NgR1 pathway antagonists. Studies using synthetic peptides established that the C-terminus (residues 31–55) of Nogo-66 is required for inhibitory activity (GrandPré et al., 2002) while high-affinity binding to NgR1-expressing COS7 cells requires residues 1–31 (GrandPré et al., 2002). A peptide comprising the first 40 residues of Nogo-66 (NEP1-40) competes with Nogo-66 for binding to NgR1 and partially blocks the inhibitory effects of CNS myelin (GrandPré 2002).

NOGO Receptor Antagonists

There have been no reports to date describing small molecular weight inhibitors of NgR1–ligand interactions. The prospects for identifying such antagonists may be improving, however, as the last several years have seen an increase in our understanding of the structural features of NgR1 and its ligands.

The LRR region of NgR1 forms a super-helical topology of repeating β-strands, as has been observed for other related LRR proteins (He et al., 2003). Two key disulfide bonds in the C-terminal cap region affect the topology of the molecule and may play an important role in positioning the C-terminus for binding to co-receptors such as p75 and LINGO-1 (Wen et al., 2005). The binding sites for each of the three myelin ligands on NgR1 have not been explicitly defined, although some progress has been made toward identifying the key regions of NgR1 involved in these interactions. Using a ribosome display approach, Schimmele and Plückthun (2005) identified key residues contained primarily within the region from LRR3 to LRR5 as critical for the interaction with Nogo66. Similarly, an alanine-scanning approach was used to demonstrate the importance of residues in the center of the concave face of the molecule, as observed with other LRR proteins, and suggested partial but not complete overlap among the binding sites for the various myelin ligands (Laurén et al., 2007). These results were consistent with the epitope of blocking antibody 7E11, which was shown to be contained within LRR3.

Recent structural studies of Nogo-66 (Li et al., 2006) and the amino terminal portion of NogoA (Zander et al., 2007) may aid in developing approaches to targeting the myelin ligands themselves with small molecules. One successful example of this involves inhibitors of MAG, in which the sialic acid-dependence of MAG-binding was used as the basis for designing potent inhibitors of myelin-mediated neurite outgrowth (Vyas et al., 2005). A likely drawback of targeting individual ligands, however, is the apparent redundancy built into the system, since blocking one ligand may be insufficient to overcome the inhibitory effects of the others.

Intracellular Signal Transduction Inhibitors

As alternatives to blocking the extracellular NgR-ligand interactions, one can imagine targeting the downstream intracellular signaling pathways mediated by myelin inhibitory ligand binding (McKerracher and Higuchi, 2006). A number of recent studies have begun to identify some of the key signaling events stimulated by the NgR complex. These include the activation of RhoA and Rho kinase (ROCK), two targets which have been explored using therapeutic agents. Two such Rho pathway inhibitors are an engineered variant of the C3 ADP-ribosyltransferase exoenzyme that blocks Rho function by ADP ribosylation of the effector domain (Winton et al., 2002) and a small molecule inhibitor of ROCK, Y27632 (Dergham et al., 2002.; Fournier et al., 2003). Both effectively block myelin-inhibitory protein-mediated neurite outgrowth inhibition in cell-based assays.

Another downstream consequence of NgR1 signal transduction is the phosphorylation of protein kinase C (PKC). An inhibitor of PKC has shown the ability to attenuate the neurite outgrowth inhibitory ability of myelin in rat cerebellar granule neurons (Sivasankaran et al., 2004).

Using a screening approach to identify small molecules with the ability to disrupt the outgrowth of rat cerebellar granule neurons, He and coworkers (Koprivica et al., 2005) found that inhibitors of EGFR were able to effectively counter the growth inhibitory effects of myelin substrates. No direct interaction of EGFR with either myelin inhibitory proteins or NgR1 was observed, suggesting that EGFR may impact neurite extension through cross-talk (transactivation) of signaling pathways that are downstream of the NgR1 complex. This unexpected finding highlights the complexity inherent to this pathway, and presents a tantalizing new target for therapeutic intervention in spinal cord injury, since EGFR inhibitors (e.g., erlotinib (Tarceva)) have already been approved for other indications.

OTHER STRATEGIES

In addition to strategies directed at using either recombinant proteins or small molecules to directly block NgR1–ligand interactions or intracellular signaling, alternative therapeutic approaches involving either immunization with myelin or the relevant inhibitory components, or gene therapy using dominant negative forms of the proteins, have shown promise. Rodents immunized with spinal cord homogenates (Ellezam et al., 2003), myelin extract (Huang et al., 1999), a mixture of recombinant Nogo66 and MAG (Sicotte et al., 2003), a NogoA-derived peptide (Hauben et al., 2001), or dendritic cells pulsed with myelin basic protein (Hauben et al., 2003) have shown signs of promoting axonal regeneration. Recombinant DNA vaccines containing either sequences encoding myelin inhibitory proteins (Xu et al., 2004) or NgR1 (Yu et al., 2007) have also shown the ability to overcome myelin inhibition. Using a gene therapy approach, Benowitz and coworkers (Fischer et al., 2004) demonstrated that delivery of a dominant negative form of NgR using recombinant adeno-associated virus stimulated axon regeneration of retinal ganglion cells when prestimulated by macrophage-derived factors. These less traditional therapeutic approaches have further strengthened the rationale for targeting myelin inhibitory proteins and their cognate receptors in the treatment of conditions in which regeneration of axons is predicted to have a therapeutic impact.

NOGO RECEPTOR INHIBITION PROMOTES AXONAL SPROUTING AND FUNCTIONAL RECOVERY IN RODENT MODELS OF SPINAL CORD INJURY

Rat and mouse models of spinal cord injury induced by transection or contusion serve as valuable means by which to study axonal regeneration. Despite the marked anatomical differences between higher species and rodents in the location of motor tracts and the relative functional importance of particular tracts (Weidner et al., 2001), individual tracts can be traced within the cord and pharmacological enhancement of regeneration or plasticity are quantifiable. Discrete differences between laboratories and investigators exist but the vast majority of NgR1-related SCI studies have been performed using models of dorsal hemisection (e.g., GrandPré et al., 2000; Li & Strittmatter, 2003; Li et al., 2004; Ji et al., 2005). This model allows complete transection of the dorsal corticospinal tract, responsible for control of fine motor function, that can be labeled via injection of an axonal tracer, for example biotin dextran amine, into the sensorimotor cortex to allow histological evaluation of corticospinal tract regeneration (e.g., Ji et al., 2005, Figure 16.2). Dorsal hemisection of the cord induces hindlimb paralysis and recovery of function is routinely quantified using the Basso Beattie Bresnahan (BBB) scoring system for rats (Basso et al., 1995), and the more

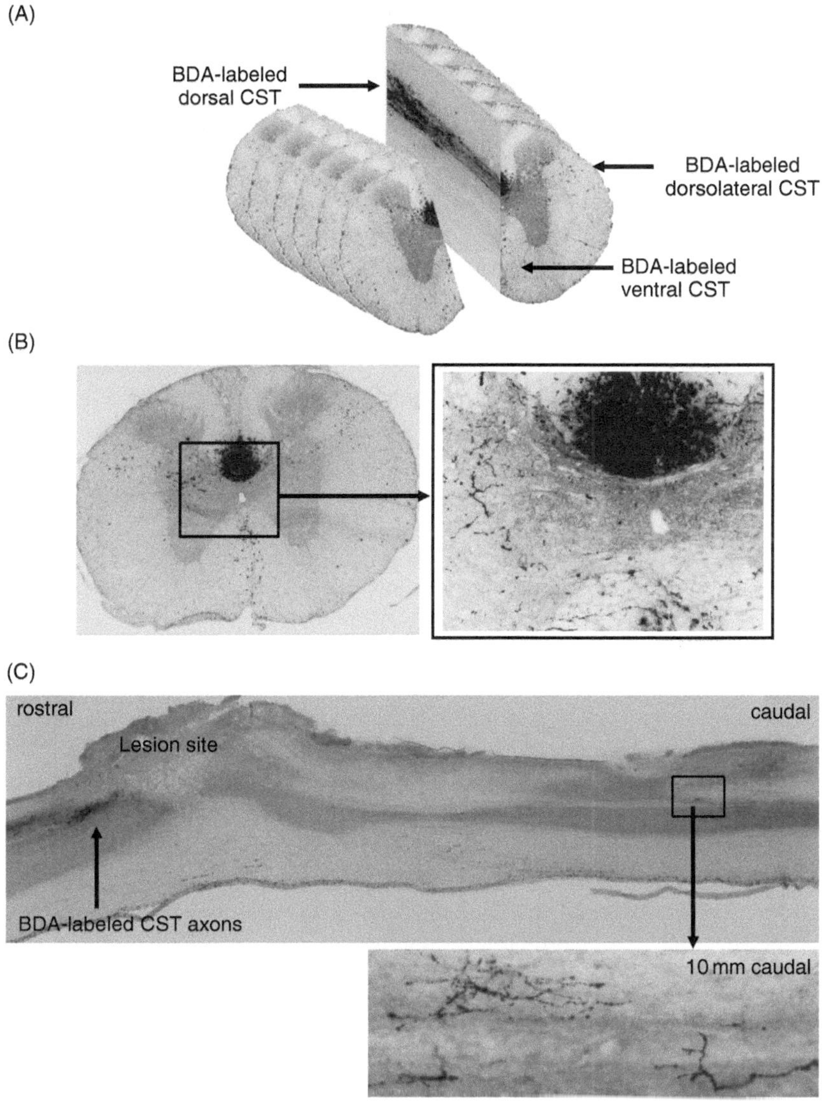

FIGURE 16.2 Biotin dextran amine labeling of the corticospinal tract of the rat after spinal cord injury. For histological tracing of the corticospinal tracts, 2 weeks after CST transection and thoracic level 6 (T6), animals were anesthetized and an incision is made in the scalp. The area around the skin incision was injected with a local anesthetic, Marcaine (Abbott Labs, Chicago, IL) and the left sensorimotor cortex exposed via a craniotomy and 7 μl 10% biotin dextran amine (BDA, 10 000 mol.wt., Molecular Probes, Eugene, OR) in PBS, injected using a nanoliter injector and micro4 controller (World Precision Instruments) at 12 points 0–3.5 mm posterior to Bregma, 0–2.5 mm lateral to the midline at a depth of 1 mm below the surface of the cortex. (A) 95% of corticospinal tract axons run in the dorsal columns and the remainder are localized to the lateral and ventral corticospinal tracts. (B) Sprouting from intact axons above the lesion can be observed in all animals after lesioning but is more pronounced in rats treated with NgR1(310)-Fc. (C) In NgR(310)-Fc treated animals, BDA-labeled axons can be observed sprouting from the intact ventral CST into the gray matter 10 mm below the level of spinal cord transection. [Previously published in Ji et al. (2005). *Eur J Neurosci* 22, 587–94].

recently developed Basso Mouse Scale (BMS) (Basso et al., 2006). These scoring methods allow uniformity across laboratories and enable direct comparison of data.

Efficacy in a model of SCI by inhibition of the Nogo-NgR1 pathway was first demonstrated several years prior to the identification of NgR1 itself. Based on the knowledge that myelin contained inhibitory components, Schwab et al., raised an antibody to myelin that inhibited neurite outgrowth in vitro (Caroni & Schwab, 1988), described in "Anti-Nogo Antibodies". Seminal in vivo studies employed a hybridoma cell line expressing this antibody (Schnell & Schwab, 1990) that was subsequently used in purified form in various models of spinal cord injury in the rat resulting in varying degrees of CST axonal regeneration and functional recovery (Bregman, et al., 1995; Brosamle et al., 2000; Baryere et al., 2002) and more recently the primate (Fouad et al., 2004; Freund et al., 2006). A peptide inhibitor of NgR1 (NEP1-40) significantly increased axonal outgrowth when administered directly to the site of SCI in the rat (GrandPré et al., 2002) or when administered systemically via an osmotic pump implanted subcutaneously in mice that had undergone spinal cord transection (Li and Strittmatter, 2003). NgR1(310)Fc, a recombinant protein comprising the ecto-domain of the rat NgR1 that acts as a decoy receptor, significantly increased axonal plasticity and/or regeneration when administered intrathecally (Li et al., 2004b) either alone or in combination with methylprednisolone (Ji et al., 2005), the current standard of care for SCI patients. Extended treatment with NgR1(310)-Fc shows continued improvement in functional and histological recovery after SCI (Figure 16.3). In this study functional recovery was significantly improved in NgR1-Fc treated rats 10 weeks after SCI (BBB score 13.8 $\pm$ 0.95) versus control animals (BBB score: 10.95 $\pm$ 0.3) and the number of BDA-labeled axons counted 15 mm caudal to the site of transection was greater in NgR1-Fc treated rats 12 weeks after SCI.

Additional clinical relevance is drawn from studies that show significant improvement in functional and histological recovery after delayed administration of NgR1-Fc after spinal cord contusion injury (Wang et al., 2006). This model is widely thought to more closely mimic human spinal cord injury indicating that this protein is acting via a pro-regenerative mechanism and supporting its utility in the prevalent SCI population.

Similarly antagonism of NgR1 signaling pathways promotes functional and histological recovery in animal models of spinal cord injury; RhoA inhibitors enhanced regeneration and reduced cell death after SCI (Dergham et al., 2002; Dubreuil et al., 2003) and the PKC inhibitor Gö6976 promotes regeneration of dorsal column axons across the lesion site in adult rats (Sivasankaran et al., 2004).

Inhibition of NgR1 co-receptor Lingo-1 using a recombinant Fc-conjugated human protein enhanced functional recovery and axonal outgrowth after spinal cord transection in the rat (Ji et al., 2006). However inhibition of p75 failed to promote axonal regeneration in vivo despite enhancing neurite outgrowth in vitro (Song et al., 2004; Zheng et al., 2005). These data demonstrate the complexity of the NgR1 system and indicate some level of redundancy in co-receptor function.

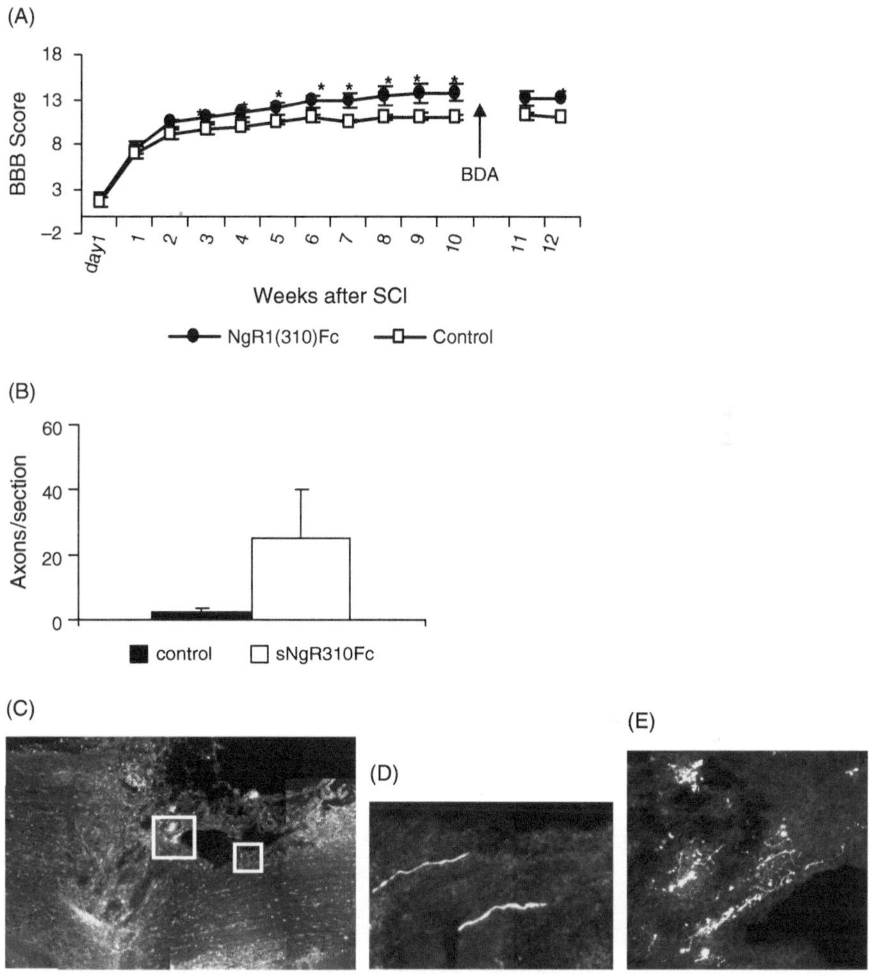

FIGURE 16.3 The effect of chronic treatment with NgR1(310)-Fc on functional and histological recovery after spinal cord injury (SCI) in the rat. Anesthetized female Long Evans rats (7 weeks old) underwent dorsal hemisection of the spinal cord, transecting the major dorsal and minor lateral corticospinal tract components at thoracic level 7 (T7). An intrathecal catheter was implanted into the subdural space at T8/T9 and PBS ($n = 11$) or NgR1(310)-Fc (50 mM, $n = 9$) infused via a subcutaneously implanted osmotic minipump at a rate of 0.25 μl/h for 8–9 weeks. Functional recovery was quantified using the BBB scoring system the day after surgery and weekly thereafter. Ten weeks after SCI rats were re-anesthetized and BDA (10 000 mol.wt) infused into the sensorimotor cortex to trace descending corticospinal axons. (A) BBB scoring of functional recovery over the 10-week period after SCI. NgR1(310)-Fc treatment significantly improved functional outcome, Two way ANOVA with Tukey's post hoc test ($p < 0.05$). (B) The number of BDA-labeled axons was increased in NgR1(310)-Fc-treated rat spinal cord 10 mm below the level of the lesion, compared to vehicle treated controls ($n = 4$ per group). Examples of BDA-labeled axons around the lesion site are shown in longitudinal section (C-E). [Previously presented in abstract form: *Ji B, Wang J, Walus L, Li M, Pepinsky RB, Lee DHS, Meier W, Relton JK. (2005). Long term effects of Nogo receptor ectodomain treatment in functional recovery and axonal regeneration after spinal cord injury in the rat. National Neurotrauma Soc.*]

NOGO RECEPTOR INHIBITION PROMOTES AXONAL PLASTICITY AND FUNCTIONAL RECOVERY IN RODENT MODELS OF STROKE

Functional recovery after stroke can be attributed in part to structural plasticity of surviving neuronal circuitry both in the spinal cord and brain (Weidner et al., 2001). After demonstrating the functional role of NgR1 signaling in mediating axonal regeneration and plasticity in the spinal cord, a natural line of investigation was to study the effects of NgR1 inhibitors in stroke. Several laboratories have now shown that blockade of the NgR1 pathway promotes functional recovery after experimental stroke using either anti-NogoA treatment (Papadopoulos et al., 2002; Markus et al., 2005) soluble NgR1-Fc (Lee et al., 2004), or genetic deletion of NogoA/B or NgR1 (Lee et al., 2004). A standard model of stroke induced by occlusion of the middle cerebral artery was employed in these studies. Treatment administered up to 1 week after stroke promoted robust functional recovery in both young and old rats (Weissner et al., 2003; Lee et al., 2004; Markus et al., 2005; Seymour et al., 2005) indicating that this is a purely regenerative effect and not due to a possible neuroprotective effect of blocking the pathway, as was seen with an anti-MAG antibody in a similar rat model of middle cerebral artery occlusion (Irving et al., 2005), and after traumatic brain injury (Thompson et al., 2006). More recent studies have demonstrated increased dendritic arborization and dendritic spine density in the cortical neurons of adult rats treated with the anti-Nogo-A antibody IN-1 after stroke (Papadopoulos et al., 2006).

CLINICAL DEVELOPMENT OF NOGO RECEPTOR INHIBITORS

Two NgR pathway inhibitors are currently in clinical trials for acute spinal cord injury; a Nogo A antibody, developed by Novartis, and the Rho inhibitor C3, developed by Boston Life Sciences Inc (now Alseres Pharmaceuticals). The exciting biology surrounding the NgR1 pathway and its potential to yield a viable therapeutic for the treatment of acute axonal injury has motivated scientists and clinicians to implement these experimental drugs into clinical settings (Buchli et al., 2007). More novel NgR1 pathway inhibitors are sure to follow.

ACKNOWLEDGEMENTS

Experiments described here were performed by Benxiu Ji and Mingwei Li, with assistance from Joy Wang and Harry Sweigard, Dept Neuropharmacology, Biogen Idec. Thanks to Blake Pepinsky, Daniel Lee, and members of the Nogo Team at Biogen Idec, and Dr Stephen Strittmater and his group at Yale University for their contributions to this work.

REFERENCES

Baptiste DC and Fehlings MG (2006). Pharmacological approaches to repair the injured spinal cord. *J Neurotrauma* 23: 318–334.

Bartsch U, Bandtlow CE, Schnell L et al. (1995). Lack of evidence that myelin-associated glycoprotein is a major inhibitor of axonal regeneration in the CNS *Neuron* 15: 1375–1381.

Baryere FM, Haudenschild B, and Schwab ME (2002). Long lasting sprouting and gene expression changes induced by the monoclonal antibody IN-1 in the adult spinal cord. *J Neurosci* 22: 7097–7110.

Basso DM, Beattie MS, and Bresnahan JC (1995). A sensitive and reliable locomotor rating scale for open field testing in rats. *J Neurotrauma* 12: 1–21.

Basso DM, Fisher LC, Anderson AJ et al. (2006). Basso Mouse Scale for locomotion detects differences in recovery after spinal cord injury in five common mouse strains. *J Neurotrauma* 23: 635–659.

Bregman BS, Kunkel-Bagden E, Schnell L et al. (1995). Recovery from spinal cord injury mediated by antibodies to neurite growth inhibitors. *Nature* 378: 498–501.

Brosamle C, Huber AB, Fiedler M et al. (2000). Regeneration of lesioned corticospinal tract fibers in the adult rat induced by a recombinant, humanized IN-1 antibody fragment. *J Neurosci* 20: 8061–8068.

Buchli AD, Rouiller E, Meuller R et al. (2007). Repair of the injured spinal cord. *Neurodegen Dis* 4: 51–56.

Cafferty WBJ, and Strittmatter SM (2006). The Nogo-Nogo receptor pathway limits a spectrum of adult CNS axonal growth. *J Neurosci* 26: 12242–12250.

Cafferty WBJ, Kim J-E, Lee J-K et al. (2007). Response to correspondence: Kim et al., "Axon regeneration in Young adult mice lacking NogoA/B" Neuron 38: 187–199. *Neuron* 54: 195–196.

Carmichael ST (2006). Cellular and molecular mechanisms of neural repair after stroke: making waves. *Ann Neurol* 59: 735–742.

Caroni P, and Schwab ME (1988). Antibody against myelin associated inhibitor of neurite growth neutralizes non permissive substrate properties of CNS white matter. *Neuron* 1: 85–96.

Chen MS, Huber AB, van der Haar ME et al. (2000). Nogo-A is a myelin associated neurite outgrowth inhibitor and an antigen for monoclonal antibody IN-1. *Nature* 403: 434–439.

Dergham P, Ellezam B, Essagian C, et al. (2002). Rho signaling pathway targeted to promote spinal cord injury repair. *J Neurosci* 22: 6570–6577.

Dimou L, Schnell L, Montani L et al. (2006). Nogo-A deficient mice reveal strain-dependent differences in axonal regeneration. *J Neurosci* 26: 5591–5603.

Dirnagl U, Iadecola C, and Moskowitz MA (1999). Pathobiology of ischemic stroke: an integrated view. *Trends in Neurosci* 22: 391–397.

Domeniconi M, Cao Z, Spencer T et al. (2002). Myelin-associated glycoprotein interacts with the Nogo-66 receptor to inhibit neurite outgrowth. *Neuron* 35: 283–290.

Dubreuil CI, Winton MJ, and McKerracher L (2003). Rho activation patterns after spinal cord injury and the role of activated Rho in apoptosis in the central nervous system. *J Cell Biol* 162: 233–243.

Ellezam B, Bertrand J, Dergham P et al. (2003). Vaccination stimulates retinal ganglion cell regeneration in the adult optic nerve. *Neurobiol Dis* 12,1–10.

Filbin MT (2003). Myelin-associated inhibitors of axonal regeneration in the adult mammalian CNS. *Nat Neurosci Rev* 4: 1–11.

Fischer D, He Z, and Benowitz LI (2004). Counteracting the Nogo receptor enhances optic nerve regeneration if retinal ganglion cells are in an active growth state. *J Neurosci* 24: 1646–1651.

Fleming JC, Norenberg MD, Ramsay DA et al. (2006). The cellular inflammatory response in human spinal cords after injury. *Brain* 129: 3249–3269.

Fouad K, Klusman I, and Schwab ME (2004). Regenerating corticospinal fibers in the Marmoset (*Callitrix jacchus*) after spinal cord lesion and treatment with the anti-Nogo-A antibody IN-1. *Eur J Neurosci* 20: 2479–2482.

Fournier AE, GrandPré T, and Strittmatter SM (2001). Identification of a receptor mediating Nogo-66 receptor inhibition of axonal regeneration. *Nature* 409: 341–346.
Fournier AE, Gould GC, Liu BP et al. (2002). Truncated soluble nogo receptor binds nogo-66 and blocks inhibition of axon growth by myelin. *J Neurosci* 22: 8876–8883.
Fournier AE, Takizawa BT, and Strittmatter SM (2003). Rho Kinase inhibition enhances axonal regeneration in the injured CNS. *J Neurosci* 23: 1416–1423.
Freund P, Schmidlin E, Wannier T et al. (2006). Nogo-A specific antibody treatment enhances sprouting and functional recovery after cervical lesion in adult primates. *Nat Med* 12: 790–792.
GrandPré T, Nakamura F, Vartanian T et al. (2000). Identification of the Nogo inhibitor of axonal regeneration as a Reticulon protein *Nature* 403: 439–444.
GrandPré T, Li S, and Strittmatter SM (2002). Nogo-66 receptor antagonist peptide promotes axonal regeneration. *Nature* 417: 547–551.
Hall A (1998). Rho GTPases and the actin cytoskeleton. *Science* 279: 509–514.
Hauben E, Ibarra A, Mizrahi T et al. (2001). Vaccination with a Nogo-A-derived peptide after incomplete spinal-cord injury promotes recovery via a T-cell mediated neuroprotective response: comparison with other myelin antigens. *Proc Natl Acad Sci* 98: 15173–15178.
Hauben E, Gothlif A, Cohen A et al. (2003). Vaccination with dendritic cells pulsed with peptides of myelin basic protein promotes functional recovery from spinal cord injury. *J Neurosci* 23: 8808–8819.
He XL, Bazan JF, McDermott G et al. (2003). Structure of the Nogo receptor ectodomain: a recognition module implicated in myelin inhibition. *Neuron* 38: 177–185.
Huang DW, McKerracher L, Braun B et al. (1999). A therapeutic vaccine approach to stimulate axon regeneration in the adult mammalian spinal cord. *Neuron* 24: 639–647.
Huber AB, Weinmann O, Brosamle C et al. (2002). Patterns of Nogo mRNA and protein expression in the developing and adult rat after CNS lesions. *J Neurosci* 22: 3553–3567.
Hunt D, Mason MRJ, Campbell G et al. (2002a). Nogo receptor mRNA expression in intact and regenerating CNS neurons. *Mol Cell Neurosci* 20: 537–552.
Hunt D, Coffin RS, and Anderson PN (2002b). The Nogo receptor, its ligands and axonal regeneration in the spinal cord; A review. *J Neurocyto* 31: 93–120.
Irving EA, Vinson M, Rosin C et al. (2005). Identification of neuroprotective properties of anti-MAG antibody: a novel approach for the treatment of stroke? *J Cereb Blood Flow Metab* 25: 98–107.
Ji B, Li M, Budel S et al. (2005). Effect of combined treatment with methylprednisolone and soluble Nogo-66 receptor after rat spinal cord injury. *Eur J Neurosci* 22: 587–594.
Ji B, Li M, Wu WT et al. (2006). LINGO-1 antagonist promotes functional recovery and axonal sprouting after spinal cord injury. *Mol Cell Neurosci* 33: 311–320.
Kastin AJ and Pan W (2005). Targeting neurite outgrowth inhibitors to induce CNS regeneration. *Curr Pharmaceutical Design* 11: 1247–1253.
Kim JE, Li S, GrandPré T et al. (2003). Axon regeneration in young adult mice lacking Nogo-A/B. *Neuron* 38: 187–199.
Kim JE, Liu BP, Park JH et al. (2004). Nogo-66 receptor prevents raphespinal and rubrospinal axon regeneration and limits functional recovery from spinal cord injury. *Neuron* 44: 439–451.
Koprivica V, Cho KS, Park JB et al. (2005). EGFR activation mediates inhibition of axon regeneration by myelin and chondroitin sulfate proteoglycans. *Science* 310: 106–110.
Laurén J, Hu F, Chin J et al. (2007). Characterization of myelin ligand complexes with neuronal Nogo-66 receptor family members. *J Biol Chem* 282: 5715–5725.
Lee DHS, Strittmatter SM, and Sah DW (2003). Targeting the Nogo receptor to treat central nervous system injuries. *Nat Rev Drug Discov* 2: 1–7.
Lee JK, Kim JE, Sivula M et al. (2004). Nogo receptor antagonism promotes stroke recovery by enhancing axonal plasticity. *J Neurosci* 24: 6209–6217.
Lehmann M, Fournier A, Selles-Navarro I et al. (1999). Inactivation of Rho signaling pathway promotes CNS axon regeneration. *J Neurosci* 19: 7537–7547.
Li M, Shibata A, Li C et al. (1996). Myelin-associated glycoprotein inhibits neurite/axon growth and causes growth cone collapse. *J Neurosci Res* 46: 404–414.

Li S, and Strittmatter SM (2003). Delayed systemic Nogo-66 receptor antagonist promotes recovery from spinal cord injury. *J Neurosci* 23, 4219–4227.

Li W, Walus L, Rabacchi SA et al. (2004a) A neutralizing anti-Nogo66 receptor monoclonal antibody reverses inhibition of neurite outgrowth by central nervous system myelin. *J Bio. Chem* 279: 43780–43788.

Li S, Liu BP, Budel S et al. (2004b) Blockade of Nogo-66, myelin-associated glycoprotein, and oligodendrocyte myelin glycoprotein by soluble Nogo-66 receptor promotes axonal sprouting and recovery after spinal injury. *J Neurosci* 24: 10511–10520.

Li M, Liu J, and Song J (2006). Nogo goes in the pure water: solution structure of Nogo-60 and design of the structured and buffer-soluble Nogo-54 for enhancing CNS regeneration. *Prot Sci* 15: 1835–1841

Liu BP, Fournier A, GrandPreT et al. (2002). Myelin associated glycoprotein as a functional ligand for the Nogo-66 receptor. *Science* 297: 1190–1193.

Markus TM, Tsai S-Y, Bollownow MR et al. (2005). Recovery and brain reorganization after stroke in adult and aged rats. *Ann Neurol* 58: 950–953.

McGee A and Strittmatter SM (2003) The Nogo-66 receptor: focusing myelin inhibition of axon regeneration. *Trends in Neurosci* 26: 193–198.

McKerracher L, David S, Jackson DL et al. (1994) Identification of myelin-associated glycoprotein as a major myelin-derived inhibitor of neurite growth. *Neuron* 13: 805–811.

McKerracher L, and Higuchi H (2006) Targeting Rho to stimulate repair after spinal cord injury. *J Neurotrauma* 23: 309–317.

Mi S, Lee X, Shao Z et al. (2004). LINGO-1 is a component of the Nogo-66 receptor/p75 signaling complex *Nat Neurosci* 7: 221–228.

Mukhopadhyay G, Doherty P, Walsh FS et al. (1994). A novel role for myelin-associated glycoprotein as an inhibitor of axonal regeneration. *Neuron* 13: 757–767.

Neideröst B, OertleT, Fritsche T et al. (2002). NogoA and myelin-associated glycoprotein mediate neurite outgrowth inhibition by antagonistic regulation of RhoA and Rac1. *J Neurosci* 22: 10368–10376.

Papadopuolos CM, Tsai SY, Alsbiei T et al. (2002). Functional recovery and neuroanatomical plasticity following middle cerebral artery occlusion and IN-1 antibody treatment in the adult rat. *Ann Neurol* 51: 433–441.

Papdopoulos CM, Tsai SY, Cheatwood JL et al. (2006). Dendritic plasticity in the adult rat following middle cerebral artery occlusion and Nogo-A neutralization. *Cereb Cortex* 16: 529–536.

Park JB, Yiu G, Kaneko S et al. (2005). A TNF receptor family member, TROY, is a co-receptor with Nogo receptor in mediating the inhibitory activity of myelin inhibitors. *Neuron* 45: 345–351.

Prinjha R, Moore SE, Vinson M et al. (2000). Inhibitor of neurite outgrowth in humans. *Nature* 403: 383–384.

Savitz S and FisherM (2007). Future of neuroprotection for acute stroke: in the aftermath of the SAINT trials. *Ann Neurol* 61: 396–402.

Scimmele A, and Plückthun A (2005). Identification of a functional epitope of the Nogo receptor by a combinatorial approach using ribosome display. *J Mol Biol* 352: 229–241.

Schnell L and Schwab ME (1990). Axonal regeneration in the rat spinal cord produced by an antibody against myelin-associated neurite growth inhibitors. *Nature* 343: 269–272.

Schwab ME (2004). Nogo and axonal regeneration. *Curr Opin. Neurobiol* 14:1–7

Seymour AB, Andrews EM, Tsai SY et al. (2005). Delayed treatment with monoclonal antibody IN-1 1 week after stroke results in recovery of function and corticorubral plasticity in adult rats. *J Cereb Blood Flow Metab* 25: 1366–1375.

Shao Z, Browning JL, Lee X et al. (2005). TAJ/TROY, an orphan TNF receptor family member, binds nogo-66 receptor 1 and regulates axonal regeneration. *Neuron* 45: 353–359.

Sicotte M, Tsatas O, Jeong SY et al. (2003). Immunization with myelin or recombinant Nogo-66/MAG in alum promotes axon regeneration and sprouting after corticospinal tract lesions in the spinal cord. *Mol Cell Neurosci* 23: 251–263.

Simonen M, Pedersen V, Weinmann O et al. (2003). Systemic deletion of the myelin-associated outgrowth inhibitor Nogo-A improves regenerative and plastic responses after spinal cord injury. *Neuron* 38: 201–211.

Sivasankaran R, Pei J, Wang KC et al. (2004). PKC mediates inhibitory effects of myelin and chondroitin sulfate proteoglycans on axonal regeneration. *Nat Neurosci* 7: 261–268.

Song XY, Zhong JH, Wang X et al. (2004). Suppression of p75NTR does not promote regeneration of injured spinal cord in mice. *J Neurosci* 24: 542–546.

Spillmann AA, Bandtlow CE, Lottspeich F et al. (1998). Identification and characterization of a bovine neurite growth inhibitor (bNI-220). *J Biol Chem* 273: 19283–19293.

Steward O, Zheng B, Banos K et al. (2007). Response to: Kim et al., "Axon regeneration in Young adult mice lacking NogoA/B" Neuron 38: 187–199. *Neuron* 54: 191–195.

Teng FYH, and Tang BL (2005). Why do Nogo/Nogo-66 receptor gene knockouts result in inferior regeneration compared to treatment with neutralizing agents? *J Neurochem* 94: 865–874.

Thompson HJ, Marklund N, LeBold DG et al. (2006). Tissue sparing and functional recovery following experimental brain injury is provided by treatment with anti-myelin-associated glycoprotein antibody. *Eur J Neurosci* 24: 3063–3072.

Vyas AA, Blixt O, Paulson JC et al. (2005). Potent glycan inhibitors of myelin-associated glycoprotein enhanced axon outgrowth in vitro. *J Biol Chem* 280: 16305–16310.

Wang KC, Koprivica V, Kim JA et al. (2002a) Oligodendrocyte myelin glycoprotein is a nogo-receptor ligand that inhibits neurite outgrowth. *Nature* 417: 941–944.

Wang KC, Kim JA, Sivasankaran R et al. (2002b). P75 interacts with the Nogo receptor as a co-receptor for Nogo, Mag and OMgp. *Nature* 420: 74–78.

Wang X, Baughman KW, Basso M et al. (2006). Delayed Nogo receptor therapy improves recovery from spinal cord contusion. *Ann Neurol* 60: 540–549.

Weidner N, Ner A, Salimi N et al. (2001). Spontaneous corticospinal axonal plasticity and functional recovery after adult central nervous system injury. *Proc Natl Acad Sci* 98: 3513–3518.

Wen D, Wildes CP, Silvian L et al. (2005). Disulfide structure of the leucine-rich repeat C-terminal cap and c-terminal stalk region of Nogo-66 receptor. *Biochemistry* 44: 16491–16501.

Wiessner C, Baryere FM, Allegrini, PR et al. (2003). Anti-nogo-A antibody infusion 24h after experimental stroke improved behavioral outcome and corticospinal plasticity in normotensive and spontaneously hypertensive rats. *J Cereb Blood Flow Metab* 23: 154–165.

Winton MJ, Dubreuil CI, Lasko D et al. (2002). Characterization of new cell permeable C3-like proteins that inactivate Rho and stimulate neurite outgrowth on inhibitory substrates. *J Biol Chem* 277: 32820–32829.

Woolf CJ (2003). No Nogo: now where do we go? *Neuron* 38: 153–156.

Xu G, Nie DY, Chen JT et al. (2004). Recombinant DNA vaccine encoding multiple domains related to inhibition of neurite outgrowth: a potential strategy for axonal regeneration. *J Neurochem* 91: 1018–1023.

Yiu G, and He Z (2003). Signaling mechanisms of the myelin inhibitors of axon regeneration. *Curr Opin Neurobiol* 13: 1–7.

Yu P, Huang L, Zhu H et al. (2007). DNA vaccine against NgR promotes functional recovery after spinal cord injury in adult rats. *Brain Res* 1147: 66–76

Zander H, Hettich E, Greiff K et al. (2007). Biochemical characterization of the recombinant human Nogo A ectodomain. *FEBS J* 274: 2603–2613.

Zheng B, Ho C, Li S et al. (2003). Lack of enhanced spinal regeneration in Nogo deficient mice. *Neuron* 38: 213–224.

Zheng B, Atwal J, Ho C et al. (2005). Genetic deletion of the Nogo receptor does not reduce neurite inhibition in vitro or promote corticospinal regeneration in vivo. *Proc Natl Acad Sci* 102: 1205–1210.

17

Cyclic AMP Modulation of Axonal Regeneration Following Spinal Cord Injury

Sari S. Hannila and Marie T. Filbin

Department of Biological Sciences, Hunter College, City University of New York, NY 10021

Abstract

Central nervous system (CNS) myelin contains several proteins that contribute to regenerative failure after spinal cord injury, and elevation of intracellular cyclic adenosine monophosphate (cAMP) has proven to be one of the most effective means of overcoming myelin inhibition. In this chapter, we describe how increasing cAMP promotes axonal regeneration in the CNS and how this finding has advanced the field of spinal cord injury repair. Upregulation of cAMP is responsible for the conditioning lesion effect observed following peripheral nerve transection, and administration of cAMP analogues promotes axonal growth in the presence of myelin both in vitro and in vivo. The phosphodiesterase (PDE) inhibitor rolipram also enhances axonal regeneration when used in combination with cell transplantation in models of spinal cord injury. Most significantly, cAMP upregulates the expression of genes such as arginase I and interleukin-6 (IL-6), which have been shown to overcome myelin inhibitors in their own right. It is therefore probable that additional study of cAMP-regulated genes will lead to the development of new agents capable of promoting axonal regeneration after spinal cord injury.

Copyright © 2008 Elsevier Inc.
All rights of reproduction in any form reserved.

THE ROLE OF THE CNS ENVIRONMENT IN REGENERATIVE FAILURE

HISTORICAL PERSPECTIVES

As the theories of Galen once dominated medicine, the observations of Santiago Ramón y Cajal have dominated the first century of research on axonal regeneration in the CNS. Like Galen, Cajal was devoted to detailed study of anatomical structure, and his findings remain relevant today thanks to his painstaking documentation of axonal morphology. Using the silver staining technique pioneered by his contemporary Camillo Golgi, Cajal was the first to describe the axonal sprouting and growth cones that characterize regenerating axons in transected peripheral nerves (Ramón y Cajal, 1928). Similar experiments in the CNS, however, produced dramatically different results. In both the brain and spinal cord, transected axons rapidly became dystrophic and extended only short distances before growth was abruptly halted (Ramón y Cajal, 1928). Additionally, the growth cones at the tips of these axons became stunted and swollen, which led Cajal to describe them as retraction bulbs (Ramón y Cajal, 1928). Cajal also used the phrase "abortive regeneration" to describe the overall response of CNS axons to injury, and concluded that, "In adult centers the nerve paths are something fixed, ended, immutable. Everything may die, nothing may be regenerated."

The findings of Cajal led many scientists to believe that CNS regeneration was impossible, and consequently, little research was conducted in this area. In the early 1980s, however, work from the laboratory of Albert Aguayo would completely change this perception and make CNS axonal regeneration one of the most widely studied topics in neuroscience. In the first of two landmark studies, small segments of peripheral nerve were grafted into the injured spinal cord and axons were then retrogradely traced by injection of horseradish peroxidase (HRP) into the graft (Richardson et al., 1980). Visualization of the tracer revealed substantial axonal growth within the graft, and more importantly, staining of neuronal cell bodies within the spinal cord, which demonstrated that these axons were CNS-derived (Richardson et al., 1980). This approach was then expanded to produce an even more dramatic demonstration of CNS axon growth. Long sciatic nerve grafts were positioned extraspinally, and the ends were inserted into the medulla oblongata and spinal cord (David and Aguayo, 1981). After a postsurgical survival period of 22–30 weeks, HRP was applied to the ends of the transected grafts to allow both anterograde and retrograde labeling of axons (David and Aguayo, 1981). Labeled axons extended up to 30 mm within the grafts and their origin in the CNS was confirmed by the presence of large numbers of HRP-labeled neurons in the medulla and spinal cord (David and Aguayo, 1981). Together, these data firmly established that CNS axons were capable of regenerating over long distances. In addition, the fact that this regeneration occurred within a permissive environment gave new prominence to a concept first proposed by Cajal: that factors in the CNS environment inhibit regeneration of injured axons.

MYELIN INHIBITORS – STRUCTURES AND SIGNALING

It is now known that CNS myelin contains several proteins that inhibit axonal growth (Figure 17.1), and the first direct evidence of this was produced by the laboratory of Martin Schwab. They first isolated 35 and 250 kD protein fractions

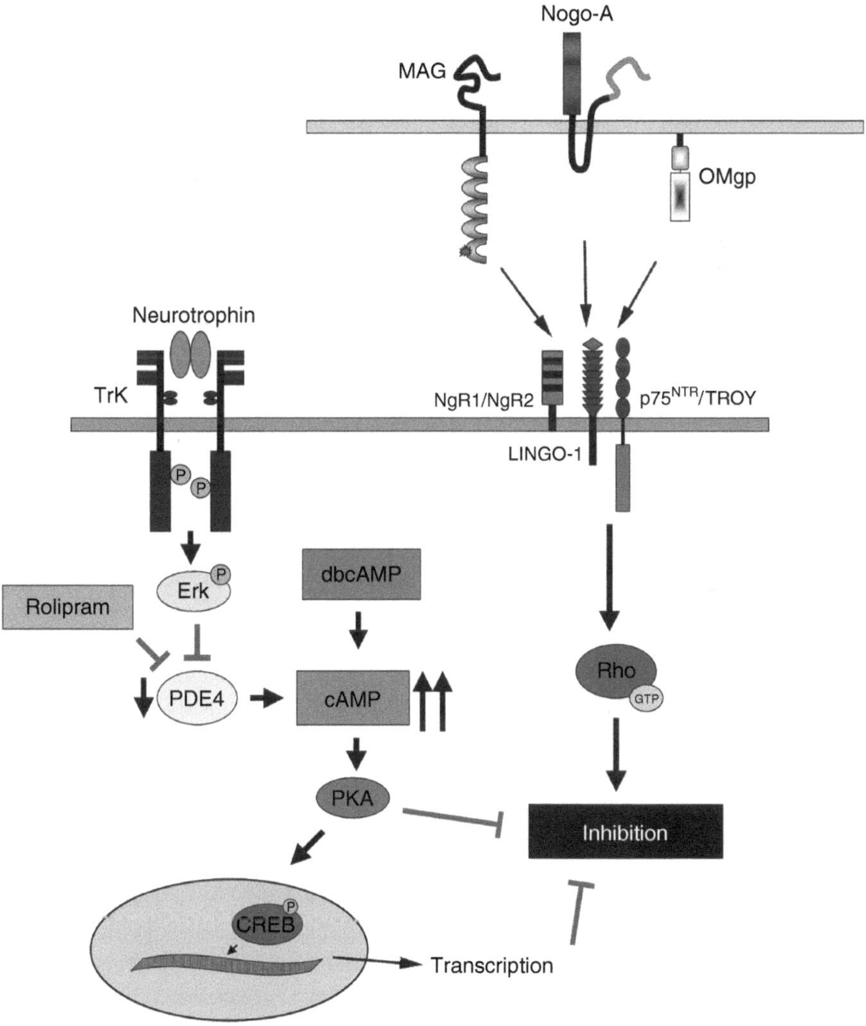

FIGURE 17.1 Schematic representation of intracellular signaling pathways activated by cAMP and their role in overcoming inhibition by CNS myelin. Neurotrophin binding to Trk receptors leads to activation of Erk, which in turn results in a transient reduction of PDE4 activity. Inhibition of PDE4 can also be achieved through administration of rolipram, and this causes cAMP levels to rise. Alternatively, intracellular cAMP can be directly increased by treatment with dbcAMP, and upon reaching a threshold level, cAMP will activate PKA and initiate transcription by CREB. These signaling events block inhibition mediated by MAG, Nogo-A and OMgp binding to the receptor complex consisting of NgR1 or NgR2, $p75^{NTR}$ or TROY and LINGO-1. (see Plate 10)

from CNS myelin and demonstrated that these proteins could inhibit neurite outgrowth in vitro (Caroni and Schwab, 1988a). Monoclonal antibodies were raised against these proteins and it was found that administration of an antibody named IN-1 not only increased neurite outgrowth on myelin but also enhanced regeneration of corticospinal axons and improved locomotor function after spinal cord injury (Caroni and Schwab, 1988b; Schnell and Schwab, 1990, 1993; Bregman et al., 1995). The antigen of the IN-1 antibody remained unknown until 2000, when the Nogo protein was independently cloned by three different groups (Chen et al., 2000; GrandPre et al., 2000; Prinjha et al., 2000). Three different isoforms of Nogo (A, B and C) are expressed in the CNS, but only Nogo-A is enriched in oligodendrocytes (Chen et al., 2000; GrandPre et al., 2000; Prinjha et al., 2000). Nogo-A inhibits neurite outgrowth by inducing growth cone collapse, and this effect is mediated through two distinct domains: a 66-residue extracellular domain (Nogo-66) that is shared by all three Nogo isoforms, and amino-Nogo, which is unique to Nogo-A (Chen et al., 2000; GrandPre et al., 2000; Prinjha et al., 2000).

The cloning of Nogo made it the second myelin inhibitor to be identified, as the growth-inhibiting properties of myelin-associated glycoprotein (MAG) had been reported several years earlier (McKerracher et al., 1994; Mukhopadhyay et al., 1994). MAG is a well-known component of both peripheral nervous system (PNS) and CNS myelin, and is a member of the immunoglobulin (Ig) superfamily, containing five extracellular Ig-like domains (Salzer et al., 1987, 1990). It is also a sialic acid-binding protein and its first four Ig-like domains are homologous to the Siglec family of sialic acid-binding Ig-like lectins (Kelm et al., 1994). In the PNS, MAG is involved in initiating myelination by Schwann cells (Owens and Bunge, 1989), but it is not required for this process as MAG null mutant mice display normal myelination of peripheral axons (Li et al., 1994; Montag et al., 1994). By contrast, MAG plays a critical role in maintaining the interaction between the axon and its myelin sheath (Martini and Schachner, 1986). This interaction is disrupted in MAG null mutant mice, leading to abnormalities in the periaxonal space and loss of compact myelin (Li et al., 1994; Montag et al., 1994). More importantly, widespread degeneration of axons and myelin is observed in MAG null mutants older than 8 months, which indicates that MAG is essential for the maintenance of axonal structure (Fruttiger et al., 1995). In 1994, it was shown that MAG also acts as an inhibitor of neurite outgrowth for several different neuronal populations, including dorsal root ganglion (DRG) and cerebellar neurons (McKerracher et al., 1994; Mukhopadhyay et al., 1994). Sialic acid binding may play a role in mediating MAG inhibition, as mutation of arginine 118, the sialic acid-binding site, abolishes inhibition by soluble MAG (Tang et al., 1997). However, inhibition by membrane-bound MAG is unaffected by this mutation, which suggests that the inhibitory domain of MAG is distinct from the sialic acid-binding site (Tang et al., 1997).

Recently published findings suggest that MAG may also play a role in axonal guidance. During development, proprioceptive axons are guided to the dorsal column through signaling by attractive and repulsive guidance cues, and these axons are excluded from the dorsal horn of the spinal cord through semaphorin 6C and 6D binding to the plexin A1 receptor (Yoshida et al., 2006). In plexin A1 null mutant mice, this axonal patterning is disrupted, and aberrant growth of proprioceptive axons into the medial dorsal horn is observed (Yoshida et al., 2006). Unmyelinated cutaneous sensory axons expressing IB4 lectin normally terminate within the medial dorsal horn; however, in plexin A1 mutants, IB4-positive axons were absent from this area. Interestingly, plexin A1 mutants also displayed a dramatic increase in the number of MAG-expressing oligodendrocytes within the medial dorsal horn from P4 onward, and these oligodendrocytes were intimately associated with the ingrowing proprioceptive axons (Yoshida et al., 2006). It was therefore proposed that these axons are accompanied by oligodendrocytes as they invade the dorsal horn, and that the expression of MAG causes IB4-positive axons to be repelled (Yoshida et al., 2006).

In the last 5 years, two more myelin-associated inhibitors have been described: oligodendrocyte myelin glycoprotein (OMgp) and ephrin B3. OMgp is a glycosyl-phosphatidylinositol (GPI)-linked protein with a leucine-rich repeat (LRR) domain and is expressed in both neurons and oligodendrocytes in the CNS (Mikol and Stefansson, 1988; Mikol et al., 1990; Habib et al., 1998; Wang et al., 2002a). It is a potent inhibitor of neurite outgrowth and induces growth cone collapse in chick DRG neurons (Wang et al., 2002a). In the spinal cord, OMgp is present in cells that ensheath the nodes of Ranvier, and it is believed that the expression of OMgp maintains nodal structure by inhibiting collateral axon sprouting (Huang et al., 2005). Ephrin B3 repels corticospinal axons from the midline of the spinal cord during development and was found to inhibit neurite outgrowth for cerebellar and cortical neurons (Benson et al., 2005).

The myelin inhibitors described above vary tremendously in structure, yet it has been found that three of them mediate inhibition by binding to a common receptor (Figure 17.1). The Nogo receptor (NgR1) was cloned from a mouse expression library using a soluble form of Nogo-66, and it was shown that binding of Nogo-66 to NgR1 was required to induce growth cone collapse (Fournier et al., 2001). OMgp binds to NgR1 in a similar manner, and exogenous expression of NgR1 conferred responsiveness to OMgp and inhibited neurite outgrowth in neurons that are normally unresponsive to myelin (Wang et al., 2002a).

MAG is unique in that it is capable of binding to both NgR1 and a structurally related receptor known as NgR2. Soluble MAG was able to precipitate NgR1 from primary neurons and it was also shown that this binding was independent of sialic acid (Domeniconi et al., 2002). Neurite outgrowth was inhibited by MAG binding to NgR1, and this inhibition could be blocked by neutralization of NgR1 function through the addition of NgR1 antibody, soluble NgR1 or dominant-negative NgR1 (Domeniconi et al., 2002; Liu et al., 2002). MAG binding to NgR2 also produced inhibition of neurite outgrowth, but unlike NgR1, this binding is sialic acid-dependent and selective for MAG, as NgR2 cannot interact with other myelin inhibitors (Venkatesh et al., 2005).

The functions of NgR1 and NgR2 are not limited to inhibition, as a newly published study describes a novel role for NgR1 and NgR2 in the clearance of macrophages from regenerating peripheral nerves. Recruitment of macrophages to the injury site is an important component of peripheral nerve regeneration, as they phagocytose the axonal and myelin debris generated by Wallerian degeneration (Mueller et al., 2003; Fry et al., 2007). These macrophages migrate out of the nerve once Wallerian degeneration is complete, but the signals that regulate this efflux are unknown. Fry and colleagues (2007) show that activated macrophages upregulate the expression of NgR1 and NgR2 as they accumulate in the injured sciatic nerve and that this upregulation is induced by phagocytosis of cellular debris. To determine the functional significance of this observation, they tested the ability of NgR-expressing macrophages to adhere to MAG and myelin substrates and found that binding was reduced. This repulsion was reversed in macrophages transfected with NgR1 or NgR2 small interfering RNAs, which confirmed that this effect was mediated by NgR (Fry et al., 2007). Ultrastructural analysis of crushed sciatic nerves revealed that the onset of macrophage efflux is correlated with the myelination of newly regenerated axons. By contrast, large numbers of macrophages persisted in ligated sciatic nerves lacking newly synthesized myelin. Macrophage migration was also impaired in sciatic nerves from NgR1 and MAG null mutant mice, which indicated that MAG binding to NgR1 is required for the expulsion of macrophages from peripheral nerve (Fry et al., 2007). Together, these findings suggest that macrophage efflux is regulated by remyelination of regenerated axons and that this process is triggered by MAG binding to NgR on macrophages (Fry et al., 2007).

Both NgR1 and NgR2 are GPI-linked proteins, which render them incapable of intracellular signaling, and they must therefore rely on co-receptors for signal transduction. The first transmembrane protein that was shown to associate with NgR1 was the p75 neurotrophin receptor ($p75^{NTR}$), a member of the tumor necrosis factor receptor superfamily (Roux and Barker, 2002). The $p75^{NTR}$ was first implicated in MAG signaling when it was shown that DRG and cerebellar neurons from $p75^{NTR}$ null mutant mice were not inhibited by MAG (Yamashita et al., 2002). This role was later confirmed in immunoprecipitation studies where MAG, Nogo-66 and OMgp each precipitated receptor complexes containing NgR1 and $p75^{NTR}$ (Wang et al., 2002b; Wong et al., 2002). Activation of the NgR1–$p75^{NTR}$ receptor complex inhibits neurite outgrowth by activating the small GTPase RhoA in a protein kinase C-dependent manner (Figure 17.1; Yamashita et al., 2002; Yamashita and Tohyama, 2003; Sivasankaran et al., 2004), and it was recently shown that regulated intramembrane proteolysis of $p75^{NTR}$ is required for MAG-mediated activation of RhoA (Domeniconi et al., 2005).

Soon after $p75^{NTR}$ was identified as the transducing component of the NgR1 receptor complex, it was noted that $p75^{NTR}$ was not widely expressed in the CNS, and so, it was suggested that another receptor was substituting for $p75^{NTR}$ in these regions. This was confirmed when two groups independently demonstrated that TROY (also known as TAJ), another member of the TNF receptor family, could associate with NgR1 and mediate myelin inhibition (Park et al., 2005; Shao et al., 2005). The role of TROY in myelin inhibition was further supported by the observation that cerebellar and DRG neurons from TROY/TAJ-deficient mice were less inhibited by OMgp and Nogo-66 (Park et al., 2005; Shao et al., 2005). In addition, inhibition could be reversed by either exogenous application of TROY or expression of truncated TROY in DRG neurons (Park et al., 2005; Shao et al., 2005).

A third component of the NgR1 receptor complex is the LRR protein LINGO-1, which was identified by Mi and colleagues in 2004. LINGO-1 is expressed exclusively in the nervous system, and neurite outgrowth was significantly reduced following transfection of cerebellar neurons with LINGO-1, while treatment with soluble LINGO-1 blocked this effect (Mi et al., 2004). The ability of LINGO-1 to transduce inhibitory signals lies in its cytoplasmic domain, as neurons expressing a truncated form of LINGO-1 were less responsive to myelin inhibitors (Mi et al., 2004).

Epidermal growth factor receptor (EGFR) is the latest protein to be ascribed a role in myelin signal transduction. Koprivica and colleagues (2005) demonstrated that Nogo-66 and OMgp phosphorylate EGFR, and that this activation is dependent on both calcium and the NgR1 receptor complex. It was found, however, that myelin inhibitors do not bind to EGFR, and so, it is unlikely that direct activation of the receptor mediates this effect. Furthermore, EGFR does not associate with NgR1 or $p75^{NTR}$, and this led the authors to conclude that phosphorylation of the receptor occurs through transactivation (Koprivica et al., 2005). Inhibition of EGFR kinase activity led to enhanced neurite outgrowth on myelin, MAG and Nogo-66, and promoted regeneration of retinal ganglion cell axons in the optic nerve, which suggests that EGFR signaling is an essential component of myelin inhibition (Koprivica et al., 2005). These findings underscore the growing complexity of myelin signaling, and as research in this field continues, new co-receptors and signaling intermediates will undoubtedly emerge.

THE CONDITIONING LESION EFFECT AND cAMP

The challenge presented by myelin inhibition has led researchers to pursue a wide range of strategies to promote axonal regeneration, and one of the most innovative of these has been to manipulate the neuron at the molecular level so that it no longer responds to myelin inhibitors. One way to accomplish this is through a procedure known as a conditioning lesion, which takes advantage of the fact that the pseudounipolar processes of DRG neurons give rise to axons that project into both the CNS and PNS. Lesioning the peripheral process of a DRG neuron enhances the neuron's intrinsic growth state, and this is reflected in the upregulated expression of regeneration-associated genes such as growth-associated protein 43 (GAP-43; Schreyer and Skene, 1993). Conversely, GAP-43 expression does not increase when DRG central processes are lesioned, and these axons fail to regenerate (Schreyer and Skene, 1993). Thus, transection of a peripheral nerve, such as the sciatic nerve, "conditions" the neurons and increases their ability to regenerate. When the conditioning lesion paradigm was applied in vivo, it was shown that concomitant lesioning of DRG central and peripheral processes enhanced regeneration of DRG central processes into a peripheral nerve graft, and that optimal regeneration was obtained when the peripheral lesion preceded the central lesion by 1 week (Richardson and Issa, 1984; Oudega et al., 1994; Chong et al., 1996).

In 1999, Neumann and Woolf used this model to determine whether a conditioning lesion could improve axonal regeneration in the spinal cord in the absence of a peripheral nerve graft. In adult animals that received only a dorsal column lesion, axons failed to regenerate into the lesion site, and this observation remained consistent when animals were examined at 2 months and 1 year after injury. Conversely, animals that received simultaneous dorsal column and sciatic nerve lesions displayed substantial growth of axons into the lesion site. This response was even more robust when the conditioning lesion was performed 1 week prior to the dorsal column transection, producing axonal regeneration that extended several millimeters beyond the site of injury.

At this point it is important to note that enhancing the growth capacity of axons does not necessarily mean that they will be able to overcome inhibition by myelin. For example, overexpression of GAP-43 is not sufficient to induce axonal regeneration following dorsal column or cerebellar lesion, even in the presence of permissive embryonic tissue grafts (Buffo et al., 1997; Neumann and Woolf, 1999; Bomze et al., 2001). Activating transcription factor-3 (ATF-3) is also strongly upregulated after conditioning lesion, and neurite outgrowth on a permissive substrate is significantly increased when ATF-3 is exogenously expressed in DRG neurons (Seijffers et al., 2006). It has therefore been proposed that ATF-3 may act as a universal regulator of axonal growth (Seijffers et al., 2006), but it is unknown if this effect is limited to general growth, or if ATF-3 can also overcome inhibition by myelin. This distinction is critical as it has been shown that neurite outgrowth on a permissive substrate such as L1 can be abolished by the addition of soluble MAG (Tang et al., 1997). Thus, observation of enhanced neurite outgrowth on a permissive substrate does not accurately reflect an agent's capacity to overcome inhibition and promote regeneration in vivo. After a conditioning lesion, many regenerating axons were observed in myelinated regions of the spinal cord (Neumann and Woolf, 1999), and this demonstrated that the conditioning lesion not only increases intrinsic growth, but also induces molecular changes that allow axons to overcome myelin inhibition in vivo. This finding has proven to be extremely valuable because prophylactic measures such as a conditioning lesion cannot be used to treat spinal cord injury in humans. Subsequent studies have therefore focused on identifying the factors that underlie the conditioning lesion effect, which in turn has led to the discovery of agents that can mimic its ability to promote axonal regeneration.

The conditioning lesion effect can now be attributed to elevation of the intracellular second messenger cAMP within the affected neurons. Our laboratory has demonstrated that cAMP levels in DRG cell bodies are increased twofold at 24 h after a conditioning lesion, and that these neurons are able to overcome inhibition by myelin (Qiu et al., 2002). It was also shown that at this time the increase in neurite outgrowth after conditioning lesion is dependent on activation of protein kinase A (PKA). At 1 week postlesion, cAMP levels have returned to baseline, but neurite outgrowth on myelin is even more extensive and has become PKA-independent (Qiu et al., 2002). To demonstrate that elevation of intracellular cAMP is sufficient to mimic the conditioning lesion effect, the non-hydrolyzable cAMP analog dibutyryl cAMP (dbcAMP) was injected into the DRG 1, 2 or 7 days prior to culturing the neurons on myelin substrates, an approach that yielded significant increases in neurite outgrowth (Neumann et al., 2002; Qiu et al., 2002). More importantly, intraganglionic injections of dbcAMP 7 days prior to a dorsal column lesion produced extensive regeneration of dorsal column axons in vivo (Neumann et al., 2002; Qiu et al., 2002). These studies demonstrated that cAMP can mimic the conditioning lesion effect both in vivo and in vitro, and that this effect occurs in sequential PKA-dependent and PKA-independent phases. More importantly, they confirmed that cAMP levels directly influence the response of neurons to myelin inhibitors and consequently, the extent of axonal regeneration.

REVERSAL OF MYELIN INHIBITION IN VITRO THROUGH ELEVATION OF cAMP

When the inhibitory properties of MAG were first described, it revealed an intriguing dichotomy. When P1 DRG neurons are plated on Chinese hamster ovary (CHO) cells expressing MAG, neurite outgrowth is promoted, whereas growth from DRG neurons that are P5 and older is inhibited (Johnson et al., 1989; Mukhopadhyay et al., 1994; Cai et al., 2001). This switch from growth promotion to inhibition has been observed in all neuronal subtypes that have been tested to date. For some neurons the switch occurs during embryonic development, while in others it occurs postnatally (Mukhopadhyay et al., 1994; De Bellard et al., 1996; Turnley and Bartlett, 1998).

In 2001, we demonstrated that there is a direct correlation between neuronal cAMP levels and inhibition of neurite outgrowth by MAG and myelin. Levels of endogenous cAMP are high in P1 DRG neurons and substantial neurite outgrowth is observed on MAG and myelin at this time point; however, at P3–4 there is a precipitous drop in neuronal cAMP, which coincides with the onset of myelin inhibition (Cai et al., 2001). These findings, in combination with those obtained in our conditioning lesion studies (Qiu et al., 2002), raised the possibility that pharmacological agents could be used to elevate cAMP levels in neurons and thereby increase their ability to regenerate.

Cell-permeable cAMP analogues are an effective means of directly elevating intracellular cAMP levels, and dbcAMP is one of the most widely used of these agents. Administration of dbcAMP overcomes inhibition by MAG and myelin for several neuronal subtypes, including DRG, cerebellar, cortical, and hippocampal neurons (Figure 17.1; Cai et al., 1999; Hannila and Filbin, unpublished observations). As mentioned previously, injection of dbcAMP into DRG can also promote regeneration of transected dorsal column axons following spinal cord injury (Neumann et al., 2002; Qiu et al., 2002). This effect is not limited to mammals, as dbcAMP can induce regeneration of transected spinal axons in zebrafish (Bhatt et al., 2004). The cAMP analog Sp-cAMPS has also been shown to reverse the inhibitory effects of MAG. In growth cone turning assays using *Xenopus* spinal neurons, MAG induces a repulsive turning response, and this can be converted to attraction by addition of Sp-cAMPS (Song et al., 1998; Henley et al., 2004).

Intracellular cAMP can be increased through indirect methods such as stimulating adenylate cyclase with forskolin, which has been shown to increase the rate of regeneration for transected axons in the sciatic nerve (Kilmer and Carlsen, 1984). Production of cAMP can also be stimulated by priming neurons with neurotrophic factors such as nerve growth factor (NGF) and brain-derived neurotrophic factor (BDNF). Priming is the term we use to describe experiments where neurons are first treated with neurotrophic factors on a permissive substrate and subsequently exposed to inhibitors such as MAG and myelin. In these experiments, cerebellar and DRG neurons were treated with BDNF, NGF or glial-derived neurotrophic factor (GDNF) overnight and then plated onto monolayers of MAG-expressing CHO cells. Cerebellar neurons treated with BDNF and GDNF were able to overcome MAG inhibition, while all three neurotrophic factors promoted neurite outgrowth for DRG neurons (Cai et al., 1999). Direct addition of BDNF and GDNF to neurons plated on MAG or myelin did not increase cerebellar neurite outgrowth, but when these neurotrophic factors were added in conjunction with pertussis toxin (PTX), an inhibitor of the GTP-binding proteins G_i and G_0, inhibition by MAG and myelin was blocked without priming (Cai et al., 1999). PTX alone did not reverse inhibition by MAG, discounting the possibility that G proteins directly mediate inhibition (Cai et al., 1999), but G_i/G_0 protein is known to inhibit adenylate cyclase and decrease intracellular levels of cAMP. It is therefore possible that G_i/G_0-mediated decreases in cAMP contribute to inhibition by MAG and myelin.

Neurotrophins are among the most potent neural growth factors known, but the fact that they cannot overcome inhibition by myelin limits their ability to promote axonal regeneration. NGF, BDNF and neurotrophin-3 (NT-3) have all been used in studies of spinal cord injury, but in most cases regeneration and functional recovery have been modest. For example, intrathecal administration of NGF after dorsal root rhizotomy induced regeneration of nociceptive axons into the spinal cord and recovery of sensory function (Ramer et al., 2000), but infusion of BDNF or NT-3 after spinal cord injury produced only moderate growth of rubrospinal and corticospinal axons (Schnell et al., 1994; Kobayashi et al.,

1997). Implantation of neurotrophin-secreting fibroblasts into the injured spinal cord produced extensive axonal regeneration, but only small improvements in functional recovery were observed (Grill et al., 1997; Tobias et al., 2003; Tuszynski et al., 2003). As shown by Cai and colleagues (1999), PTX allows neurotrophin-treated neurons to overcome inhibition by myelin, and thereby introduces an additional growth-promoting element that neurotrophins alone do not possess. The abilities of PTX and other G protein inhibitors to enhance the efficacy of neurotrophins merit further study, and it is possible that combining these agents could greatly improve axonal regeneration and functional recovery after spinal cord injury.

Neurotrophins do not directly affect cAMP levels, but rather initiate intracellular signaling by binding to Trk tyrosine kinase receptors (Kaplan and Miller, 2000). In 2003, Gao and colleagues showed that Trk receptor signaling and Trk-mediated activation of extracellular signal-regulated kinase (Erk) were required for BDNF and NGF to overcome inhibition by MAG (Figure 17.1). It was then found that activated Erk caused a transient inhibition of phosphodiesterase 4 (PDE4), the main enzyme responsible for cAMP degradation (Gao et al., 2003). The inhibition of PDE4 allowed intracellular cAMP to accumulate and eventually promote neurite outgrowth on MAG. This suggested that a threshold level of cAMP had to be reached in order to overcome inhibition and this was supported by the observation that Erk activation works synergistically with dbcAMP to achieve the required levels of cAMP (Gao et al., 2003). The regulation of PDE4 activity by Erk therefore represents a unique mechanism in which cross talk between the neurotrophin and cAMP signaling pathways allows primed neurons to overcome inhibition by myelin.

ELEVATION OF cAMP IN MODELS OF SPINAL CORD INJURY

Elevating cAMP levels through inhibition of PDE activity has proven to be a highly effective strategy for promoting axonal regeneration in the spinal cord. PDE4 is the major source of PDE activity in the CNS (Iona et al., 1998), making it a logical target for therapeutic intervention. Rolipram is a specific PDE4 inhibitor that has the added advantage of being able to cross the blood–brain barrier (Figure 17.1; Krause and Kuhne, 1988), which makes subcutaneous and oral administration possible. Rolipram was initially developed as an antidepressant (Horowski and Sastre y Hernandez, 1985) but has recently been found to enhance neurite outgrowth and axonal regeneration in the presence of myelin inhibitors.

The therapeutic potential of cAMP and rolipram was recently assessed in three separate studies, each using different treatment strategies and models of spinal cord injury (Figure 17.2). In the first of these studies, neurite outgrowth on MAG-expressing CHO cells or CNS myelin was significantly increased when DRG neurons received rolipram through priming or subcutaneous delivery in vivo (Nikulina et al., 2004). To build on these findings, adult rats received spinal cord hemisections, and this was followed by transplantation of embryonic spinal cord

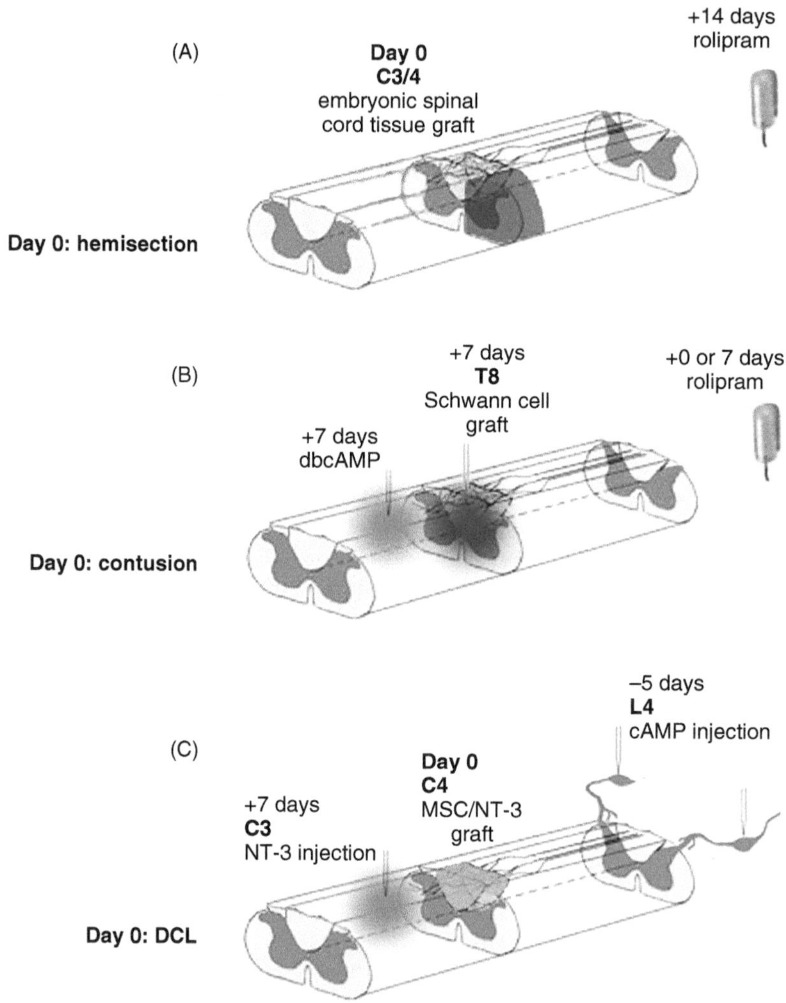

FIGURE 17.2 Comparison of spinal cord injury models that have used elevated cAMP to promote axonal regeneration. (A) In the study by Nikulina and colleagues (2004), spinal cord hemisection was performed at C3/4 and this was immediately followed by implantation of embryonic spinal cord tissue into the lesion site. Two weeks after surgery, rolipram was administered subcutaneously via osmotic minipump for 10 days. (B) Pearse and colleagues (2004) performed thoracic (T8) spinal cord contusions, which were followed by Schwann cell grafting and intraspinal injection of dbcAMP 1 week later. Some animals also received subcutaneous delivery of rolipram for 2 weeks, beginning either at the time of injury or 1 week after injury. (C) In the experimental model used by Lu and colleagues (2004), dbcAMP was injected into the L4 DRG 5 days prior to performing a dorsal column lesion (DCL). Animals received MSC grafts and NT-3 injections at the time of injury, and at 1 week after injury, an additional intraspinal injection of NT-3 was made rostral to the lesion site. (Copyright 2004 by the Society for Neuroscience.) (see Plate 11)

tissue into the lesion site and subcutaneous delivery of rolipram (Figure 17.2). When regeneration was assessed, few axons were observed within the transplant for vehicle-treated animals, but animals that received rolipram displayed significantly more growth of serotonergic fibers into the transplant (Figure 17.3; Nikulina et al., 2004). Rolipram-treated animals also had greater functional recovery as measured by forelimb paw placement, which suggests that the regenerating axons may be forming synaptic connections (Nikulina et al., 2004).

When these animals received rolipram, decreased expression of glial-fibrillary acidic protein (GFAP) was unexpectedly observed within the transplant, and this is indicative of a reduction in glial scarring (Figure 17.3; Nikulina et al., 2004). The glial scar is another key factor in regenerative failure, forming both a physical and biochemical barrier to regenerating axons. It is the result of astrogliosis, which is characterized by injury-induced morphological changes in astrocytes and elevated expression of GFAP and extracellular matrix molecules such as chondroitin sulfate proteoglycans (CSPGs), keratin sulfate proteoglycan and cytotactin/tenascin (McKeon et al., 1991; Jones et al., 2002; Silver and Miller, 2004). CSPGs are major components of the glial scar and are believed to inhibit axonal regeneration after spinal cord injury (Jones et al., 2002, 2003; Tang et al., 2003). Our results demonstrated that elevation of cAMP by rolipram reduces astrogliosis, which could enhance axonal regeneration by rendering the CNS environment more permissive. It is not known how elevated cAMP mediates this effect, but inhibition of reactive astrocyte proliferation and downregulation of CSPG expression are two possibilities that should be investigated.

Rolipram and dbcAMP have also been very effective when used in combination with cell transplantation techniques in other models of spinal cord injury. In a study by Pearse and colleagues (2004), adult rats received Schwann cell grafts and intraspinal injections of dbcAMP 1 week after a moderate spinal cord contusion (Figure 17.2), a model of injury that is considered more clinically relevant than spinal cord transection. Animals that received acute delivery of rolipram combined with Schwann cell grafts and dbcAMP displayed enhanced sparing of myelinated axons, greater myelination of spinal cord axons by the engrafted Schwann cells and an overall increase in the number of axons within the grafts (Pearse et al., 2004). Importantly, there was also significant regeneration of serotonergic axons across the lesion site and recovery of hind limb locomotor function as measured on the Basso–Beattie–Bresnahan scale (Figure 17.4; Basso et al., 1995). These findings demonstrate that elevation of cAMP increases the regenerative capacity of CNS axons following spinal cord contusion. The tissue sparing observed when rolipram and dbcAMP were administered in conjunction with Schwann cell grafting indicates that this approach is also highly neuroprotective.

While this combination produced the best results, arguably the most important finding of this study was that acute administration of rolipram alone was sufficient to increase axonal sparing and myelination, and significantly improve functional outcome (Figure 17.4; Pearse et al., 2004). The fact that a single agent could so dramatically impact functional recovery after spinal cord injury is

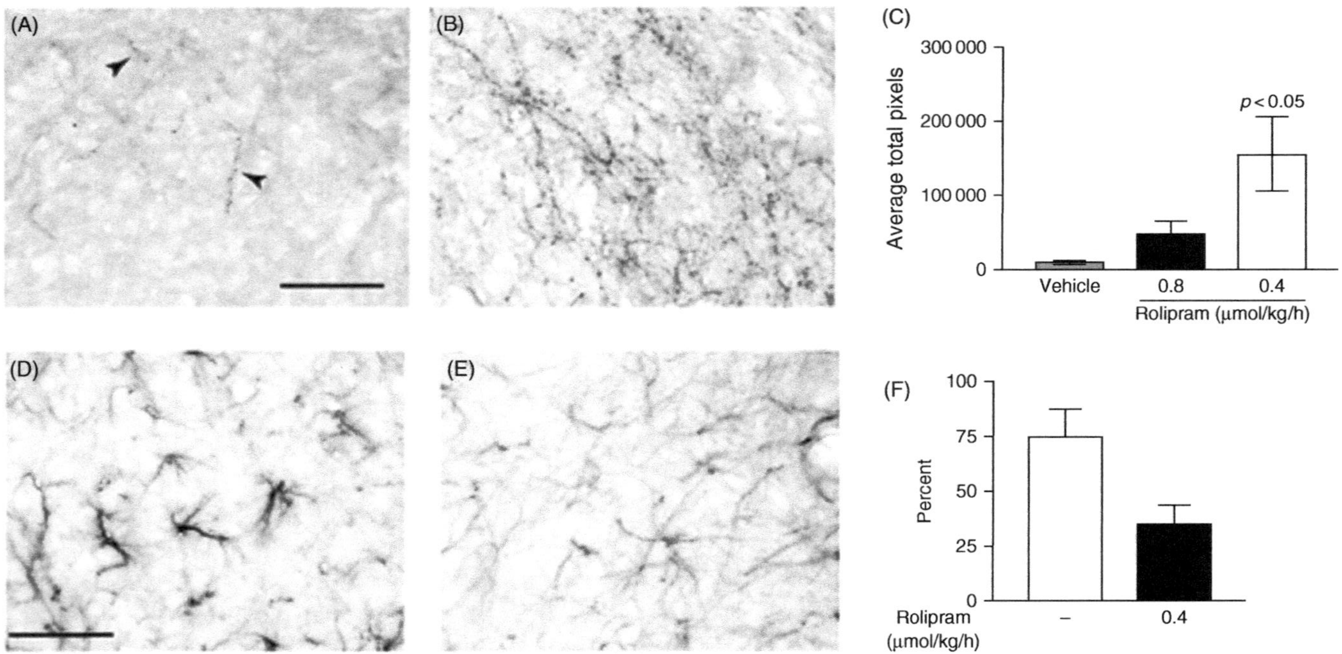

FIGURE 17.3 Rolipram promotes growth of serotonergic axons, reduces astrogliosis and improves functional recovery after spinal cord injury. Few serotonergic fibers (arrowheads) are present within the embryonic spinal cord tissue grafts of vehicle-treated animals (A) at 6–8 weeks after spinal cord hemisection, but in animals that received rolipram, growth of serotonergic axons is significantly increased (B, C). Animals treated with rolipram (E) also show significantly less GFAP staining within the tissue grafts when compared to animals that received vehicle alone (D), and this is indicative of a reduction in astrogliosis. Lastly, animals treated with rolipram show improved functional recovery, as demonstrated by a significant decrease in paw placement errors (F). Scale bars = 50 μm. (Copyright 2004 National Academy of Sciences, USA.)

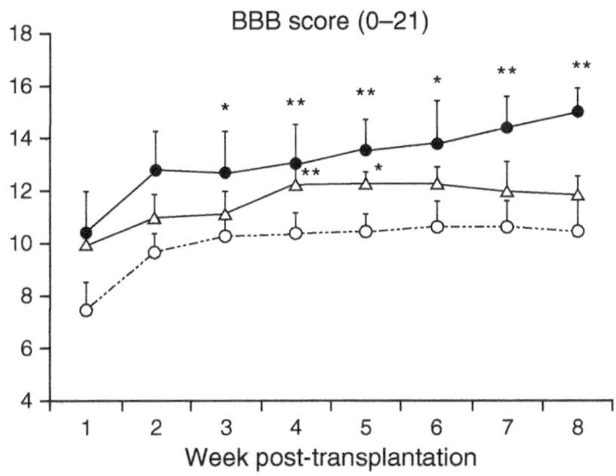

FIGURE 17.4 Rolipram promotes functional recovery after spinal cord contusion injury. BBB scores of animals that received rolipram alone (Δ) are significantly higher than those of control animals (○) at both 4 and 5 weeks after injury (* $p < 0.05$, ** $p < 0.01$). Animals that received rolipram in conjunction with Schwann cell transplantation and dbcAMP (●) show the greatest recovery of motor function, as BBB scores steadily increase beginning at 3 weeks after injury. (Adapted by permission from Macmillan Publishers Ltd., Pearse DD, Pereira FC, Marcillo AE, Bates ML, Berrocal YA, Filbin MT, and Bunge MB, cAMP and Schwann cells promote axonal growth and functional recovery after spinal cord injury. *Nat Med* 10: 610–616, copyright 2004.)

nothing less than remarkable, and attests to the tremendous therapeutic potential of this drug. The ability of rolipram to cross the blood–brain barrier is equally important, as this allows it to be delivered subcutaneously rather than to the lesion site. From a clinical perspective, this feature is invaluable because it eliminates the need for invasive surgery that could cause further damage to the spinal cord. Methylprednisolone is no longer considered the standard of care for spinal cord injury (Hurlbert, 2000; Short, 2001; Hugenholtz, 2003), and new agents are urgently needed to fill this void. Because of its efficacy in animal models and ease of administration, rolipram should be considered a leading candidate for this role and its use in human clinical trials should be expedited. It is very possible that rolipram would only produce modest functional recovery following spinal cord injury in humans, but even small improvements in sensory or motor function would make a tremendous difference in a patient's quality of life.

A similar study combined administration of dbcAMP and NT-3 with transplantation of bone marrow stromal cells (MSCs), which have been shown to physically support axonal regeneration in the spinal cord (Hofstetter et al., 2002; Lu et al., 2004, 2005). dbcAMP was injected into the L4 DRG prior to injury, while NT-3 was injected into the injury site and caudal spinal cord immediately after dorsal column lesion (Figure 17.2; Lu et al., 2004). MSCs were also transplanted after injury. The goal of this approach was to replicate the conditioning lesion effect in vivo by stimulating neuronal cell bodies with dbcAMP, while simultaneously providing trophic support to the transected axons. Like a conditioning lesion, the injections of dbcAMP made prior to injury are prophylactic, and so, this approach should be considered a means of providing proof of principle rather than a clinically applicable treatment strategy. Animals that received a combination of dbcAMP, NT-3 and MSCs were the only treatment group that displayed extensive axonal regeneration beyond the lesion site, more than ever reported before in this lesion paradigm (Figure 17.5; Lu et al.,

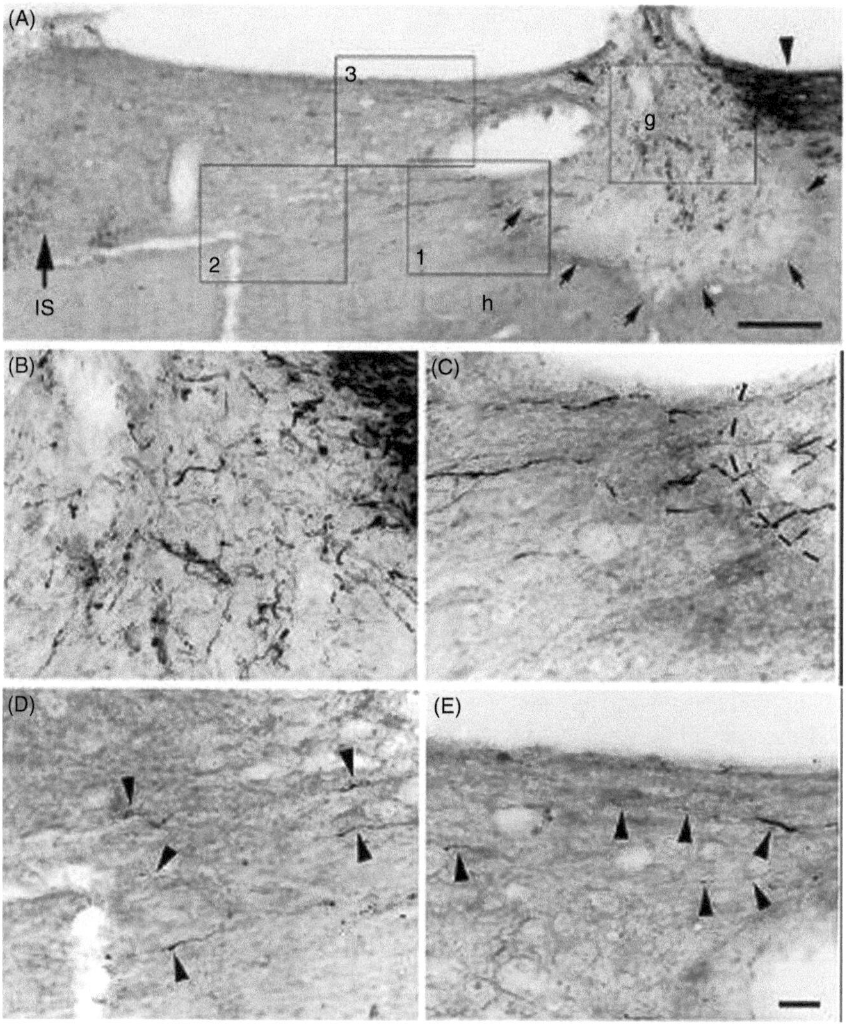

FIGURE 17.5 Axonal regeneration induced by a combination of elevated cAMP, NT-3 and MSC transplantation. At 1–3 months after dorsal column lesion, animals that received intraganglionic injection of dbcAMP, MSC grafts (g) and intraspinal injection of NT-3 (IS) displayed regeneration of axons within and beyond the lesion site (A). Axons labeled with cholera toxin B are present within the MSC graft (B), and re-enter the host tissue rostral to the graft (C). These axons extend up to 0.7 mm beyond the lesion site (D) in both the ventral and dorsal (E) aspects of the spinal cord. (Copyright 2004 by the Society for Neuroscience.)

2004). Even though functional recovery was not observed, this study still provides another clear example of axonal regeneration mediated by dbcAMP, and also emphasizes the benefits of combinatorial approaches to spinal cord injury repair.

DOWNSTREAM EFFECTORS OF cAMP IN SPINAL CORD INJURY

The studies that first described cAMP's ability to reverse inhibition by MAG and myelin also determined that this effect occurs in two phases that are PKA- and transcription-dependent (Figure 17.1). In the first phase, PKA activation by cAMP is required to overcome inhibition, and this is supported by the observation that PKA inhibitors block the effects of a conditioning lesion or priming with neurotrophins (Cai et al., 1999; Qiu et al., 2002). Subsequently, the effects of cAMP become PKA-independent but transcription-dependent, and evidence of this was provided by experiments that used 5,6-dichloro-1-b-D-ribo-furanosyl-benzimidazole (DRB) to inhibit transcription. Cerebellar neurons that received DRB in conjunction with dbcAMP or priming with BDNF were unable to overcome inhibition by MAG, thereby confirming that transcription is an essential component of cAMP signaling (Cai et al., 2002).

Elevated levels of cAMP induce gene transcription through activation of the transcription factor cAMP response element binding protein (CREB) (Lonze and Ginty, 2002), and in 2004, Gao and colleagues showed that activation of CREB is necessary to overcome inhibition by myelin. Adenoviral expression of dominant negative CREB in DRG and cerebellar neurons blocked the ability of dbcAMP to overcome inhibition, whereas constitutively active CREB promoted neurite outgrowth on MAG in the absence of elevated cAMP (Gao et al., 2004). To determine whether activation of CREB is sufficient to promote axonal regeneration in vivo, adenoviruses expressing constitutively active CREB were injected into the L4 DRG of adult rats 4 days prior to a dorsal column lesion. The extent of axonal regeneration was assessed by anterograde tracing with cholera toxin B subunit (CTB), and in animals that expressed constitutively active CREB regeneration was significantly increased compared to animals that received injections of control adenovirus (Gao et al., 2004). These findings suggested that cAMP induces CREB-mediated transcription of genes that are involved in overcoming myelin inhibition.

The first cAMP-regulated gene to be examined in the context of myelin inhibition was arginase I (ArgI), a key enzyme in the synthesis of polyamines (Lange et al., 2004). Developmental expression of ArgI is intimately linked to intracellular cAMP levels, as ArgI expression in DRG neurons suddenly decreases between P3 and P5, coinciding with the reduction in cAMP and the onset of myelin inhibition (Cai et al., 2002). ArgI expression and polyamine synthesis are increased in cerebellar neurons in response to dbcAMP or BDNF

(Cai et al., 2002), and it was subsequently shown that the upregulation of ArgI is mediated by activation of PKA and CREB (Gao et al., 2004). Neurite outgrowth on MAG was significantly increased when ArgI was overexpressed in cerebellar neurons, but reduced when inhibitors of polyamine synthesis were administered in conjunction with dbcAMP and BDNF (Cai et al., 2002). These results suggested that polyamines play an important role in overcoming inhibition by MAG, and this hypothesis was supported by the observation that neurite outgrowth could be enhanced by priming with the polyamine putrescine (Cai et al., 2002). In vivo studies evaluating the ability of polyamines to promote axonal regeneration after spinal cord injury are currently underway.

Microarray analysis of DRG neurons that received a conditioning lesion or treatment with dbcAMP revealed increased expression of the cytokine IL-6 (Cao et al., 2006), making it a second potential target for therapeutic intervention in spinal cord injury. When tested in vitro, recombinant IL-6 overcame inhibition by MAG and myelin in a dose-dependent manner and this effect was also dependent on transcription (Cao et al., 2006). IL-6 mediates its effects through binding to the IL-6 receptor (IL-6R) and activation of Janus kinase (JAK), and blocking the function of either of these proteins abolished the ability of IL-6 to promote neurite outgrowth on MAG. Neurite outgrowth mediated by dbcAMP, however, was not affected by blockade of IL-6R or JAK, which suggested that IL-6 is not an essential component of cAMP signaling (Cao et al., 2006). This finding was supported by studies of the conditioning lesion effect in IL-6 null mutant mice. DRG neurons from IL-6 mutants retained the ability to extend neurites on MAG following a conditioning lesion, and when the conditioning lesion was followed by dorsal column transection, IL-6 mutants and wild-type mice displayed comparable regeneration of dorsal column axons (Cao et al., 2006). These results suggested that IL-6 is sufficient but not necessary to overcome myelin inhibition (Cao et al., 2006). It must be noted that Cafferty and colleagues (2004) had previously presented evidence that IL-6 is required for the conditioning lesion effect, as they observed no regeneration of dorsal column axons in IL-6 mutant mice. The reasons for this discrepancy are unclear, but it should not detract from the fact that substantial axonal regeneration was observed in adult rats that received intrathecal administration of IL-6 after dorsal column lesion (Cao et al., 2006).

While these results are encouraging, the potentially detrimental effects of IL-6 must also be acknowledged. IL-6 is a pro-inflammatory cytokine and is strongly upregulated in response to CNS injury and peripheral axotomy (Gadient and Otten, 1997). Increased expression of IL-6 has been correlated with accelerated peripheral nerve regeneration (Hirohata et al., 1996), but high levels of IL-6 could also exacerbate the inflammation observed after spinal cord injury and lead to increased cell death. This is supported by observations made in IL-6 transgenic mice, which display extensive astrogliosis, neurodegeneration and breakdown of the blood–brain barrier when IL-6 is overexpressed in the CNS (Campbell et al., 1993; Chiang et al., 1994; Brett et al., 1995). Clearly, additional work is needed to determine whether IL-6 is a viable option for the treatment of spinal cord injury, and it would be particularly beneficial to elucidate the mechanism underlying the growth-promoting properties of IL-6. Identification of the signaling pathways and intermediates activated by IL-6 could lead to the development of agents that promote axonal regeneration without stimulating an inflammatory response.

CONCLUSIONS

The enhanced regeneration observed following CREB activation and administration of IL-6 raises the possibility that other cAMP-regulated genes may yield additional agents capable of promoting axonal regeneration. Cao and colleagues (2006) noted that 11 genes were upregulated in response to dbcAMP, and these included neuropeptide Y, CREM (cAMP response element modulator) and VGF (NGF-inducible growth factor). These and other genes merit further study in models of axonal regeneration, and it is possible that a combination of cAMP-induced factors could prove to be an effective therapeutic strategy for the treatment of spinal cord injury.

These findings are just one example of the many promising strategies that have resulted from the identification of cAMP as a modulator of axonal regeneration. For their potential to be realized, however, they must be tested in a clinical setting. Some agents, such as rolipram, have already been approved by the Food and Drug Administration for clinical use in humans, and so long as adequate funding is available, there is no reason to delay clinical trials any longer. Rigorous clinical testing of new therapies will not only advance research but also better serve the needs of spinal cord injury patients.

REFERENCES

Basso DM, Beattie MS, and Bresnahan JC (1995). A sensitive and reliable locomotor rating scale for open field testing in rats. *J Neurotrauma* 12: 1–21.

Benson MD, Romero MI, Lush ME et al. (2005). Ephrin-B3 is a myelin-based inhibitor of neurite outgrowth. *Proc Natl Acad Sci U S A* 102, 10694–10699.

Bhatt DH, Otto SJ, Depoister B et al. (2004). Cyclic AMP-induced repair of zebrafish spinal circuits. *Science* 305: 254–258.

Bomze HM, Bulsara KR, Iskandar BJ et al. (2001). Spinal axon regeneration evoked by replacing two growth cone proteins in adult neurons. *Nat Neurosci* 4: 38–43.

Bregman BS, Kunkel-Bagden E, Schnell L et al. (1995). Recovery from spinal cord injury mediated by antibodies to neurite growth inhibitors. *Nature* 378: 498–501.

Brett FM, Mizisin AP, Powell HC et al. (1995). Evolution of neuropathologic abnormalities associated with blood-brain barrier breakdown in transgenic mice expressing interleukin-6 in astrocytes. *J Neuropathol Exp Neurol* 54: 766–775.

Buffo A, Holtmaat AJ, Savio T et al. (1997). Targeted overexpression of the neurite growth-associated protein B-50/GAP-43 in cerebellar Purkinje cells induces sprouting after axotomy but not axon regeneration into growth-permissive transplants. *J Neurosci* 17: 8778–8791.

Cafferty WB, Gardiner NJ, Das P et al. (2004). Conditioning injury-induced spinal axon regeneration fails in interleukin-6 knock-out mice. *J Neurosci* 24: 4432–4443.

Cai D, Shen Y, De Bellard M et al. (1999). Prior exposure to neurotrophins blocks inhibition of axonal regeneration by MAG and myelin via a cAMP-dependent mechanism. *Neuron* 22: 89–101.

Cai D, Qiu J, Cao Z et al. (2001). Neuronal cyclic AMP controls the developmental loss in ability of axons to regenerate. *J Neurosci* 21: 4731–4739.

Cai D, Deng K, Mellado W et al. (2002). Arginase I and polyamines act downstream from cyclic AMP in overcoming inhibition of axonal growth MAG and myelin in vitro. *Neuron* 35: 711–719.

Campbell IL, Abraham CR, Masliah E et al. (1993). Neurologic disease induced in transgenic mice by cerebral overexpression of interleukin 6. *Proc Natl Acad Sci U S A* 90: 10061–10065.

Cao Z, Gao Y, Bryson JB et al. (2006). The cytokine interleukin-6 is sufficient but not necessary to mimic the peripheral conditioning lesion effect on axonal growth. *J Neurosci* 26: 5565–5573.

Caroni P and Schwab ME (1988a). Two membrane protein fractions from rat central myelin with inhibitory properties for neurite growth and fibroblast spreading. *J Cell Biol* 106: 1281–1288.

Caroni P and Schwab ME (1988b). Antibody against myelin-associated inhibitor of neurite growth neutralizes nonpermissive substrate properties of CNS white matter. *Neuron* 1: 85–96.

Chen MS, Huber AB, van der Haar ME et al. (2000). Nogo-A is a myelin-associated neurite outgrowth inhibitor and an antigen for monoclonal antibody IN-1. *Nature* 403: 434–439.

Chiang C, Stalder A, Samimi A et al. (1994). Reactive gliosis as a consequence of interleukin-6 expression in the brain: studies in transgenic mice. *Dev Neurosci* 16: 212–221.

Chong MS, Woolf CJ, Turmaine M et al. (1996). Intrinsic versus extrinsic factors in determining the regeneration of the central processes of rat dorsal root ganglion neurons: the influence of a peripheral nerve graft. *J Comp Neurol* 370: 97–104.

David S and Aguayo AJ (1981). Axonal elongation into peripheral nervous system "bridges" after central nervous system injury in adult rats. *Science* 214: 931–933.

De Bellard ME, Tang S, Mukhopadhyay G et al. (1996). Myelin-associated glycoprotein inhibits axonal regeneration from a variety of neurons via interaction with a sialoglycoprotein. *Mol Cell Neurosci* 7: 89–101.

Domeniconi M, Cao Z, Spencer T et al. (2002). Myelin-associated glycoprotein interacts with the Nogo66 receptor to inhibit neurite outgrowth. *Neuron* 35: 283–290.

Domeniconi M, Zampieri N, Spencer T et al. (2005). MAG induces regulated intramembrane proteolysis of the p75 neurotrophin receptor to inhibit neurite outgrowth. *Neuron* 46: 849–855.

Fournier AE, GrandPre T, and Strittmatter SM (2001). Identification of a receptor mediating Nogo-66 inhibition of axonal regeneration. *Nature* 409: 341–346.

Fruttiger M, Montag D, Schachner M et al. (1995). Crucial role for the myelin-associated glycoprotein in the maintenance of axon-myelin integrity. *Eur J Neurosci* 7: 511–515.

Fry EJ, Ho C, and David S (2007). A role for Nogo receptor in macrophage clearance from injured peripheral nerve. *Neuron* 53: 649–662.

Gadient RA and Otten UH (1997). Interleukin-6 (IL-6)–a molecule with both beneficial and destructive potentials. *Prog Neurobiol* 52: 379–390.

Gao Y, Nikulina E, Mellado W et al. (2003). Neurotrophins elevate cAMP to reach a threshold required to overcome inhibition by MAG through extracellular signal-regulated kinase-dependent inhibition of phosphodiesterase. *J Neurosci* 23: 11770–11777.

Gao Y, Deng K, Hou J et al. (2004). Activated CREB is sufficient to overcome inhibitors in myelin and promote spinal axon regeneration in vivo. *Neuron* 44: 609–621.

GrandPre T, Nakamura F, Vartanian T et al. (2000). Identification of the Nogo inhibitor of axon regeneration as a Reticulon protein. *Nature* 403: 439–444.

Grill RJ, Blesch A, and Tuszynski MH (1997). Robust growth of chronically injured spinal cord axons induced by grafts of genetically modified NGF-secreting cells. *Exp Neurol* 148: 444–452.

Habib AA, Marton LS, Allwardt B et al. (1998). Expression of the oligodendrocyte-myelin glycoprotein by neurons in the mouse central nervous system. *J Neurochem* 70: 1704–1711.

Henley JR, Huang KH, Wang D et al. (2004). Calcium mediates bidirectional growth cone turning induced by myelin-associated glycoprotein. *Neuron* 44: 909–916.

Hirohata H, Kiyama H, Kishimoto T et al. (1996). Accelerated nerve regeneration in mice by upregulated expression of interleukin (IL) 6 and IL-6 receptor after trauma. *J Exp Med* 183: 2627–2634.

Hofstetter CP, Schwarz EJ, Hess D et al. (2002). Marrow stromal cells form guiding strands in the injured spinal cord and promote recovery. *Proc Natl Acad Sci U S A* 99: 2199–2204.

Horowski R and Sastre y Hernandez M (1985). Clinical effects of the neurotrophic selective cAMP phosphodiesterase inhibitor rolipram in depressed patients: global evaluation of the preliminary reports. *Curr Ther ResClin Exp* 38: 23–29.

Huang JK, Phillips GR, Roth AD et al. (2005). Glial membranes at the node of Ranvier prevent neurite outgrowth. *Science* 310: 1813–1817.

Hugenholtz H (2003). Methylprednisolone for acute spinal cord injury: not a standard of care. *CMAJ* 168: 1145–1146.

Hurlbert RJ (2000). Methylprednisolone for acute spinal cord injury: an inappropriate standard of care. *J Neurosurg* 93: 1–7.

Iona S, Cuomo M, Bushnik T et al. (1998). Characterization of the rolipram-sensitive, cyclic AMP-specific phosphodiesterases: identification and differential expression of immunologically distinct forms in the rat brain. *Mol Pharmacol* 53: 23–32.

Johnson PW, Abramow-Newerly W, Seilheimer B et al. (1989). Recombinant myelin-associated glycoprotein confers neural adhesion and neurite outgrowth function. *Neuron* 3: 377–385.

Jones LL, Yamaguchi Y, Stallcup WB et al. (2002). NG2 is a major chondroitin sulfate proteoglycan produced after spinal cord injury and is expressed by macrophages and oligodendrocyte progenitors. *J Neurosci* 22: 2792–2803.

Jones LL, Margolis RU, and Tuszynski MH (2003). The chondroitin sulfate proteoglycans neurocan, brevican, phosphacan and versican are differentially regulated following spinal cord injury. *Exp Neurol* 182: 399–411.

Kaplan DR and Miller FD (2000). Neurotrophin signal transduction in the nervous system. *Curr Opin Neurobiol* 10: 381–391.

Kelm S, Pelz A, Schauer R et al. (1994). Sialoadhesin, myelin-associated glycoprotein and CD22 define a new family of sialic acid-dependent adhesion molecules of the immunoglobulin superfamily. *Curr Biol* 4: 965–972.

Kilmer SL and Carlsen RC (1984). Forskolin activation of adenylate cyclase in vivo stimulates nerve regeneration. *Nature* 307: 455–457.

Kobayashi NR, Fan DP, Giehl KM et al. (1997). BDNF and NT-4/5 prevent atrophy of rat rubrospinal neurons after cervical axotomy, stimulate GAP-43 and Talpha1-tubulin mRNA expression, and promote axonal regeneration. *J Neurosci* 17: 9583–9595.

Koprivica V, Cho KS, Park JB et al. (2005). EGFR activation mediates inhibition of axon regeneration by myelin and chondroitin sulfate proteoglycans. *Science* 310: 106–110.

Krause W and Kuhne G (1988). Pharmacokinetics of rolipram in the rhesus and cynomolgus monkeys, the rat and the rabbit. Studies on species differences. *Xenobiotica* 18: 561–571.

Lange PS, Langley B, Lu P et al. (2004). Novel roles for arginase in cell survival, regeneration, and translation in the central nervous system. *J Nutr* 134: 2812S–2817S.

Li C, Tropak MB, Gerlai R et al. (1994). Myelination in the absence of myelin-associated glycoprotein. *Nature* 369: 747–750.

Liu BP, Fournier A, GrandPre T et al. (2002). Myelin-associated glycoprotein as a functional ligand for the Nogo-66 receptor. *Science* 297: 1190–1193.

Lonze BE and Ginty DD (2002). Function and regulation of CREB family transcription factors in the nervous system. *Neuron* 35: 605–623.

Lu P, Yang H, Jones LL et al. (2004). Combinatorial therapy with neurotrophins and cAMP promotes axonal regeneration beyond sites of spinal cord injury. *J Neurosci* 24: 6402–6409.

Lu P, Jones LL, and Tuszynski MH (2005). BDNF-expressing marrow stromal cells support extensive axonal growth at sites of spinal cord injury. *Exp Neurol* 191: 344–360.

Martini R and Schachner M (1986). Immunoelectron microscopic localization of neural cell adhesion molecules (L1, N-CAM, and MAG) and their shared carbohydrate epitope and myelin basic protein in developing sciatic nerve. *J Cell Biol* 103: 2439–2448.

McKeon RJ, Schreiber RC, Rudge JS et al. (1991). Reduction of neurite outgrowth in a model of glial scarring following CNS injury is correlated with the expression of inhibitory molecules on reactive astrocytes. *J Neurosci* 11: 3398–3411.

McKerracher L, David S, Jackson DL et al. (1994). Identification of myelin-associated glycoprotein as a major myelin-derived inhibitor of neurite growth. *Neuron* 13: 805–811.

Mi S, Lee X, Shao Z et al. (2004). LINGO-1 is a component of the Nogo-66 receptor/p75 signaling complex. *Nat Neurosci* 7: 221–228.

Mikol DD and Stefansson K (1988). A phosphatidylinositol-linked peanut agglutinin-binding glycoprotein in central nervous system myelin and on oligodendrocytes. *J Cell Biol* 106: 1273–1279.

Mikol DD, Gulcher JR, and Stefansson K (1990). The oligodendrocyte-myelin glycoprotein belongs to a distinct family of proteins and contains the HNK-1 carbohydrate. *J Cell Biol* 110: 471–479.

Montag D, Giese KP, Bartsch U et al. (1994). Mice deficient for the myelin-associated glycoprotein show subtle abnormalities in myelin. *Neuron* 13: 229–246.

Mueller M, Leonhard C, Wacker K et al. (2003). Macrophage response to peripheral nerve injury: the quantitative contribution of resident and hematogenous macrophages. *Lab Invest* 83: 175–185.

Mukhopadhyay G, Doherty P, Walsh FS et al. (1994). A novel role for myelin-associated glycoprotein as an inhibitor of axonal regeneration. *Neuron* 13: 757–767.

Neumann S and Woolf CJ (1999). Regeneration of dorsal column fibers into and beyond the lesion site following adult spinal cord injury. *Neuron* 23: 83–91.

Neumann S, Bradke F, Tessier-Lavigne M et al. (2002). Regeneration of sensory axons within the injured spinal cord induced by intraganglionic cAMP elevation. *Neuron* 34: 885–893.

Nikulina E, Tidwell JL, Dai HN et al. (2004). The phosphodiesterase inhibitor rolipram delivered after a spinal cord lesion promotes axonal regeneration and functional recovery. *Proc Natl Acad Sci U S A* 101: 8786–8790.

Oudega M, Varon S, and Hagg T (1994). Regeneration of adult rat sensory axons into intraspinal nerve grafts: promoting effects of conditioning lesion and graft predegeneration. *Exp Neurol* 129: 194–206.

Owens GC and Bunge RP (1989). Evidence for an early role for myelin-associated glycoprotein in the process of myelination. *Glia* 2: 119–128.

Park JB, Yiu G, Kaneko S et al. (2005). A TNF receptor family member, TROY, is a coreceptor with Nogo receptor in mediating the inhibitory activity of myelin inhibitors. *Neuron* 45: 345–351.

Pearse DD, Pereira FC, Marcillo AE et al. (2004). cAMP and Schwann cells promote axonal growth and functional recovery after spinal cord injury. *Nat Med* 10: 610–616.

Prinjha R, Moore SE, Vinson M et al. (2000). Inhibitor of neurite outgrowth in humans. *Nature* 403: 383–384.

Qiu J, Cai D, Dai H et al. (2002). Spinal axon regeneration induced by elevation of cyclic AMP. *Neuron* 34: 895–903.

Ramer MS, Priestley JV, and McMahon SB (2000). Functional regeneration of sensory axons into the adult spinal cord. *Nature* 403: 312–316.

Ramón y Cajal S (1928). *Degeneration and regeneration of the nervous system.* J DeFelipe and EG Jones, eds. Oxford University Press, Oxford.

Richardson PM, McGuinness UM, and Aguayo AJ (1980). Axons from CNS neurons regenerate into PNS grafts. *Nature* 284: 264–265.

Richardson PM and Issa VM (1984). Peripheral injury enhances central regeneration of primary sensory neurones. *Nature* 309: 791–793.

Roux PP and Barker PA (2002). Neurotrophin signaling through the p75 neurotrophin receptor. *Prog Neurobiol* 67: 203–233.

Salzer JL, Holmes WP, and Colman DR (1987). The amino acid sequences of the myelin-associated glycoproteins: homology to the immunoglobulin gene superfamily. *J Cell Biol* 104: 957–965.

Salzer JL, Pedraza L, Brown M et al. (1990). Structure and function of the myelin-associated glycoproteins. *Ann N Y Acad Sci* 605: 302–312.
Schnell L and Schwab ME (1990). Axonal regeneration in the rat spinal cord produced by an antibody against myelin-associated neurite growth inhibitors. *Nature* 343: 269–272.
Schnell L and Schwab ME (1993). Sprouting and regeneration of lesioned corticospinal tract fibres in the adult rat spinal cord. *Eur J Neurosci* 5: 1156–1171.
Schnell L, Schneider R, Kolbeck R et al. (1994). Neurotrophin-3 enhances sprouting of corticospinal tract during development and after adult spinal cord lesion. *Nature* 367: 170–173.
Schreyer DJ and Skene JH (1993). Injury-associated induction of GAP-43 expression displays axon branch specificity in rat dorsal root ganglion neurons. *J Neurobiol* 24: 959–970.
Seijffers R, Allchorne AJ, and Woolf CJ (2006). The transcription factor ATF-3 promotes neurite outgrowth. *Mol Cell Neurosci* 32: 143–154.
Shao Z, Browning JL, Lee X et al. (2005). TAJ/TROY, an orphan TNF receptor family member, binds Nogo-66 receptor 1 and regulates axonal regeneration. *Neuron* 45: 353–359.
Short D (2001). Is the role of steroids in acute spinal cord injury now resolved? *Curr Opin Neurol* 14: 759–763.
Silver J and Miller JH (2004). Regeneration beyond the glial scar. *Nat Rev Neurosci* 5: 146–156.
Sivasankaran R, Pei J, Wang KC et al. (2004). PKC mediates inhibitory effects of myelin and chondroitin sulfate proteoglycans on axonal regeneration. *Nat Neurosci* 7: 261–268.
Song H, Ming G, He Z et al. (1998). Conversion of neuronal growth cone responses from repulsion to attraction by cyclic nucleotides. *Science* 281: 1515–1518.
Tang S, Shen YJ, De Bellard ME et al. (1997). Myelin-associated glycoprotein interacts with neurons via a sialic acid binding site at ARG118 and a distinct neurite inhibition site. *J Cell Biol* 138: 1355–1366.
Tang X, Davies JE, and Davies SJ (2003). Changes in distribution, cell associations, and protein expression levels of NG2, neurocan, phosphacan, brevican, versican V2, and tenascin-C during acute to chronic maturation of spinal cord scar tissue. *J Neurosci Res* 71: 427–444.
Tobias CA, Shumsky JS, Shibata M et al. (2003). Delayed grafting of BDNF and NT-3 producing fibroblasts into the injured spinal cord stimulates sprouting, partially rescues axotomized red nucleus neurons from loss and atrophy, and provides limited regeneration. *Exp Neurol* 184: 97–113.
Turnley AM and Bartlett PF (1998). MAG and MOG enhance neurite outgrowth of embryonic mouse spinal cord neurons. *Neuroreport* 9: 1987–1990.
Tuszynski MH, Grill R, Jones LL et al. (2003). NT-3 gene delivery elicits growth of chronically injured corticospinal axons and modestly improves functional deficits after chronic scar resection. *Exp Neurol* 181: 47–56.
Venkatesh K, Chivatakarn O, Lee H et al. (2005). The Nogo-66 receptor homolog NgR2 is a sialic acid-dependent receptor selective for myelin-associated glycoprotein. *J Neurosci* 25: 808–822.
Wang KC, Koprivica V, Kim JA et al. (2002a). Oligodendrocyte-myelin glycoprotein is a Nogo receptor ligand that inhibits neurite outgrowth. *Nature* 417: 941–944.
Wang KC, Kim JA, Sivasankaran R et al. (2002b). P75 interacts with the Nogo receptor as a co-receptor for Nogo, MAG and OMgp. *Nature* 420: 74–78.
Wong ST, Henley JR, Kanning KC et al. (2002). A p75(NTR) and Nogo receptor complex mediates repulsive signaling by myelin-associated glycoprotein. *Nat Neurosci* 5: 1302–1308.
Yamashita T, Higuchi H, and Tohyama M (2002). The p75 receptor transduces the signal from myelin-associated glycoprotein to Rho. *J Cell Biol* 157: 565–570.
Yamashita T and Tohyama M (2003). The p75 receptor acts as a displacement factor that releases Rho from Rho-GDI. *Nat Neurosci* 6: 461–467.
Yoshida Y, Han B, Mendelsohn M et al. (2006). PlexinA1 signaling directs the segregation of proprioceptive sensory axons in the developing spinal cord. *Neuron* 52: 775–788.

18

Multiple Sclerosis: Remyelination

Jeffery D. Kocsis, Masanori Sasaki, Karen L. Lankford, and Christine Radtke

Department of Neurology, Yale University School of Medicine, and Neuroscience Rehabilitation Research and Development Center, VA Connecticut Healthcare System, West Haven, CT 06516

Abstract

Demyelinating diseases can arise in a variety of forms in the central and as well as in the peripheral nervous system, and loss of myelin results in conduction abnormalities and functional deficits. Remyelination can occur from activation of endogenous progenitor cells and by exogenously transplanted myelin-forming cells. Here we discuss conduction abnormalities associated with demyelination and myelin repair by transplanted myelin-forming cells. Axonal repair and proper impulse conduction following remyelination also requires appropriate ion channel organization on the remyelinated axon. Axonal sodium and potassium channel distribution on remyelinated axons is discussed. Axon damage and transection is another prominent pathological feature of demyelinated lesions in multiple sclerosis (MS). This chapter also discusses potential neuroprotective effects of transplantation of myelin-forming cells into sites of axonal transection.

Introduction

Demyelination in the central nervous system (CNS) occurs in a variety of pathophysiological conditions. Perhaps the most notable is demyelination associated with multiple sclerosis (MS). MS is an inflammatory disease characterized

Copyright © 2008 Elsevier Inc.
All rights of reproduction in any form reserved.

by white matter plaques of demyelination in the brain and the spinal cord (Charcot, 1868; Lumsden, 1970). In addition to demyelination, MS plaques are often associated with axonopathy (Trapp et al., 1999). Impulse conduction is either blocked or slowed in these lesion sites resulting in various neurological symptoms depending on the plaque site. Demyelination can also occur in traumatic spinal cord injury (SCI) and following cerebral infarction. In contusive SCI, the spinal cord often presents with a central necrotic core, but areas of demyelinated axons are present outside of this region. Interestingly, apoptotic oligodendrocytes have been observed in experimentally induced SCI models at considerable distances of the injury sites (Crowe et al., 1997). Thus, interventional approaches to encourage CNS remyelination have relevance to many immunologic and traumatic CNS disorders. Experimental cellular transplantation has proven successful in a number of demyelination and injury models to remyelinate and improve functional outcome. Here we discuss the remyelination and neuroprotective potential of several myelin-forming cell types and their behavior in different demyelination and injury models. Better understanding of experimental cell-based strategies for remyelination and neuroprotection offer exciting opportunities to develop strategies for clinical studies.

CONDUCTION ABNORMALITIES IN DEMYELINATED AXONS

The most prominent conduction abnormalities in demyelinated axons are decreased conduction velocity, reduced ability to transmit high frequency trains of action potentials and conduction block (McDonald, 1963; McDonald and Sears 1970). Internodal conduction time can increase to nearly 500 ms in demyelinated fibers as compared to 20 μs in a normal fiber (Rasminsky and Sears, 1972). Thus, if multiple internodal regions of an axon are demyelinated this could result in considerable slowing of conduction even if the demyelination is focal. If a number of axons are demyelinated within a given tract, loss of synchrony occurs with temporal dispersion of the impulses (Figure 18.1C), resulting in inappropriate processing of information. Temporally dispersed and slowed afferent inputs in demyelinating polyneuropathies can result in loss of deep tendon reflexes.

Conduction block in demyelinated axons can take on various forms. It can be frequency-dependent, a form of conduction block in which high frequency impulse trains fail to conduct, but low frequency trains can conduct albeit with a time delay (Figure 18.1D). Conduction block can be complete for even single impulses (Figure 18.1D). High frequency conduction block may result from hyperpolarization due to electrogenic pump (Na^+, K^+ ATPase) activity (Bostock and Grafe, 1985). Moreover, increased intracellular Na^+ concentration at the "driving node" (Rasminsky and Sears, 1972) and accumulation of extracellular K^+ in the demyelinated areas could lead to Na^+ channel inactivation and conduction block (Brismar, 1981).

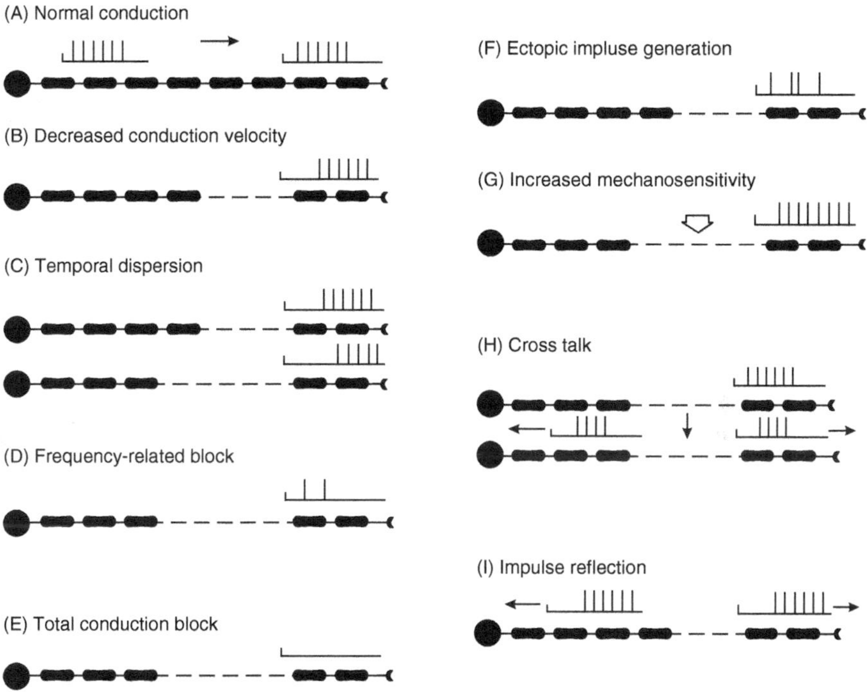

FIGURE 18.1 Conduction abnormalities in demyelinated axons. Demyelinated regions are schematically shown as dashed lines. Cell bodies are represented as circles on the left, and axon terminals are on the right. Arrows indicate the direction of conduction.

Hyperexcitable or positive events also occur in demyelinated axons. Ectopic action potential generation was reported for demyelinated cat dorsal column axons (Smith and McDonald, 1980). There are changes in the accommodative properties of demyelinated sensory vs. motor fibers with sensory fibers showing reduced accommodation. This may account for the more frequent appearance of paresthesia than increased motor activity in MS patients. Demyelinated fibers also display increased mechanosensitivity (Figure 18.1G) which may contribute to clinical phenomenon such as Lhermitte's sign; paresthesia induced by flexion of the neck in MS patients with cervical involvement (Smith and McDonald, 1980). Electrical cross talk (Figure 18.1H) may occur between demyelinated axons and contribute to hyperexcitability (Rasminsky, 1981; Devor and Seltzer, 1999). Impulse reflection (Figure 18.1I) has been reported in some demyelinated axons (Burchiel, 1980) and may result in paresthesia, pain or tonic spasms. Reflected (antidromic) impulses could collide with normal orthodromic impulse flow and further reduce the fidelity of impulse signaling. Thus, a key question is whether remyelination of demyelinated axons restores normal physiological function without inducing hyperexcitable events.

ENDOGENOUS MYELIN REPAIR AND POTENTIAL STIMULATION OF ENDOGENOUS PROGENITORS

Endogenous remyelination occurs in most rodent models of demyelination. When rat spinal cord white matter is chemically demyelinated, substantial endogenous remyelination occurs by a few weeks (Gledhill and McDonald, 1977; Smith et al., 1979). The remyelinated spinal cord axons show patterns of both central- and peripheral-like myelin. Remyelination has also been reported in animals where a focal experimental autoimmune encephalomyelitis (EAE) lesion was induced in the spinal cord (Kerschensteiner et al., 2004). The origin of cells responsible for this endogenous repair of myelin is thought to derive from spinal cord progenitor cells and invasion of peripheral Schwann cells. However, recent work suggests that Schwann cells can be derived from central progenitor cells and at least some of the peripheral-like myelin observed in remyelinated spinal cord may be centrally derived (Mujtaba et al., 1998; Keirstead et al., 1999; Akiyama et al., 2001).

The MS plaques in patients can become repaired by remyelination (Prineas and Connell, 1979). Oligodendrocytes reappear within active lesions in patients with early MS and are associated with early stages of remyelination (Lassmann, 1983; Prineas et al., 1984; Raine and Wu, 1993; Bruck et al., 1994). These areas of remyelination in MS patients are characterized by thinly myelinated, oligodendrocyte-like myelinated axons and are referred to as shadow plaques (Lassmann, 1983; Prineas, 1985). However, in late stages of progressive disease, remyelination is thought to be sparse and localized to outer margins of inactive plaques (Suzuki et al., 1969). Interestingly, Patrikios et al. (2006) recently reported from a post-mortem histopathological analysis that in 20% of MS patients studied 60–96% of the lesion area was remyelinated. Thus, spontaneous remyelination occurs in humans as well as in animal models, and strategies to improve spontaneous repair are quite important.

Oligodendrocyte precursors have been identified in some chronic MS lesions (Scolding et al., 1995; Wolswijk, 1998; Chang et al., 2002), but these cells appear to be quiescent. Recently, Nait-Oumesmar et al. (2007) found that activation of gliogenesis occurs in the subventricular zone (SVZ) of MS patients. The SVZ around the lateral ventricles is one of the largest germinative areas in the adult brain including humans, and is imbued with multipotent cells that have the potential to proliferate and differentiate into neurons and glia (Luskin, 1993; Lois and Alvarez-Buylla, 1994). Their data (Nait-Oumesmar et al., 2007) suggest the mobilization of SVZ-derived early glial progenitors to periventricular lesions where they might give rise to oligodendrocyte precursors. Thus, these cells could be a target to promote myelin repair in MS. However, to date, pharmacological approaches, such as application of neurotrophic factors and mitogens [e.g. basic fibroblast growth factor (bFGF) and epidermal growth factor (EGF)], to enhance remyelination have not achieved extensive myelin repair. An alternative approach to induce remyelination may be to isolate and expand human myelin-forming progenitor cells in vitro, and to reintroduce them into the demyelinated human CNS via cell transplantation approaches. While myelin-forming progenitors may be present in the adult CNS, their activation may be problematic, Thus, in vitro manipulation, expansion and re-introduction into the CNS (cell transplantation) of myelin-forming cells may be required to achieve an appropriate therapeutic outcome (Keirstead et al., 2005).

REMYELINATION AS A CELL THERAPY APPROACH TO IMPROVE CONDUCTION IN MS

While cellular transplantation approaches are being considered as an attempt to remyelinate demyelinated axons in MS, and to improve impulse conduction and functional outcome, a concern with such an approach is that MS is an immunological disease with central myelin-presenting antigens that elicit white matter inflammation. Introduction of exogenous oligodendrocytes or their precursors could exacerbate the disease. However, with recent developments in immunological therapies for MS, it is conceivable that in the future the immune process in MS could be controlled. These patients could have residual demyelinated plaques that associated with neurological symptoms. An effective cell therapy that targets sites of demyelination and induces remyelination in these patients could improve axonal function and reduce neurological symptom. Moreover, in traumatic diseases such as contusive SCI, demyelination can occur in long fiber tracts and remyelination of these tracts could contribute to improved functional outcome.

Experimental issues that need to be addressed in consideration of a clinical cell transplantation study include determination of an appropriate and safe donor cell type, demonstration of stable anatomical myelination and appropriate functional reorganization of the axons including axonal ion channel organization. If this reorganization fails in remyelinated axons, this would result in abnormal impulse conduction or even in the generation of ectopic or spontaneous action potential generation which could lead to pain or paresthesia. These events would severely limit the use of transplantation of myelin-forming cells as a therapeutic approach. However, much experimental work reviewed below indicates that axons remyelinated by transplanted cells generally show improvement in impulse conduction without adverse effects.

REMYELINATION BY ENDOGENOUS CELLS AND CONDUCTION IMPROVEMENT

Robust remyelination occurs following experimentally induced demyelination in spinal cord of rodents even after a few weeks (Gledhill and McDonald, 1977; Smith et al., 1979; Akiyama et al., 2004). Characteristic of remyelination the CNS are thinner and shorter internodes implying that new nodes of Ranvier have regenerated on these axons (Harrison et al., 1972; Gledhill and McDonald, 1977). Endogenous remyelination in rat spinal cord is accompanied by restoration of rapid and secure conduction (Smith et al. 1979). While structurally different from normal axons (thinner myelin and shorter internodal lengths), the remyelinated axons were able to conduct at rapid velocities and to follow high frequency stimuli. Functional improvement has been reported after endogenous remyelination of dorsal column axons (Jeffery and Blakemore, 1997). In agreement with these early physiological studies, recent work indicates that nodal sodium channels (Nav1.6) and paranodal potassium channels (Kv1.2) are restored on the endogenously remyelinated axons (Black et al., 2006).

REMYELINATION OF AXONS BY CELLULAR TRANSPLANTS

Remyelination in genetic and chemically induced demyelination models has been demonstrated by transplantation of oligodendrocyte lineage cells (Duncan et al., 1988; Lachapelle et al., 1992; Utzschneider et al., 1994; Keirstead et al., 2005), Schwann cells (Blakemore and Crang, 1985; Baron-Van Evercooren et al., 1992; Honmou et al., 1996; Kohama et al., 2001), olfactory ensheathing cells (OECs) (Franklin et al., 1996; Imaizumi et al., 1998) and various of stem cell types (Brustle et al., 1999; Keirstead et al., 1999; Akiyama et al., 2001; Akiyama et al., 2002). The first demonstration of anatomical myelination by cell transplantation was in the shiverer mouse (Lachapelle et al., 1983) which has a defect in the myelin basic protein (*MBP*) gene that results in no or poor CNS myelination (Roach et al., 1985). The demonstration of improved conduction in axons remyelinated by cell transplantation was carried out using neonatal optic nerve-derived glial cells transplanted into the spinal cord of the myelin-deficient (*md*) rat (Utzschneider et al., 1994). The *md* rat has a point mutation in proteolipid protein and does not form central myelin. Conduction velocity of axons in the amyelinated *md* rat is about 1 m/s or less regardless of axon diameter as compared to conduction velocities of over 10 m/s in normal controls (Utzschneider et al., 1994). Thus, while virtually all of the CNS lacks myelin in the *md* rat, their axons do conduct nerve impulses albeit at a very slow velocity.

Remyelination of the *md* spinal cord was observed after transplantation into the spinal cord of glial progenitors harvested from optic nerve (Utzschneider et al., 1994) (Figure 18.2). Conduction velocity and the ability to follow high frequency stimulation are significantly improved in the remyelinated axons. The improvement in conduction velocity is shown in Figure 18.3. Thus, remyelination by transplanted glial cells restores conduction velocity and possibly the integrative properties of the axons. A number of cell types including neural precursors can remyelinate the shiverer brain after intraventricular or direct injection into the brain (Learish et al., 1999; Yandava et al., 1999; Mitome et al., 2001; Windrem et al., 2004). A recent study demonstrated very extensive remyelination

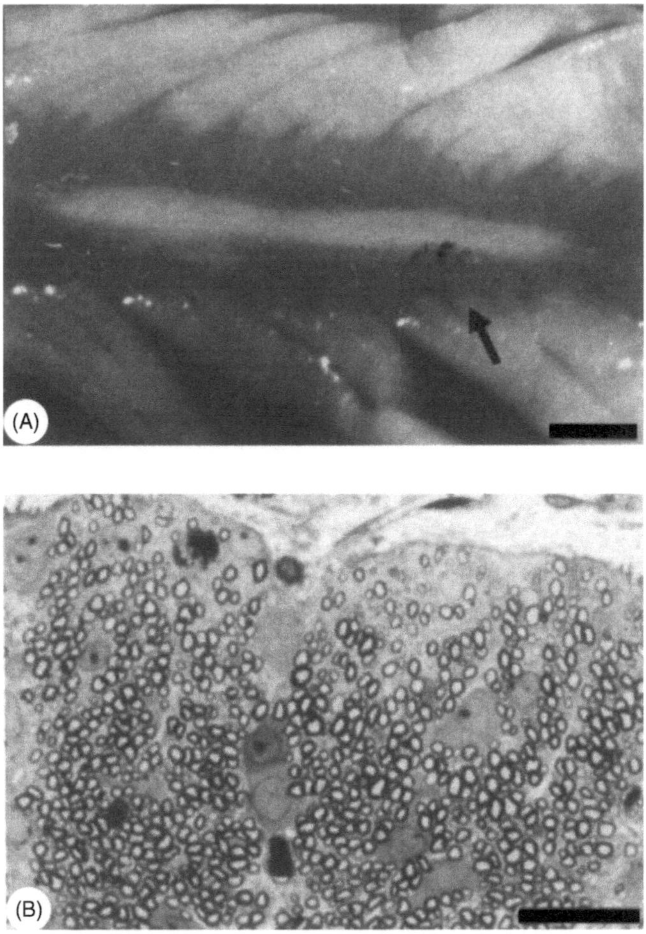

FIGURE 18.2 Oligodendrocyte precursor cells remyelinate the myelin-deficient (*md*) rat spinal cord. (A) Three weeks after intraspinal injection of glial precursors into the dorsal columns a longitudinal white stripe can be seen. Arrow points to site of cell injection marked by charcoal. (B) A section through this regions indicates extensive remyelination. Calibration in A corresponds to 0.5 mm and 20 μm in B. Modified from Utzschneider et al., 1994.

in the shiverer following transplantation of human oligodendrocyte progenitors (Windrem et al., 2004). This remyelination following multiple cell injections into brain increased the longevity of the animals. As encouraging as these results are, these genetic myelin mutants are not true demyelinating disorders but developmental dysmyelinating disorders. Therefore, it is important to determine whether demyelinated adult CNS axons could be remyelinated by cell transplantation and whether adult axons could reconstruct appropriate myelin and nodal organization to support normal impulse conduction.

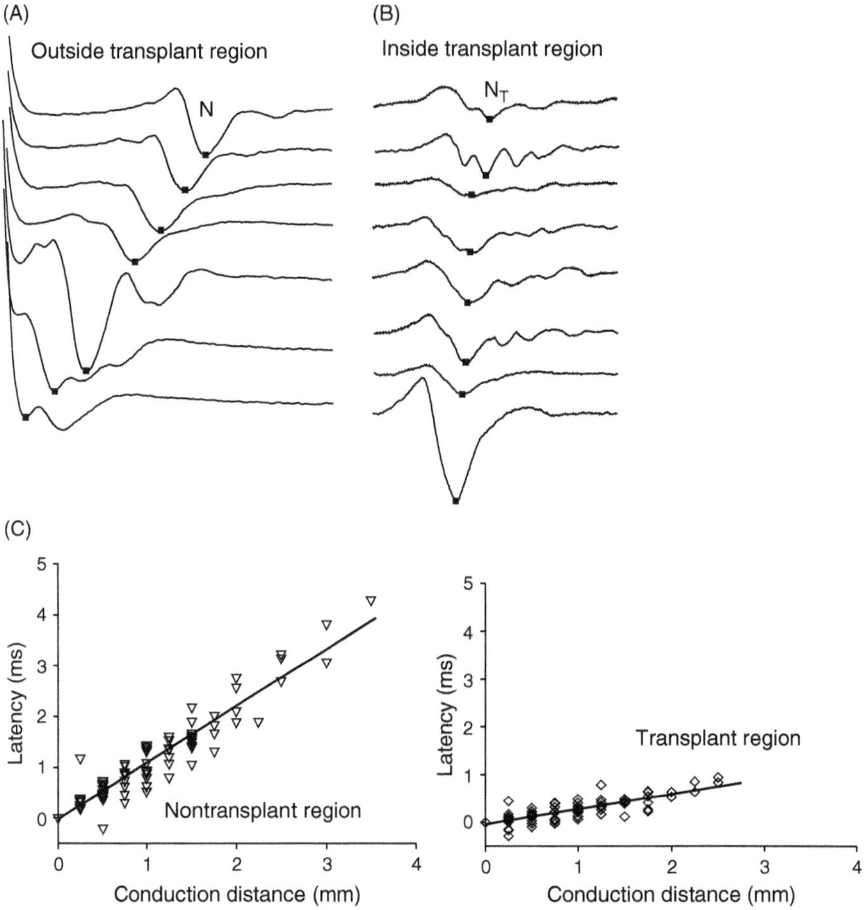

FIGURE 18.3 Conduction velocity is improved following remyelination of the spinal cord in the myelin-deficient rat. (A) Field potential recordings at 0.5 mm increments showing conduction outside of the transplant zone in the myelin-deficient rat. (B) Recordings from the transplant zone indicate a reduction in latency, i.e., an increase in conduction velocity. (C) Latency plots of compound action potentials outside of lesion zone and inside of lesion zone. Note the increase in conduction velocity in the transplant zone. Modified from Utzschneider et al., 1994.

Cell therapy approaches for MS may include mechanisms other than remyelination. Improved neurological function in mouse EAE has been reported following intravenous infusion of mesenchymal stem cells (Zhang et al., 2005) and neurosphere-derived multipotent precursors (Pluchino et al., 2003, 2005). Suggested mechanisms to account for the improvement in these EAE animals are reduction of inflammatory infiltration and thus reduction in demyelination and axonal pathology, and elevation of trophic factors that may be neuroprotective or stimulate oligodendrogliosis.

TRANSPLANTATION OF PERIPHERAL MYELIN-FORMING CELLS FOR REMYELINATION OF ADULT CNS AXONS

While Schwann cells normally remyelinated peripheral nerve, transplantation of Schwann cells into the demyelinated adult spinal cord results in remyelination with a characteristic peripheral pattern (Blakemore and Crang, 1985; Baron-Van Evercooren et al., 1992; Honmou et al., 1996). Near-normal conduction velocity is restored on these Schwann cell remyelinated central axons (Honmou et al., 1996). Transplantation of human sural nerve Schwann cells can remyelinate spinal cord axons in the immunosuppressed rat (Kohama et al., 2001). In these experiments, a focal demyelinated lesion was created in the dorsal column of the spinal cord of 12-week-old rats by X-irradiation and ethidium bromide injection (X-EB) (see Kohama et al., 2001 for technical details). This lesion presents as a persistent area of demyelination which lacks astrocytes. An example of a remyelinated axon region of the spinal cord 3 weeks after focal injection of reconstituted cryopreserved human Schwann cells is shown in Figure 18.4. Note the relatively large number of myelinated axons with typical Schwann cell morphology, i.e. large cytoplasmic and nuclear regions (signet ring configuration). Electron micrographs (not shown) reveal the presence of a basement membrane and extracellular collagen deposition. The conduction velocity of the axons remyelinated by the human Schwann cells was improved (Figure 18.5), indicating that electrophysiological function of the remyelinated axons was improved. Importantly, from intra-axonal recordings from the remyelinated axons, no hyperexcitability or ectopic impulse generation was observed, further suggesting that appropriate impulse conduction was restored by the transplant-induced remyelination.

The demonstration of anatomical and electrophysiological repair of demyelinated axons by adult human Schwann cells is an important prerequisite for future consideration of these cells as candidates for autologous transplantation studies in humans. A potential advantage of Schwann cells for transplantation into MS patients is that they are not antigenically predisposed to the immunological attack, as are oligodendrocytes. One potential problem with the use of Schwann cells to remyelinate lesions in MS patients is the presence of a glial scar in MS lesion sites which could limit cell migration and remyelination potential. In the X-EB lesion in the rat where relatively extensive remyelination is observed, it is important to note that the lesion is agliotic and thus the potential impediment of gliosis is not an issue.

The OEC is another cell type that has attracted much attention over the past decade as a cell candidate to encourage both axonal remyelination and regeneration. OECs are a specialized glia cell located in olfactory mucosa and nerves, and the outer nerve layer of the olfactory bulb. Adult olfactory receptor neurons continually undergo turnover from an endogenous progenitor pool, and their axons, which are ensheathed by OECs, grow through the olfactory nerves and cross the PNS–CNS interface where they form new synaptic connections in

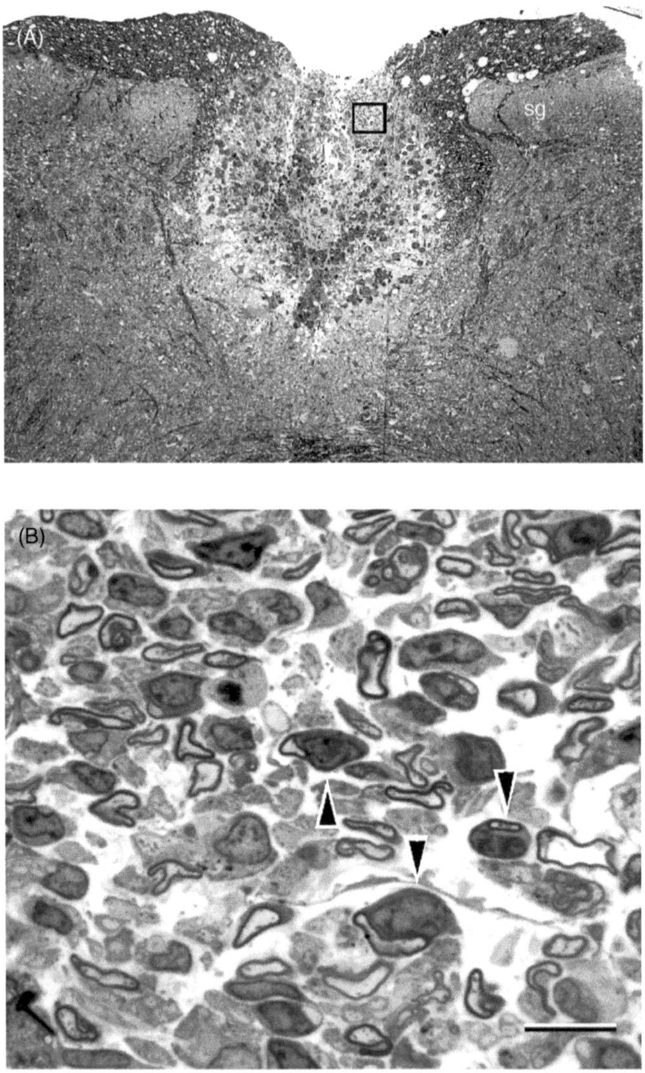

FIGURE 18.4 Remyelinated axons following human Schwann cell transplantation show a peripheral pattern of myelination. Photomicrographs were obtained from spinal cords placed in fixative after *in vitro* electrophysiological recordings. (A) Lesion area of dorsal columns 3 weeks after induction of the X-EB lesion. Sg refers to substantia gelatinosa in the dorsal horn. (B) Higher power micrograph from the boxed region of the lesion showing remyelinated axons. Arrows in (B) indicate examples of axons myelinated by cells with large nuclear and cytoplasmic domains characteristic of peripheral myelin. Calibration in (B) corresponds to 100 μm in (A) and 10 μm in (B). (Modified from Kohama et al., 2001).

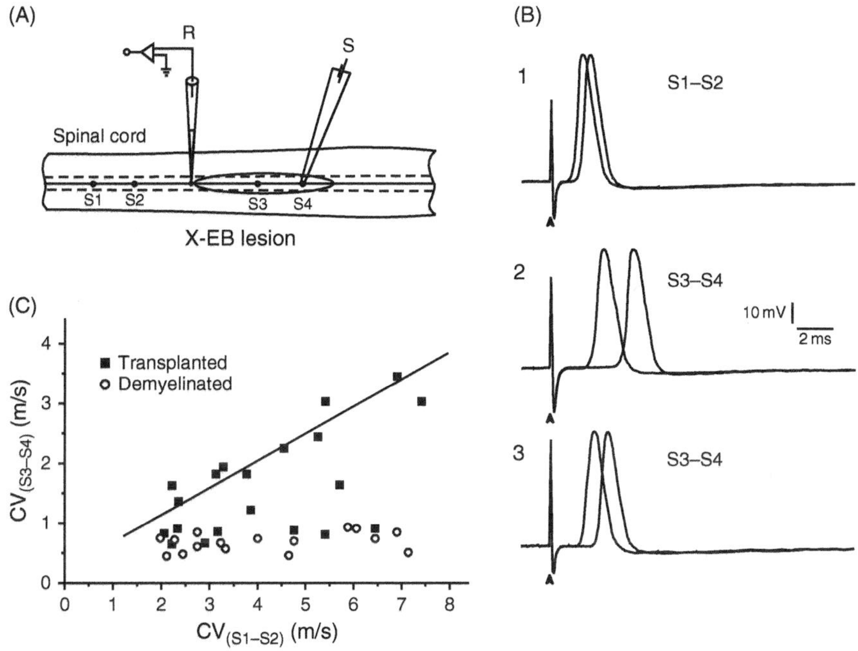

FIGURE 18.5 Transplantation of human Schwann cells into the demyelinated spinal cord improves conduction. (A) Schematic showing arrangement of intra-axonal recording and stimulation sites. Intra-axonal recordings were obtained from dorsal column axons outside of the lesion where the axons were normally myelinated. Stimulating electrodes were positioned outside (S1–S2) and within (S3–S4) the X-EB lesion zone to assess single axon conduction velocity over both the demyelinated or remyelinated axon segment and the normally myelinated axon segment of the same axon. (B) Pairs of action potentials recorded from S1–S2 stimulation (1), S3–S4 in the demyelinated dorsal columns (2) and S3–S4 following cell transplantation (3). Recordings were obtained at comparable conduction distances. (C) Plot of the conduction velocity of axon segments within the lesion (S3–S4) vs. conduction velocity of the axon segment outside of the lesion (S1–S2) for X-EB lesioned spinal cord without (open circles) and with (closed squares) transplantation. Modified from Kohama et al., 2001. (see Plate 12)

the olfactory bulb (Graziadei et al., 1978; Graziadei and Graziadei, 1979). This support role of OECs in axonal growth in the adult CNS has spawned extensive research aimed at studying the potential of OEC transplants to encourage axonal regeneration and to potentially remyelinate axons in the CNS (Li et al., 1997, 1998; Ramon-Cueto and Avila, 1998; Ramon-Cueto et al., 1998; Imaizumi et al., 2000a,b; Ramon-Cueto et al., 2000). OECs normally do not form myelin, but can do so when transplanted into the CNS (Franklin et al., 1996; Imaizumi et al., 1998). OECs are an unusual population of glial cells in that they share characteristics with both astrocytes in the CNS and Schwann cells in the PNS (Ramon-Cueto and Valverde, 1995), and are the only glial cells known to cross the PNS–CNS transitional zone, accompanying the axons that they ensheath (Doucette, 1991). Schwann cells and OECs have been reported to enhance axonal regeneration in the spinal cord when transplanted into spinal cord transection lesion sites (Li et al., 1997; Ramon-Cueto and Avila, 1998; Imaizumi et al., 2000a,b; Sasaki et al., 2004). OECs can produce a number of neurotrophic factors, and neuroprotection has been reported following OEC transplantation into transected spinal cord (Sasaki et al., 2006b).

Myelination was not observed in an in vitro co-culture experiment with dorsal root ganglion neurons and immunoselected (p75) OECs under culture conditions permissive for myelination by Schwann cells (Plant et al., 2002). These authors suggest that OECs might induce or enhance the migration of endogenous Schwann cells into the transplantation site (Brook et al., 1998). Moreover, while numerous reports suggest that OECs can form myelin when transplanted into the demyelinated (Franklin et al., 1996; Imaizumi et al., 1998; Barnett et al., 2000; Kato et al., 2000; Akiyama et al., 2004; Radtke et al., 2004) or injured spinal cord (Li et al., 1998; Imaizumi et al., 2000a,b), a recent study was unable to find evidence of OEC myelination in the injured spinal cord (Boyd et al., 2004). However, Franklin et al. (1996) used an OEC line and found remyelination in the spinal cord by identified OECs. Moreover, Akiyama et al. (2004) prepared cell suspensions of OECs from the olfactory bulb of alkaline phosphatase expressing adult transgenic rats (Kisseberth et al., 1999). The marker gene, *human placental alkaline phosphatase* (*hPAP*), is linked to the ubiquitous active R26 gene promoter, and its stable expression has been demonstrated by neural precursor cells in culture and after transplantation into the CNS (Han et al., 2002; Mujtaba et al., 2002). Transplantation of cell suspensions enriched in adult OECs (>95% $p75^+$ and $S100^+$) derived from hPAP (Akiyama et al., 2004) or eGFP transgenic rats (Sasaki et al., 2004; Sasaki et al., 2006a, b) can be easily identified in vivo and are associated with myelin formation (Figure 18.6). The remyelination by identified OECs and strongly suggests that OECs under appropriate conditions can form myelin.

The majority of the experimental work showing remyelination using OECs was in rodent systems and the OEC preparations were of varying purity and cellular composition. Both endogenous and cell transplant-induced remyelination may be different in primates as compared to rodents. Unlike the rodent, very little endogenous repair was observed after EB lesions in the nonhuman primate spinal cord at 6 weeks post-injection (Radtke et al., 2004). However, after grafting of OECs derived from a transgenic pig model expressing H-transferase to alter carbohydrate structure of the cells to mimic that of the human Type O blood group, considerable peripheral-like myelin was observed in the primate spinal cord at 3–4 weeks after transplantation. Bachelin et al. (2005) also demonstrated remyelination of the nonhuman primate spinal cord by transplantation of Schwann cells. These results indicate that transplanted peripheral-myelin forming

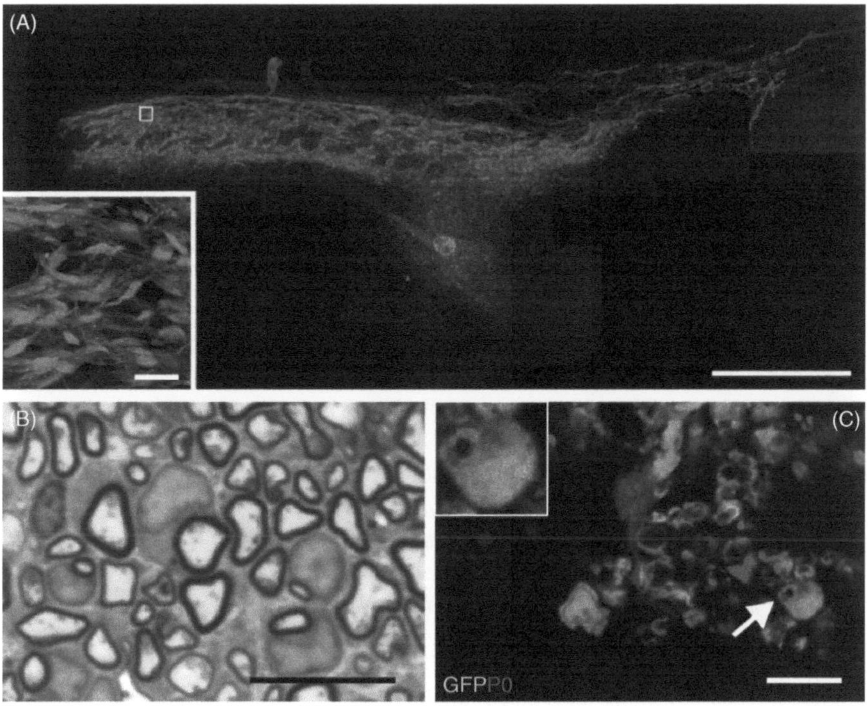

FIGURE 18.6 (A) Transplanted green fluorescent protein (GFP) expressing OECs transplanted into a dorsal transected spinal cord integrate and span the transection site (sagittal section). The transplanted cells are primarily confined to the lesion site. The dashed line demarcates lesion edge. (B) High-magnification micrograph of semithin plastic sections stained with methylene blue/Azure II through the OEC-transplanted lesion demonstrating that the transplanted dorsal funiculus was extensively myelinated. (C). Red-P0 rings are associated with green cellular elements, indicating that transplanted OECs remyelinated the demyelinated axons. Inset, expansion of the cell indicated by an arrow. Scale bars: 1 mm (A), 10 μm (B, C) and 20 μm (Inset A). Modified from Sasaki et al., 2004. (see Plate 13)

cells are capable of remyelinating primate spinal cord axons, a preclinical observation that is important for consideration of potential future cell therapy studies in humans.

NODAL RECONSTRUCTION OF REMYELINATED SPINAL CORD AXONS

The restoration of rapid and secure impulse conduction after demyelination is dependent on the acquisition of myelin sheaths and the clustering of specific molecules within discrete domains of the myelinated axon membrane. In myelinated axons, voltage-gated TTX-sensitive sodium (Na_v) channels are aggregated in high density at nodes of Ranvier, whereas *Shaker*-type potassium (K_v1) channels are separated from nodal Na_v channels by septate-like paranodal junctions (Peles and Salzer, 2000; Rasband and Trimmer, 2001b; Girault and Peles, 2002). Of the multiple Na_v channel isoforms expressed in nervous tissue (Goldin et al., 2000), $Na_v1.6$ is the predominant one at mature nodes in both the PNS and CNS (Caldwell et al., 2000; Boiko et al., 2001). The channel clustering (Vabnick et al., 1997; Rasband et al., 1999a) is dependent on interaction of the axon with myelinating cells (Kaplan et al., 1997; Eshed et al., 2005). Remyelinated axons display inappropriately short internodal lengths (Gledhill and McDonald, 1977; Weiner et al., 1980; Blakemore and Murray, 1981; Hildebrand et al., 1985), indicating that new nodes are formed. Despite their location at formerly internodal sites, remyelinated PNS axons have been shown to display high densities of Na_v channels at nodes (Novakovic et al., 1996) and K_v1 aggregations within juxtaparanodal domains (Rasband et al., 1999b). The fact that conduction velocity is restored in remyelinated axons argues for proper ion channel reorganization of remyelinated axons.

Recently we reported that ultrastructual analysis of spinal cords performed at three weeks after GFP–OEC transplantation demonstrated distinct nodes of Ranvier, and immunohistochemical analysis showed $Na_v1.6$ staining at nodes of Ranvier of the remyelanted axons, whereas detectable $Na_v1.2$ immunostaining was not apparent at nodes (Figure 18.4). In dorsal columns 3 weeks after transplantation, virtually all nodes bounded by GFP–OEC myelin sheaths exhibited $Na_v1.6$ staining (Figure 18.7); similar to control spinal cord axons, $Na_v1.2$ immunolabeling was not observed at any nodes (Figure 18.7). The $Na_v1.6$ labeling was localized to the nodal domain and was not observed in paranodal or juxtaparanodal regions or beneath the myelin sheath in remyelinated axons, suggesting that the transplanted GFP–OECs are competent to contribute to the specific clustering of Na_v channels at nodes. As an additional determinant of the ability of axons myelinated by GFP–OECs to support the asymmetric organization of ion channels within remyelinated nodal regions, we examined the distribution of $K_v1.2$ which have been shown previously to be aggregated in juxtaparanodal regions of most spinal cord axons (Rasband et al., 1999b; Rasband and Trimmer, 2001a,b). At 3 weeks (Figure 18.7), $K_v1.2$ is aggregated within juxtaparanodal regions of the remyelinated axons. A similar re-establishment of appropriate sodium and potassium channel organization is observed after endogenous remyelination (Black et al., 2006) and following remyelination in the shiverer mouse secondary to transplantation of adult neural precursor cells (Eftekharpour et al., 2007). Thus, it appears that sodium and potassium channel deployment on remyelinated axons is appropriate regardless of the type of myelinating cell that forms myelin. This result is encouraging for future clinical studies aimed at achieving remyelination in that anatomical repair of myelin appears co-terminous with axonal ion channel reconstruction and appropriate impulse conduction.

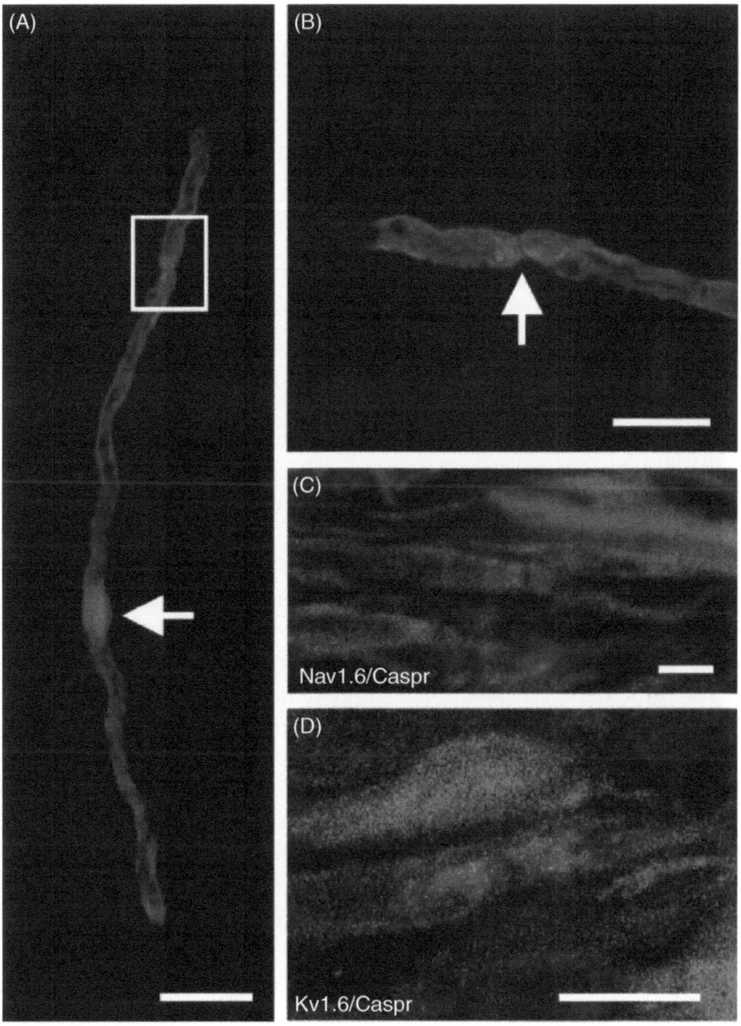

FIGURE 18.7 (A) Nodal formation of spinal cord axons remyelinated by transplanted OECs. OECs oriented longitudinally with periodic bulbous regions of intense green (GFP) fluorescence at the level of the cell body and nucleus (arrow). The elongated processes showed fluorescence on their outer margins corresponding to the outer cytoplasmic slips of the myelin forming cells. (B) Nodal-like regions were observed as thin breaks between adjacent green cell regions from boxed area in A. (C) Sodium channel $Na_v1.6$ at node of Ranvier of remyelinated dorsal columns at 3 weeks after OEC transplantation into demyelinating lesion in rat spinal cord. (D) Juxtaparanodal $K_v1.2$ immuno-labeling after GFP–OEC transplantation dorsal columns. Paranodes display Caspr staining that is flanked by $K_v1.2$ aggregations within juxtaparanodal regions. Scale bars: 100 μm (A), 20 μm (B) and 10 μm (C, D). Modified from Sasaki et al., 2006b. (see Plate 14)

NEUROPROTECTION OF CORTICOSPINAL TRACT NEURONS BY OEC TRANSPLANTATION

In addition to demyelination, axonopathy occurs in MS as recently highlighted by (Trapp et al., 1998; Woodhall et al., 2001). It is not clear whether myelin itself is trophic for axonal survival, or if the axon loss observed in some MS patients is secondary to the inflammation. Regardless, strategies to protect as well as to remyelinate axons are important. Primary motor cortex (M1) pyramidal neurons become atrophic and can undergo apoptotic cell death after axotomizing spinal cord transection (McBride et al., 1989; Hains et al., 2003). OECs express a number of neurotrophic factors including NGF and brain-derived neurotrophic factor (BDNF), but not NT-3 (Boruch et al., 2001; Woodhall et al., 2001), and several studies suggest that exposure to neurotrophins can limit death of axotomized neurons in the CNS after injury (for review see Giehl, 2001). To study the possibility that OECs are neuroprotective for injured corticospinal tract (CST) neurons, we transplanted OECs into the dorsal transected spinal cord (T9) which includes the CST and examined M1 to assess apoptosis and neuronal loss at 1 and 4 weeks post-transplantation (Sasaki et al., 2006b). Our results indicate that apoptosis of primary motor cortical neurons is reduced and that cortical neuronal density is increased after OEC transplantation. Enhanced levels of BDNF were observed in the OEC transplanted lesion and functional outcome was improved. Thus, transplantation of OECs into injured spinal cord has a neuroprotective effect on corticospinal neurons. The relative contribution of this effect to the observed functional improvement after OEC transplantation is uncertain, but this data indicates that OEC transplantation results in a larger pool of surviving corticospinal neurons. Thus, OEC transplantation into the injured spinal cord has distant neuroprotective effects on descending cortical projection neurons as well (Sasaki et al., 2006b). Consistent with this conclusion, paracrine BDNF support has been shown to be necessary for survival of CST neurons after proximal axotomy (Giehl and Tetzlaff, 1996; Schutte et al., 2000; Giehl, 2001), and BDNF promotes sprouting of CST axons after axotomy in the spinal cord (Namiki et al., 2000; Hiebert et al., 2002; Zhou and Shine, 2003).

CLINICAL STUDIES USING OECS IN SPINAL CORD INJURY

A number of clinical studies transplanting OECs into SCI patients are ongoing (Senior, 2002; Watts, 2005; Ibrahim et al., 2006). While these studies are not directed at MS, given the prospect of demyelination in SCI, important information relevant to MS could be derived from these studies. Feron et al. (2005) have conducted a phase I safety study using suspensions of OECs cultured from biopsied tissue from the patients' own olfactory mucosa. They report no adverse effects at 12 months post-transplantation, but no neurological improvement. Carlos Lima and colleagues at the Egaz Moniz Hospital (Lisbon, Portugal) have reported an experimental approach where the cavity of the SCI site is packed with acutely prepared minced olfactory mucosa tissue, which includes a number of cell types in addition to OECs. They report that the procedure was safe and potentially beneficial, but efficacy was not clearly established (Lima et al., 2006).

In China, Huang and colleagues (Chaoyang Hospital, Beijing, China) report that several hundred patients have received transplants of cultures from human embryonic olfactory bulbs obtained from 14 to 16 fetuses (Huang et al., 2003; Curt and Dietz, 2005; Dobkin et al., 2006). Modest functional improvement was reported beginning as early as a day after transplantation. This early effect could not be the result of axonal regeneration or remyelination. It is important to note that the Lima and Huang studies have not carried out control studies nor have these observations been independently confirmed (Dobkin et al., 2006; Ibrahim et al., 2006). Dobkin et al. (2006) independently studied several chronic SCI patients undergoing surgery by the Huang group in Beijing. For assessment they used MRI, the protocol of the American Spinal Injury Association (ASIA) for change in disability, and detailed history of the perioperative course. They identified shortcomings which include (1) the phenotype and fate of the cells referred to as OECs are unknown, (2) perioperative morbidity and (3) lack of functional benefit. They also emphasize no attempt to meet international standards for safety and efficacy. From their observations they urge physicians not to recommend this procedure to patients at this point in time. However, plans to establish careful guidelines for these potentially important clinical studies on SCI in China are being discussed and organized (Cyranoski, 2007).

"Spontaneous" functional improvement occurs in most SCI patients making assessment of efficacy of therapeutic interventions in SCI including cell therapy approaches difficult. Surgical intervention necessary to transplant cells can in and of itself lead to modest functional improvement. Issues related to methods of assessment of SCI patients in clinical studies are currently being discussed with an emphasis on assessing the degree of functional recovery of an individual patient (Curt et al., 2004). The complexity of SCI and the difficulty of accurately assessing functional recovery will be a challenge for all interventional clinical studies for SCI. These considerations are also critical in MS clinical studies were exacerbations and remissions are common.

Reconstruction of appropriate spinal circuits by cell-based therapies is the ultimate long-term goal of cell transplantation research. However, laboratory work to date suggests that more immediate therapeutic benefits will come from neuroprotective effects and remyelination. The most extensive reported functional recovery in animal models of SCI with cell transplantation is following treatment of acute and subacute SCI prior to extensive spinal cord scarring. Early cell intervention may reduce scar formation and secondary cell death by release of appropriate trophic factors by engrafted cells. Angiogenic factors released by transplanted cells could increase neovascularization which would be critical for tissue preservation. If long-term SCI patients have preservation of long tracts in the spinal cord that are demyelinated, remyelination of these tracts by cell transplantation might lead to some degree of functional improvement.

CONCLUDING REMARKS

Demyelinated axons in the CNS have the intrinsic potential to be remyelinated by endogenous progenitor cells and by invasion of Schwann cells from the periphery. This potential has been observed in a subset of MS patients where there is extensive endogenous repair of myelin. Experimental cell transplantation studies indicate oligodendrocyte progenitor cells, a variety of stem cell types, and peripheral myelin-forming cells can remyelinate central axons. Demyelination in MS is also often associated with axon loss of varying degrees. The mechanism of the axon loss is not fully established, but trophic support of axons by myelin and possible axonal damage from inflammation are being considered. Experimental cell therapy strategies in animal models have achieved remyelination, neuroprotection, enhanced axonal outgrowth and immunomodulation of T-cell infiltration in EAE. An important challenge that remains for cell-based therapies for MS and SCI is to determine the optimal cell type, method of delivery and timing of cellular intervention.

REFERENCES

Akiyama Y, Honmou O, Kato T et al. (2001). Transplantation of clonal neural precursor cells derived from adult human brain establishes functional peripheral myelin in the rat spinal cord. *Exp Neurol* 167: 27–39.

Akiyama Y, Radtke C, and Kocsis JD (2002). Remyelination of the rat spinal cord by transplantation of identified bone marrow stromal cells. *J Neurosci* 22: 6623–6630.

Akiyama Y, Lankford KL, Radtke C et al. (2004). Remyelination of spinal cord axons by olfactory ensheathing cells and Schwann cells derived from a transgenic rat expressing alkaline phosphatase marker gene. *Neuron Glia Biol* 1: 1–9.

Bachelin C, Lachapelle F, Girard C et al. (2005). Efficient myelin repair in the macaque spinal cord by autologous grafts of Schwann cells. *Brain* 128: 540–549.

Barnett SC, Alexander CL, Iwashita Y et al. (2000). Identification of a human olfactory ensheathing cell that can effect transplant-mediated remyelination of demyelinated CNS axons. *Brain* 123: 1581–1588.

Baron-Van Evercooren A, Gansmuller A, Duhamel E et al. (1992). Repair of a myelin lesion by Schwann cells transplanted in the adult mouse spinal cord. *J Neuroimmunol* 40: 235–242.

Black JA, Waxman SG, and Smith KJ (2006). Remyelination of dorsal column axons by endogenous Schwann cells restores the normal pattern of Nav1.6 and Kv1.2 at nodes of Ranvier. *Brain* 129: 1319–1329.

Blakemore WF and Crang AJ (1985). The use of cultured autologous Schwann cells to remyelinate areas of persistent demyelination in the central nervous system. *J Neurol Sci* 70: 207–223.

Blakemore WF and Murray JA (1981). Quantitative examination of internodal length of remyelinated nerve fibres in the central nervous system. *J Neurol Sci* 49: 273–284.

Boiko T, Rasband MN, Levinson SR et al. (2001). Compact myelin dictates the differential targeting of two sodium channel isoforms in the same axon. *Neuron* 30: 91–104.

Boruch AV, Conners JJ, Pipitone M et al. (2001). Neurotrophic and migratory properties of an olfactory ensheathing cell line. *Glia* 33: 225–229.

Bostock H and Grafe P (1985). Activity-dependent excitability changes in normal and demyelinated rat spinal root axons. *J Physiol* 365: 239–257.

Boyd JG, Lee J, Skihar V et al. (2004). LacZ-expressing olfactory ensheathing cells do not associate with myelinated axons after implantation into the compressed spinal cord. *Proc Natl Acad Sci U S A* 101: 2162–2166.

Brismar T (1981). Electrical properties of isolated demyelinated rat nerve fibres. *Acta Physiol Scand* 113: 161–166.

Brook GA, Plate D, Franzen R et al. (1998). Spontaneous longitudinally orientated axonal regeneration is associated with the Schwann cell framework within the lesion site following spinal cord compression injury of the rat. *J Neurosci Res* 53: 51–65.

Bruck W, Schmied M, Suchanek G et al. (1994). Oligodendrocytes in the early course of multiple sclerosis. *Ann Neurol* 35: 65–73.

Brustle O, Jones KN, Learish RD et al. (1999). Embryonic stem cell-derived glial precursors: a source of myelinating transplants. *Science* 285: 754–756.

Burchiel KJ (1980). Abnormal impulse generation in focally demyelinated trigeminal roots. *J Neurosurg* 53: 674–683.

Caldwell JH, Schaller KL, Lasher RS et al. (2000). Sodium channel Na(v)1.6 is localized at nodes of ranvier, dendrites, and synapses. *Proc Natl Acad Sci U S A* 97: 5616–5620.

Chang A, Tourtellotte WW, Rudick R et al. (2002). Premyelinating oligodendrocytes in chronic lesions of multiple sclerosis. *N Engl J Med* 346: 165–173.

Charcot J (1868). Histologie de la sclerose en plaques. *Gazette Hopitaux* 41: 554–566.

Crowe MJ, Bresnahan JC, Shuman SL et al. (1997). Apoptosis and delayed degeneration after spinal cord injury in rats and monkeys. *Nat Med* 3: 73–6.

Curt A and Dietz V (2005). Controversial treatments for spinal-cord injuries. *Lancet* 365: 841.

Curt A, Schwab ME, and Dietz V (2004). Providing the clinical basis for new interventional therapies: refined diagnosis and assessment of recovery after spinal cord injury. *Spinal Cord* 42: 1–6.

Cyranoski D (2007). Chinese network to start trials of spinal surgery. *Nature* 446: 476–477.

Devor M and Seltzer Z (1999). Pathophysiology of damaged nerves in relation to chronic pain. In PD Wall and R Melzack, eds. *Textbook of Pain*. Churchill Livingstone, London, vol. 4, pp. 129–164.

Dobkin BH, Curt A, and Guest J (2006). Cellular transplants in China: observational study from the largest human experiment in chronic spinal cord injury. *Neurorehabil Neural Repair* 20: 5–13.

Doucette R (1991). PNS-CNS transitional zone of the first cranial nerve. *J Comp Neurol* 312: 451–466.

Duncan ID, Hammang JP, Jackson KF et al. (1988). Transplantation of oligodendrocytes and Schwann cells into the spinal cord of the myelin-deficient rat. *J Neurocytol* 17: 351–360.

Eftekharpour E, Karimi-Abdolrezaee S, Wang J et al. (2007). Myelination of congenitally dysmyelinated spinal cord axons by adult neural precursor cells results in formation of nodes of Ranvier and improved axonal conduction. *J Neurosci* 27: 3416–3428.

Eshed Y, Feinberg K, Poliak S et al. (2005). Gliomedin mediates Schwann cell-axon interaction and the molecular assembly of the nodes of Ranvier. *Neuron* 47: 215–229.

Feron F, Perry C, Cochrane J et al. (2005). Autologous olfactory ensheathing cell transplantation in human spinal cord injury. *Brain* 128: 2951–2960.

Franklin RJ, Gilson JM, Franceschini IA et al. (1996). Schwann cell-like myelination following transplantation of an olfactory bulb-ensheathing cell line into areas of demyelination in the adult CNS. *Glia* 17: 217–224.

Giehl KM (2001). Trophic dependencies of rodent corticospinal neurons. *Rev Neurosci* 12: 79–94.

Giehl KM and Tetzlaff W (1996). BDNF and NT-3, but not NGF, prevent axotomy-induced death of rat corticospinal neurons in vivo. *Eur J Neurosci* 8: 1167–1175.

Girault JA and Peles E (2002). Development of nodes of Ranvier. *Curr Opin Neurobiol* 12: 476–485.

Gledhill RF and McDonald WI (1977). Morphological characteristics of central demyelination and remyelination: a single-fiber study. *Ann Neurol* 1: 552–560.

Goldin AL, Barchi RL, Caldwell JH et al. (2000). Nomenclature of voltage-gated sodium channels. *Neuron* 28: 365–368.

Graziadei PP and Graziadei GA (1979). Neurogenesis and neuron regeneration in the olfactory system of mammals. I. Morphological aspects of differentiation and structural organization of the olfactory sensory neurons. *J Neurocytol* 8: 1–18.

Graziadei PP, Levine RR, and Graziadei GA (1978). Regeneration of olfactory axons and synapse formation in the forebrain after bulbectomy in neonatal mice. *Proc Natl Acad Sci U S A* 75: 5230–5234.

Hains BC, Black JA, and Waxman SG (2003). Primary cortical motor neurons undergo apoptosis after axotomizing spinal cord injury. *J Comp Neurol* 462: 328–341.

Han SS, Kang DY, Mujtaba T et al. (2002). Grafted lineage-restricted precursors differentiate exclusively into neurons in the adult spinal cord. *Exp Neurol* 177: 360–375.

Harrison BM, McDonald WI, and Ochoa J (1972). Remyelination in the central diphtheria toxin lesion. *J Neurol Sci* 17: 293–302.

Hiebert GW, Khodarahmi K, McGraw J et al. (2002). Brain-derived neurotrophic factor applied to the motor cortex promotes sprouting of corticospinal fibers but not regeneration into a peripheral nerve transplant. *J Neurosci Res* 69: 160–168.

Hildebrand C, Kocsis JD, Berglund S et al. (1985). Myelin sheath remodelling in regenerated rat sciatic nerve. *Brain Res* 358: 163–170.

Honmou O, Felts PA, Waxman SG et al. (1996). Restoration of normal conduction properties in demyelinated spinal cord axons in the adult rat by transplantation of exogenous Schwann cells. *J Neurosci* 16: 3199–3208.

Huang H, Chen L, Wang H et al. (2003). Influence of patients' age on functional recovery after transplantation of olfactory ensheathing cells into injured spinal cord injury. *Chin Med J (Engl)* 116: 1488–1491.

Ibrahim A, Li Y, Li D et al. (2006). Olfactory ensheathing cells: ripples of an incoming tide? *Lancet Neurol* 5: 453–457.

Imaizumi T, Lankford KL, Waxman SG et al. (1998). Transplanted olfactory ensheathing cells remyelinate and enhance axonal conduction in the demyelinated dorsal columns of the rat spinal cord. *J Neurosci* 18: 6176–6185.

Imaizumi T, Lankford KL, Burton WV et al. (2000a). Xenotransplantation of transgenic pig olfactory ensheathing cells promotes axonal regeneration in rat spinal cord. *Nat Biotechnol* 18: 949–953.

Imaizumi T, Lankford KL, and Kocsis JD (2000b). Transplantation of olfactory ensheathing cells or Schwann cells restores rapid and secure conduction across the transected spinal cord. *Brain Res* 854: 70–78.

Jeffery ND and Blakemore WF (1997). Locomotor deficits induced by experimental spinal cord demyelination are abolished by spontaneous remyelination. *Brain* 120: 27–37.

Kahoma et al. (2001). Transplantation of cryopreserved adult human Schwann cells enhances axonal conduction in demyelinated spinal cord. *Journal of Neuroscience* 21: 944–950.

Kaplan MR, Meyer-Franke A, Lambert S et al. (1997). Induction of sodium channel clustering by oligodendrocytes. *Nature* 386: 724–728.

Kato T, Honmou O, Uede T et al. (2000). Transplantation of human olfactory ensheathing cells elicits remyelination of demyelinated rat spinal cord. *Glia* 30: 209–218.

Keirstead HS, Ben-Hur T, Rogister B et al. (1999). Polysialylated neural cell adhesion molecule-positive CNS precursors generate both oligodendrocytes and Schwann cells to remyelinate the CNS after transplantation. *J Neurosci* 19: 7529–7536.

Keirstead HS, Nistor G, Bernal G et al. (2005). Human embryonic stem cell-derived oligodendrocyte progenitor cell transplants remyelinate and restore locomotion after spinal cord injury. *J Neurosci* 25: 4694–4705.

Kerschensteiner M, Stadelmann C, Buddeberg BS et al. (2004). Targeting experimental autoimmune encephalomyelitis lesions to a predetermined axonal tract system allows for refined behavioral testing in an animal model of multiple sclerosis. *Am J Pathol* 164: 1455–1469.

Kisseberth WC, Brettingen NT, Lohse JK et al. (1999). Ubiquitous expression of marker transgenes in mice and rats. *Dev Biol* 214: 128–138.

Kohama I, Lankford KL, Preiningerova J et al. (2001). Transplantation of cryopreserved adult human Schwann cells enhances axonal conduction in demyelinated spinal cord. *J Neurosci* 21: 944–950.

Lachapelle F, Gumpel M, Baulac M et al. (1983). Transplantation of CNS fragments into the brain of shiverer mutant mice: extensive myelination by implanted oligodendrocytes. I. Immunohistochemical studies. *Dev Neurosci* 6: 325–334.

Lachapelle F, Lapie P, and Gumpel M (1992). Oligodendrocytes from jimpy and normal mature tissue can be 'activated' when transplanted in a newborn environment. *Dev Neurosci* 14: 105–113.

Lassmann H (1983). Comparative neuropathology of chronic experimental allergic encephalomyelitis and multiple sclerosis. *Schriftenr Neurol* 25: 1–135.

Learish RD, Brustle O, Zhang SC et al. (1999). Intraventricular transplantation of oligodendrocyte progenitors into a fetal myelin mutant results in widespread formation of myelin. *Ann Neurol* 46: 716–722.

Li Y, Field PM, and Raisman G (1997). Repair of adult rat corticospinal tract by transplants of olfactory ensheathing cells. *Science* 277: 2000–2002.

Li Y, Field PM, and Raisman G (1998). Regeneration of adult rat corticospinal axons induced by transplanted olfactory ensheathing cells. *J Neurosci* 18: 10514–10524.

Lima C, Pratas-Vital J, Escada P et al. (2006). Olfactory mucosa autografts in human spinal cord injury: a pilot clinical study. *J Spinal Cord Med* 29: 191–203; discussion 204–206.

Lois C and Alvarez-Buylla A (1994). Long-distance neuronal migration in the adult mammalian brain. *Science* 264: 1145–1148.

Lumsden CE (1970). The neuropathology of multiple sclerosis and other demyelinating diseases. In P Vinken and GW Bruyn, eds. Handbook of clinical neurology. North Holland Publishing, Amsterdam, vol. 9, pp. 217–309.

Luskin MB (1993). Restricted proliferation and migration of postnatally generated neurons derived from the forebrain subventricular zone. *Neuron* 11: 173–189.

McBride RL, Feringa ER, Garver MK et al. (1989). Prelabeled red nucleus and sensorimotor cortex neurons of the rat survive 10 and 20 weeks after spinal cord transection. *J Neuropathol Exp Neurol* 48: 568–576.

McDonald WI (1963). The effects of experimental demyelination on conduction in peripheral nerve: a histological and electrophysiological study. ii. Electrophysiological observations. *Brain* 86: 501–524.

McDonald WI and Sears TA (1970). The effects of experimental demyelination on conduction in the central nervous system. *Brain* 93: 583–598.

Mitome M, Low HP, van den Pol A et al. (2001). Towards the reconstruction of central nervous system white matter using neural precursor cells. *Brain* 124: 2147–2161.

Mujtaba T, Mayer-Proschel M, and Rao MS (1998). A common neural progenitor for the CNS and PNS. *Dev Biol* 200: 1–15.

Mujtaba T, Han SS, Fischer I et al. (2002). Stable expression of the alkaline phosphatase marker gene by neural cells in culture and after transplantation into the CNS using cells derived from a transgenic rat. *Exp Neurol* 174: 48–57.

Nait-Oumesmar B, Picard-Riera N, Kerninon C et al. (2007). Activation of the subventricular zone in multiple sclerosis: evidence for early glial progenitors. *Proc Natl Acad Sci U S A* 104: 4694–4699.

Namiki J, Kojima A, and Tator CH (2000). Effect of brain-derived neurotrophic factor, nerve growth factor, and neurotrophin-3 on functional recovery and regeneration after spinal cord injury in adult rats. *J Neurotrauma* 17: 1219–1231.

Novakovic SD, Deerinck TJ, Levinson SR et al. (1996). Clusters of axonal Na^+ channels adjacent to remyelinating Schwann cells. *J Neurocytol* 25: 403–412.

Patrikios P, Stadelmann C, Kutzelnigg A et al. (2006). Remyelination is extensive in a subset of multiple sclerosis patients. *Brain* 129: 3165–3172.

Peles E, and Salzer JL (2000). Molecular domains of myelinated axons. *Curr Opin Neurobiol* 10: 558–565.

Plant GW, Currier PF, Cuervo EP et al. (2002). Purified adult ensheathing glia fail to myelinate axons under culture conditions that enable Schwann cells to form myelin. *J Neurosci* 22: 6083–6091.

Pluchino S, Quattrini A, Brambilla E et al. (2003). Injection of adult neurospheres induces recovery in a chronic model of multiple sclerosis. *Nature* 422: 688–694.

Pluchino S, Zanotti L, Rossi B et al. (2005). Neurosphere-derived multipotent precursors promote neuroprotection by an immunomodulatory mechanism. *Nature* 436: 266–271.

Prineas JW (1985). The neuropathology of multiple sclerosis. In JC Koetsier, ed. *Handbook of neurology: demeylinating diseases*. Elsevier, Amsterdam, vol. 47, pp. 337–395.

Prineas JW and Connell F (1979). Remyelination in multiple sclerosis. *Ann Neurol* 5: 22–31.

Prineas JW, Kwon EE, Cho ES et al. (1984). Continual breakdown and regeneration of myelin in progressive multiple sclerosis plaques. *Ann N Y Acad Sci* 436: 11–32.

Radtke C, Akiyama Y, Brokaw J et al. (2004). Remyelination of the nonhuman primate spinal cord by transplantation of H-transferase transgenic adult pig olfactory ensheathing cells. *Faseb J* 18: 335–337.

Raine CS and Wu E (1993). Multiple sclerosis: remyelination in acute lesions. *J Neuropathol Exp Neurol* 52: 199–204.

Ramon-Cueto A and Avila J (1998). Olfactory ensheathing glia: properties and function. *Brain Res Bull* 46: 175–187.

Ramon-Cueto A and Valverde F (1995). Olfactory bulb ensheathing glia: a unique cell type with axonal growth-promoting properties. *Glia* 14: 163–73.

Ramon-Cueto A, Plant GW, Avila J et al. (1998). Long-distance axonal regeneration in the transected adult rat spinal cord is promoted by olfactory ensheathing glia transplants. *J Neurosci* 18: 3803–3815.

Ramon-Cueto A, Cordero MI, Santos-Benito FF et al. (2000). Functional recovery of paraplegic rats and motor axon regeneration in their spinal cords by olfactory ensheathing glia. *Neuron* 25: 425–435.

Rasband MN and Trimmer JS (2001a). Developmental clustering of ion channels at and near the node of Ranvier. *Dev Biol* 236: 5–16.

Rasband MN and Trimmer JS (2001b). Subunit composition and novel localization of K^+ channels in spinal cord. *J Comp Neurol* 429: 166–176.

Rasband MN, Peles E, Trimmer JS et al. (1999a). Dependence of nodal sodium channel clustering on paranodal axoglial contact in the developing CNS. *J Neurosci* 19: 7516–7528.

Rasband MN, Trimmer JS, Peles E et al. (1999b). K^+ channel distribution and clustering in developing and hypomyelinated axons of the optic nerve. *J Neurocytol* 28: 319–331.

Rasminsky M (1981). Hyperexcitability of pathologically myelinated axons and positive symptoms in multiple sclerosis. *Adv Neurol* 31: 289–297.

Rasminsky M and Sears TA (1972). Internodal conduction in undissected demyelinated nerve fibres. *J Physiol* 227: 323–350.

Roach A, Takahashi N, Pravtcheva D et al. (1985). Chromosomal mapping of mouse myelin basic protein gene and structure and transcription of the partially deleted gene in shiverer mutant mice. *Cell* 42: 149–155.

Sasaki M, Lankford KL, Zemedkun M et al. (2004). Identified olfactory ensheathing cells transplanted into the transected dorsal funiculus bridge the lesion and form myelin. *J Neurosci* 24: 8485–8493.

Sasaki M, Black JA, Lankford KL et al. (2006a). Molecular reconstruction of nodes of Ranvier after remyelination by transplanted olfactory ensheathing cells in the demyelinated spinal cord. *J Neurosci* 26: 1803–1812.

Sasaki M, Hains BC, Lankford KL et al. (2006b). Protection of corticospinal tract neurons after dorsal spinal cord transection and engraftment of olfactory ensheathing cells. *Glia* 53: 352–359.

Schutte A, Yan Q, Mestres P et al. (2000). The endogenous survival promotion of axotomized rat corticospinal neurons by brain-derived neurotrophic factor is mediated via paracrine, rather than autocrine mechanisms. *Neurosci Lett* 290: 185–188.

Scolding NJ, Rayner PJ, Sussman J et al. (1995). A proliferative adult human oligodendrocyte progenitor. *Neuroreport* 6: 441–445.

Senior K (2002). Olfactory ensheathing cells to be used in spinal-cord repair trial. *Lancet Neurol* 1: 269.

Smith KJ and McDonald WI (1980). Spontaneous and mechanically evoked activity due to central demyelinating lesion. *Nature* 286: 154–155.

Smith EJ, Blakemore WF, and McDonald WI (1979). Central remyelination restores secure conduction. *Nature* 280: 395–396.

Suzuki JI, Tokumasu K Goto K (1969). Eye movements from single utricular nerve stimulation in the cat. *Acta Otolaryngol* Oct, 68(4): 350–362.

Trapp BD, Peterson J, Ransohoff RM et al. (1998). Axonal transection in the lesions of multiple sclerosis. *N Engl J Med* 338: 278–285.

Trapp BD, Ransohoff R, and Rudick R (1999). Axonal pathology in multiple sclerosis: relationship to neurologic disability. *Curr Opin Neurol* 12: 295–302.

Utzschneider DA, Archer DR, Kocsis JD et al. (1994). Transplantation of glial cells enhances action potential conduction of amyelinated spinal cord axons in the myelin-deficient rat. *Proc Natl Acad Sci U S A* 91: 53–57.

Vabnick I, Messing A, Chiu SY et al. (1997). Sodium channel distribution in axons of hypomyelinated and MAG null mutant mice. *J Neurosci Res* 50: 321–336.

Watts J (2005). Controversy in China. *Lancet* 365: 109–110.

Weiner LP, Waxman SG, Stohlman SA et al. (1980). Remyelination following viral-induced demyelination: ferric ion-ferrocyanide staining of nodes of Ranvier within the CNS. *Ann Neurol* 8: 580–583.

Windrem MS, Nunes MC, Rashbaum WK et al. (2004). Fetal and adult human oligodendrocyte progenitor cell isolates myelinate the congenitally dysmyelinated brain. *Nat Med* 10: 93–97.

Wolswijk G (1998). Chronic stage multiple sclerosis lesions contain a relatively quiescent population of oligodendrocyte precursor cells. *J Neurosci* 18: 601–609.

Woodhall E, West AK, and Chuah MI (2001). Cultured olfactory ensheathing cells express nerve growth factor, brain-derived neurotrophic factor, glia cell line-derived neurotrophic factor and their receptors. *Brain Res Mol Brain Res* 88: 203–213.

Yandava BD, Billinghurst LL, and Snyder EY (1999). "Global" cell replacement is feasible via neural stem cell transplantation: evidence from the dysmyelinated shiverer mouse brain. *Proc Natl Acad Sci U S A* 96: 7029–7034.

Zhang J, Li Y, Chen J et al. (2005). Human bone marrow stromal cell treatment improves neurological functional recovery in EAE mice. *Exp Neurol* 195: 16–26.

Zhou L and Shine HD (2003). Neurotrophic factors expressed in both cortex and spinal cord induce axonal plasticity after spinal cord injury. *J Neurosci Res* 74: 221–226.

PART

VI

DESIGN OF CLINICAL TRIALS

19

Clinical Trials in Rehabilitation With Neural Repair

BRUCE H. DOBKIN

University of California Los Angeles, Los Angeles, CA 90095

ABSTRACT

Pre-clinical experiments that demonstrate neural repair must be interpreted with caution by clinicians who wish to apply studies of animal models of disease and regeneration to their patients. Ideally, in vivo models bear parallels and recognizable differences from the cascades of post-injury and pro-restorative mechanisms that may evolve in patients. Experience-dependent learning may also drive neural and behavioral adaptations differently in caged rodents and people. With caution, biological experiments can lead to well-designed clinical trials to test new interventions.

INTRODUCTION

Basic studies in developmental neurobiology and pre-clinical experiments in models of stroke, spinal cord injury (SCI), traumatic brain injury (TBI), multiple sclerosis, peripheral nerve injury, and other acute neurological diseases suggest the growing feasibility of strategies for neural repair to lessen the physical and cognitive impairments and disabilities in patients. Trials for these diseases differ from those for chronic degenerative diseases such as Alzheimer's and Parkinson's, which may aim to lessen symptoms and slow progression or stop the disease, as well as reverse impairments and disability, which is the main focus of regenerative strategies. Media reports of successful experiments in

Copyright © 2008 Elsevier Inc.
All rights of reproduction in any form reserved.

brain or spinal cord injured rodents raise hope for imminent cures of paralysis, sensory loss, bladder dysfunction, neurogenic pain, and cognitive impairments. The translation from models of injury and repair to patient care, however, includes many potential confounders.

Pre-clinical data must reveal safety and identify the effects of the repair intervention on neural networks and on meaningful behaviors. Prior chapters have emphasized these elements. Potential strategies include manipulations of the promoters, inhibitors and modulators of endogenous neurogenesis, exogenous cell transplants, axonal transport and guidance systems needed for regeneration, cell–cell signaling in myelinated fibers, re-expression of developmental genes, dendritic and spine sprouting, the assembly of synapses, remyelination, and activity-dependent synaptic reorganization. Well-planned pre-clinical investigations set the scientific and regulatory course toward establishing the conceptual basis, methods, safety, and potential for efficacy of restorative interventions in patients. From this work and an understanding of the goals of an intervention in patients, the design of a clinical trial moves forward. Trials that involve attempts at repair may be even more complex to design than those for neuroprotection, but much can be learned from the difficulties encountered in failed studies that attempted to limit injury. Both positive and negative studies of biologic approaches for neurodegenerative diseases also offer help in the design of neurorestorative trials. This chapter examines some of the confounders that clinicians who develop trials can keep in mind as they interpret information drawn from pre-clinical studies and then discusses design features that are especially applicable to using neural repair to augment neurorehabilitation.

CONFOUNDERS IN THE TRANSLATION OF PRE-CLINICAL TRIALS

The Food and Drug Administration in the United States requires the safety of a drug at a range of dosages to be demonstrated in animal models prior to clinical use. Pre-clinical trials are not required to demonstrate physiologic or behavioral improvement for an experimental intervention compared to a placebo, but it would be most difficult to design a human trial without this knowledge. Highly focused pre-clinical mechanistic studies, however, do not necessarily serve as steps for the development of clinical therapies. They can suggest bench-to-bedside applications.

Pre-clinical testing should aim to set the conditions for meeting ethical standards for a subsequent phase 1 or 2 clinical trial (Table 19.1). These initial trials are not expected to benefit an individual patient, but should offer a reasonable opportunity to document scientifically important and clinically relevant information. From an ethical view, prior to subjecting patients to an experimental intervention, rigorous pre-clinical experiments ought to employ an animal model

TABLE 19.1 Human clinical trial experimental phases

Phase 1: The first stage of clinical testing assesses the safety, tolerability, pharmacokinetics, pharmacodynamics, and maximum tolerated dose of a therapy in healthy and affected subjects. Dose–response safety curves are developed. The number treated is too small to assess efficacy.
Phase 2: The first clinical studies aim to gather preliminary data on the effectiveness of the experimental therapy by testing one or more doses in a modest number of subjects. The design for these trials also assesses the feasibility of the inclusion and exclusion criteria for participation (location of lesion, age, gender, time after injury, severity of deficits, other medical complications, concomitant medications, ability to give informed consent, family support, etc.) the utility of chosen outcome measures, and the likely robustness of the intervention to help determine the number of subjects that will need to be treated successfully to best power a larger trial. Phase 2 studies test the standardization procedures that have been adopted. Trials include a control group that receives a specified intervention (such as a placebo medication, manipulation or physical therapy) to equal the experimental group's intensity of care, blinded outcome measures, and external monitoring by a safety committee for short-term adverse reactions and risks.
Phase 3: Subjects with the disease and entry criteria of interest are randomly assigned to the experimental or control intervention in sufficient numbers to reject the null hypothesis that the experimental treatment is no better than the control therapy. The design and management of the trial are performed so as to eliminate bias – random allocation to a placebo, sham or other control intervention, predefined primary and secondary outcomes for statistical analyses, masking at least the person who conducts the primary outcome measures, and the subjects and investigators when feasible, a data management group that receives and analyses all data independent of the investigators, and a safety monitoring committee that evaluates risks, benefits, and ethical issues as the trial proceeds.
Phase 4: Post-marketing safety surveillance collects voluntarily offered data sent to the commercial company that licenses the intervention. Formal studies may add information about the long-term risks, benefits, and optimal indications for use of the approved treatment.

designed to be relevant to the human disease, use valid and reliable histologic, physiologic, and behavioral outcome measures that have parallel outcomes in patients, and then interpret the results in keeping with its magnitude and the applicability of the response to patients.

Pre-clinical modeling in rodents must try to solve a complex matrix if translation into a clinical trial is to be feasible. Experiments should emphasize the search for safety in a disease model that is similar to the human condition in its pathology, location, and induced deficits; seek the optimal dose, delivery schedule, and timing for the intervention; employ task-related activities to maximize learning and neural adaptations; collect reproducible measures of efficacy that can be related to human impairments and disabilities, and follow up the animals for at least 6 months. In human trials, however, heterogeneity in genes, age, sex, medication taken, premorbid health, severity of injury, precise timing after injury to onset of intervention, and other variables will differ from the perfect breeding and controlled environment of animal experiments.

Rodent models are an approximation to the human condition and response. Differences in surface area and volume of the brain and spinal cord between mice and humans are in the order of 10^3, an important factor when considering the distance an axon may need to regenerate or a cell to migrate. Highly inbred rats or mice of the same species, strain, age, weight, and gender serve as the usual experimental population. Variation in any one of these factors can alter the results of an experiment. Different inbred murine strains, for example, may respond quite differently to ischemia or trauma and most mice respond differently than most rats in terms of injury, inflammatory, immune, and regenerative cascades (Steward et al., 1999). Some rodents are much less likely than others to develop a glial scar after the same SCI that produces a large barrier to axonal regeneration in another species (Hagg and Oudega, 2006). Rats develop cystic cavities after a cord contusion whereas mice generally do not. Differences in genetic injury-induced immune responses or in the potential for endogenous regeneration of neurons in inbred mouse strains, such as Nogo-A-specific knock-outs (Dimou et al., 2006), can account for variations in experimental outcomes for different rodent species and strains.

Different strains of rats may vary in important ways when morphologic, locomotor and sensory measures are examined (Webb et al., 2003). An induced physical and cognitive impairment in a rodent does not induce the disability that patients with stroke, SCI, or TBI may experience. Behavioral tests, then, have uncertain parallels in laboratory rodents and patients living in their homes.

Pre-clinical studies would better serve the decision to go forward with a human trial if they examined the statistical significance of changes in the expression of a regenerative substance, the length and number of sprouting axons, and number of new neurons in light of the behavioral significance of the induced adaptation. The interpretation of functional losses and gains in quadrupedal locomotion or reaching for a food pellet is complicated in models of repair by not knowing the anatomic tracts necessary to perform the activity. Physiologic, histologic, and molecular markers may serve as surrogates for behavioral outcomes in pre-clinical models to assess the biologic activity of the repair intervention. Differences between rodents that do better than controls may be statistically significant, but the difference may not be behaviorally significant.

In addition, pre-clinical studies of an intervention often do not attempt to replicate the results of the same injury and repair model in different laboratories. The biologic significance of a manipulation for repair would increase if results were reproduced in another rodent species or strain and in two different models of injury. For example, an axonal regeneration strategy for SCI, if reproduced in both a severe contusion and a hemicord section model or in a SCI and a stroke white matter injury model, would suggest a robust biologic mechanism for pursuing a human intervention. Replication of the responses in larger animals such as pigs or nonhuman primates may further secure the biologic relevance, safety, and most feasible dose–response curve for clinical trials.

Pre-clinical experiments that inform the design of human trials ideally include a lesion of etiology, volume, and location that reflects the human disease; examine changes induced by injury and by repair procedures both near and remote from the lesion; distinguish between reactive molecular and histologic changes vs. changes critical to repair cascades; employ explicit training paradigms for the reacquisition of testable skills; correlate morphologic and physiologic measures of repair with behavioral measures of task reacquisition; reproduce key results in more than one laboratory, in different strains or species of rodent, and in a larger mammal; and generalize the results across several disease models, such as axonal regeneration in a stroke and SCI platform. These studies also permit the development of drugs and cells with the purity, potency, stability, best dose, and delivery route that will be necessary for use in patients. To optimize pre-clinical modeling, collaborations between basic and clinical scientists in the development of translational animal models of injury and repair may best move bench-to-bedside therapies forward for the rehabilitation of disabled patients.

Studies in animal models have led to a growing number of safety studies of human interventions for stroke. These include intravenous injection of autologous mesenchymal stem cells about 40 days after a hemispheric stroke (Bang et al., 2005) and implantation of human neuronal cells (LBS-Neurons; Layton Bioscience, Inc.) into the edge of deep infarcts near the basal ganglia (Kondziolka et al., 2005). Complications were modest. Safety studies in SCI have proceeded with injection of human fetal spinal cord tissue into a syrinx (Anderson et al., 2005). Trials of autologous activated macrophages injected into the acute injury (Proneuron Biotechnologies, Israel) and of porcine oligodendrocyte progenitors (www.diacrin.com) were halted. Autologous olfactory ensheathing glia injected into chronic spinal injuries (Feron et al., 2005), as well as recent dural/intrathecal infusions of Rho-kinase (Dubreuil et al., 2003) and Nogo inhibitors (Schwab, 2004; Markus et al., 2005), are in progress.

At conferences and on Web pages, but not in peer-reviewed published reports, clinicians at hospitals in China (www.stemcellschina.com), Russia (www.stemedica.com), and other countries have described fetal, olfactory ensheathing glia, hematopoietic, and other unspecified cell implantations into the cord and brain, as well as peripheral nerve bridging experiments in uncontrolled experiments on patients for stroke, SCI, TBI, and degenerative diseases (Dobkin et al., 2006; Enserink, 2006). These unpublished interventions were based on a broad interpretation of the applicability of pre-clinical experiments in rodents. For example, one Chinese neurosurgeon has verbally reported that he has injected over 500 patients with chronic SCI and amyotrophic lateral sclerosis with olfactory ensheathing cells. The cells have been injected into the cystic cavity of the cord or through burr holes into the frontal lobes (Dobkin et al., 2006). The only published characterization of the transplants, which are cultured from two aborted fetuses per patient, suggested that the cells in that sample were probably not ensheathing glia (Guest et al., 2006). Nothing, however, has been learned from these human experiments, because no subject inclusion and exclusion criteria exist, no anticipated effects of the intervention are pre-specified, no standardized measures are performed before or at intervals after the surgical intervention to assess for possible complications or for effectiveness of the implant procedure, no unbiased observers carry out behavioral tests, and patients and families draw their own conclusions about whether any change is related to the surgery (Dobkin et al., 2006). Transplantation and other invasive or noninvasive biologic interventions that proceed without basic elements of an ethical design cannot inform clinicians or patients about actual risks and benefits.

CLINICAL TRIAL DESIGNS

If pre-clinical studies achieve the level of evidence and promise that warrant a clinical trial, the design and conduct of the phases of human trials can go forward (Table 19.1). No set standards exist for when to forge ahead, but many valuable suggestions have been offered (Redmond and Freeman, 2001; Steeves et al., 2004, 2006; Anderson et al., 2005; Fisher, 2005; Blight and Tuszynski, 2006; Dobkin, 2007a,b). The goals are to demonstrate safety of the intervention, show that the strategy is better than what is available, which in general will mean task-related neurorehabilitation training, and demonstrate a gain that entails a clinically meaningful change in impairment and disability. Safety and the protection of human subjects is the most critical element in a neural repair intervention, especially because such interventions are likely to stir the imagination of highly disabled persons and clinicians and may lead to excessive risk-taking behavior. The basic rules for informed human consent to participate in a trial have been codified by the reports in Table 19.2.

Resources for the organization of a clinical trial for a biologic intervention can be found at www.ich.org (International Conference on Harmonization) and www.ICORD.org (Guidelines for Clinical Trials in SCI). One approach to the development of a randomized clinical trial that compares two interventions is for investigators to consider their responses to each aspect of organization of the experiment. Many of the key details are listed in Table 19.3. The following

TABLE 19.2 Protection of human subjects in clinical Trials

Helsinki Declaration. The Declaration of Helsinki, developed by the World Medical Association, is a set of ethical principles for the medical community regarding human experimentation. It was originally adopted in June 1964 and has since been amended multiple times. The recommendations concerning the guidance of physicians involved in medical research may be found at: http://www.wma.net/e/policy/b3.htm
Belmont Report The Belmont report is a report created by the former United States Department of Health, Education, and Welfare (which was renamed to Health and Human Services) entitled "Ethical Principles and Guidelines for the Protection of Human Subjects of Research." The text is available at http://www.hhs.gov/ohrp/humansubjects/guidance/belmont.htm

TABLE 19.3 Organizational checklist for a randomized clinical trial

Specific aims
Importance of trial – background literature and rationale
Experimental intervention and comparison treatment
Hypotheses to be tested
Specific primary and secondary outcomes
Methods
Define study populations and criteria for inclusion and exclusion
Define potential risks and benefits
Provide training in the ethical conduct of research
Define descriptors about subjects
Rationale for specific outcome measures
Appropriateness to intervention
Reliability, validity, and sensitivity for the study population and intervention
Training plan to provide uniform experimental and control treatments
Specify duration and intensity of all interventions
Schedule of interventions
Schedule of assessment measures
Methods for blinding subjects, personnel, or those who collect outcomes
Training plan for blinded observers to uniformly collect and measure outcomes
Plan to monitor adequacy of blinding
Plan to monitor adverse reactions
Informed consent
Meet Institutional Review Board requirements
Establish common approaches across sites for presentation, advertising, maintaining privacy, and offering payments and reimbursements to subjects
Explicitly state whether the experimental intervention is offered outside of the trial
Calculate sample size based on pilot studies
Expected entry and dropout rates
Demonstrate access to this sample size
Subject randomization assignment
Check eligibility of potential subjects
Allocation schedule and stratification criteria
Method to generate and conceal allocation
Flow diagram for subjects by assignment, assessments, interventions, outcome measures
Develop procedures to assure compliance by subjects and therapists with the assigned intervention
Establish data management group's responsibilities
Develop practical data forms and transmission of forms
Training for data entry
Data collection assuring completeness, quality and privacy
Procedures for adverse reactions, dropouts, and missing data
Interim reports
Establish an administrative oversight committee
Establish an external safety committee and set procedures
Maintain an operations manual

(*Continues*)

TABLE 19.3 (*Continued*)

Analytic plan
Rationale for statistical methods for primary, subgroup, or covariate analyses
Intention-to-treat analysis vs. efficacy analysis
Uses of descriptor data
Plan interim analyses for safety and efficacy; specify study stop rules
Describe protocol and deviations
Results and publication
Provide summary data including means, standard deviations, and effect size
State results in relative and absolute numbers, not just in percentages
State estimated effect on each outcome measure, including confidence intervals
Interpret findings in terms of internal validity and generalizability
Put study into the context of all available evidence

Adapted from B. Dobkin, *The Clinical Science of Neurologic Rehabilitation*, Oxford University Press, 2003.

discussion is primarily about biologic interventions that are of uncertain safety and efficacy, may alter the expression of genes and substances in neural tissue, may alter neural networks, may evoke seizures, tumors and maladaptive neural plasticity by their adverse effects, and are considered to carry more than minimal risk whether given by oral, parenteral or intrathecal means and especially if injected into parenchyma. This review emphasizes enrichment strategies (Table 19.4) that identify subjects who are most likely to respond and whose treatment is optimized to maximize the chance for a successful trial.

TABLE 19.4 Enrichment strategies for clinical trials

Perform pre-trial studies that develop realistic numbers of subjects needed to recruit and randomize (effect size).
Determine the most ideal timing for the intervention.
Determine the most feasible dose that is likely to be therapeutic.
Select the population most likely to respond to the specific repair technique.
Provide a baseline course of rehabilitation for management of the clinical endpoints.
Consider anatomic, physiologic, functional neuromaging, and gene studies to enter subjects who are most likely to respond or as potential biomarkers of efficacy.
Maximize the dose of task-specific rehabilitation to promote activity-dependent plasticity.
Use the most sensitive and specific outcome measures.
Minimize the variability of the assessment of outcomes.
Consider unique methods of analysis of outcomes.

AIMS

The goal of a neural repair intervention is to lessen impairments and disabilities beyond the effects of the natural history of recovery and the best possible techniques of rehabilitation. Modest benefits need to be appreciated through an array of outcome measures. Minor effects that reveal some positive biologic activity may be of value when added to another future intervention. The changes that increase the functional abilities of patients, however, are the primary goal.

An initial demonstration of safety for an intrathecal drug or intraparenchymal cellular strategy may enter patients with a chronic, complete thoracic SCI or a stroke with hemiplegia due to a chronic subcortical infarct. The dose of the treatment could be escalated over intervals of time and concentration of the dose to demonstrate relative safety. The investigation might proceed to determine safety in subjects with a subacute injury at the time that further clinically important gains were unlikely and that time could be based on prospective clinical studies and the likely interactions of the intervention on the status of the injury milieu.

For example, for at least 3 weeks, the peri-infarct milieu (Weller et al., 2006) into which substances or cells are exposed expresses a rise or fall in a range of neuronal growth-promoting and growth-inhibiting genes that affect the making of cytoskeletal proteins, direct axon growth cones, and produce angiogenesis and other molecular signals for the proliferation and migration of endogenous stem cells from the subventricular zone (Zhang et al., 2005; Carmichael, 2006). Exogenous stem cell interventions, including bone marrow stromal cells, may be most useful within this timeframe to extend and sculpt these mechanisms for axonal regeneration, cellular signaling, and synaptogenesis (Zhang et al., 2006). Clinically, the likelihood of achieving useful function of a hand is minimal if no finger or wrist extension has recovered by 2–4 weeks after an infarct. Thus, there would be a reasonable conceptual basis for a clinical trial aimed at recovery of the hand for grasping and pinching starting at about 3–4 weeks after stroke, if pre-clinical animal model findings were consistent with this.

After complete cervical SCI, functional use of the elbow, wrist, or hand at two levels below the lesion becomes very unlikely if no gains have occurred by 4–8 weeks. The biologic and clinical aim would be to achieve regeneration and reinnervation of spinal motor pools and limb muscles at two levels below the SCI, starting as soon as a month after onset (Steeves et al., 2006). If the population had an incomplete SCI with some movement below the site of trauma, a much larger number of subjects would be needed to try to demonstrate efficacy, because many subjects are still evolving toward their maximal gains. Although patients with paralysis at entry may seem the most definitive group to use to demonstrate gains, interventions may require some level of network intactness and clinical function for remodeling to occur. This would be especially true for an intervention that altered synaptic efficacy and Hebbian plasticity. Severely impaired subjects with chronic stroke or SCI and only slight motor control beyond 6–12 months would tend to maintain a stable baseline, unlikely to make further gains. The concern about using subjects with chronic deficits and pathology is that the neural milieu may include physical barriers such as scar and cysts and be less responsive to an exogenous intervention. Also, patients may have evolved compensatory and synergistic movements and pathologies such as muscle atrophy and fatigability that may interfere with the effects of the new strategy. Insight into this problem can come from experiments in subacute and chronic animal models of injury.

The primary outcome for these two stroke and SCI scenarios could be a gain in muscle strength to at least motion against gravity, accompanied by a secondary outcome or combined outcome that includes functional use of the hand for several self-care actions.

METHODS

ENTRY CRITERIA

Entry criteria ought to be as inclusionary as possible, if generalization of the benefit is to be expected. In addition, most sites will have to screen large numbers of subjects to enter what is typically about 10% into the trial who meet inclusion and exclusion criteria, so more open entry criteria may allow the trial to meet its goal for entries. A priori exclusion of subjects requires a firm rationale. Subjects entered ought to be as typical of the target population as possible. Time after onset of neurologic impairment, type and degree of impairment and disability that is being targeted, location of lesion, medical stability, understanding by patient and family of the full nature of the experiment, home support system, and availability for long-term monitoring are among the most important entry criteria. For example, patients may need to take drugs for hypertension, diabetes, pain, seizures, spasticity, and for management of other diseases. These medications may alter the absorption, metabolism or bioavailability of the experimental intervention. Some may affect regeneration, such as drugs that alter cAMP levels, or block dopamine and other neurotransmitters and possibly enhance or diminish learning and repair. Routinely taken medications may confound the early effects on an intervention for repair. The need for immunosuppressants or antibiotics after a cellular transplant may have adverse effects on regenerative capacity. Even the anesthetic given at the time of a surgical transplantation strategy may alter inhibitory and excitatory neurotransmission for days in animals and in human. It will not be possible to eliminate patients from joining a trial based on the need for medications used to manage the disease at hand; so assays of the possible interactions of the experimental intervention and commonly taken classes of drugs are an important pre-clinical step.

DESIGN

In trials that involve rehabilitation, sampling error, the diversity of impairments and related disabilities, and measurability of outcomes can confound the best intentions of investigators. A multicenter, prospective, randomized trial with outcomes measured by an assessor who is blinded to the treatment, along with additional blinding of subjects and treating physicians and therapists when feasible, is most likely to create unbiased treatment groups that are similar with respect to other possible determinants of outcome. For biologic repair studies, a parallel group design would usually be most fitting. Subjects are divided into the number of groups equal to the number of interventions. If the balance between groups in important variables is uncertain, then stratification by these variables can be employed at the time of randomization. This may increase the number of entries required. Data analyses can also adjust for discrepancies between groups. A crossover design may be employed when concern arises that the variability between subjects is greater than within subjects at entry or in response to the intervention. However, in repair trials, giving participants one treatment with a 6- to 12-month followup and then switching those who received the placebo to the experimental intervention and switching those who received the latter to a placebo may have substantial effects related to timing of the intervention and carry over that blur the final outcomes.

Phase 2 trials can be combined with a phase 3 trial. A safety run-in might include escalation of the dose and duration of the therapy using, for example, patients 2 months after a persistent, complete thoracic SCI, followed by entry of subjects 2 months after a complete cervical SCI, for which the intervention may carry more risk and bear more obvious benefit. Adaptive designs can serve as enrichment strategies (Fisher, 2005). This approach permits real-time learning about the optimal dose–response relationship, the safety of increasing doses, and a continuous reassessment of the futility of the experimental intervention. This can be a complicated strategy, however, and may require a large number of subjects unless adverse effects occur early or gains are very robust. Other alternatives include (1) testing a series of fixed doses in a small number of subjects, (2) using a large number of treatment arms that are given different doses, or (3) employing different stratification criteria at entry for a particular dose. Adaptive designs can also compare, for example, a single intrathecal dose to repeated doses each at higher concentrations for a drug therapy or multiples of dose concentrations in a wider region of interest for implantation interventions.

In another enrichment strategy, subjects can be titrated with an increasing dose of the experimental intervention to identify potential responders. These responders are then randomized to receive a placebo or the experimental treatment. Care must be taken to account for any carry-over effects of the first stage on the placebo group. The results of course cannot then be generalized to all patients with the problem of interest with regard to adverse effects and efficacy.

INTERVENTIONS

A great number of neural repair interventions are in the offing. Every one ought to be combined with a task-specific set of neurorehabilitation physical and cognitive therapies that can optimize the effects of the intervention. These therapies can also be added to trials of interventions aimed at slowing neurodegenerative diseases as well to maximize the contrast in any benefits of the treatment on impairments and disabilities. One basis for this approach arises from animal models. Exercise, environmental enrichment that permits activity in cages, and skills training enhance synaptogenesis, neurotrophin levels, mechanisms of learning, and recovery in most (Johansson and Belichenko, 2002; Vaynman and Gomez-Pinilla, 2005), but not in all, models of brain and spinal cord injury. The optimal intensity and duration of therapy to maximize gains after stroke or SCI is rarely examined in animals or prior to undertaking clinical trials (Dobkin, 2005a,b). Skills training, procedures to make practice progressively more difficult, and reinforcement paradigms to maximize learning are rarely considered in animal and human studies. Indeed, most studies of augmented practice during rehabilitation for stroke, for example, have offered about 16 h of additional treatment for a specified problem in patients with mild-to-moderate impairments or disabilities; these trials have revealed only a 10% improvement with the additional treatment time (Kwakkel et al., 2004). The intensity of task-related training that aims to incorporate the cells, axons, dendrites, and synapses of a partially regenerated pathway is no less important to the success of a regeneration strategy than the type and timing of the repair intervention. Therapies with a strong conceptual basis that are applicable to the behavioral and network targets for the repair intervention should be incorporated into the rehabilitation component of the trial (Kimberley et al., 2006).

Another enrichment strategy aims to start with a stable baseline of function across subjects. For example, subjects can be assessed at baseline on a study's motor and functional scales if better use of the arm or walking were the goal. Then, a period of rehabilitation that targets the expected behavioral effects of the biologic intervention on motor control is undertaken. Three weeks with 18 h of therapy would be a reasonable dose. Scores on a repeated assessment become the new baseline. If the change exceeds a predefined level on any of the primary outcome assessments, for example $\geq$10% better, the rehabilitation intervention would continue for six additional sessions. If no further change occurred and the subject still met the entry criteria for the level of impairment and disability, the medication, transplant or placebo, would be given. Progressive practice would continue for at least 3 contact hours a week for 24 weeks, along with a daily home program. This approach minimizes the experimental noise of gains that are not likely to be related to the experimental intervention. A stable behavioral baseline may also decrease the number of subjects needed to power a randomized clinical trial. Most important, this strategy aims to maximize the potential benefits of the regenerative therapy as an adjunct to rehabilitation.

OUTCOME MEASURES

Trials should include descriptors of the population that allow both a comparison between the experimental and control group at entry and a means by which readers can relate the studied group to their own patients and to other clinical trial data. Information includes age, gender, ethnicity, lesion location, time from onset of injury to entry into the trial, severity of initial impairment and disability, amount of prior rehabilitation, medications, and co-morbidities. Most important, investigators must measure the expected goal and other potential outcomes for the intervention. One concern in trials for neural repair is that the clinical science of mechanisms for repair may benefit from relative success, such as in restoration of hand movement, even if this change provides no gain in functional use of the hand. Improvement in voluntary wrist extension of the hemiplegic arm without the ability to grasp and release for holding and manipulating objects will have little impact on the daily activities of the patient with stroke, because the subject will have to continue to use the unaffected hand for all gross and dexterous actions. An intervention that carries risk, especially by invasive cell implantation, must aim to enable better function.

Outcome measurement tools for trials include reliable and valid ordinal and interval scales for disease-related impairments and disabilities, as well as for health-related quality of life and home and community participation. Ordinal scales are common in neurorehabilitation and include the Fugl–Meyer Motor Assessment for selective arm and leg movements, the Functional Independence Measure (FIM) for assistance needed in performing activities of daily living, and the Stroke Impact Scale to assess participation in activities that require mobility, use of the affected hand, and cognition and mood. They measure magnitude by a predetermined ranked order among possible responses in a classification, but the 3–10 responses for each domain are not of equal intervals. Walking speed, on the other hand, would be an interval measure and a ratio scale in that the numerical differences between two measures are interpretable as being of equal magnitude. The scales employed in trials must have been proven reliable, in that each time within a period of no treatment that a score is tallied it is the same. They must have face and content validity for the disease and the impairments and disabilities being examined. They must be responsive to changes that are meaningful to patients and clinicians. Ideally, they tell us not only whether a treatment works (efficacy), but also how well it works (effectiveness). Outcome measures should reflect the expected main effects of the intervention and look for potential effects that may provide insights for future trials. Although the primary outcome may be recovery of hand function, some measure of bladder control, autonomic function, and pain may be appropriate for a trial in SCI. In addition, explanatory measures, such as motor-evoked potentials in relation to a change in motor control, may be of value in understanding why the intervention proved useful. Most important, the tools employed must capture the magnitude of any effect so that risks, costs, and benefits can be considered in the future care of patients.

Functional outcome scales such as the FIM and Barthel Index have ceiling effects, because patients can have marked impairments on the affected side but relative independence in physical functioning. A quality of life measure such as the SF-36 may reveal patient preferences and interpretations about changes over the course of a treatment, but may not reflect the biologic activity of a regenerative strategy on the affected arm or leg as well as it might reflect a change in pain and satisfaction. Physical limitations in using the affected limb or in a range of cognitive functions must be captured in a study of a biologic intervention. Thus, one or more of the reliable and valid outcome measures may need to be refined. Single highly relevant items drawn from an existing scale may need to be employed, rather than the entire measure, as an outcome. The entire measure may, however, serve as a global descriptor of the patient population for the reader. A study may assess the change in the distribution of an outcome measure across its entire range, which may be a more sensitive approach for measurement when patients are heterogeneous and the variability of outcomes is likely to be broad. Scales can also be made dichotomous. For example, the effectiveness of a walking intervention after stroke or SCI can be revealed by the level of ability to walk with less assistance on the FIM scale. For repair, a change from stepping with maximal assistance to walking with any lesser level of help would be a valuable gain, but each higher level of independence on the 7-item FIM locomotor scale would have a different clinical impact. At the same time, a gain in walking speed from a baseline of 0.3 m/s to 0.4 m/s would be an increase of 33%, but not clinically useful. If the control group gained <0.5 m/s and the experimental group improved to >0.5 m/s, a dichotomous outcome measure, greater participation in walking-related activities would be likely for the latter, which would be very meaningful clinically. Investigators can also plan a responder analysis in which subjects are randomized into groups based on severity, each with a predefined outcome measure, then a positive effect for each stratum is selected for statistical analysis.

A global statistic, rather than a single ordinal or interval measure, incorporates several relevant measures. For example, in ASIA A and B subjects with lesions at C4–C6 begun 2 months after SCI, the combination of enabling patients to make gains toward a threshold of 3/5 for strength of the wrist and fingers and toward a 50% increase in the number of blocks moved on the Box and Blocks test of hand dexterity may best reveal the impact of the treatment for the sample of subjects.

Physiological studies in animal models provide insights for human studies. Functional magnetic resonance imaging (fMRI), positron emission tomography, diffusion tensor imaging of axons, evoked potentials, optical intrinsic signal imaging, and other methods could be used in parallel in animal and human trials to explore residual connectivity, the best site for a repair intervention, monitor remyelination, follow the movement of implanted cells, determine whether the motor network is engaged by task-specific rehabilitation, and monitor the dose of therapy needed to maximize cerebral reorganization after a biologic therapy (Dobkin et al., 2004; Dobkin, 2005a; Carey et al., 2006; Lotze et al., 2006). Several studies of rehabilitation interventions for the upper extremity and for walking monitored by repeated fMRI or transcranial magnetic stimulation studies

TABLE 19.5 Rules used to determine the causal relationship of a serious adverse event (SAE) to the study agent

The investigators should assess whether there is a reasonable possibility that study drug caused or contributed to the serious adverse event. The statement of relationship to study drug will be located on of the CRF SAE forms. The investigator shall characterize the he relatedness of a serious adverse event to study drug as unrelated, unlikely, possible, probable, or definite. In primary analysis, a serious adverse event will be considered associated with the use of the study drug if the attribution is possible, probable, or definite.

The investigator will employ the following guide in making attributions of relatedness of an SAE to study agent:

Unrelated
Adverse event is clearly due to extraneous causes (e.g., underlying disease, environment)

Unlikely (must have 2)
May or may not have temporal relationship to intervention.

1. Could readily have been produced by the subject's clinical state,or environmental or other interventions.
2. Does not follow known pattern of response to intervention.
3. Does not reappear or worsen with reintroduction of intervention.

Possible (must have 2)

1. Has a reasonable temporal relationship to intervention.
2. Could not readily have been produced by the subject's clinical state or environmental or other interventions.
3. Follows a known pattern of response to intervention.

Probable (must have 3)

1. Has a reasonable temporal relationship to intervention.
2. Could not readily have been produced by the subject's clinical state or have been due to environmental or other interventions.
3. Follows a known pattern of response to intervention.
4. Disappears or decreases with reduction in dose or cessation of intervention.

Definite (must have all 4)

1. Has a reasonable temporal relationship to intervention.
2. Could not readily have been produced by the subject's clinical state or have been due to environmental or other interventions.
3. Follows a known pattern of response to intervention.
4. Disappears or decreases with reduction in dose or cessation of intervention and recurs with re-exposure.

Based on: Adverse Events Reporting Requirements SOP. NIH-NIAID. http://www.niaid.nih.gov/ncn/sop/adverseevents.htm, accessed 30 November 2004.

at regular intervals during therapy suggest that this approach may give insight into gradually adapting brain–behavior and dose–response relationships (Koski et al., 2004; Dobkin, 2005a; Dong et al., 2006).

ADVERSE EFFECTS

Pre-clinical animal studies may identify some of the potential side effects of a biologic intervention, but almost certainly will not predict all of the potentially serious neuromedical complications. These may include meningoencephalitis (chemical, related to rejection, bacterial, and viral if a viral agent is part of the treatment paradigm), a toxic-metabolic encephalopathy, hemorrhage at an injection site, seizures, transient hypotonia if the cord is traumatized, hypertonicity with spasms, central pain, and exacerbations of bowel and bladder dysfunction. A list of all possible adverse effects ought to be developed into a checklist and filled in for each subject as they arise. Table 19.5 shows the algorithm used by the trial's safety committee to determine whether a serious adverse event is related to the intervention.

DATA ANALYSIS

The results of a trial include the planned primary analysis by intention-to-treat, as well as an efficacy analysis on subjects who met some minimal level of treatment but dropped out. This may be followed by an analysis of interesting subgroups from which no conclusions can be drawn, but this fishing in the database may offer insight into the futility of the intervention or the potential for a particular subgroup to be used for future studies. The results can also contribute to future trials by examining the sensitivity of the outcome measures for the sample studied.

Investigators should not overly rely on p values in statistical tests of the null hypothesis. Confidence intervals must be calculated. Also, the p value does not tell how large the effect was of a successful intervention. To better assess the clinical meaning, the effect size (EF) should be calculated as the

EF = mean (treatment 1) – mean (treatment 2)/standard deviation of control. An effect size of 0.8 is large, 0.5 moderate, and 0.2 small.

ETHICAL CONCERNS IN CLINICAL TRIALS OF REPAIR

EMBRYONIC STEM CELLS

Scientific, legal, cultural, religious, and ethical differences among patients, biomedical scientists, clinicians, the public, and nations have a profound effect on the conduct of clinical trials that involve invasive placebo procedures, on randomization with blinding of the subject, on oversight about safety and efficacy, and especially about the use of human embryonic stem cells and cells that can generate complex tissues. Experiments must be conducted in a transparent and responsible fashion within the conventions and regulations of a society. Special attention must be given to stem cell research. Guidelines from national and international scientific and legislative groups must be followed for the procurement, derivation, banking, distribution, and use of cells and tissues taken from human fetal material, embryos, and blastocysts. Researchers should ensure that both the written and verbal components of the informed consent process are clear about the material that is being used. Nonexclusive access by the research community for human stem cell lines is also necessary to assure progress and verification of the findings from any single laboratory during pre-clinical animal and initial human studies. Cells can be made available through national or international depositories. Local, national, and international review of procedures may be necessary to reach a consensus on volatile subjects such as the use of stem cells obtained during human development.

SHAM SURGERY

An invasive procedure that injects cells or releases substances into the brain or cord needs to develop a procedure for control subjects. Possible effects of a simple manipulation on the spinal cord from injection of cells into the surrounds of a cystic lesion include decompression, detethering, or improved cerebrospinal fluid flow dynamics around the lesion site, local inflammation that releases cytokines or alters excitability, and the potentially powerful effects of knowledge about surgery that raises subject and investigator expectations. It is quite possible that a simple surgical procedure itself, independent of the injected test substance, may improve certain outcomes and alter local biology in a way that suggests that the experimental material is the cause. For example, five of seven subjects who were injected with cells derived from the brain tissue of aborted fetuses into the cystic region of their chronic SCI reported a transient decrease in spasms and hypertonicity (Dobkin et al., 2006). The change allowed several of them to make slightly greater movements at a joint in which they had had some motor control and lasted for about 10 days. All appeared to have had this change from either manipulation of the cord or possibly in association with fever and a CSF pleocytosis that suggested local inflammation from rejection of the cells. Thus, without a placebo procedure, and careful monitoring of behavioral changes, it is impossible to learn anything from this experiment on patients.

Surgery that applies all of the invasive procedures except for use of the experimental agent will most ideally serve as the placebo comparison. If administration of a placebo agent that contains all but the active substance is not possible during a surgical procedure already planned as part of standard care or as a requirement for participation in a trial, then a sham control could include a skin incision and subcutaneous manipulation under anesthesia, which serves to keep the subject blinded.

PARTICIPATION IN CLINICAL TRIALS

Some highly disabled patients may choose to enter a study not because they have weighed the potential risks against a small chance of benefit, but because their desperation leads them to disregard anything other than the possibility of a benefit. They are willing to risk any adverse reaction for the most meager prospect of benefit. This problem places a high ethical obligation on researchers when recruiting subjects for a trial. Transparency of information about the state of the science is essential for trials in neural repair. Scientists, clinicians, and biotechnology commercial interests must guard against unrealistic expectations and premature human trials. The specific goals and limitations of results must be made manifest for every basic research finding and pre-clinical animal and human study that precedes a clinical trial. Oversight is imperative in the review of the data that is used to support a clinical trial design. Patients must know that uninvolved experts in developmental neurobiology, clinicians with experience in clinical trials, rehabilitation, and neurological disease, statisticians, and informed laymen have examined the readiness of the strategy and the risks and benefits of the trial.

The search for therapies must be based on scientific methods. It is most disappointing that cellular and other medical interventions are being provided without being subjected to the gold standard of a scientific trial in which patients meet a specified criteria for entry, are randomly assigned to one of two treatments, are carefully examined for the duration of the likely benefit of the experimental and other intervention, and are examined at the end of the trial by a person who does not know which treatment was given (Enserink, 2006). This is the Western standard for obtaining an unbiased measure of possible efficacy of a new treatment. Anything less is fool's gold. No one learns whether a treatment is of value without this standard approach. In an all too typical example, a Chinese neurosurgeon obtained his only follow up about his injection of cells into the cord or brain for SCI by e-mail, if at all (Dobkin et al., 2006). Patients sent greetings and sometimes said that some aspect of their function seemed better, but no one had examined them to determine just what that benefit may have been or whether that benefit was related to a plausible mechanism. In this example, patients and the surgeon described some detail after surgery that seemed favorable but could easily fall within ordinary physiologic fluxes – a toe that sometimes wiggled or a wrist that changed from a twitch of motion to slight movement, but not against gravity, which was still not useful for any activity. The most pertinent outcome will be in regard to functional gains. Will the procedure allow patients whose legs are paralyzed to improve enough to stand or walk? Will the implanted cells allow patients who cannot flex their elbow regain the strength to bring a spoon to their mouth? How many people must be treated to get one person to be able to perform a new function–10, 100, 500? Clinicians must help patients ask these questions and understand that participation in well-designed trials is the only way to help the field of neural repair make the advances that will lessen their disability or that of future patients.

REFERENCES

Anderson D, Beattie M, Blesch A et al. (2005). Recommended guidelines for studies of human subjects with spinal cord injury. *Spinal Cord* 43: 453–458.

Bang O, Lee J, Lee P et al.(2005). Autologous mesenchymal stem cell transplantation in stroke patients. *Ann Neurol* 57: 874–882.

Blight A and Tuszynski M. (2006). Clinical trials in spinal cord injury. *J Neurotrauma* 23: 586–593.

Carey L, Abbott D, Egan G et al. (2006). Evolution of brain activation with good and poor motor recovery after stroke. *Neurorehabil Neural Repair* 20: 42–48.

Carmichael S (2006). Cellular and molecular mechanisms of neural repair after stroke: making waves. *Ann Neurol* 59: 735–742.

Dimou L, Schnell L, Montani L et al. (2006). Nogo-A-deficient mice strain-dependent differences in axonal regeneration. *J Neurosci* 26: 5591–5603.

Dobkin B (2005a). Rehabilitation after stroke. *New Engl J Med* 352: 1677–1684.

Dobkin B (2005b). Rehabilitation and functional neuroimaging dose-response trajectories for clinical trials. *Neurorehabil Neural Repair* 19(4): 276–282.

Dobkin B (2007a). Behavioral, temporal, and spatial targets for cellular transplants as adjuncts to rehabilitation for stroke. *Stroke* 38: 832–839.

Dobkin B (2007b). Curiosity and cure: translational research strategies for neural repair-mediated rehabilitation. *J Neurobiol* 67: 1133–1147.

Dobkin B, Firestine A, West M et al. (2004). Ankle dorsiflexion as an fMRI paradigm to assay motor control for walking during rehabilitation. *Neuroimage* 23: 370–381.

Dobkin BH, Curt A, and Guest J (2006). Cellular transplants in China: observational study from the largest human experiment in chronic spinal cord injury. *Neurorehabil Neural Repair* 20(1): 5–13.

Dong Y, Dobkin BH, Cen SY et al. (2006). Motor cortex activation during treatment may predict therapeutic gains in paretic hand function after stroke. *Stroke* 37: 1552–1555.

Dubreuil C, Winton M, and McKerracher L (2003). Rho activation patterns after spinal cord injury and the role of activated Rho in apoptosis in the central nervous system. *J Cell Biol* 162(2): 233–243.

Enserink M (2006). Selling the stem cell dream. *Science* 313: 160–163.

Feron F, Perry C, Cochrane J et al. (2005). Autologous olfactory ensheathing cell transplantation in human spinal cord injury. *Brain* 128: 2951–2960.

Fisher M (2005). Enhancing the development and approval of acute stroke therapies. *Stroke* 36: 1808–1813.

Guest J, Herrera L, and Qian T (2006). Rapid recovery of segmental neurological function in a tetraplegic patient following transplantation of fetal olfactory bulb-derived cells. *Spinal Cord* 44: 135–142.

Hagg T and Oudega M (2006). Degenerative and spontaneous regenerative processes after spinal cord injury. *J Neurotrauma* 23: 264–280.

Johansson B and Belichenko P (2002). Neuronal plasticity and dendritic spines: effect of environmental enrichment on intact and postischemic rat brain. *J Cereb Blood Flow Metab* 22: 89–96.

Kimberley T, Khandekar G, Skraba L et al. (2006). Neural substrates for motor imagery in severe hemiparesis. *Neurorehabil Neural Repair* 20: 268–277.

Kondziolka D, Steinberg G, Wechsler L et al. (2005). Neurotransplantation for patients with subcortical motor stroke: a Phase 2 randomized trial. *J Neurosurg* 103: 38–45.

Koski L, Mernar T, and Dobkin B. Immediate and long-term changes in corticomotor output response to rehabilitation: correlation with functional improvements in chronic stroke. *Neurorehabil Neural Repair* 18: 230–249.

Kwakkel G, van Peppen R, Wagenaar R et al. (2004). Effects of augmented exercise therapy time after stroke: a meta-analysis. *Stroke* 35: 2529–2539.

Lotze M, Grodd W, Rodden F et al. (2006). Neuroimaging patterns associated with motor control in traumatic brain injury. *Neurorehabil Neural Repair* 20: 14–23.

Markus T, Tsai S-Y, Bollnow M et al. (2005). Recovery and brain reorganization after stroke in adult and aged rats. *Ann Neurol* 58: 950–953.

Redmond D and Freeman T (2001). The American Society of Neural Transplantation and Repair considerations and guidelines for studies of human subjects: the practice committee of the society. *Cell Transplant* 10(8): 661–664.

Schwab ME (2004). Nogo and axon regeneration. *Curr Opin Neurobiol* 14(1): 118–24.

Steeves J, Lammertse D, Curt A et al. (2006). Guidelines for the conduct of clinical trials for spinal cord injury as developed by the ICCP panel. *Spine* (in press).

Steeves JD, Fawcett J, and Tuszynski M (2004). Report of international clinical trials workshop on spinal cord injury February 20–21, 2004. *Spinal Cord* 42: 591–597.

Steward O, Schauwecker P, Guth L et al. (1999). Genetic approaches to neurotrauma research: opportunities and potential pitfalls of murine models. *Exp Neurol* 157: 19–42.

Vaynman S and Gomez-Pinilla F (2005). License to run: exercise impacts functional plasticity in the intact and injured CNS by using neurotrophins. *Neurorehabil Neural Repair* 19: 283–295.

Webb A, Gowribai K, and Muir G (2003). Fischer rats have different morphology, sensorimotor and locomotor capabilities compared to Lewis, Long-Evans, Sprague-Dawley and Wistar rats. *Behav Brain Res* 144: 143–156.

Weller P, Wittsack H-J, Siebler M et al. (2006). Motor recovery as assessed with isometric finger movements and perfusion magnetic resonance imaging after acute ischemic stroke. *Neurorehabil Neural Repair* 20: 390–397.

Zhang C, Li Y, Chen J et al. (2006). Bone marrow stromal cells upregulate expression of bone morphogenetic proteins 2 and 4, GAP junction protein connexin-43 and synaptophysin after stroke in rats. *Neurosci* 141: 687–695.

Zhang RL, Zhang ZG, and Chopp M. Neurogenesis in the adult ischemic brain: generation, migration, survival, and restorative therapy. *Neuroscientist* 11(5): 408–416.

INDEX

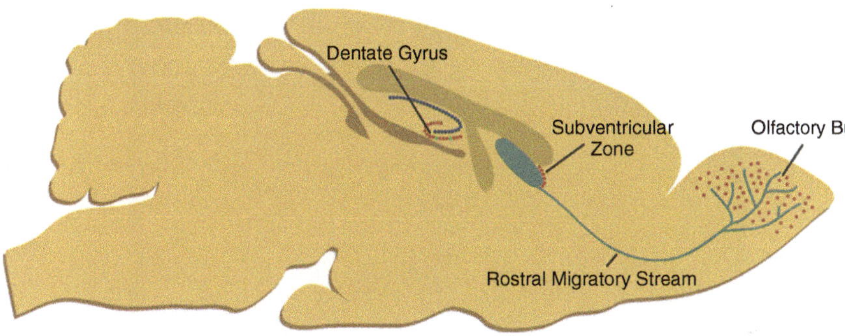

PLATE 1 Neurogenesis occurs in two distinct regions of the adult brain. A schematic drawing of a sagittal section through a rat brain shows these two neurogenic regions of the adult brain. NPCs in the dentate gyrus generate new granule cells throughout life. In addition, new neuronal cells are born in the subventricular zone and migrate through the rostral migratory stream into the olfactory bulb, where they differentiate into olfactory granule cells and periglomerular neurons. (see Figure 7.1 on p. 182)

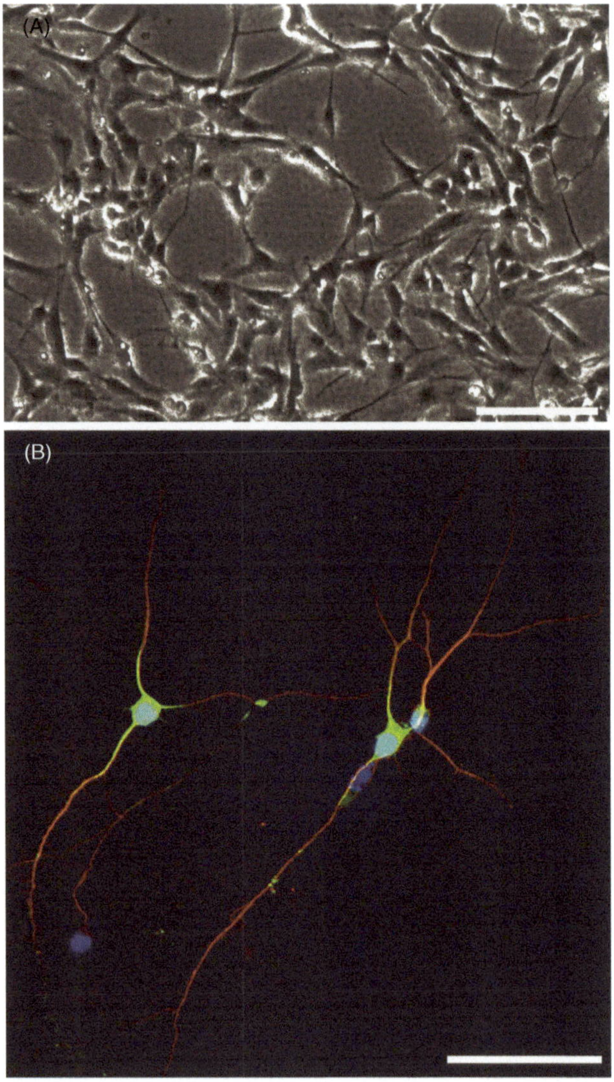

PLATE 2 The NPCs can be propagated and differentiated in vitro. (A) Example of NPCs under proliferating conditions derived from the mouse SVZ and grown as a monolayer. (B) Neuronal differentiation of rat NPCs derived from the DG. Shown is an example of cells that differentiated into MAP2ab-expressing neuronal cells (red) following retroviral overexpression of the neurogenic bHLH transcription factor, NeuroD1 (transduced cells are green). Scale bars in A and B: 100 μm. (see Figure 7.2 on p. 184)

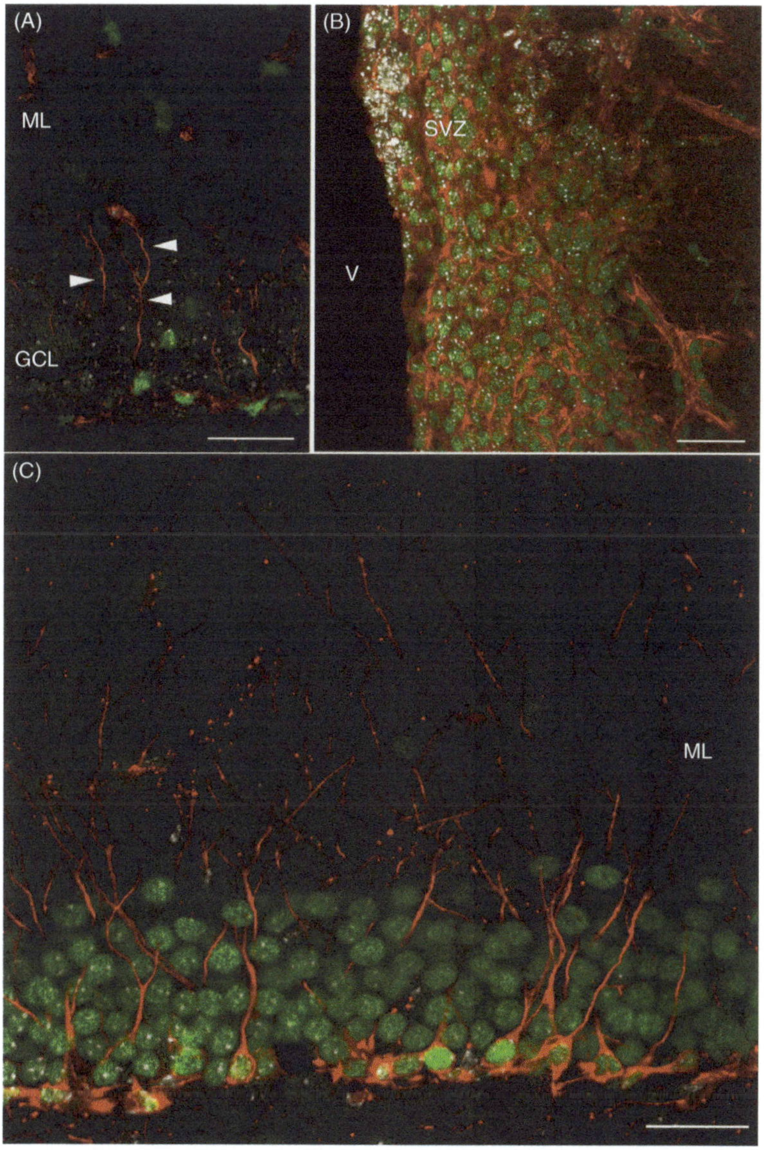

PLATE 3 Expression of proteins associated with adult neurogenesis. (A) The somata of dividing NPCs in the adult hippocampus are localized in the SGZ and express a distinct set of marker proteins (Kempermann et al., 2004). Sox-2 (green in A, B) is expressed in the SGZ and labels cells that also co-label with the intermediate filament nestin (red in A, B) showing a radial glia-like morphology (arrowheads). (B) Similar markers are expressed in the adult SVZ, where a large number of cells express Sox-2 and nestin. (C) Dividing progenitors and immature newborn granule cells in the DG transiently express the microtubuli-associated protein DCX (red in B). The granule cell-specific transcription factor, Prox1 (green in C), is expressed early on (note the partial co-expression with DCX) but continues to label granule cells once they mature. GCL: granule cell layer, ML: molecular layer. V: ventricle. Scale bars in A–C are 50 μm. (see Figure 7.3 on p. 187)

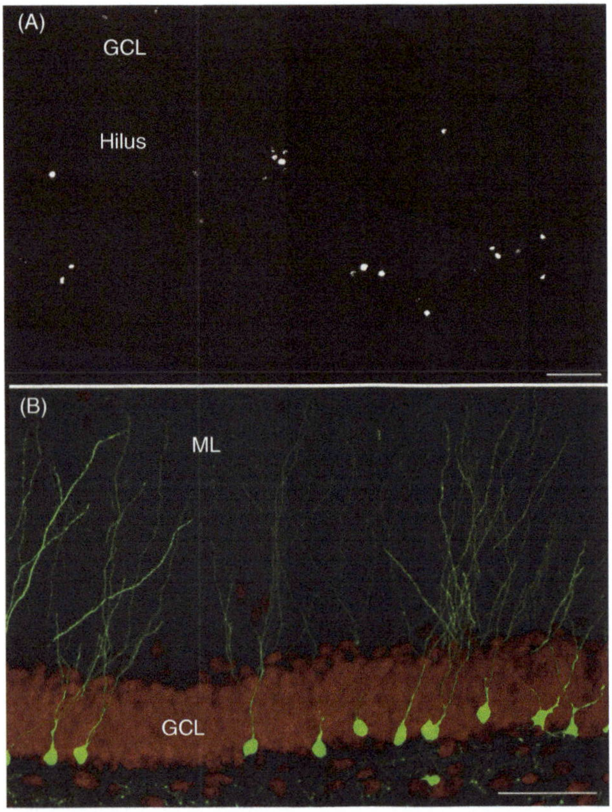

PLATE 4 Labeling of newborn cells in the adult brain using thymidine-analogues or retroviral vectors. (A) BrdU is integrated into the DNA during s-phase and can be later visualized using specific antibodies. The example shown is from a mouse DG 6 weeks following a series of BrdU injections. Note that the majority of the cells are still localized in the inner third of the DG. (B) Intracerebral injections of fluorescence label-expressing retroviral vectors (GFP in green) that specifically integrate into dividing cells and label their progeny allow the whole cell visualization (e.g., Zhao et al., 2006), physiological characterization (e.g., Laplagne et al., 2006), and genetic manipulation (e.g., Tashiro et al., 2006) of newborn cells in the adult brain. Note the highly polarized morphology of newborn granule cells extending an apical dendrite into the ML. GCL: granule cell layer, ML: molecular layer. Scale bars in A and B: 100 μm. (see Figure 7.4 on p. 188)

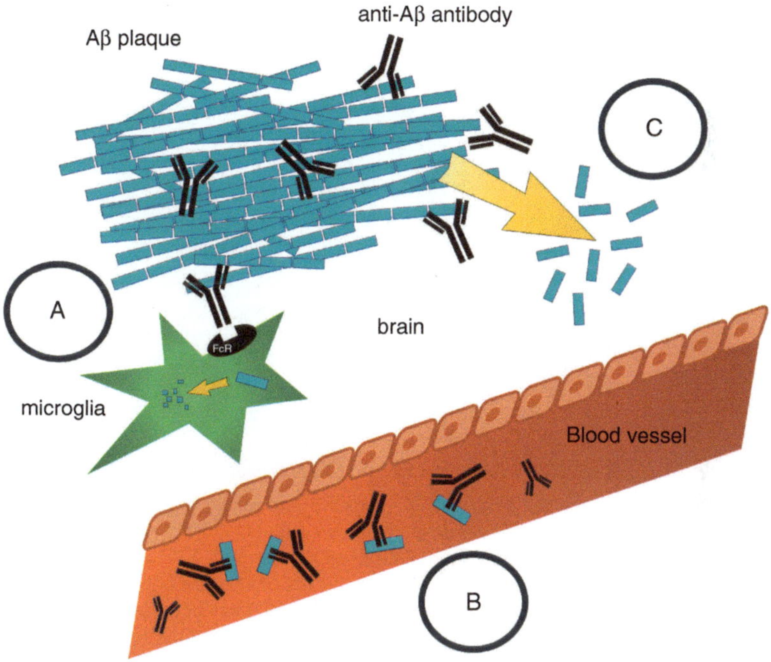

PLATE 5 Mechanisms of immunization-mediated clearance of Aβ. (A) Opsonization mechanism. Anti-Aβ antibodies cross the blood–brain barrier and bind to Aβ with amyloid plaques. Microglia recognize the antibody via the Fc receptor and phagocytose the amyloid plaque. (B) Peripheral sink mechanism. Antibodies within the blood bind to Aβ, thus preventing it crossing into the brain or increasing the net flux of Aβ from brain into blood. (C) Direct dissolution mechanism. Antibodies bind to amyloid plaque within the brain, thereby changing the conformation of the Aβ aggregate and facilitating Aβ re-entry into the soluble phase. (see Figure 12.2 on p. 302)

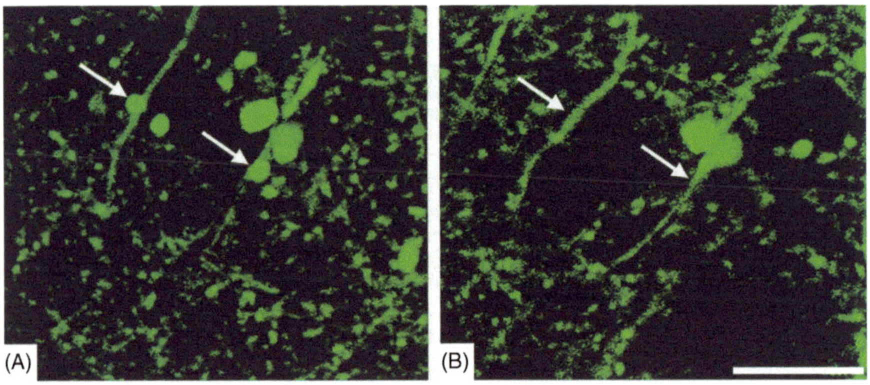

PLATE 6 Morphological recovery of amyloid-associated neuritic dystrophy in a *PDAPP;Thy1:YFP* double-transgenic mouse after anti-Aβ antibody (10D5) treatment. PDAPP transgenic mice overexpressing human APP were bred to mice expressing YFP within a subset of their cortical neurons and processes. The two panels show the same population of YFP-labeled neurites, which are associated with a neuritic plaque imaged on the initial day of surgery (A) and 72 h later (B). The antibody was administered directly to the surface of the brain during the cranial-window surgical procedure on day 0. The arrows in the day 0 image (A) point to 2 enlarged dystrophic neurites, which are absent in the day 3 image (B). The associated amyloid is not shown. Scale bar: 20 μm. Reprinted with permission from Brendza et al. (2005). *J Clin Invest* 115: 428–433. (see Figure 12.3 on p. 305)

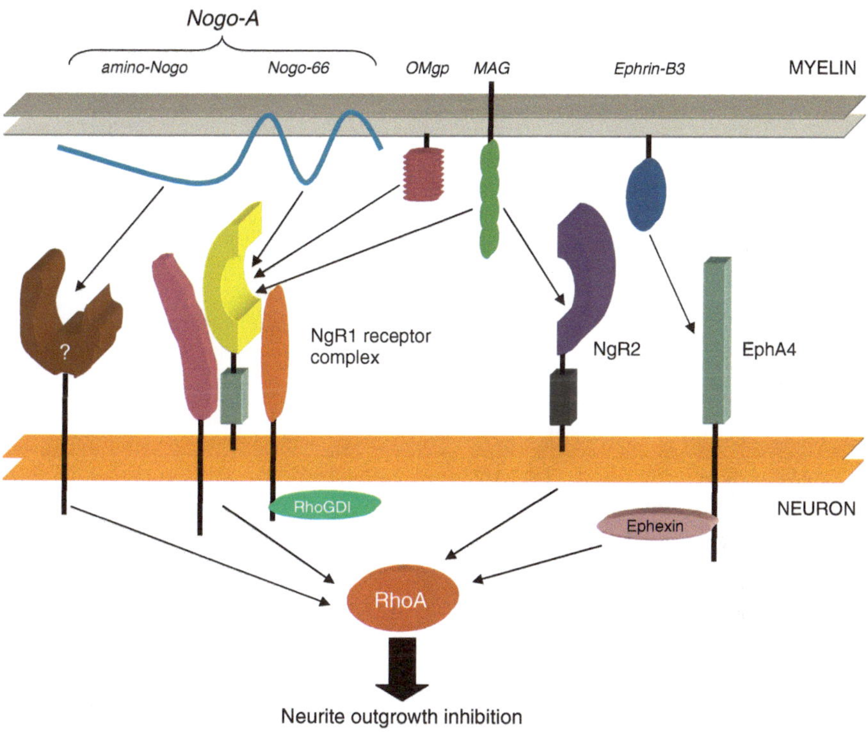

PLATE 7 Myelin-associated inhibitors of neurite outgrowth. The CNS myelin membrane proteins MAG, OMgp and the Nogo-66 domain of Nogo-A bind to overlapping sites on the neuronal GPI-anchored receptor NgR1 (yellow) which signals to the inside of the cell via the transmembrane proteins LINGO-1 (purple) and $p75^{NTR}$ or TROY/TAJ (orange). MAG also binds to the NgR1 homologue, NgR2, the transmembrane co-receptors of which remain to be identified. Apart from Nogo-66, Nogo-A has additional inhibitory regions in the amino-Nogo-A domain which signal through an unknown receptor complex (brown) independent of NgR1. All these ligands activate the small GTPase RhoA in the neuron leading to actin rigidification and neurite outgrowth inhibition. MAG and Nogo-66 appear to activate RhoA by enhancing the binding of the $p75^{NTR}$ intracellular domain to RhoGDI which precludes its association with RhoA allowing the activation of RhoA by GEFs. Myelin-associated ephrin-B3 activates neuronal EphA4 receptors and mediates neurite outgrowth inhibition possibly via the activation of RhoA by the GEF ephexin. (see Figure 15.1 on p. 356)

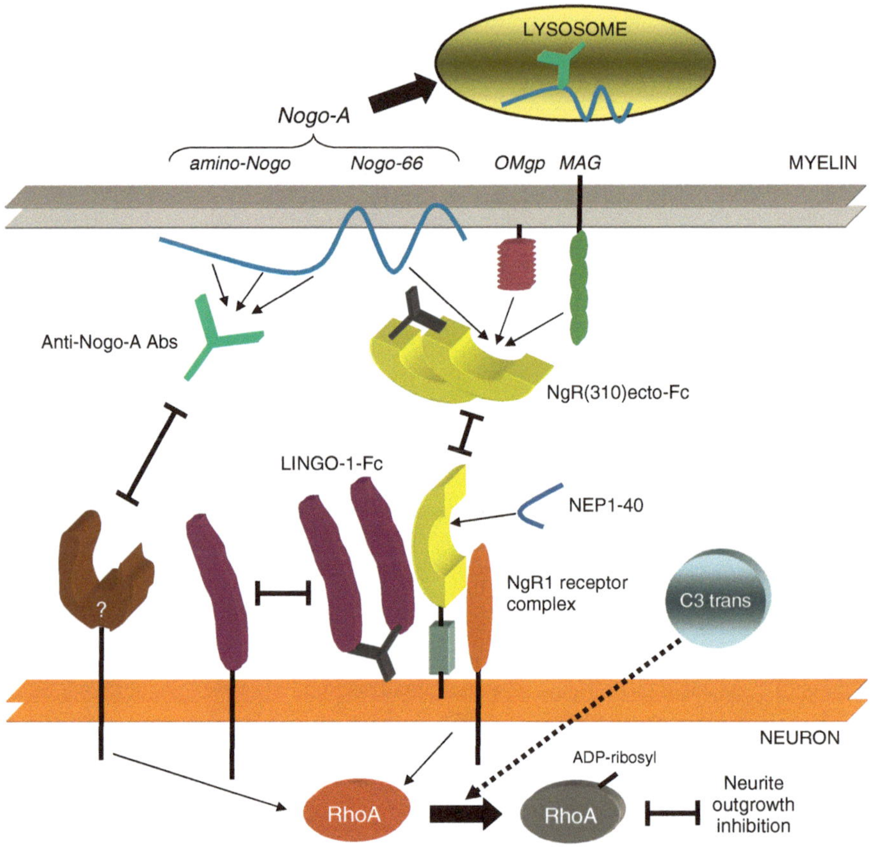

PLATE 8 Approaches to block myelin-mediated neurite outgrowth inhibition Anti-Nogo-A antibodies to disparate epitopes in the amino-Nogo-A region of Nogo-A are likely to sterically hinder the binding of Nogo-A to its cognate neuronal receptors. The anti-Nogo-A antibody: Nogo-A complex is also internalized into lysosomal structures leading to a downregulation of the Nogo-A protein in myelin. NgR(310)ecto-Fc is a soluble form of the ligand-binding domain of NgR1 that binds NgR1 ligands but cannot signal as it lacks the C-terminal domain necessary to form an interaction with the transmembrane co-receptors of the complex. Similarly, LINGO-1-Fc is a soluble form of the LINGO-1 ectodomain that binds to NgR1 but cannot signal as it lacks an intracellular domain. The NEP1-40 peptide comprises the first 40 amino acids of Nogo-66 that binds to NgR but does not activate the receptor hence blocking its activation by Nogo-66. NEP1-40, however, fails to block the activation of NgR1 by MAG, possibly because the binding sites for MAG and Nogo-66 on the receptor are overlapping but distinct. The exoenzyme C3 transferase irreversibly inactivates RhoA by ADP-ribosylation of the effector domain and has now been engineered to yield membrane permeable variants. (see Figure 15.2 on p. 362)

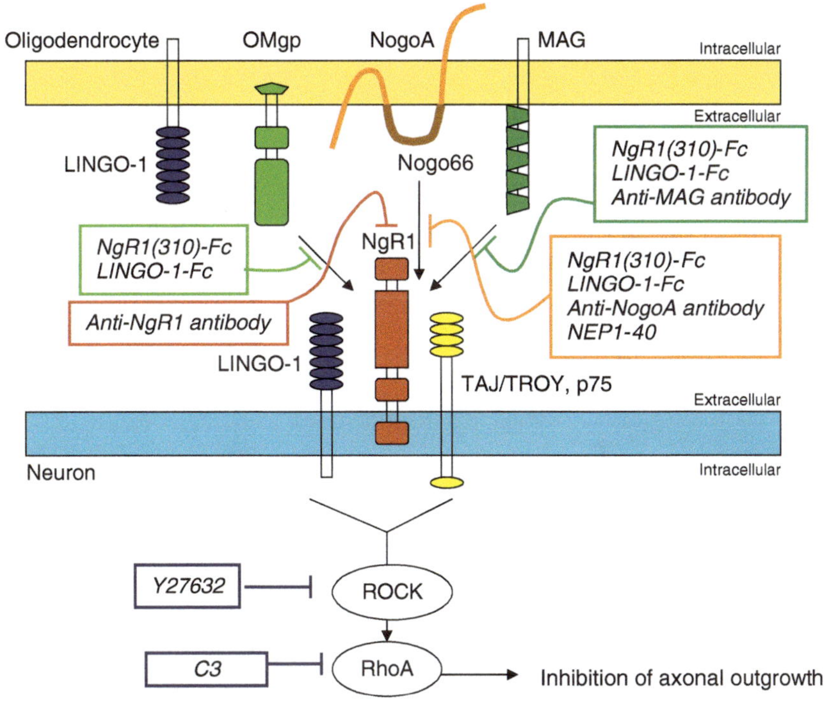

PLATE 9 Schematic representation of NgR1 and its co-receptors and signaling mechanisms. The three characterized myelin ligands for NgR1 are Nogo-A, myelin-associated glycoprotein (MAG) and oligodendrocyte myelin glycoprotein (OMgp). The known co-receptors for NgR1 are p75, TAJ/TROY and LINGO-1. The signaling molecules are ROCK and RhoA. NgR1 and its co-receptors are expressed on neurons and its ligands are expressed on oligodendrocytes. The leucine rich repeats are required for interaction with Nogo-66, MAG and OMgp. The amino terminal of NogoA can be resent either in the cytosol or in the extracellular space. Shown in boxes are the potential sites of intervention and known inhibitors of signaling through NgR1. (see Figure 16.1 on p. 376)

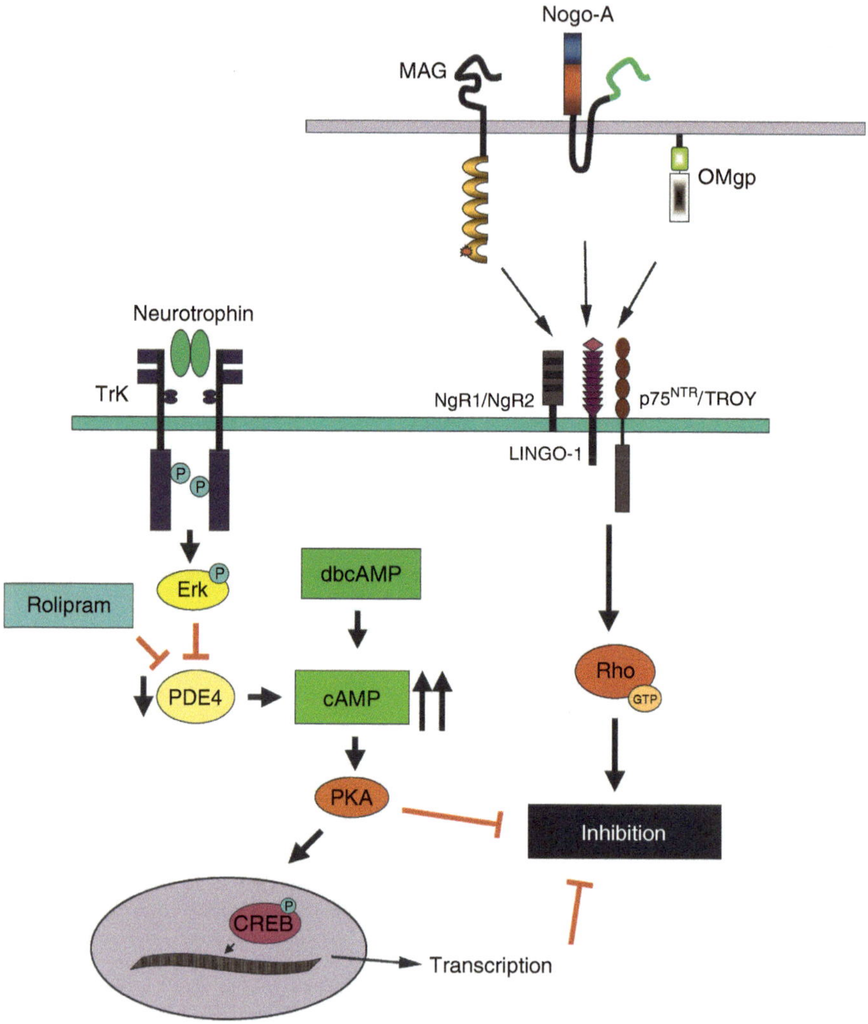

PLATE 10 Schematic representation of intracellular signaling pathways activated by cAMP and their role in overcoming inhibition by CNS myelin. Neurotrophin binding to Trk receptors leads to activation of Erk, which in turn results in a transient reduction of PDE4 activity. Inhibition of PDE4 can also be achieved through administration of rolipram, and this causes cAMP levels to rise. Alternatively, intracellular cAMP can be directly increased by treatment with dbcAMP, and upon reaching a threshold level, cAMP will activate PKA and initiate transcription by CREB. These signaling events block inhibition mediated by MAG, Nogo-A and OMgp binding to the receptor complex consisting of NgR1 or NgR2, $p75^{NTR}$ or TROY and LINGO-1. (see Figure 17.1 on p. 391)

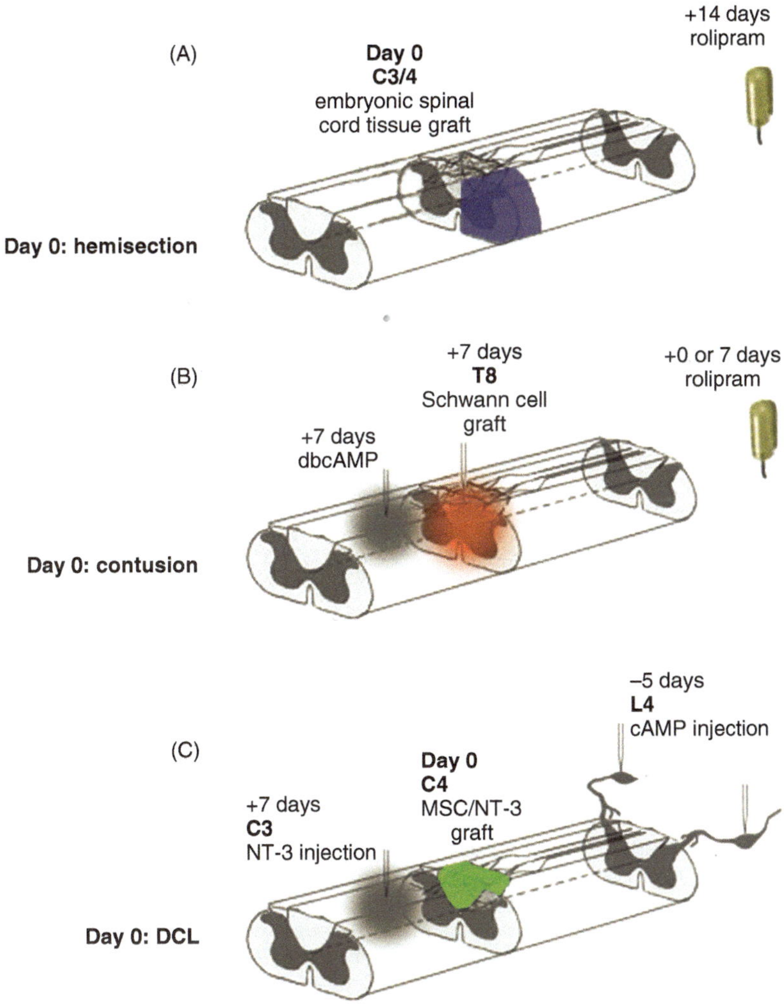

PLATE 11 Comparison of spinal cord injury models that have used elevated cAMP to promote axonal regeneration. (A) In the study by Nikulina and colleagues (2004), spinal cord hemisection was performed at C3/4 and this was immediately followed by implantation of embryonic spinal cord tissue into the lesion site. Two weeks after surgery, rolipram was administered subcutaneously via osmotic minipump for 10 days. (B) Pearse and colleagues (2004) performed thoracic (T8) spinal cord contusions, which were followed by Schwann cell grafting and intraspinal injection of dbcAMP 1 week later. Some animals also received subcutaneous delivery of rolipram for 2 weeks, beginning either at the time of injury or 1 week after injury. (C) In the experimental model used by Lu and colleagues (2004), dbcAMP was injected into the L4 DRG 5 days prior to performing a dorsal column lesion (DCL). Animals received MSC grafts and NT-3 injections at the time of injury, and at 1 week after injury, an additional intraspinal injection of NT-3 was made rostral to the lesion site. (Copyright 2004 by the Society for Neuroscience.) (see Figure 17.2 on p. 400)

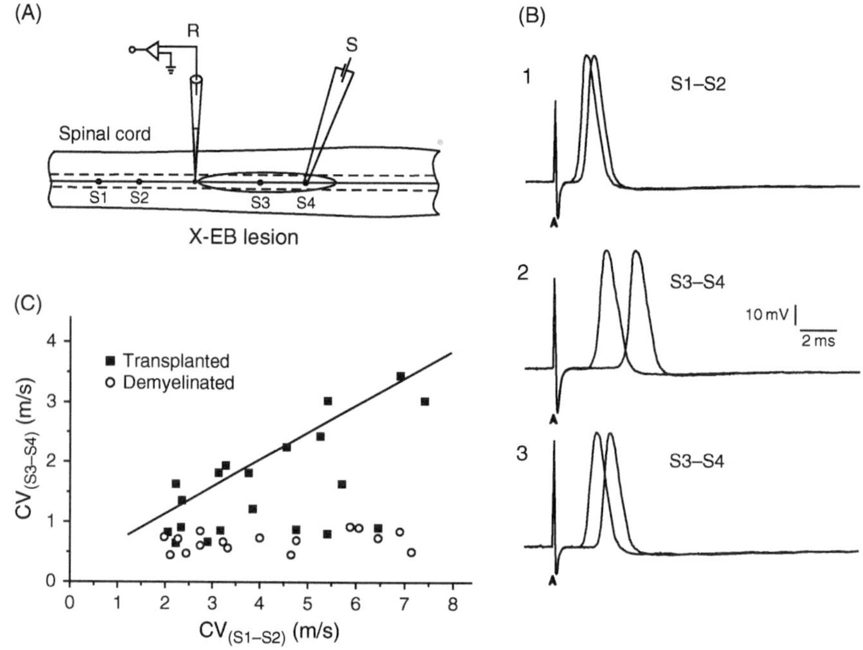

PLATE 12 Transplantation of human Schwann cells into the demyelinated spinal cord improves conduction. (A) Schematic showing arrangement of intra-axonal recording and stimulation sites. Intra-axonal recordings were obtained from dorsal column axons outside of the lesion where the axons were normally myelinated. Stimulating electrodes were positioned outside (S1–S2) and within (S3–S4) the X-EB lesion zone to assess single axon conduction velocity over both the demyelinated or remyelinated axon segment and the normally myelinated axon segment of the same axon. (B) Pairs of action potentials recorded from S1–S2 stimulation (1), S3–S4 in the demyelinated dorsal columns (2) and S3–S4 following cell transplantation (3). Recordings were obtained at comparable conduction distances. (C) Plot of the conduction velocity of axon segments within the lesion (S3–S4) vs. conduction velocity of the axon segment outside of the lesion (S1–S2) for X-EB lesioned spinal cord without (open circles) and with (closed squares) transplantation. Modified from Kohama et al., 2001. (see Figure 18.5 on p. 423)

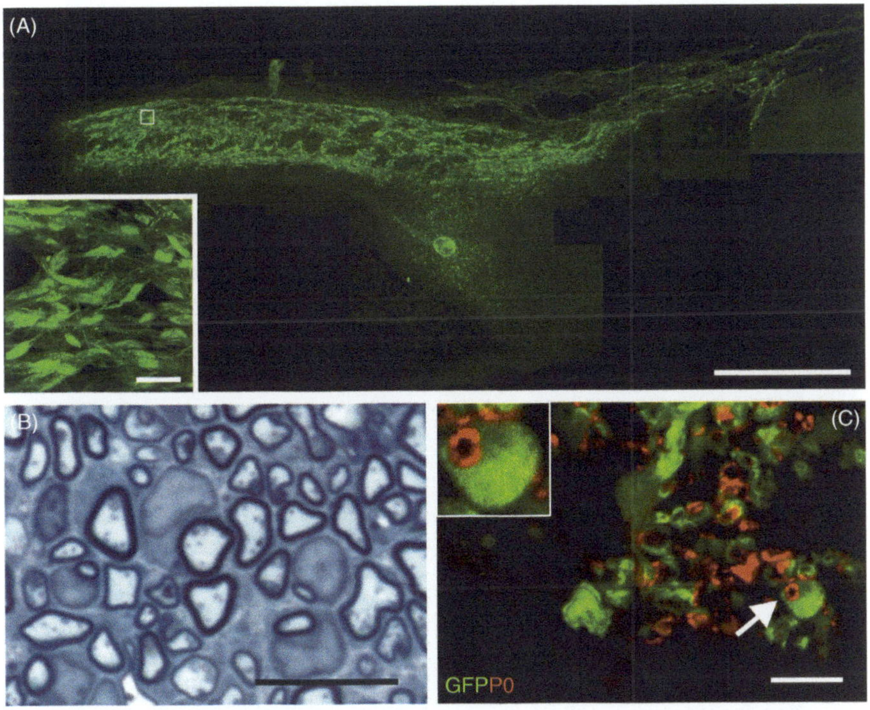

PLATE 13 (A) Transplanted green fluorescent protein (GFP) expressing OECs transplanted into a dorsal transected spinal cord integrate and span the transection site (sagittal section). The transplanted cells are primarily confined to the lesion site. The dashed line demarcates lesion edge. (B) High-magnification micrograph of semithin plastic sections stained with methylene blue/Azure II through the OEC-transplanted lesion demonstrating that the transplanted dorsal funiculus was extensively myelinated. (C). Red-P0 rings are associated with green cellular elements, indicating that transplanted OECs remyelinated the demyelinated axons. Inset, expansion of the cell indicated by an arrow. Scale bars: 1 mm (A), 10 μm (B, C) and 20 μm (Inset A). Modified from Sasaki et al., 2004. (see Figure 18.6 on p. 425)

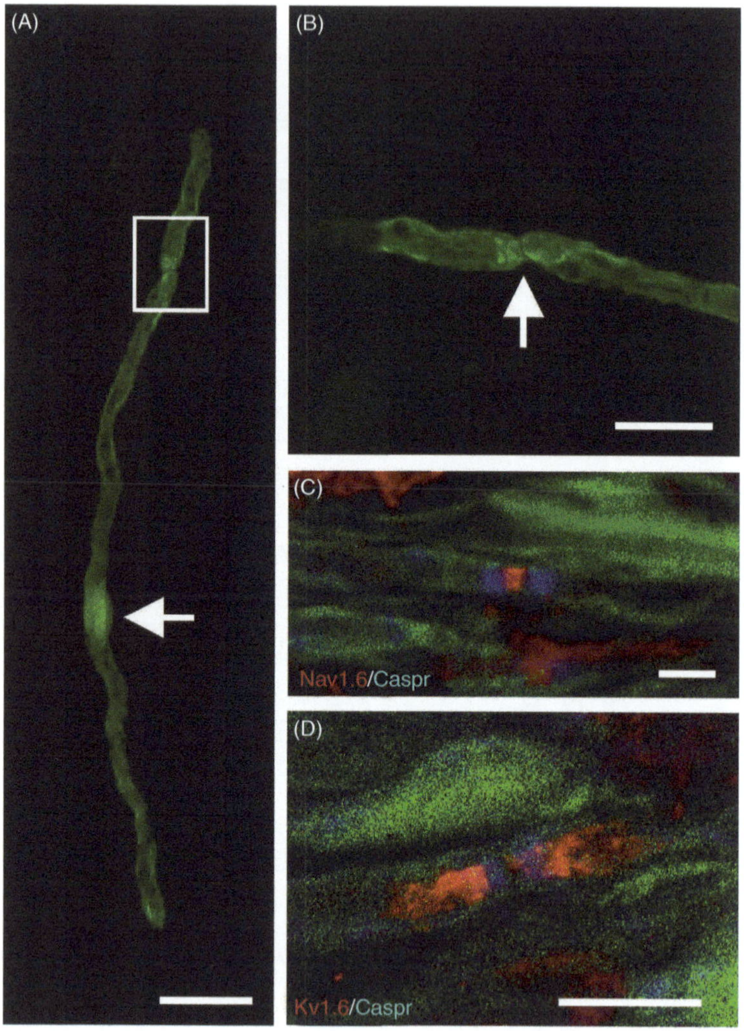

PLATE 14 (A) Nodal formation of spinal cord axons remyelinated by transplanted OECs. OECs oriented longitudinally with periodic bulbous regions of intense green (GFP) fluorescence at the level of the cell body and nucleus (arrow). The elongated processes showed fluorescence on their outer margins corresponding to the outer cytoplasmic slips of the myelin forming cells. (B) Nodal-like regions were observed as thin breaks between adjacent green cell regions from boxed area in A. (C) Sodium channel $Na_v1.6$ at node of Ranvier of remyelinated dorsal columns at 3 weeks after OEC transplantation into demyelinating lesion in rat spinal cord. (D) Juxtaparanodal $K_v1.2$ immuno-labeling after GFP–OEC transplantation dorsal columns. Paranodes display Caspr staining that is flanked by $K_v1.2$ aggregations within juxtaparanodal regions. Scale bars: 100 μm (A), 20 μm (B) and 10 μm (C, D). Modified from Sasaki et al., 2006b. (see Figure 18.7 on p. 427)

FSC
www.fsc.org
MIX
Papier | Fördert gute Waldnutzung
FSC® C083411

Kolibri 360 GmbH
Ferdinand-Jühlke-Straße 7
99095 Erfurt, Deutschland
produktsicherheit@kolibri360.de